ADVANCED MATHEMATICS

Christine McRae
Esther Gawac Ehava
Mac Dandava
John Gesa Junior
Phyl Haydock
John Watson
Rory Barrett

OXFORD

Oxford University Press is a department of the University of Oxford. It furthers the University's objective of excellence in research, scholarship, and education by publishing worldwide. Oxford is a registered trademark of Oxford University Press in the UK and in certain other countries.

Published in Australia by
Oxford University Press
Level 8, 737 Bourke Street, Docklands, Victoria 3008, Australia

First published 2013
Reprinted 2013, 2014 (twice), 2016, 2017(three times), 2019 (twice), 2020, 2021, 2023, 2024, 2025, 2026

This book was originally published by ESA Publications, Auckland, New Zealand. This edition, specially adapted for the Grade 11 syllabus in Papua New Guinea, is published by arrangement with ESA Publications. Authors of the original work were Phyl Haydock, John Watson and Rory Barrett and this adaptation has been produced by Christine McRae with the assistance of Esther Gawac Ehava, Mac Dandava and John Gesa Junior.

ISBN 978 0 19 557871 3

Illustrated by diacriTech, Chennai, India
Typeset by diacriTech, Chennai, India
Printed in China by Golden Cup Printing Co. Ltd

Oxford University Press Australia & New Zealand is committed to sourcing paper responsibly.

Contents

Acknowledgments

There are many people to be thanked for their help and assistance in enabling the publication of this series to happen. First, we acknowledge the cooperation and generosity of Mark Sayes at ESA Publications in New Zealand, who responded with interest and support when the proposal to adapt his Study Guide series was put to him.

The main author of this book is Christine McRae, with input from Esther Gawac Ehava, Mac Dandava and John Gesa Junior, but with the approval of ESA Publications, she drew on material from *NCEA Level 1 Mathematics* by Phyl Haydock and John Watson, and *NCEA Level 2 Mathematics* by Rory Barrett. The objective has been to provide Grade 11 students in PNG with a book that they can use as a compact summary of content and skills related to the PNG Grade 11 Advanced Mathematics syllabus.

We would also like to acknowledge many other individuals who have been happy to advise and assist in different ways: Ms Vagi Hanua Bino, Ms Betty Pulpulis, Ms Joy Sahumlal, Mr Greg Kapanombo, Mrs Anne Sangi and Mr Safak Deliismail.

Unit 11.1 Number and Application

Topic 1: Basic numeracy—the real number system

This Unit focuses on the mathematics we use in our daily lives. The Topic covers:

- The development of the real number system.
- Different types of numbers.
- Properties of whole numbers: closure, associative, commutative and distributative laws.

Real numbers

The numbers and their symbols that we use in today's society have developed over many centuries. The original use of numbers in many cultures was to count objects, and, as communities became farmers and builders, the system and the way of writing numbers became more sophisticated. The **decimal** system of numbers that we use today was thought to originate with the Hindus.

Types of numbers

- **Natural numbers (Counting numbers):** The set of numbers 1, 2, 3, 4, 5, ... is called the set of natural (or counting) numbers.
- **Whole numbers:** At some point in the development of numbers it was realised that there was a need for a number that represented 'none' and so the number zero was added to the natural numbers to form the set of whole numbers
- **Integers:** Later, the notion of 'owing' led to the development of negative numbers and so the set of integers was created : ... −4, −3, −2, −1, 0, 1, 2, 3, 4, 5, ...
- **Fractions and decimals:** The notion of 'part of a whole' led to the development of fractions and later their decimal representation.

Rational numbers

For many centuries whole numbers and fractions were thought to be the only types of numbers that we would ever have to deal with. They were collectively called the set of **rational numbers**.

Rational numbers:

- Have the special property that they can all be written as a **ratio**, ie as one integer divided by another.
- Can be positive or negative.
- Cannot have the number zero as the number in the denominator, eg $\frac{8}{0}$; these numbers are **undefined**.
- Can have the number zero in the numerator, eg $\frac{0}{8}$, as this number is equal to zero.
- In set notation, are described as $\frac{m}{n}$, $n \neq 0$ where $m, n \in J$.
- Were called **ratio**nal numbers because the first five letters spell the word '**ratio**'.

Irrational numbers

It became clear about 2 500 years ago that there existed another type of number that could not be written as a ratio.

This was in the time of the Greek philosopher and mathematician Pythagoras (580–500 BC) who belonged to a powerful brotherhood whose motto was 'All is number'. They believed that all things in the universe could be explained in terms of rational numbers. However, using a **right-angled triangle** and Pythagoras' rule they discovered and **proved** that there existed another type of number that could not be written as a ratio. These were numbers such as $\sqrt{2}$ $\sqrt{7.69}$, and π which appear as **infinite, non-recurring decimals**. Today these numbers are called **irrational numbers**. Pythagoras's brotherhood kept the existence of these irrational numbers secret for about 400 years as it ruined their 'All is number' theory.

Today, the set of **rational numbers** together with the set of **irrational numbers** make up the set of **real numbers**.

The set of real numbers is displayed on the following diagram. The capital letter next to the number headings represents commonly used letters for that set of numbers.

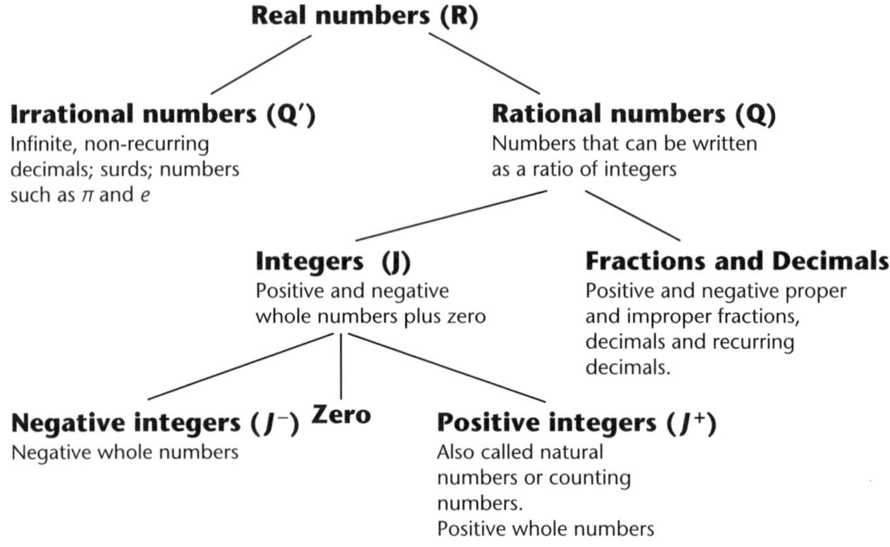

Showing that numbers are rational by changing them into a ratio of integers

i. **Whole numbers** and **integers** can be expressed as a ratio by using the whole number in the numerator and 1 in the denominator.

The number 3 can be written as $\frac{3}{1}$ which is a ratio of the whole numbers 3 and 1.

The integer –7 can be written as the ratio $\frac{-7}{1}$.

ii. **Proper fractions** and **improper fractions** are already given as a ratio; it is usual to give these in their simplest form.

The proper fraction $\frac{5}{12}$ is the ratio of 5 and 12.

The improper $\frac{24}{9}$ fraction is a ratio but should be written as $\frac{8}{3}$.

iii. Mixed numbers need to be changed to improper fractions.

For example: $5\frac{3}{4}$ can be written as $\frac{23}{4}$; the numerator of the improper fraction is found by multiplying the whole part (5) by the denominator (4) and adding the numerator (3) to give 23.

iv. Decimals: In the denominator, place a 1 below the decimal point and add as many zeros as needed to cover the decimal parts. Remove the decimal point from the numerator.

For example:

a. 0.83 can be written as $\frac{0\ 83}{100} = \frac{83}{100}$.

b. 25.6679 is written as $\frac{25\ 6679}{10000} = \frac{256\ 679}{10\ 000}$.

c. 0.0013 is written as $\frac{0\ 0013}{10000} = \frac{13}{10\ 000}$.

v. Recurring decimals can be written in several ways:

$2.\dot{6}$ = 2.666666... the 6 is the only repeated number.

$4.\underline{5027}$ = 4 502750275027... a dot above the beginning and the end of the repeated **digits**.

$0.5\overline{237}$ = 0.5237237237...... ; a bar above the digits that are repeated.

Example A

Q. Express each of the following recurring decimals as a ratio of integers:

1. $0.\dot{6}$

2. $2.\overline{105}$

A. 1. $0.\dot{6} = 0.666666...$

Let $x = 0.666666...$ (1)

so $10x = 6.6666...$ (2)

Subtract (1) from (2)

$10x - x = 6.6666... - 0.6666...$

$9x = 6$: Divide both sides by 9

$x = \frac{6}{9} = \frac{2}{3}$

So $0.6 = \frac{2}{3}$

2. $2.\overline{105} = 2.105105105......$

Let $x = 2.105105105...$ (1)

$1\,000x = 2\,105.105105...$ (2)

Subtract (1) from (2)

$1\,000x - x = 2\,105.105105... - 2.105105...$

$999\,x = 2103$: divide both sides by 999

$x = \frac{2\,103}{999}$

So $2.105 = \frac{2103}{999} = \frac{701}{333}$ (check this on your calculator by dividing 701 by 333)

Properties of rational numbers

The properties of whole numbers called axioms specifically apply to addition and multiplication only. An axiom does not need to be proved as it is assumed to be true.

Axiom 1: The closure law of addition

The closure law of addition states that when adding two whole numbers the sum is another whole number. It can be shown algebraically as, $a + b = c$ where a, b and c are all whole numbers. Example: 2 + 3 = 5

Axiom 2: Commutative law of addition

The law states that the whole numbers can be added in any order but their sum will be the same. That is shown algebraically as $a + b = b + a$. Example: $4 + 5 = 5 + 4 = 9$

Axiom 3: Associative law of addition

The order of association of whole numbers does not change their sum.
$(a + b) + c = a + (b + c)$ Example: $(3 + 4) + 5 = 3 + (4 + 5)$

Axiom 4: Identity law of addition

When adding zero to any whole number the sum is that number. The number did not lose its identity therefore zero is known as the identity element. That is algebraically represented as $a + 0 = 0 + a = a$. Example: $6 + 0 = 0 + 6 = 6$

Axiom 5: Closure law of multiplication

The law states that when multiplying any two whole numbers, the **product** is another whole number. That is shown algebraically as $a \times b = c$, where a, b *and* c are whole numbers. Example: $2 \times 3 = 6$

Axiom 6: Commutative law of multiplication

Whole numbers can be multiplied in any order but their product is unchanged. That is shown algebraically as $a \times b = b \times a$. Example: $3 \times 5 = 5 \times 3 = 15$

Axiom 7: Associative law of multiplication

The order in which numbers are associated in multiplication does not alter the product. That is shown algebraically as $(a \times b) \times c = a \times (b \times c)$. Example: $(2 \times 3) \times 4 = 2 \times (3 \times 4) = 24$

Axiom 8: Identity law of multiplication

One is the identity element for multiplication. When it is multiplied by any whole number, the product is the same as that number. That is, $a \times 1 = 1 \times a = a$. Example: $2 \times 1 = 1 \times 2 = 2$

Unit 11.1 Activity 1A: The real number system

1. By writing them in the form $\frac{m}{n} (n \neq 0)$, show that the following numbers are rational.

a. 15 **b.** -4 **c.** 0 **d.** 12.5 **e.** 0.045

f. $2\frac{3}{8}$ **g.** 23.224 **h.** $-6\frac{8}{9}$ **i.** -0.00005 **j.** $3\frac{23}{25}$

2. Show that the following recurring decimals are rational by converting them to the form $\frac{m}{n} (n \neq 0)$. Give your answers in simplest fractional form.:

a. 0.333333… **b.** $-0.\dot{7}$ **c.** 0.565656… **d.** $0.\overline{67}$

e. 2.555555… **f.** $12.\overline{45}$ **g.** $62.\overline{822}$

3. Explain the identity law of addition, giving an example. What is the identity element of addition?

4. Explain the identity law of multiplication, giving an example. What is the identity element of multiplication?

5. We know that $5 \times (3 \times 2) = 5 \times 6 = 30$ and $(5 \times 3) \times 2 = 15 \times 2 = 30$. Which of the axioms of rational numbers is this illustrating?

Unit 11.1 Number and Application
Topic 2: Basic numeracy—working with real numbers

The material in this Topic is about working with real numbers and solving straightforward number problems in context, and covers:
- Integers.
- Making sensible estimates and checking the reasonableness of results.
- Solving number problems in context, involving manipulation, several steps or reversing processes.

Introduction

Solving problems in context

The General Mathematics syllabus emphasises the importance of solving problems *in context* so that problems arising from practical situations can be correctly solved and the answers meaningfully **interpreted**.

Many straightforward calculations can be carried out mentally, or using pen and paper. It is most important to *understand* the processes involved in operations with numbers, and not to simply rely on a calculator. Unless these processes are understood thoroughly, it is difficult to succeed with algebra problems involving these skills (when a calculator can no longer be used).

Sets of numbers

In this Achievement Standard, various types of numbers are used.

Natural numbers	1, 2, 3, 4, 5, 6, 7, 8, 9, 10, …
Whole numbers	0, 1, 2, 3, 4, 5, 6, 7, 8, 9, 10, …
Integers	… –4, –3, –2, –1, 0, 1, 2, 3, 4, …
Rational numbers	Numbers which can be written in the form $\frac{\text{integer}}{\text{integer}}$. Rational numbers include fractions and finite or recurring decimals as well as integers.
Irrational numbers	Numbers which are not rational, such as $\sqrt{2}$ and π.
Real numbers	The combined set of all rational and irrational numbers.

Note: The Natural numbers are also called **Counting numbers**.

To solve number problems, numbers can be combined using **operations** such as addition, subtraction, multiplication and division.

Order of operations

In expressions involving more than one operation, there is an accepted order in which operations are performed. The *mnemonic* **BEMA** is useful for remembering the correct order:

B	Brackets first
E	Exponents (powers) next
M	Multiplication/division next
A	Addition/subtraction last

Note: **1.** Multiplication and division are inverses and are therefore the same 'type' of operation, eg division by 4 is the same as multiplication by $\frac{1}{4}$. Since neither operation 'outranks' the other, multiplication and division are carried out in order from left to right. (Remember 'of' is also used to mean multiplication.)

2. Similarly addition and subtraction are inverses, and are carried out in order from left to right.

Example A

Q. Calculate: **1.** $3 \times 2 - 16 \div 4$ **2.** $15 + 3 \times 2 - 4^2$ **3.** $5 + 3(15 - 3 \times 2)$

A. **1.** In $3 \times 2 - 16 \div 4$, there are no **brackets** or **exponents** so multiplication and division are done first. 3×2 is calculated before $16 \div 4$ because the multiplication appears before the division.

$3 \times 2 - 16 \div 4 = 6 - 16 \div 4$ [multiplication first]
$= 6 - 4$ [division next]
$= 2$ [subtraction next]

2. $15 + 3 \times 2 - 4^2 = 15 + 3 \times 2 - 16$ [the exponent 4^2 is done first]
$= 15 + 6 - 16$ [then multiplication]
$= 21 - 16$ [15 + 6 is done, not 6 – 16]
$= 5$

3. Here the bracket is done first. Within the bracket, the multiplication is done before the subtraction.

$5 + 3(15 - 3 \times 2) = 5 + 3(15 - 6)$ [multiplication in brackets first]
$= 5 + 3 \times 9$ [subtraction in brackets next]
$= 5 + 27$
$= 32$

Calculators

Scientific **calculators** are inexpensive and efficient tools for longer, more complex calculations. With scientific calculators such as the *Casio fx*-82, brackets may be entered, the correct order of operations is automatically followed and the **memory** is useful for storing parts of the answer as you go.

Example B

Q. Use a calculator to evaluate the expressions in Example A.

A. **1.** Press 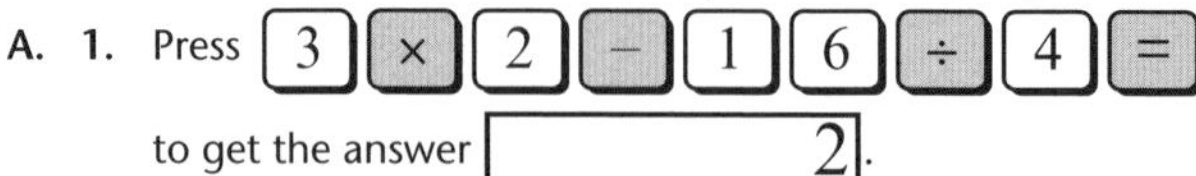

to get the answer 2.

2. Press 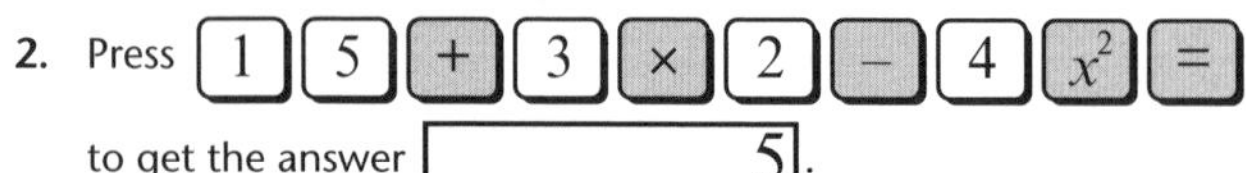

to get the answer 5.

3. Press 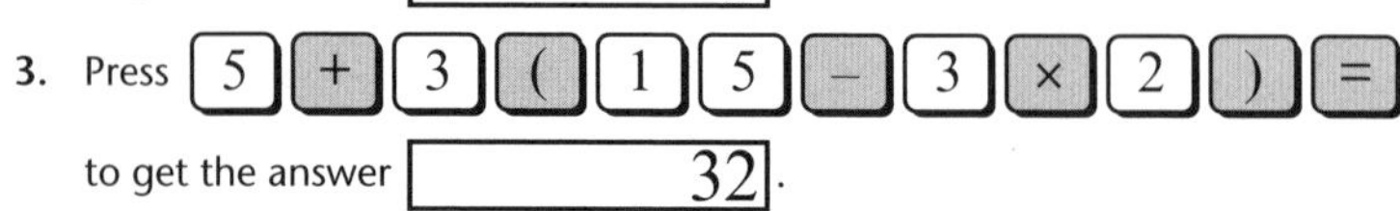

to get the answer 32.

Note: In expressions where brackets are 'understood' they may not be shown.

For example, the fraction $\dfrac{47 + 53}{12 - 7}$ is understood to mean $\dfrac{(47 + 53)}{(12 - 7)}$, so press

[( ] [4] [7] [+] [5] [3] [ )] [÷] [( ] [1] [2] [−] [7] [ )] [=]

to get answer [20].

Calculators vary so it is important that students become familiar with their own brand of calculator (ie read the instruction manual thoroughly) so that calculations are handled correctly.

Integers

The set of integers is the set of whole numbers and their 'opposites'.

> The integers are ..., –3, –2, –1, 0, 1, 2, 3, ...

Adding and subtracting integers

A number line is useful for illustrating addition and subtraction of integers.

Example C

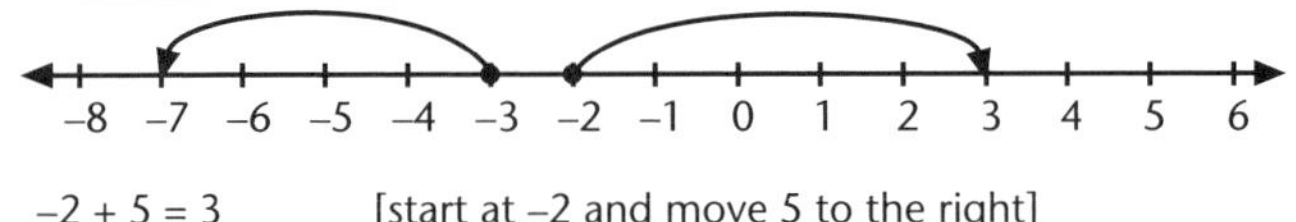

–2 + 5 = 3 [start at –2 and move 5 to the right]

–3 – 4 = –7 [start at –3 and move 4 to the left]

Two consecutive signs can be simplified to a single sign according to the following rule:

> If two consecutive signs are the same change to '+'.
>
> If two consecutive signs are different change to '–'.

Example D

1. 4 – –3 = 4 + 3 = 7 [subtracting a negative means adding]
2. 5 + –8 = 5 – 8 = –3 [adding a negative means subtracting]

Using a number line works well when adding or subtracting small integers. To add or subtract larger integers, the **size** and **sign** of the integers should be taken into account (for example –45 has a *size* of 45 and a *sign* that is negative).

When adding two integers:

- If the signs are the *same*, then *add* the sizes. The sign of the answer is the same as the signs of the two numbers, eg 40 + 60 = 100 and –40 + –60 = –100, ie –40 – 60 = –100.
- If the signs are *opposite* then *subtract* the sizes. The sign of the answer matches the sign of the number whose size is larger, eg –40 + 60 = 20 [60 is larger in size so answer is positive] and –80 + 30 = –50 [–80 is larger in size so answer is negative].

Problems expressed in words may be set out as numerical expressions and simplified.

Example E

Q. Vagi's bank balance is overdrawn by K153. She deposits K47. What is the current state of her bank balance?

A. Vagi still owes money, but not as much as before. Her account is now overdrawn by K(153 – 47) = K106.

Writing this in terms of integers:

$-153 + 47 = -106$ [negative since –153 is larger in size]

Example F

Q. The **temperature** at the top of Mt Wilhelm was –5°C at midnight, but by noon had risen by 11°C. Find the temperature at noon.

A. $-5 + 11 = 6$ [positive integer has larger size, so answer positive]

The temperature is 6°C at noon.

Multiplying and dividing integers

The rules for multiplying and dividing integers are more straightforward. When multiplying or dividing a *pair* of integers, 'like' signs give a positive answer, 'unlike' signs a negative answer. This is shown in the table:

+ x +	gives	+		+ ÷ +	gives	+
– x +	gives	–		– ÷ +	gives	–
+ x –	gives	–		+ ÷ –	gives	–
– x –	gives	+		– ÷ –	gives	+

Example G

1. $-40 \times 5 = -200$ [since signs are unlike answer is negative]
2. $-6 \times -3 = 18$ [since signs are like answer is positive]
3. $\dfrac{120}{-40} = -3$ [since signs are unlike answer is negative]
4. $\dfrac{-45}{-9} = 5$ [since signs are like answer is positive]

The rules for multiplying and dividing integers can be generalised as follows:

> When an **even number** of negative numbers are multiplied together or divided, the result is a *positive* number.
>
> When an **odd number** of negative numbers are multiplied together or divided, the result is a *negative* number.

Example H

Q. Calculate each of the following:

1. $-2 \times 4 \times -5$ **2.** $(-2)^3$ **3.** $\dfrac{-4 \times 15}{-3 \times -2}$ **4.** $\dfrac{-5 \times -6}{-2 \times -3}$

A. **1.** $-2 \times 4 \times -5 = 40$ [the answer is positive because there are two negative numbers in the calculation (two is even)]

2. $(-2)^3 = -2 \times -2 \times -2$
$= -8$ [the answer is negative because there are three negative numbers in the calculation (three is odd)]

3. $\dfrac{-4 \times 15}{-3 \times -2} = -\dfrac{60}{6}$
$= -10$ [the three negative numbers in the calculation give a negative answer]

4. $\dfrac{-5 \times -6}{-2 \times -3} = \dfrac{30}{6}$
$= 5$ [the four negative numbers in the calculation give a positive answer]

Note: Division by zero is **undefined**. This means that it is impossible to divide by zero (attempting to do so on a calculator will result in an error message). Thus, expressions such as $\frac{3}{0}$, or $\frac{-4}{0}$ cannot be simplified in any way to a finite value.

Integers and the calculator

Use the negative key [(–)] to enter negative numbers into the calculator; eg to enter –5 press

Note: On most calculators the subtraction key [–] may be used instead of the negative key to enter negative numbers. Some calculators may have a [+/–] key which can be used to change the sign of a number. For example [2] [+/–] gives –2

Example I

Q. Calculate $-45 \times 78 \times -23$.

A. Enter the calculation by pressing

to get the answer 80 730 [positive since two negative integers in the calculation]

Unit 11.1 Activity 2A: Order of operations and integers

1. Use the correct order of operations to evaluate (where possible) each of the following *without* using a calculator:

a. $4 + 3 \times 2$ **b.** $4 \times 3 + 2$ **c.** $(5 + 3) \times 4$ **d.** $2(6 + 3)$

e. $12 - 6 \div 3$ **f.** $18 \div (6 + 3)$ **g.** $8 \times 5 - 3 \times 4$ **h.** $3^2 \times 2^2$

i. $5^2 - 4^2$ **j.** $8 - 5 + 3 \times 2$ **k.** $2 + 3^2 - 9 + 2$ **l.** $3 + 4^2 - 6 + 2$

m. $10 - (12 - 4 \times 2)$ **n.** $\dfrac{6 + 9}{7 - 4}$ **o.** $\dfrac{15 - 3}{3 \times 2}$ **p.** $\dfrac{5 \times 2 - 3 \times 2}{5 - 3}$

q. $\dfrac{3 + 12 \times 5}{9 - 2 \times 3}$ **r.** $\dfrac{4 - 4}{4 + 4}$ **s.** $\dfrac{4 + 4}{4 - 4}$ **t.** $\dfrac{6 \times 4 - 8 \times 3}{12 \times 2 - 8 \times 3}$

2. Peter purchases a phone for K157. He pays K35 and returns the next day to complete the payment on his phone. He wishes to buy an K18 phone case at the same time and to cash in a voucher for 'K10 off any purchase over K50'. How much does he need to pay the shop to settle his account?

3. Jane goes on a diet. She starts with a weight of 61 kg, and loses 2 kg every 3 weeks for 12 weeks. She abandons her diet at *the start* of a 2-week holiday, and gains back 1 kg per week. How much does she weigh at *the end* of her holiday?

4. Evaluate each of the following:

a. $8 - 12$ **b.** $5 - -2$ **c.** $9 - (7 - 10)$ **d.** $\frac{-12}{4}$

e. 5×-4 **f.** -3×-4 **g.** $-9 - 15$ **h.** $3 \times (-2)^2$

i. $-6 - -6$ **j.** $11 - (3 + 5)$ **k.** $\frac{35}{-5}$ **l.** $\frac{-3 \times -2 \times 4 \times -7}{8}$

m. $\frac{-2 \times 2 \times 4 \times -5}{16}$ **n.** $14 \div 7 \div -2$ **o.** $8 - 2(4 - 6)$ **p.** $\frac{2 - 3 - 4}{-2 - 3}$

q. $(6 - 11)^2$ **r.** $9 - (8 - 4)^2$ **s.** $\frac{(-4)^2 - 1}{-4 - 1}$ **t.** $((-7 + 4)^2 - 10)^2$

5. The 7 am temperature reading at Mt Wilhelm is –8°C. The noon reading is 10°C higher than this, but by midnight the temperature has fallen to 2°C below the 7 am reading.

a. What is the temperature at noon?

b. What is the temperature at midnight?

c. By how many degrees Celsius has the temperature fallen between noon and midnight?

6. Temperatures in Enga on three consecutive days are:

Day	Monday	Tuesday	Wednesday
Noon	2°C	–1°C	–3°C
Midnight	–5°C	–8°C	–9°C

a. By how many degrees did the temperature fall between noon and midnight on:
i. Monday? **ii.** Tuesday?

b. What was the change in temperature between Tuesday midnight and Wednesday noon?

c. The temperature at noon on the following Thursday was 7°C above the temperature at midnight on Wednesday. What was the Thursday noon temperature?

7. Pius had K32.50 in his cheque account. He wrote cheques for K25.60 and K45.80. He deposited K24. What was the balance after these transactions?

8. A bank offers an 'overdrawn' service in which customers are permitted a negative balance, usually with some maximum limit to the amount the customer may be overdrawn.

Paul has an account with this service, with a maximum 'overdrawn' limit of –K2 000.

Paul has his monthly salary of K1 980 credited automatically on the first of each month. On the first day of each month he automatically pays bills of K250 and withdraws K675 spending money as well as his monthly rent of K712. A friend who bought a car from him is repaying the debt at K425 per month by automatically crediting Paul's account on the first of each month.

At the end of May, Paul has just bought a new car and is overdrawn by K1 875.

a. If Paul spends as described above, how long will it be before Paul has at least K1 200 balance? Show all working clearly.

b. Paul wants to have a balance of at least K1 200 by the end of August. To do this he will reduce his monthly spending money. What is the maximum amount he should spend per month? Set out your working clearly.

Squares and square roots

The **square** of a number is found by multiplying that number by itself, eg the square of 5 is $5 \times 5 = 25$.

The square of a number, a, is $a \times a$, and is written a^2

It is useful to be familiar with the squares of the first dozen or so counting numbers: $1^2 = 1$, $2^2 = 4$, $3^2 = 9$, $4^2 = 16$, $5^2 = 25$, $6^2 = 36$, $7^2 = 49$, $8^2 = 64$, $9^2 = 81$, $10^2 = 100$, $11^2 = 121$, $12^2 = 144$, ... (These numbers are often called **perfect squares**.)

Scientific calculators have a squaring key 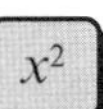.

Example J

1. $13^2 = 13 \times 13$
$= 169$

2. $(-4)^2 = -4 \times -4$
$= 16$

3. $-4^2 = -1 \times 4^2$
$= -1 \times 16$
$= -16$

Note: There is an important difference in meaning between the expressions $(-4)^2$ and -4^2 in part **2** and part **3** of the example. Use brackets carefully so that correct values are calculated.

The inverse of squaring a number is finding the **square root** of a number. For example, the square of 7 is 49 so the square root of 49 is 7 (ie, 7 is the number which squares to make 49). Similarly the square root of 100 is 10 since $10^2 = 100$.

The positive square root of a number a is written $\sqrt{a}$

For example, $\sqrt{49} = 7$ and $\sqrt{100} = 10$. Thus, even though $-4 \times -4 = 16$, the value of $\sqrt{16}$ is +4 (not –4).

Scientific calculators have a square root key $\sqrt{\ }$ for evaluating square roots.

Since squaring and taking the square root are inverses it follows that, for non-negative real numbers, a,

$$\left(\sqrt{a}\right)^2 = \sqrt{a} \times \sqrt{a} = a$$

For example $\sqrt{49} \times \sqrt{49} = 49$ and $\sqrt{3} \times \sqrt{3} = 3$.

Note: Since squares of numbers are always positive, the square root of a negative number is undefined (no real number multiplies by itself to make a negative number). For example, $\sqrt{-4}$ is undefined.

Example K

Q. 1. Without using a calculator,

a. Find $\sqrt{64}$. **b.** Find $\sqrt{81}$. **c.** Estimate $\sqrt{70}$.

2. Use a calculator to evaluate $\sqrt{70}$ to 3 dp.

3. Find $-\sqrt{144}$.

A. 1. a. $\sqrt{64} = 8$ [since $8 \times 8 = 64$]

b. $\sqrt{81} = 9$ [since $9 \times 9 = 81$]

c. 70 lies between the perfect squares 64 and 81 so $\sqrt{70}$ lies between $\sqrt{64}$ and $\sqrt{81}$, ie between 8 and 9. A reasonable approximation to $\sqrt{70}$ would be 8.5.

2. Using the square root key, $\sqrt{70} = 8.367$ (3 dp).

3. $-\sqrt{144} = -1 \times \sqrt{144} = -1 \times 12 = -12$

Order of operations with squares and square roots

Some expressions involving squares and square roots involve other operations as well. In the correct order of operations (mnemonic BEMA), brackets are done before exponents (**powers**). Thus, squaring and its **inverse operation**, taking the square root, are done *after* any expressions in brackets are evaluated. Note that expressions under the square root symbol are assumed to be in brackets (ie the brackets are implicit).

Example L

Q. Find the value of **1.** $(5 + 6)^2$ **2.** $\sqrt{100 + 44}$ **3.** $\sqrt{5^2 - 3^2}$

A. 1. $(5 + 6)^2 = 11^2 = 121$ [adding inside bracket first]

2. $\sqrt{100 + 44}$ means $\sqrt{(100 + 44)}$ [brackets under square root symbol are implicit]

$\sqrt{100 + 44} = \sqrt{144} = 12$ [add before taking square root]

3. $\sqrt{5^2 - 3^2} = \sqrt{(25 - 9)} = \sqrt{16} = 4$
[implicit brackets so subtract before taking square root]

Higher powers and roots of numbers

In a similar way to squaring, numbers can be raised to higher powers. For example, to **cube** a number (raise it to the power of three), the number is multiplied by itself then by itself again, so that it appears *three* times in the product. Thus, $5^3 = 5 \times 5 \times 5 = 125$.

In general, to raise a number to any whole number power, *n*, the following definition holds:

$$x^n = \underbrace{x.x.x. \ldots .x}_{n \text{ factors}}$$

For example, $2^6 = \underbrace{2 \times 2 \times 2 \times 2 \times 2 \times 2}_{6 \text{ factors}} = 64$ and $3^5 = \underbrace{3 \times 3 \times 3 \times 3 \times 3}_{5 \text{ factors}} = 243$.

Using calculators to evaluate higher powers and roots

Calculators are very useful tools for evaluating powers of numbers. Most scientific calculators have a cube key [x^3] for finding the cubes of numbers. To evaluate higher powers, the **power key** is used. On your calculator this may look like [x^y] or [^]. For example, to calculate 9^4, press [9] [^] [4] [=] (or [9] [x^y] [4] [=]) to get 6 561.

The inverse of *raising a number to the power of n is finding the nth root of a number*. For example, the cube of 5 is 125, so the **cube root** of 125 is 5 (ie, 5 is the number which cubes to make 125). Many scientific calculators have a cube root key [$\sqrt[3]{\ }$] for evaluating cube roots.

In general the **nth root** of a number *a* is written $\sqrt[n]{a}$, eg the 7th root of 2 187 is written $\sqrt[7]{2\ 187}$.

To evaluate this number, the *n*th root key [$\sqrt[x]{\ }$] is used on the calculator (press SHIFT [x^y]). For example, to calculate $\sqrt[7]{2\ 187}$, press [7] [$\sqrt[x]{\ }$] [2] [1] [8] [7] [=] to get the answer 3.

Note: Since numbers to even powers are positive, the *n*th root of an odd number is undefined when *n* is even. For example $\sqrt[4]{-16}$ is undefined since the fourth power of any number is always positive. Similarly $\sqrt[6]{-1}$, or $\sqrt[8]{-256}$ are undefined.

Example M

Q. Find **1.** The **volume** of a cube whose side length is 16 cm.

2. The side length of a cube whose volume is 343 cm^3.

A. **1.** Since volume = $(\text{side length})^3$ [**formula** for volume of cube]

Volume of cube of side length 16 = 16^3 = 4 096 cm^3

2. Since side length = $\sqrt[3]{\text{volume}}$ [cube root is inverse of cubing]

Side length of a cube whose volume is 343 = $\sqrt[3]{343}$ = 7 cm

Unit 11.1 Activity 2B: Powers and roots of numbers

1. Evaluate each of the following to a maximum of 3 dp:

a. 5^2 **b.** 9^2 **c.** 25^2 **d.** 43^2 **e.** $\sqrt{36}$

f. $\sqrt{169}$ **g.** $\sqrt{3\ 136}$ **h.** $\sqrt{9\ 604}$ **i.** $\sqrt{346}$ **j.** $\sqrt{1\ 234}$

k. $\sqrt{850}$ **l.** $\sqrt{60}$ **m.** $(\sqrt{3})^2$ **n.** $\sqrt{4^2}$ **o.** $(\sqrt{5})^2$

p. $\sqrt{9^2}$ **q.** $-\sqrt{121}$ **r.** $-\sqrt{400}$ **s.** $-\sqrt{600}$ **t.** $\sqrt{(-5)^2}$

2. Calculate each of the following using a calculator:

a. $8^2 - 4 \times 5^2$ **b.** $(2 - 3)^2$ **c.** $(-2 - 4)^2$

d. $5^2 - 5 \times 5^2 + 5^2$ **e.** $\sqrt{25^2 - 24^2}$ **f.** $8 - \sqrt{59 + 5}$

3. Use your knowledge of perfect squares to give two consecutive numbers these square roots lie between. Confirm using a calculator.

a. $\sqrt{140}$ **b.** $\sqrt{87}$ **c.** $\sqrt{27}$ **d.** $\sqrt{410}$

4. A square has an area of 1 369 cm^2.

a. What is its side length?

b. How much longer would the side of the square need to be to create a square with double the original area?

5. Two square gardens are planted. The first has a side length of 12 m. The second garden has a side length of 15 m.

a. How many **square metres** greater is the area of the second garden than the first?

b. A third square garden has an area equal to the difference between the areas of the first and second gardens. What is the side length of this garden?

6. Evaluate where possible to a maximum of 3 dp.

a. 8^3 **b.** 9^4 **c.** 3^5 **d.** 4^6 **e.** $\sqrt[3]{27}$

f. $\sqrt[4]{14\,641}$ **g.** $\sqrt[5]{1\,024}$ **h.** $\sqrt[7]{128}$ **i.** $\sqrt[6]{6}$ **j.** $\sqrt[4]{1}$

k. $\sqrt[10]{50}$ **l.** $\sqrt[3]{510}$ **m.** $(\sqrt[3]{2})^3$ **n.** $\sqrt[4]{5^4}$ **o.** $\sqrt[3]{7^3}$

p. $\sqrt[4]{2^8}$ **q.** $-\sqrt[3]{216}$ **r.** $\sqrt[3]{-216}$ **s.** $\sqrt[4]{-64}$ **t.** $\sqrt[7]{-9}$

7. Evaluate (where possible) to a maximum of 3 dp.

a. $\sqrt[4]{416 - 200}$ **b.** $(1 + 2)^4$ **c.** $\sqrt[4]{2^4 - 1^4}$ **d.** $\dfrac{\sqrt[5]{32}}{\sqrt[6]{64}}$

e. $\sqrt[3]{(8 - 6)^3}$ **f.** $\dfrac{1 + \sqrt[3]{27}}{\sqrt[3]{-64}}$

8. A cube has side length 5 cm.

a. What is its volume?

A second cube has twice the volume of the first cube.

b. What is its side length?

9. A lump of clay has volume 1 000 cm^3. Four cubes are made from it. The first three cubes have equal volumes, but the third cube is double the volume of each of the other cubes.

a. What is the side length of the small cubes?

b. What is the side length of the larger cube?

c. It is suggested that the side length of the larger cube is equal to the product of the side length of the small cube and $\sqrt[3]{k}$ for some k. Find the value of k.

10. A block of chocolate has volume 2^{10} **cubic centimetres**. It is melted and poured into four identical moulds in the shape of cubes. There is 160 cm^3 quantity of chocolate left. What is the side length of each of these cube-shaped moulds?

Unit 11.1 Number and Application

Topic 3: Basic numeracy—revision of fractions

This Topic presents a revision of fractions and looks at problem solving in context. It covers:

- Fractions and mixed numbers.
- Solving fraction problems in context.
- Sharing quantities in a given ratio.
- Solve number problems in context, involving manipulation, several steps or reversing processes.

Introduction

The aim of this Topic is to solve practical problems involving fractions. Many operations with fractions should be familiar by now, and scientific calculators with a fraction key can be used to carry out fraction calculations rapidly. However, the processes involved must also be understood (without the use of a calculator) so that algebraic problems involving fractions can be solved.

Fractions

A **fraction** is part of a whole. Fractions can be illustrated with diagrams which show the part compared to the whole. A fraction is commonly expressed as a **numerator** divided by a **denominator**. The numerator is the number on top of the fraction line and the denominator is the number below the fraction line. (The fraction line should be horizontal, not sloping.)

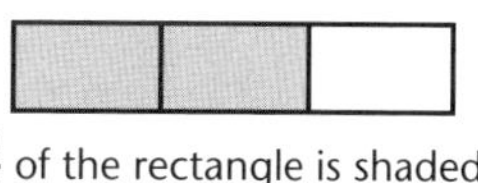

$\frac{2}{3}$ of the rectangle is shaded

Equivalent fractions

Fractions representing the same quantity are called **equivalent fractions**. For example, $\frac{3}{4}, \frac{6}{8}, \frac{12}{16}$ are equivalent fractions, as the following diagrams show:

$\frac{3}{4}$

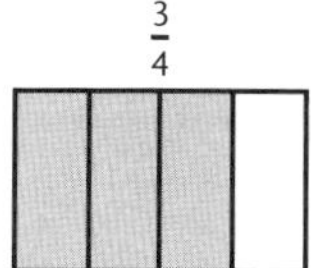

$\frac{6}{8}$

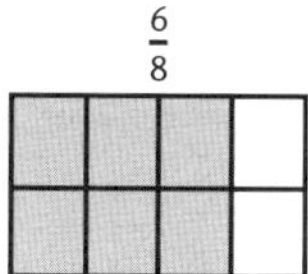

$\frac{12}{16}$

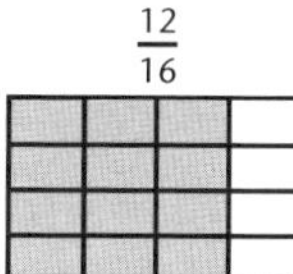

Forming equivalent fractions

Equivalent fractions can be formed by *multiplying* the numerator *and* denominator of a fraction by the *same* number.

Thus $\frac{3}{4}$ and $\frac{6}{8}$ are equivalent fractions because $\frac{3}{4} \overset{\times 2}{=} \frac{6}{8}$ (numerator × 2, denominator × 2)

Equivalent fractions may also be formed by *dividing* the numerator *and* denominator of a fraction by the *same* number.

For example, $\frac{12}{16}$ and $\frac{3}{4}$ are equivalent fractions because

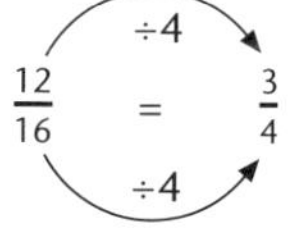

Simplifying fractions

Simplifying a fraction means writing it as an equivalent fraction with the smallest possible denominator. Simplifying involves **cancelling**, which means *dividing* the numerator *and* denominator ('top and bottom') by the *same* number. Cancelling continues until the fraction is in its **simplest form** (the numerator and denominator have no more common **factors**).

Example A

Q. Express in simplest form: **1.** $\frac{6}{8}$ **2.** $\frac{16}{24}$

1. $\frac{6}{8}$ can be simplified to $\frac{3}{4}$ by dividing both 6 and 8 by 2.

This is written $\frac{6}{8} = \frac{6 \div 2}{8 \div 2} = \frac{3}{4}$

2. $\frac{16}{24} = \frac{4}{6}$ [dividing top and bottom by 4]

$= \frac{2}{3}$ [dividing top and bottom by 2]

Note: In part **2**, $\frac{4}{6}$ is *not* the answer because it can be simplified further.

An alternative procedure for part **2** involves dividing the top and bottom of the fraction by 8, the highest common factor of 16 and 24.

Thus $\frac{16}{24} = \frac{16 \div 8}{24 \div 8} = \frac{2}{3}$

Equivalent fractions are useful for **comparing** fractions (putting fractions in order of size). Express each fraction with the same denominator, and then compare numerators.

Example B

Q. Put in order of size from smallest to largest: $\frac{1}{3}, \frac{2}{5}, \frac{9}{25}$.

A. The **lowest common multiple** of 3, 5 and 25 is 75.

Expressing each fraction with denominator 75:

$\frac{1 \times 25}{3 \times 25} = \frac{25}{75}$ $\frac{2 \times 15}{5 \times 15} = \frac{30}{75}$ $\frac{9 \times 3}{25 \times 3} = \frac{27}{75}$

gives $\frac{1}{3}, \frac{9}{25}, \frac{2}{5}$ as the required order. [since $25 < 27 < 30$]

Types of fractions

Fractions can be classified as shown below:

Proper fraction	Improper fraction	Mixed number
The size of the numerator is *smaller* than the size of the denominator. Examples: $\frac{1}{2}, \frac{-3}{4}, \frac{15}{16}$	The size of the numerator is *bigger* than the size of the denominator. Examples: $\frac{3}{2}, \frac{4}{3}, \frac{-17}{9}$	An integer and a proper fraction are combined and written next to each other. Examples: $2\frac{3}{4}, 3\frac{1}{2}, -4\frac{1}{5}$

Mixed numbers and improper fractions

To change an **improper fraction** to a **mixed number**, divide the numerator by the denominator.

- The **quotient** is the integer part of the mixed number.
- The remainder is written over the divisor (which is the bottom part of the improper fraction).

Example C

The improper fraction $\frac{75}{8}$ is changed to a mixed fraction by dividing 75 by 8:

$$\begin{array}{r} 9 \\ 8\overline{)75} \\ \underline{72} \\ 3 \end{array}$$

$\therefore \frac{75}{8} = 9\frac{3}{8}$ [the quotient 9 is the integer part, the remainder 3 is written over the **divisor** 8]

Note: The mixed number $9\frac{3}{8}$ means $9 + \frac{3}{8}$.

To change a mixed number to an improper fraction, multiply the *integer* part of the mixed number by the denominator of the fraction and add the result to the numerator of the fraction.

- The result of this calculation gives the numerator of the improper fraction.
- The denominator of the fraction remains unchanged.

Example D

Q. **1.** Show that $3\frac{4}{5}$ is equivalent to the improper fraction $\frac{19}{5}$.

2. Illustrate the relationship between $3\frac{4}{5}$ and $\frac{19}{5}$ diagrammatically.

3. Change $7\frac{3}{8}$ to an improper fraction.

A. **1.** The mixed number $3\frac{4}{5}$ becomes $\frac{3 \times 5 + 4}{5} = \frac{19}{5}$.

[multiply integer 3 by denominator 5 and add the numerator 4]

2. $3 + \frac{4}{5} = \frac{19}{5}$

3. $7\frac{3}{8} = \frac{7 \times 8 + 3}{8}$

$= \frac{59}{8}$

Unit 11.1 Activity 3A: Equivalent fractions and mixed numbers

1. Express the shaded parts as a fraction of the whole:

a. 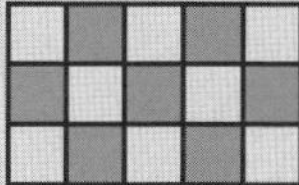

b. 

c. 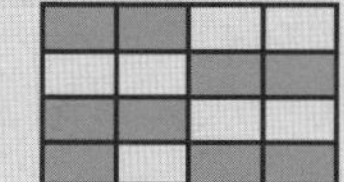

2. Jeanne has a bag containing 8 red, 4 orange and 5 green sweets.

a. What fraction of the sweets are red?

b. What fraction of the sweets are green?

c. What fraction of the sweets are *not* green?

3. Eri purchased a K25 hammer, K10 worth of nails and a K15 metal ruler. What fraction of the total cost was

a. The hammer? **b.** The nails? **c.** The ruler?

Simplify your answers.

4. Write each of the following fractions in their simplest form:

a. $\frac{4}{8}$ **b.** $\frac{6}{24}$ **c.** $\frac{28}{60}$ **d.** $\frac{125}{1\,000}$ **e.** $\frac{72}{132}$ **f.** $\frac{5}{10}$

g. $\frac{18}{24}$ **h.** $\frac{75}{125}$ **i.** $\frac{175}{250}$ **j.** $\frac{625}{1\,000}$

5. Change each of the following to mixed numbers:

a. $\frac{19}{2}$ **b.** $\frac{48}{5}$ **c.** $\frac{79}{6}$ **d.** $\frac{87}{10}$ **e.** $\frac{100}{12}$ **f.** $\frac{11}{2}$

g. $\frac{17}{3}$ **h.** $\frac{28}{5}$ **i.** $\frac{47}{6}$ **j.** $\frac{95}{10}$

6. Change each of the following to improper fractions:

a. $2\frac{1}{2}$ **b.** $4\frac{2}{3}$ **c.** $5\frac{3}{4}$ **d.** $5\frac{7}{8}$ **e.** $12\frac{7}{20}$ **f.** $2\frac{1}{3}$

g. $3\frac{4}{5}$ **h.** $2\frac{11}{20}$ **i.** $11\frac{5}{8}$ **j.** $12\frac{11}{15}$

7. Put in order of size (from smallest to largest):

a. $\frac{3}{8}, \frac{2}{7}, \frac{1}{3}$ **b.** $\frac{1}{6}, \frac{3}{20}, \frac{2}{15}$ **c.** $\frac{7}{8}, 1\frac{1}{5}, \frac{4}{3}$ **d.** $2\frac{3}{4}, \frac{12}{5}, \frac{8}{3}$

8. Shift workers in a factory were offered part shifts as overtime. Anne worked $\frac{1}{3}$ of a shift, Bob worked $\frac{3}{8}$ of a shift and Carol worked $\frac{5}{14}$ of a shift as overtime. Who did

a. The most overtime? **b.** The least overtime?

9. Three out of every five employees in a clinic are male. If there are 36 males, how many workers are there in the clinic?

10. Takis is a triathlete. Each day he runs, swims and cycles. Takis runs 12 km each day.

a. After he has run 8 km what fraction of his run remains?

Takis swims 3 km each day.

b. What fraction of his total running and swimming distance is the run?

Takis cycles 25 km each day.

c. What fraction of the total training distance is the cycling?

Give all fractions in simplest form.

Addition and subtraction of fractions

When adding (or subtracting) fractions with the *same denominators*:

- The numerators are added together (or subtracted).
- The denominator of the answer is the same as the original denominators.

Example E

Q. Calculate **1.** $\frac{1}{5} + \frac{2}{5}$ **2.** $\frac{3}{5} - \frac{1}{5}$

A. **1.** $\frac{1}{5} + \frac{2}{5} = \frac{1+2}{5}$ [add numerators (1 and 2), keep the denominator the same]

$= \frac{3}{5}$

This can be shown in the diagram shown alongside.

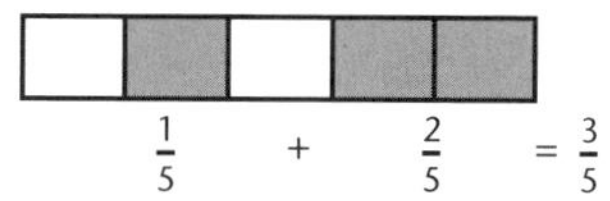

2. $\frac{3}{5} - \frac{1}{5} = \frac{3-1}{5}$ [subtracting numerators, keeping denominator the same]

$= \frac{2}{5}$

If the *denominators are different*, the fractions must first be changed into equivalent fractions with the same denominators. Then proceed as before.

Example F

Q. Calculate: **1.** $\frac{2}{3} + \frac{1}{4}$ **2.** $\frac{7}{8} - \frac{2}{5}$ **3.** $1 - \left(\frac{2}{3} + \frac{1}{4}\right)$

A. **1.** Since denominators are different, change into equivalent fractions with a common denominator of 12. [12 is the LCM of 3 and 4]

$$\frac{2}{3} + \frac{1}{4} = \frac{2 \times 4}{3 \times 4} + \frac{1 \times 3}{4 \times 3}$$

$$= \frac{8}{12} + \frac{3}{12}$$

$$= \frac{11}{12}$$

2. $$\frac{7}{8} - \frac{2}{5} = \frac{7 \times 5}{8 \times 5} - \frac{2 \times 8}{5 \times 8}$$

$$= \frac{35}{40} - \frac{16}{40}$$

$$= \frac{19}{40}$$

Note: An alternative setting out is:

$$\frac{7}{8} - \frac{2}{5} = \frac{7 \times 5 - 2 \times 8}{40}$$

[changing $\frac{7}{8}$ and $\frac{2}{5}$ to fractions with a denominator of 40, since 40 is the LCM of 5 and 8]

$$= \frac{35 - 16}{40}$$

$$= \frac{19}{40}$$

3. $1 - \left(\frac{2}{3} + \frac{1}{4}\right) = 1 - \frac{11}{12}$ **[from part 1]**

$= \frac{12}{12} - \frac{11}{12}$ [since $1 = \frac{12}{12}$]

$= \frac{1}{12}$

Mixed numbers can be added or subtracted by changing them to improper fractions and continuing as before. The final answer is expressed in mixed number form, where appropriate.

Example G

Q. Calculate: $8\frac{2}{3} - 2\frac{5}{12}$

A. $8\frac{2}{3} - 2\frac{5}{12} = \frac{26}{3} - \frac{29}{12}$ [changing to improper fractions]

$= \frac{104}{12} - \frac{29}{12}$ [change $\frac{26}{3}$ to a fraction with a denominator of 12]

$= \frac{75}{12}$

$= 6\frac{3}{12}$ [change to mixed number]

$= 6\frac{1}{4}$ [simplifying]

Note: Many calculators have a fraction key [$a\frac{b}{c}$] so that addition, subtraction, multiplication and division of fractions may be carried out with the answer given as a fraction.

For example, to find the answer to Example G, press

[8] [$a\frac{b}{c}$] [2] [$a\frac{b}{c}$] [3] [−] [2] [$a\frac{b}{c}$] [5] [$a\frac{b}{c}$] [1] [2] [=] to get the answer

[6 ⌟ l ⌟ 4] which is $6\frac{1}{4}$.

Multiplying fractions

To multiply proper or improper fractions:

- Multiply numerators together.
- Multiply denominators together.

Note that the word 'of' is used to mean multiplication eg $\frac{1}{2}$ of $\frac{5}{12}$ means $\frac{1}{2} \times \frac{5}{12} = \frac{5}{24}$.

Example H

Q. One fifth of a garden is planted in crops. Two thirds of the crops are taro. What fraction of the garden is planted with taro?

A. As the diagram illustrates, $\frac{2}{3}$ of $\frac{1}{5}$ of the garden is in taro.

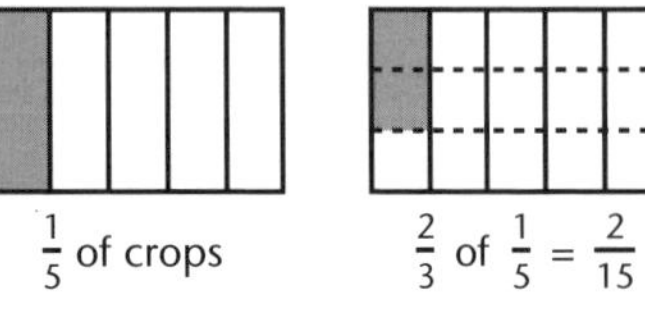

$\frac{2}{3} \times \frac{1}{5} = \frac{2 \times 1}{3 \times 5}$ ['of' means multiply]

$= \frac{2}{15}$

So $\frac{2}{15}$ of the garden is planted with taro.

Often the multiplication process can be simplified by cancelling first.

Example I

Q. Twenty-eight boxes of fruit are delivered of which five boxes contain guavas. $\frac{4}{15}$ of the guavas are Indian variety. What fraction of the delivery is Indian?

A. $\frac{5}{28}$ of the delivery are guavas. $\frac{4}{15}$ of $\frac{5}{28}$ are Indian.

$$\frac{4}{15} \times \frac{5}{28} = \frac{{}^{1}\cancel{4}}{\cancel{15}_{3}} \times \frac{\cancel{5}^{1}}{\cancel{28}_{7}}$$

[dividing left numerator and right denominator by 4; dividing left denominator and right numerator by 5]

$$= \frac{1 \times 1}{3 \times 7}$$

$$= \frac{1}{21}$$

To multiply mixed numbers, change them to improper fractions and then proceed as in the previous example.

Example J

$$2\frac{2}{3} \times 5\frac{3}{4} = \frac{{}^{2}\cancel{8}}{3} \times \frac{23}{\cancel{4}_{1}}$$

[changing to improper fractions and cancelling]

$$= \frac{2 \times 23}{3 \times 1}$$

$$= \frac{46}{3}$$

$$= 15\frac{1}{3}$$

Fractions may be raised to powers, using the rules for multiplication of fractions. Again, change mixed numbers to improper fractions first.

Example K

Q. Calculate **1.** $\left(\frac{3}{4}\right)^3$ **2.** $\left(2\frac{2}{3}\right)^2$

A. **1.** $\left(\frac{3}{4}\right)^3 = \frac{3 \times 3 \times 3}{4 \times 4 \times 4}$

$$= \frac{27}{64}$$

2. $\left(2\frac{2}{3}\right)^2 = \left(\frac{8}{3}\right)^2$ [changing to improper fraction]

$$= \frac{8 \times 8}{3 \times 3}$$

$$= \frac{64}{9} \text{ or } 7\frac{1}{9}$$

Division of fractions

The inverse of multiplying by a fraction $\frac{a}{b}$ is either

- Dividing by $\frac{a}{b}$ (since multiplication and division are inverses), or
- Multiplying by $\frac{b}{a}$, (since $\frac{a}{b} \times \frac{b}{a} = \frac{ab}{ba} = 1$).

Thus, dividing by a fraction $\frac{a}{b}$ is the same as multiplying by the inverted fraction $\frac{b}{a}$.

> To divide by a fraction $\frac{a}{b}$, multiply by $\frac{b}{a}$

For example, division by $\frac{2}{3}$ is changed to multiplication by $\frac{3}{2}$.

Example L

$\frac{2}{3} \div \frac{1}{5} = \frac{2}{3} \times \frac{5}{1}$ [division by $\frac{1}{5}$ becomes multiplication by $\frac{5}{1}$]

$= \frac{2 \times 5}{3 \times 1}$ [multiplying numerators and multiplying denominators]

$= \frac{10}{3}$ or $3\frac{1}{3}$

To invert a mixed number, express it as an improper fraction first. For example to invert $2\frac{1}{2}$, change it to $\frac{5}{2}$ then invert it to get $\frac{2}{5}$.

Example M

$2\frac{3}{4} \div 3\frac{2}{3} = \frac{11}{4} \div \frac{11}{3}$ [change mixed numbers to improper fractions]

$= \frac{^{1}\cancel{11}}{4} \times \frac{3}{\cancel{11}_{1}}$ [changing '$\div \frac{11}{3}$' to '$\times \frac{3}{11}$' and cancelling]

$= \frac{3}{4}$ [multiplying numerators and multiplying denominators]

Unit 11.1 Activity 3B: Operations with fractions

1. Calculate each of the following, giving answers as fractions or mixed numbers in simplest form.

a. $\frac{3}{5} + \frac{1}{5}$ **b.** $\frac{3}{10} + \frac{4}{10}$ **c.** $\frac{5}{8} + \frac{1}{8}$ **d.** $\frac{1}{3} + \frac{1}{3}$ **e.** $\frac{2}{3} + \frac{1}{5}$

f. $\frac{5}{8} + \frac{2}{5}$ **g.** $\frac{1}{4} + \frac{1}{5}$ **h.** $\frac{2}{5} + \frac{1}{6}$ **i.** $\frac{4}{5} + \frac{3}{4}$ **j.** $1\frac{1}{5} + \frac{1}{5}$

k. $2\frac{1}{2} + 1\frac{1}{4}$ **l.** $1\frac{3}{4} + 2\frac{5}{6}$

2. Calculate each of the following, giving answers as fractions or mixed numbers in simplest form.

a. $\frac{5}{8} - \frac{2}{8}$ **b.** $\frac{7}{10} - \frac{4}{10}$ **c.** $\frac{11}{12} - \frac{4}{12}$ **d.** $\frac{1}{2} - \frac{1}{3}$ **e.** $\frac{2}{3} - \frac{1}{4}$

f. $\frac{5}{9} - \frac{3}{8}$ **g.** $\frac{1}{4} - \frac{1}{5}$ **h.** $\frac{7}{8} - \frac{5}{6}$ **i.** $\frac{5}{12} - \frac{3}{8}$ **j.** $1\frac{1}{3} - \frac{2}{3}$

k. $2\frac{4}{5} - 1\frac{2}{3}$ **l.** $3\frac{1}{4} - 1\frac{3}{8}$

3. Calculate each of the following, giving answers as fractions or mixed numbers in simplest form.

a. $\frac{1}{2} \times \frac{1}{3}$ **b.** $\frac{5}{8} \times \frac{4}{15}$ **c.** $\frac{7}{12} \times \frac{2}{5}$ **d.** $\frac{1}{4} \times \frac{1}{5}$ **e.** $\frac{7}{8} \times \frac{16}{21}$

f. $\frac{8}{15} \times \frac{25}{36}$ **g.** $2\frac{2}{3} \times 4$ **h.** $6\frac{1}{2} \times 3\frac{4}{5}$ **i.** $4\frac{3}{8} \times 2\frac{4}{5}$ **j.** $3\frac{2}{5} \times 5$

k. $5\frac{1}{3} \times 1\frac{1}{4}$ **l.** $10\frac{2}{3} \times 5\frac{3}{8}$

4. Calculate each of the following, giving answers as fractions or mixed numbers in simplest form.

a. $\frac{1}{2} \div \frac{1}{3}$ **b.** $\frac{3}{4} \div \frac{1}{2}$ **c.** $\frac{5}{8} \div \frac{2}{3}$ **d.** $\frac{1}{4} \div \frac{1}{5}$ **e.** $\frac{2}{5} \div \frac{6}{11}$

f. $\frac{7}{15} \div \frac{6}{35}$ **g.** $1\frac{1}{2} \div \frac{3}{4}$ **h.** $\frac{4}{5} \div 4$ **i.** $4\frac{1}{6} \div 3\frac{1}{3}$ **j.** $\frac{2}{3} \div 3$

k. $2\frac{3}{5} \div 3\frac{1}{2}$ **l.** $4\frac{2}{3} \div 3\frac{4}{7}$

5. Evaluate each of the following as fractions.

a. $\left(\frac{2}{3}\right)^2$ **b.** $\left(\frac{5}{9}\right)^3$ **c.** $\frac{2}{3}$ of 24 **d.** $\frac{3}{4}$ of 30 **e.** $\left(1\frac{1}{2}\right)^2$

f. $\left(-2\frac{1}{3}\right)^2$ **g.** $\left(3\frac{4}{5}\right)^3$ **h.** $\left(-1\frac{2}{3}\right)^4$

6. Give your answers as mixed numbers in simplest form.

a. A square has side length $3\frac{3}{4}$ m. What is its area?

b. A cube has side length $2\frac{1}{4}$ cm. What is its volume?

7. An area of land is divided into blocks of each area $\frac{5}{8}$ hectare. If the total area is 25 hectares, how many blocks are there?

8. At the Port Moresby swimming pool the children's pool is divided into six lanes, five for swimming lessons and one for free play. In the free play area a slide takes up $\frac{2}{5}$ of the lane. What fraction of the pool is taken up by the slide?

9. Sai did $\frac{3}{5}$ of her project on Saturday and another $\frac{1}{4}$ on Monday. What fraction of her project is now complete?

10. Timi and Pala order a pizza. Together they eat $\frac{7}{8}$ of the pizza. If Timi ate $\frac{1}{2}$ of the pizza, what fraction of the pizza did Pala eat?

11. Carol sat a multichoice test of 60 questions. She got $\frac{11}{15}$ of them correct. How many did she get wrong?

12. Susan used $\frac{2}{3}$ cup of plain flour and $\frac{1}{2}$ cup of wholemeal flour in a recipe. How much flour did she use?

Fractions and order of operations

Many fraction calculations involve more than one operation. The expression must be evaluated following the correct order of operations (BEMA).

Example N

Q. Calculate $3\frac{3}{4} \times (\frac{2}{3} - \frac{1}{2})$

A. $3\frac{3}{4} \times (\frac{2}{3} - \frac{1}{2}) = 3\frac{3}{4} \times (\frac{4}{6} - \frac{3}{6})$ [subtracting inside the brackets first]

$= 3\frac{3}{4} \times \frac{1}{6}$ [subtracting]

$= \frac{15}{4} \times \frac{1}{6}$ [changing to a mixed number]

$= \frac{15}{24}$ [multiplying]

$= \frac{3}{8}$ [simplifying]

Alternatively, enter the calculation into a scientific calculator (using fraction key and brackets) and the correct order of operations will be followed automatically.

Solving word problems

When solving problems involving fractions, be sure to identify clearly the operation(s) involved. Remember that all the fractions of a quantity total 1 (the whole quantity).

Example O

Q. Max spends $\frac{2}{5}$ of his money on rent, and $\frac{1}{8}$ on petrol. What fraction of his money remains?

A. Fraction spent $= \frac{2}{5} + \frac{1}{8}$

$= \frac{16 + 5}{40}$ [40 is the LCM of 5, 8]

$= \frac{21}{40}$

Fraction remaining $= 1 - \frac{21}{40}$

$= \frac{40}{40} - \frac{21}{40}$ [creating equivalent fractions with the same denominator]

$= \frac{19}{40}$

Note: A statement about what is being calculated should be included with each separate calculation.

Alternatively the calculation is worked out to be $1 - (\frac{2}{5} + \frac{1}{8})$ or $1 - \frac{2}{5} - \frac{1}{8}$ and this is entered into a scientific calculator to arrive at the answer.

Unit 11.1 Activity 3C: Problem-solving involving fractions

1. Evaluate the following. Give answers as fractions or mixed numbers.

a. $\frac{3}{7} + \frac{1}{2} \times \frac{3}{7}$ **b.** $2\frac{4}{5} - 6\frac{1}{2} \times 3$ **c.** $3\frac{1}{2} \times 2\frac{3}{4} - 4\frac{1}{4} \times 5\frac{1}{5}$

d. $\left(\frac{1}{7} \times \frac{2}{3}\right) \times \frac{3}{2}$ **e.** $\frac{4}{5} \times \left(\frac{5}{4} \times \frac{3}{25}\right)$ **f.** $5 - \left(\frac{7}{8} + \frac{3}{4}\right)$

g. $\left(1\frac{1}{2} + \frac{3}{4}\right)^2$ **h.** $\dfrac{\frac{7}{10} - \frac{2}{5}}{\frac{1}{3} + \frac{2}{5}}$ **i.** $\frac{3}{4} + 1\frac{1}{2} \times 2\frac{2}{3}$

j. $\left(5\frac{3}{4} - 4\frac{2}{5}\right)^2$ **k.** $8\frac{5}{8} - 5\frac{1}{4} \div \frac{3}{4}$ **l.** $\sqrt[3]{\frac{3}{16} + \frac{5}{64} + \frac{5}{32}}$

2. Kila brings $1\frac{1}{4}$ kg of sausages and $2\frac{1}{2}$ kg of steak to a function. If $\frac{4}{5}$ of the meat is cooked and eaten, what weight of meat is that?

3. David competed in a triathlon which involved swimming, cycling, and running. If the length of the swim was $\frac{1}{12}$ of the total distance of the triathlon and the cycling was $\frac{2}{3}$ of the total distance, what fraction of the total distance was the run?

4. In her will, Mrs Diwai leaves $\frac{1}{8}$ of her money to her son, twice as much to her daughter, and half as much as her son receives to her sister. The remainder is left to her husband. What fraction does her husband get?

5. Rawa ate $\frac{3}{5}$ of a pizza. The next day he ate a third of the remaining pizza. What fraction of the pizza has Rawa eaten altogether?

6. Sharon spends $\frac{2}{3}$ of her money at a circus, then half of what is left on afternoon tea. What fraction of her money remains?

7. Mrs Karam brings a dozen fruit pies to a function. At the end $1\frac{1}{2}$ pies remain. What fraction of the total number of pies was eaten?

8. Two friends go shares in a car. Brian pays $\frac{1}{4}$ and Gabriel pays $\frac{3}{8}$ of the total price of the car. The balance of K300 is loaned to them by Gabriel's mother. How much does the car cost?

9. A water tank is $\frac{4}{5}$ full at the beginning of a holiday. During the holiday 600 L of water are used, leaving the tank $\frac{1}{2}$ full. How much water does the tank hold when full?

10. A survey found that two out of every five school students listen to music while doing their homework. If there were 18 students in the survey who listened to music while doing their homework, how many students were there in the survey?

Unit 11.1 Activity 3D: Multiple choice—fractions

1. Which one of the following is **not** equal to $6\frac{1}{4}$?

A. $6+\frac{1}{4}$ **B.** 6.25 **C.** $\frac{75}{12}$ **D.** $\frac{61}{4}$

2. $\left(2\frac{2}{3}\right)^2$ is equal to:

A. $4\frac{4}{9}$ **B.** $7\frac{1}{9}$ **C.** $\frac{64}{3}$ **D.** $\frac{16}{9}$

3. $2\frac{1}{3}-\frac{1}{3}\times\frac{4}{5}=$

A. $\frac{8}{5}$ **B.** $\frac{31}{15}$ **C.** $\frac{26}{15}$ **D.** $\frac{6}{5}$

4. $4\frac{8}{9}\div 2\frac{2}{3}=$

A. $1\frac{5}{6}$ **B.** 2 **C.** $\frac{41}{8}$ **D.** $2\frac{4}{3}$

5. $3\frac{3}{4}+2\frac{1}{3}=$

A. $5\frac{3}{7}$ **B.** $\frac{22}{7}$ **C.** $6\frac{1}{4}$ **D.** $6\frac{1}{12}$

6. Manus weeded $\frac{3}{5}$ of his garden in the morning and another $\frac{1}{3}$ after school. The remainder of his garden that is not weeded is:

A. $\frac{1}{5}$ **B.** $\frac{1}{2}$ **C.** $\frac{1}{15}$ **D.** $\frac{1}{8}$

7. John and Amos buy 16 mandarins. If John eats 5 mandarins and Amos eats a quarter of the total, what fraction of the total is remaining?

A. $\frac{7}{16}$ **B.** $\frac{9}{16}$ **C.** $\frac{2}{9}$ **D.** $\frac{7}{9}$

8. Josie spent two-fifths of her money on food and then one-third of the remainder on clothing. If she had K36.00 to start with, how much does she have left?

A. K9.60 **B.** K9.00 **C.** K14.40 **D.** K12.00

9. At the beginning of the week $\frac{4}{9}$ of a garden is planted with sweet potato and $\frac{1}{3}$ with taro. At the end of the week half of the sweet potato and $\frac{3}{7}$ of the taro is harvested. What fraction of the garden now contains no plants?

A. $\frac{23}{63}$ **B.** $\frac{37}{63}$ **C.** $\frac{40}{63}$ **D.** $\frac{107}{252}$

Unit 11.1 Number and Application
Topic 4: Basic numeracy—decimals and approximations

This topic is also about solving number problems in context. It covers:

- Decimals.
- Making sensible estimates and checking the reasonableness of results.
- Using numbers expressed in standard form.
- Solve number problems in context, involving manipulation, several steps or reversing processes.

Introduction

In a decimal number such as 16.345, the **decimal point** separates the *whole part* of the number, 16, from the *fractional part* of the number, 345 thousandths. Numbers with a decimal point are frequently referred to as **decimals**. In certain calculations it is useful to treat a whole number as a decimal, so a decimal point and extra zeros are added, eg 25 may be written as 25.0 or 25.00, etc.

Changing fractions to decimals

To change a fraction to a decimal the **numerator** of the fraction is divided by the **denominator** of the fraction.

Example A

Q. Change $\frac{1}{4}$ to a decimal.

A. The division 1 ÷ 4 is done, as the following sequence shows:

$$\frac{1}{4} \longrightarrow 4\overline{)1} \longrightarrow \begin{array}{r} 0.25 \\ 4\overline{)1.00} \end{array}$$ [inserting decimal point and extra zeros]

$\therefore \frac{1}{4} = 0.25$

This can be carried out using the division key on the calculator:

press [1] [÷] [4] [=] to get = 0.25.

Notes: 1. Once a fraction is entered into a scientific calculator and [=] is pressed, pressing the fraction key [$a^{b/c}$] again will automatically convert the fraction to a decimal.

2. Changing fractions to decimals is a useful alternative method for comparing the size of fractions, eg $\frac{3}{8} = 0.375$, $\frac{2}{5} = 0.4$. So $\frac{2}{5}$ is bigger than $\frac{3}{8}$.

3. It is useful to be familiar with the decimal form of simple fractions such as $\frac{1}{2}, \frac{1}{4}, \frac{1}{5}, \frac{3}{4}$ etc; and vice versa. [$\frac{1}{2} = 0.5$, $\frac{1}{4} = 0.25$, $\frac{1}{5} = 0.2$, $\frac{3}{4} = 0.75$]

To convert a mixed number to a decimal, convert the fraction to a decimal first, then add the whole number, eg $15\frac{1}{4} = 15.25$.

Changing decimals to fractions

To change a decimal number to a fraction:

- The numerator of the fraction is equal to the whole number formed by removing the decimal point from the decimal.
- The denominator is a power of ten equal to the number of **digits** after the decimal point in the original decimal.

The resulting fraction should be simplified if possible.

Example B

Q. Change to fractions: **1.** 0.3 **2.** 0.47 **3.** 0.0125

A. **1.** $0.3 = \frac{3}{10}$ [numerator is 03 = 3, denominator = 10^1]

2. $0.47 = \frac{47}{100}$ [numerator is 047 = 47, denominator = 10^2]

3. $0.0125 = \frac{125}{10\,000}$ [numerator is 00125 = 125, denominator = 10^4]

$= \frac{1}{80}$

Note: **1.** The denominators in the fractions above have as many zeros as there are digits after the decimal point: 0.3 has one digit after the decimal place and therefore one zero in the denominator, 0.47 has two digits, so two zeros, 0.0125 has four digits, so four zeros, and so on.

2. By pressing the fraction key [$a^{b/c}$] after a decimal is entered and [=] is pressed, many scientific calculators will automatically convert a decimal to a fraction in its simplest form. For example, pressing [•] [2] [=] enters the decimal 0.2 into the calculator. Pressing [$a^{b/c}$] after this automatically converts the calculator display to [1 ⌟ 5] which is the fraction $\frac{1}{5}$.

Recurring decimals

When using division to change a fraction to a decimal, the same digit or group of digits after the decimal point may repeat. This type of decimal is called a **recurring** or **repeating decimal**. To indicate a recurring part, the repeating part is written once with a *dot above the digits at the beginning and end of it*.

Example C

1. 0.123123 ... is written $0.\dot{1}2\dot{3}$

2. $\frac{1}{3} = 0.333$ ... and is written $0.\dot{3}$

3. $\frac{5}{6} = 0.8333$ ... and is written $0.8\dot{3}$

4. $15\frac{4\,508}{9\,990} = 15.4512512$ ... and is written $15.4\dot{5}1\dot{2}$

Decimals and the basic operations

A calculator makes short work of decimal calculations involving the basic operations of addition, subtraction, multiplication and division.

Multiplying by powers of 10

It is important to understand the rule for multiplying decimals by powers of ten. To multiply a decimal number by a power of 10, the decimal point is moved the same *number* of places as there are zeros in the power of 10.

- The decimal point is moved to the *right* for positive powers.
- The decimal point is moved to the *left* for negative powers.

Note: Powers of ten include: $10^0 = 1$, $10^1 = 10$, $10^2 = 100$, $10^3 = 1\,000$, $10^4 = 10\,000$ etc

$10^{-1} = \frac{1}{10} = 0.1$, $10^{-2} = \frac{1}{10^2} = \frac{1}{100} = 0.01$, $10^{-3} = \frac{1}{10^3} = \frac{1}{1\,000} = 0.001$, etc

Further explanation on powers can be found under *Indices* on pages 5–8 in Chapter 1.

Example D

1. $16.23 \times 10 = 162.3$ [decimal point is moved one place to the right]
2. $2.46 \times 1\,000 = 2\,460$ [decimal point is moved three places to the right]
3. $0.2437 \times 10^2 = 24.37$ [decimal point is moved two places to the right]
4. $1.5 \times 10^{-2} = 0.015$ [decimal point is moved two places to the left]
5. $34.86 \times 10^{-4} = 0.003486$ [decimal point is moved four places to the left]

Note: The '0' in 2 460 is required as a **place holder** in **2**. (Zeros are also used in **4** and **5**.)

Unit 11.1 Activity 4A: Calculations involving decimals

1. Change the following fractions to decimals:

a. $\frac{4}{5}$ **b.** $3\frac{19}{25}$ **c.** $\frac{5}{6}$ **d.** $18\frac{7}{12}$ **e.** $\frac{13}{99}$

Change the following decimals to fractions (or mixed numbers) in their simplest form:

f. 0.546 **g.** 3.02 **h.** $4.\dot{3}$ **i.** $0.\dot{4}$ **j.** −11.001

2. Calculate each of the following:

a. $1.2 \times 10 + 3.2$ **b.** $3.2 \times 10^{-1} + 1.5$ **c.** $\frac{0.3 + 4.6}{13 \times 0.2}$ **d.** $\frac{0.4 \times 0.2^2}{0.8}$

e. $\frac{5.8 + 4.6}{13 \times 0.2}$ **f.** $\frac{4.9 \times 2.5}{1.5 - 0.25}$ **g.** Subtract 9.4 from 6.3^2.

h. Multiply (5.2×10^3) by (1.2×10^{-4}).

3. Evaluate each of the following (without a calculator):

a. 10×5.4 **b.** 0.7×100 **c.** 4.2×10^2 **d.** 23.8×10^{-2}

e. 5.763×10^4 **f.** 2.958×10^{-3} **g.** 13.581×10^3

4. Four towns lie along a straight road: A B C D

The distance between town A and town C is 18.5 km and the distance between town B and town D is twice that distance. If the distance between town A and town D is 45 km, find the distance between town B and town C.

5. I have K40 to spend during Monday, Tuesday and Wednesday. On Monday and Tuesday I spend a total of K26.45. On Tuesday and Wednesday I spend a total of K15.30. If I have K5.20 left at the end of Wednesday, how much did I spend on Tuesday?

6. Using his calculator, Silas inputs a number, multiplies it by 4.8, presses = and gets 16.896. What number was input?

7. In Sam's short answer test he got 13 marks out of 20. In the long answer section he got 49 out of 80.

In which part of the test did Sam do better?

8. Janet is 1.5 m tall and her brother is 1.2 times taller. Their father is half their combined height, which is 0.1 m more than their mother's height.

a. What is the combined height of Janet's mother and father?

b. Is Janet more or less than half of her parents' combined height?

9. To calculate Peter's mark, the average of 2.4, 5.76 and 3.09 is added to the (positive) difference between 9.76 and 5.81. The answer is then multiplied by 10. What is Peter's mark?

10. Beth's telephone bill is made up of a charge for line rental and a charge for calls. Beth budgets K70 per month for her phone bill. In July she made K27.45 worth of calls and her bill was K12.72 above her budgeted amount. How much is Beth charged for line rental?

Significant figures

- Every non-zero digit in a number is a **significant figure** (**sf**).
- Any zero that is not at the end of a whole number or at the beginning of a decimal fraction is also a significant figure.

To count the number of significant figures in a number, start counting (from the left) at the first non-zero digit and continue to the end of the number (excluding zeros on the end of a whole number). All zeros within the number and at the end of a decimal are counted.

Example E

Q. State the number of significant figures in each of the following numbers:

1. 641 **2.** 0.0036 **3.** 3 425 000 **4.** 5.009 **5.** 89.0

A.

1.	641	3 significant figures	[the first is 6, the second is 4 and the third is 1]
2.	0.0036	2 significant figures	[count from 3 (leading zeros are not significant)]
3.	3 425 000	4 significant figures	[zeros at the end of a whole number are not significant – see note]
4.	5.009	4 significant figures	[zeros within a number are significant]
5.	89.0	3 significant figures	[zeros at the end of a decimal are significant (the final zero indicates that the number is accurate to the nearest tenth)]

Note: In part **3**, the number 3 425 000 may be accurate to 4, 5, 6 or 7 significant figures, depending on how many of the final zeros are actual (counted) values, instead of placeholder zeros resulting from rounding. For example, there may be exactly 3 425 000 people on a mailing list (the figure is accurate to 7 sf). Alternatively, a country's population may be given to the nearest hundred as 3 425 000 in which case the figure is accurate to 5 sf. Consequently, all zeros at the end of a whole number are assumed to be not significant unless otherwise stated.

Rounding

Rounding a number means writing the number with fewer significant figures. The rounded number will be approximately equal in size to the original number.

To **round** a number to a particular number of significant figures:

- Start from the first significant figure in the number and count to the right the number of significant figures required.
- If the next digit is *5 or greater*, add a 1 to the last significant figure and use zeros where necessary as place holders.
- If the next digit is *less than* 5, leave the last significant figure unchanged and use zeros where necessary as place holders.

Example F

1. 3 486 = 3 500 (2 sf) [The first two significant figures are 3 and 4. The next digit is 8, which is bigger than 5, so 1 is added to the second significant figure, 4. Two zeros are used as place holders.]

2. 0.0642 = 0.064 (2 sf) [The first two significant figures are 6 and 4. The next digit is 2, which is less than 5, so the 4 is left unchanged. Two zeros are used as place holders.]

3. 20.850 = 20.9 (3 sf) [The first three significant figures are 2, 0 and 8. The next digit is 5, so 1 is added to the third significant figure 8. No zeros are needed as place-holders.]

Note: **1.** A common error is to forget place holder zeros when rounding whole numbers. For example in part **1** the rounded number 3 500 is similar in size to 3 486. A rounded value of 35 would be absurd.

2. Do not write the answer in part **3** as 20.900 since this would imply that the answer is accurate to the nearest one thousandth (ie 20.900 has 5 sig figs).]

When rounding a number to a particular number of **decimal places**, the digits *after* the decimal point are counted.

Example G

1. 18.346 = 18.35 (2 dp) [The first two digits after the decimal point are 3 and 4. The next is 6, which is bigger than 5, so 1 is added to the 4 in the second decimal place.]

2. 4.0523 = 4.052 (3 dp) [The first three digits after the decimal point are 0, 5, and 2. The next is 3, which is less than 5, so .052 is left unchanged.]

3. 1.98 = 2.0 (1 dp) [The first digit after the decimal point is 9 and the next is 8, which is greater than 5, so 1 must be added to 9.]

Note: 1. The answer to part **1** is 18.35, *not* 18.350, because the zero in 18.350 shows three digits after the decimal point, not two.

2. In part **3**, the 0 in the 2.0 is essential because it shows the rounding to 1 dp has been done. In other words, the 0 is a significant figure.

Rounding answers to calculations

Calculator answers are often given to too many decimal places when considering the accuracy of the figures started with. For example, in the calculation $16.2 \div 1.72 = 9.418604651$, numbers with 3 sf give rise to an answer with 10 sf.

Answers to calculations should be rounded off to a *sensible* number of significant figures or decimal places. (A problem may specify exactly what accuracy is required eg round off to 2 decimal places.)

Whenever rounding is done, the accuracy of the rounding should be stated: 'to 3 sf', 'to 2 dp', etc.

Generally, rounding is done so that *the answer has the same number of significant figures as the number in the calculation with the least number of significant figures.*

This rule is flexible.

Example H

In the calculation $16.2 \div 1.72$, the answer should be rounded to 3 sf.

Thus $16.2 \div 1.72 = 9.42$ (3 sf)

Example I

Q. Calculate $\frac{19.35 \times 4.2}{6.81}$, rounding your answer sensibly.

A. $\frac{19.35 \times 4.2}{6.81} = 12$ (2 sf)

[Using a calculator, the answer is 11.9339207. Noting that 19.35 has 4 sf, 4.2 has 2 sf and 6.81 has 3 sf, the answer should have 2 sf. However, rounding to 2 sf is not a 'sensible' answer in some situations because it may suggest that the answer is *exactly* 12. The number may therefore be rounded to 3 sf to give 11.9.]

Estimates

An exact answer to a calculation is not always required. In many practical situations, an **estimate** (an **approximate answer**) may be sufficient.

An estimate is made by rounding the numbers in a calculation to one (sometimes two) significant figures. When a calculation involves **division**, the divisor is rounded to 1 sf, and the numerator is rounded to a multiple of the divisor (which may be to 2 sf if necessary). An estimate can be used to check the accuracy of a calculator answer.

Example J

Q. Find estimates for the following calculations:

1. 28.3×846.7
2. $\dfrac{47.56}{6.34}$
3. $\dfrac{734 \times 43.7}{8.87}$

A. 1. $28.3 \times 846.7 \approx 30 \times 800$ [rounding to 1 sf]

$= 24\,000$

2. $\dfrac{47.56}{6.34} \approx \dfrac{48}{6}$ [6.34 is rounded to 6 (1 sf) and 47.56 is rounded to 48, which is the closest multiple of 6 to 47.56]

$= 8$

3. $\dfrac{734 \times 43.7}{8.87} \approx \dfrac{700 \times 40}{9}$ [rounding all to 1 sf]

$= \dfrac{28\,000}{9}$ [simplifying]

$\approx \dfrac{27\,000}{9}$ [rounding numerator to a multiple of 9]

$= 3\,000$

Note: These estimates are not unique. They vary according to how the numbers involved are rounded.

Unit 11.1 Activity 4B: Significant figures and rounding

1. Write each of the following numbers correct to the number of significant figures shown in brackets:

a. 115 (2) **b.** 5 914 (3) **c.** 31 786 (3)

d. 14.713 (3) **e.** 89.863 (3) **f.** 0.0487 (2)

2. Write each of the following numbers correct to the number of decimal places indicated in brackets:

a. 1.537 (2) **b.** 11.705 (1) **c.** 0.4137 (3)

d. 89.8632 (2) **e.** 0.00413 (4) **f.** 9.97 (1)

3. Estimate answers to each of the following (without a calculator) giving answers to 1 or 2 significant figures.

a. 86×149 **b.** 43.4×847 **c.** 348×0.843

d. $\dfrac{855}{28}$ **e.** $\dfrac{98.7}{5.37}$ **f.** $\dfrac{3\,584}{84.7}$

g. $\dfrac{15.7 \times 2.8}{0.34}$ **h.** $\dfrac{437 \times 75.4}{29.1}$ **i.** $\dfrac{431 \times 682.5}{74}$

4. The volume of a cylinder is given by the formula $V = \pi r^2 h$, where r is the **radius** of the circular base, and h is the height. If $r = 1.89$, $h = 59.28$ and $\pi = 3.14159$ (6 sf):

a. Use appropriate whole number approximations to estimate the value of V.

b. Evaluate V correct to 2 decimal places, using the values given for r and h and your calculator value for π.

5. Sila used her calculator and the formula $A = \pi r^2$ to find the area of a circle with a radius of 6.2 cm. She used her calculator value for π. The answer her calculator gave was 120.76282.

   a. Explain why she should not record her answer as 120.76282 cm^2.

   b. Write down a sensible answer that Sila might use.

6. Christine was laying lino tiles on her kitchen floor. The area of the floor is 11.48 m^2 and the area of each tile is 0.0625 m^2. Tiles come in packets of 20. Assuming no tiles need to be cut, how many packets should Christine buy? Explain your answer.

7. The volume of a sphere is $\frac{4}{3}\pi r^3$. Use this formula:

   a. To estimate the volume of a sphere whose radius is 2.75 cm. Give your answer to 1 sf.

   b. To calculate the volume of a sphere whose radius is 2.75 cm, rounding your answer sensibly.

8. A rectangular garden is 4.75 m long and 3.9 m wide.

   a. Estimate its area to 1 sf.

   b. Calculate its area, rounding sensibly.

Standard form

A number is said to be in **standard form** if it is expressed in the form

(number between 1 and 10) × (power of ten)

For example the numbers 1.4×10^4 and 3.6743×10^{-8} are in standard form.

Writing numbers in standard form has a number of advantages:

- Large numbers, particularly those with many zeros, occupy considerably less space when written in standard form.
- It is often easier to round a number if it is written in standard form and to determine the number of significant figures it has.
- The sizes of very large, or of very small numbers are more readily compared when expressed in standard form.

To express a number in standard form, move the decimal point to directly after the first significant digit of the number, then multiply by the appropriate power of ten (so that the value of the original number is kept the same).

Example K

Q. Write each of the following in standard form:

1. 340 **2.** 0.0045 **3.** 16.3×10^2

A. **1.** $340 = 3.4 \times 10^2$ [Place the decimal point after the first significant figure (3). The decimal point moved 2 places so 3.4 needs to be multiplied by 10^2 in order to equal the original number, 340.]

2. $0.0045 = 4.5 \times 10^{-3}$ [The decimal point is placed after the first significant figure (4). The point moved three places, so 4.5 needs to be multiplied by 10^{-3}, in order to equal the original number 0.0045.]

3. $16.3 \times 10^2 = 1.63 \times 10 \times 10^2$ [expressing 16.3 in standard form]]

$= 1.63 \times 10^3$ [since $10 \times 10^2 = 10^3$]

Alternatively, $16.3 \times 10^2 = 1\ 630 = 1.63 \times 10^3$

Note: A good rule to remember is that when numbers are expressed in standard form, large numbers (greater than 10) have a positive power of ten, whereas small numbers (less than 1) have a negative power of ten.

To change a number written in standard form back to an 'ordinary' number, move the decimal point in the usual manner for multiplying by powers of ten. Often zeros must be inserted as place-holders.

Example L

Q. Write each of the following as ordinary numbers:

1. 1.34×10^2 **2.** 2.4×10^4 **3.** 3.6×10^{-3}

A. **1.** $1.34 \times 10^2 = 134$ [The 2 in 10^2 shows that the decimal point needs to move two places to the right.]

2. $2.4 \times 10^4 = 24\ 000$ [The 4 in 10^4 shows that the decimal point is moved four places to the right (insert three zeros as place-holders.]

3. $3.6 \times 10^{-3} = 0.0036$ [The –3 in 10^{-3} indicates the decimal point needs to move three places to the left (insert two zeros as place-holders).]

Standard form and calculators

Scientific calculators automatically express very large or very small answers to calculations in standard form. The answer is displayed as a number between one and ten, followed by a gap, then the power of ten required. Since a calculator has a fixed number of digits in its display, this enables a much larger range of answers to be displayed.

Numbers can be entered into the calculator using the exponent [EXP] key. The calculator will leave the number in standard form if it is a very large or very small number. Otherwise it will convert it to an 'ordinary' number.

Example M

1. To enter 2.598×10^{-6}, press [2] [•] [5] [9] [8] [EXP] [(–)] [6].

Pressing [=] will give [2.598 $^{-06}$] or similar.

2. To enter 3.54×10^4 press [3] [•] [5] [4] [EXP] [4] [=] to get a display [35400].

The calculator has converted the number since it is in the acceptable range of values for display in the 'ordinary' form.

Note: On some calculators [+/–] is pressed instead of [(–)].

Scientific calculators and the [EXP] key make fast work of calculations done with numbers in standard form.

Example N

Q. Calculate the following, leaving your answer in standard form.

1. $2.6 \times 10^{-4} \times 4 \times 10^{8}$ **2.** $\dfrac{2.6 \times 10^{-4}}{4 \times 10^{8}}$

A. 1. By calculator, press [2] [•] [6] [EXP] [(–)] [4] [×] [4] [EXP] [8] [=]

to get a display [104000].

This can be converted to standard form in the usual way to 1.04×10^{5}.

2. By calculator, press [2] [•] [6] [EXP] [–] [4] [÷] [4] [EXP] [8] [=]

to get 6.5×10^{-13}.

Note: The calculator treats a number such as 2.6 EXP – 4 as a single number so no brackets are needed around this expression when dividing on a calculator.

Rounding numbers in standard form

Numbers in standard form are rounded by rounding the 'number between 1 and 10' in the usual way (the power of ten is unchanged).

Example O

Q. Calculate $(765\ 258)^2$, rounding your answer appropriately.

A. Enter [765258], then press [x^2] and [=].

Depending upon the type of calculator used, the calculator displays

[$5.856198066 \times 10^{11}$] *or* [5.856198066 11] *or* [5.856198066 E +11]

which is the number 5.85620×10^{11} (6 sf) [rounding]

Note: The number of significant figures is clear from the number 5.85620 whose final zero would not be included unless it was significant. (6 sf is chosen since 765 258 has 6 sf.)

Comparing numbers in standard form

To compare numbers in standard form, compare their powers of ten first. The larger the number, the higher the power of ten it has. If powers of ten are the same, compare their initial numbers (between one and ten).

Example P

Q. Put in order of size from smallest to largest:

3.56×10^{4}, 1.9×10^{-2}, 3.8×10, 4.6×10^{-2}, 9.019×10^{5}

A. The lowest power of 10 is –2, then 1, 4, 5.

Comparing the two numbers with the same power of 10, note that
$1.9 \times 10^{-2} < 4.6 \times 10^{-2}$ [since $1.9 < 4.6$]

The required order is, therefore:
1.9×10^{-2}, 4.6×10^{-2}, 3.8×10, 3.56×10^{4}, 9.019×10^{5}

Unit 11.1 Activity 4C: Standard form

1. Write each of the following in standard form:

a. 487 **b.** 9 120 **c.** 763 000 **d.** 0.0743

e. 0.00518 **f.** 0.00000451 **g.** 69 500 000 **h.** 495.8

i. 16.401 **j.** 126.5×10^2 **k.** 0.0032×10^4 **l.** 48.59×10^{-3}

2. Write each of the following as ordinary numbers:

a. 2.51×10^2 **b.** 6.39×10^3 **c.** 8.412×10^4 **d.** 3.57×10^{-1}

e. 1.7×10^{-3} **f.** 9.75×10^{-4} **g.** 3.48×10 **h.** 4.6×10^4

i. 7.83×10^{-2} **j.** 18.45×10^3 **k.** 126.4×10^{-2} **l.** 0.57×10^{-3}

3. Put in order of size, from smallest to largest, each of the following sets of numbers:

a. 2.4×10^3, 3.65×10^2, 1.901×10^4, 3.7×10^2

b. 4.1×10^{-2}, 0.0014, 2.6×10^{-1}, 0.002

c. 12.6×10^3, 1.3×10^4, 12 965, 9.9×10^3

4. Write each of the following correct to the number of significant figures shown in brackets.

a. 4.754×10^3 (2) **b.** 5.6648×10^{-3} (4) **c.** 8.98×10^4 (2)

d. 7.841×10^{-2} (2) **e.** 3.76×10^5 (1) **f.** 9.95×10^{-3} (2)

5. Find approximate answers to each of the following, then verify by calculation, leaving your answer in standard form, correct to three significant figures.

a. $8.5 \times 10^2 \times 2.6 \times 10^3$ **b.** $(7.48 \times 10^5) - (4.7 \times 10^4)$

c. $(6.92 \times 10^4)^2$ **d.** $\dfrac{8.74 \times 10^3}{6.5 \times 10^2} \times \dfrac{43 \times 10^{-2}}{7.3 \times 10^{-3}}$

6. A website gave the estimated population of New Zealand on 12 October 2004 as 4 071 829. The population of Auckland, the largest city in New Zealand, was given as 1.05×10^6. Using these data:

a. Express the population of New Zealand in standard form to 3 significant figures.

b. Approximately how many New Zealanders do not live in Auckland?

c. Approximately 8 out of every 37 people in New Zealand live in the South Island. Estimate the population of the South Island on 12 October 2004. Give your answer in standard form to 3 significant figures.

7. The mass of an electron at rest is 9.109×10^{-31} kg. The mass of a proton at rest is 1.673×10^{-27} kg, both figures correct to 4 significant figures. How many times greater is the mass of the proton, when compared with the mass of the electron?

8. The planet Pluto is 5 913 000 000 km from the sun. Neptune, another planet, is 4.498×10^9 km from the sun. How much further from the sun is Pluto than Neptune? Give your answer in standard form.

9. In 2003 the population of China was approximately 1.29×10^9 and the population of New Zealand was 3.95×10^6.

a. How much greater was the population of China than the population of New Zealand. Give your answer in standard form to 4 significant figures. Explain why an answer to 3 significant figures would be misleading.

b. How many times greater was the population of China than the population of New Zealand? Give your answer correct to 3 significant figures.

10. A confectionery manufacturer produces 1.76×10^6 individual sweets and 2.41×10^5 individual chocolates per year. He has been producing chocolates for 15 years and sweets for 10 years.

a. Which type of confectionery has the manufacturer produced more of since he has been in business, sweets or chocolates? How many more? (Give your answer in standard form to 3 significant figures.)

b. $\frac{4}{5}$ of the sweets produced by the manufacturer are unwrapped. How many wrapped sweets are produced per year? Give answer in standard form.

c. The manufacturer produces twice as many soft-centred chocolates as hard-centred ones. How many soft-centred chocolates are produced annually? Give answer in standard form to 3 significant figures.

Unit 11.1 Activity 4D: Multiple choice—decimals

1. The number 8476 written correct to 2 significant figures and 3 significant figures, respectively, is:

A. 8500 and 8470 **B.** 85 and 848

C. 8500 and 8480 **D.** 8400 and 8480

2. For the number 654386, which one of the following is **not** a correct rounding?

A. 654000 to 3 significant figures **B.** 654300 to 4 significant figures

C. 700000 to 1 significant figure **D.** 654390 to 5 significant figures

3. 963.44×10^5 written in standard form is

A. 963.44×10^5 **B.** 9.6344×10^3

C. 9.6344×10^7 **D.** 96344000

4. $8 \times 10^2 \times 12 \times 10^{-3}$ is equal to:

A. 96×10^{-6} **B.** 9.6×10^{-1}

C. 96×10^1 **D.** 9.6

5. $\frac{2.4 \times 10^3}{0.00008}$ written in standard form is:

A. 3×10^7 **B.** 0.3×10^8

C. 3×10^8 **D.** 3×10^{-3}

Unit 11.1 Number and Application

Topic 5: Basic numeracy—irrational numbers and surds

This Topic covers the content as set out on p. 13 of the syllabus: apply the properties of surds, simplify and rationalise surds. It includes:

- The number π.
- The number e.
- Simplifying surds.
- Addition, subtraction and multiplication of surds.
- Conjugation of surds.
- Rationalising the denominator of fractional surds.

The number π

The ratio of the circumference to the **diameter** of any **circle** is a **constant** number called **pi**, usually represented by the Greek letter π.

Ancient civilisations knew that there was a ratio between the circumference and the diameter of a circle and that it was approximately equal to 3. Pi proved to be a number that fascinated mathematicians and the approximation to pi was improved over the centuries until the mathematician Lambert, in1761, proved that pi was an **irrational number**; it could *not* be written as a ratio of integers.

- $\pi = 3.141\ 593\ 654\ 589\ 793\ 238\ 46$; correct to 20 decimal places. It is an infinite non-recurring decimal.
- Archimedes calculated pi as a number between $\frac{223}{71}$ and $\frac{22}{7}$ in 250 BC.
- The Chinese used the approximation $\frac{355}{113}$ in AD 450.
- Many famous mathematicians have developed formulae to calculate pi (these can be found if you research pi on the Internet).
- Computers have been put to the task of calculating pi and it can now be written correct to millions of decimal places.

The number e

It was noticed, when calculating compound interest, that the formula $\left(1+\frac{1}{n}\right)^n$ converged to a constant number between 2 and 3 as n became larger. This number, denoted by e, is commonly called Euler's number; it is named after the mathematician Leonhard Euler, who first referred to it in 1727. The number e also fascinated many mathematicians who developed formulae for its calculation.

- $\boldsymbol{e}$ is an irrational number; an infinite non-recurring decimal.
- $\boldsymbol{e} = 2.718\ 281\ 828\ 459\ 045\ 235\ 36$; correct to 20 decimal places.
- You can find the number e on your calculator by calculating e^1.

Unit 11.1 Activity 5A: The numbers π and e

1. Use your calculator to determine each of the following approximations to pi. State the accuracy correct to the relevant number of decimal places.

a. $\frac{223}{71}$ **b.** $\frac{22}{7}$ **c.** $\sqrt{2}+\sqrt{3}$ **d.** $\frac{355}{113}$

e. $4\left(\frac{1}{1}-\frac{1}{3}+\frac{1}{5}-\frac{1}{7}\ \ldots\ldots -\frac{1}{15}\right)$ Try extending this pattern further.

f. The square root of the following number: $6\left(\frac{1}{1^2}+\frac{1}{2^2}+\frac{1}{3^2}\ \ldots\ldots\ \frac{1}{10^2}\right)$. Try extending this pattern further.

2. Use your calculator to determine each of the following approximations to *e*. State the accuracy correct to the relevant number of decimal places.

a. $\left(1+\frac{1}{10}\right)^{10}$ **b.** $\left(1+\frac{1}{100}\right)^{100}$

c. $\left(1+\frac{1}{1000}\right)^{1000}$ **d.** $1+\frac{1}{1!}+\frac{1}{2!}+\frac{1}{3!}+\frac{1}{4!}+\frac{1}{5!}=1+\frac{1}{1}+\frac{1}{2}+\frac{1}{6}+\frac{1}{24}+\frac{1}{120}$

(Note: $4! = 1 \times 2 \times 3 \times 4 = 24$ etc.)

e. Extend the formula from part **d** until you get an approximation that is correct to 4 decimal places. Write down the fractions that you need to add to your answer from part **d**.

Surds

Surds are **irrational** numbers. Surds cannot be written as a ratio of integers.

Surds like $\sqrt{2}$ or $\sqrt[3]{7}$ are **infinite, non-recurring decimals**. This means that the decimal parts go on forever, not repeating the pattern of decimals.

The **exact value of a surd** is written as $\sqrt{2}, \sqrt[3]{5}, \sqrt{17}, \sqrt{6.85}, \sqrt{0.0258}$ … etc although approximations can be found for all of them.

There are some square roots that are not surds, ie they are **rational numbers**.

- $\sqrt{4} \times \sqrt{4} = 4$ but we know that $2 \times 2 = 4$ so $\sqrt{4} = 2$, which is a rational number.
- $13 \times 13 = 169$ so $\sqrt{169} = 13$; a rational number.
- $1.1 \times 1.1 = 1.21$ so $\sqrt{1.21} = 1.1 = \frac{11}{10}$; a rational number.
- The numbers 4, 9, 16, 25, 36, 49, 1.21 etc are called **perfect squares** because they have **rational** square roots.

Note: It is not possible to find the square root of a negative number, ie it is impossible to find a number n such that $n \times n$ = a negative number.

Some numbers are perfect cubes and so their cube root is not a surd.

- $2^3 = 8$ so $\sqrt[3]{8} = 2$; a rational number.
- $\sqrt[3]{-27} = -3$; a rational number, because $-3 \times -3 \times -3 = -27$.

Common perfect cubes are 8, 27, 64, 125, 216, 343 …

Simplifying surds

$\sqrt{x}$ is called a **simple surd** if none of the factors (excluding 1) of x are perfect squares; eg $\sqrt{6}$ is a simple surd because it has factors (1), 2, 3 and 6 and none of these are perfect squares.

Surds that have one or more factors that are perfect squares can be **simplified**.

$\sqrt{12}$ can be simplified because it has a factor, 4, that is a perfect square

$\sqrt{12} = \sqrt{4 \times 3} = \sqrt{4} \times \sqrt{3} = 2 \times \sqrt{3} = 2\sqrt{3}$.

$\sqrt{12}$ is called a **single surd** and $2\sqrt{3}$ is the **simplified** version of $\sqrt{12}$.

Unit 11.1 Activity 5B: Surds

1. Which of the following numbers are surds?

$\sqrt{101}$, 81, $\sqrt{36}$, $\sqrt{17}$, 1.732, $\sqrt[3]{0.125}$, $\sqrt{8}$, $\sqrt{83.4}$, $\sqrt[3]{216}$, $\sqrt{1.44}$, $\sqrt{4900}$

2. a. Use your calculator to find a decimal value for the following numbers, rounding to two decimal places where necessary.

$\sqrt{64}, \sqrt{0.25}, \sqrt[3]{81}, \sqrt{400}, \sqrt{4000}, \sqrt[3]{-0.216}, \sqrt{-0.36}, \sqrt{0.0016}, \sqrt[3]{-0.8}, \sqrt{250000}$

b. Which of the numbers in part **a**. are perfect squares?

3. Identify the following numbers as irrational (Q') or rational (Q):

a. $\sqrt{0.49}$ **b.** $-\sqrt{0.4}$ **c.** $\sqrt{0}$ **d.** $1 + \sqrt{1.6}$

e. $5\sqrt[3]{8}$ **f.** $4 - 5\sqrt[3]{0.08}$ **g.** $\sqrt[5]{-32}$ **h.** $7 - 3\sqrt[3]{64}$

4. Simplify the following surds:

a. $\sqrt{50}$ **b.** $\sqrt{48}$ **c.** $\sqrt{98}$ **d.** $\sqrt{320}$

e. $\sqrt{675}$ **f.** $\sqrt{108}$ **g.** $\sqrt{324}$ **h.** $\sqrt{8}$

5. Express as a single surd:

a. $5\sqrt{3}$ **b.** $-2\sqrt{7}$ **c.** $10\sqrt{10}$ **d.** $5\sqrt{6}$

e. $-3\sqrt{5}$ **f.** $4\sqrt{2}$ **g.** $2\sqrt{11}$ **h.** $0.2\sqrt{14}$

Addition and subtraction of surds

Only 'like' surds can be added or subtracted. Like surds have the same number under the root sign. For example $\sqrt{3}, -2\sqrt{3}$, and $5\sqrt{3}$ are like surds.

Example A

Q. Simplify:

1. $\sqrt{3} - 2\sqrt{3} + 5\sqrt{3}$
2. $6\sqrt{5} + 5\sqrt{3} - 4\sqrt{5}$
3. $\sqrt{12} + \sqrt{8} - \sqrt{2}$

A. 1. $\sqrt{3} - 2\sqrt{3} + 5\sqrt{3} = (1 - 2 + 5)\sqrt{3} = 4\sqrt{3}$

2. $6\sqrt{5} + 5\sqrt{3} - 4\sqrt{5} = 2\sqrt{5} + 5\sqrt{3}$

3. $\sqrt{12} + \sqrt{8} - \sqrt{2} = \sqrt{4 \times 3} + \sqrt{4 \times 2} - \sqrt{2}$

$= 2\sqrt{3} + 2\sqrt{2} - \sqrt{2}$

$= 2\sqrt{3} + \sqrt{2}$

Unit 11.1 Activity 5C: Simplifying surds

1. Simplify where possible.
   - **a.** $5\sqrt{7} + 7\sqrt{3} - 2\sqrt{7}$ **b.** $3\sqrt{3} - \sqrt{2} + 4\sqrt{2} - \sqrt{3}$
   - **c.** $\sqrt{6} + 2\sqrt{3} - 6\sqrt{3} + 3$ **d.** $5\sqrt{2} + 3\sqrt{5} - 2\sqrt{10}$
2. Simplify:
   - **a.** $\sqrt{24} - \sqrt{6}$ **b.** $\sqrt{45} + \sqrt{80}$
   - **c.** $\sqrt{75} - \sqrt{147}$ **d.** $2\sqrt{3} + \sqrt{300}$
   - **e.** $\sqrt{75} - 2\sqrt{12}$ **f.** $\sqrt{125} - \sqrt{5}$
   - **g.** $\sqrt{242} + \sqrt{99}$ **h.** $\sqrt{32} + \sqrt{50} - 3\sqrt{2}$
3. Simplify the following surd expressions:
   - **a.** $2 + \sqrt{12} - 3 + \sqrt{8}$ **b.** $3\sqrt{5} - 2\sqrt{18} + 3\sqrt{8} - \sqrt{75}$
   - **c.** $4 - 3\sqrt{24} - \sqrt{144} + 4\sqrt{48}$ **d.** $\sqrt{6} + 2\sqrt{3} - 5\sqrt{12} - \sqrt{24} + 2\sqrt{72}$
   - **e.** $\sqrt{2} + \sqrt{4} + \sqrt{8} + \sqrt{16} + \sqrt{32} + \sqrt{64} + \sqrt{128}$
   - **f.** $\sqrt{3} - \sqrt{9} + \sqrt{27} - \sqrt{81} + \sqrt{243} - \sqrt{729}$

Multiplication of surds

Surd terms can be multiplied together:

$a\sqrt{m} \times b\sqrt{n} = ab\sqrt{mn}$, with the surd $\sqrt{mn}$ being simplified if possible.

Example B

Q. Simplify:

1. $\sqrt{5} \times \sqrt{3}$
2. $3\sqrt{6} \times 5\sqrt{3}$
3. $\sqrt{3} \times 2\sqrt{3} \times 5\sqrt{3}$

A. 1. $\sqrt{5} \times \sqrt{3} = \sqrt{5 \times 3} = \sqrt{15}$

2. $3\sqrt{6} \times 5\sqrt{3} = 3 \times 5\sqrt{6 \times 3} = 15\sqrt{18} = 15 \times 3\sqrt{2} = 45\sqrt{2}$

3. $\sqrt{3} \times 2\sqrt{3} \times 5\sqrt{3} = 2 \times 5\sqrt{3 \times 3 \times 3} = 10\sqrt{27} = 10 \times 3\sqrt{3} = 30\sqrt{3}$

Expanding brackets containing surds

The **distributive law** can be used to **expand** surd expressions:

$a(b + c) = ab + ac$

$(a + b)(c + d) = ac + ad + bc + bd$

Example C

Q. Expand and simplify:

1. $3\sqrt{2}(2\sqrt{6}+5)$
2. $\left(2\sqrt{5}-\sqrt{2}\right)\left(\sqrt{15}+7\right)$
3. $\left(\sqrt{7}+\sqrt{5}\right)\left(\sqrt{7}-\sqrt{5}\right)$

A. 1.
$$\begin{aligned}3\sqrt{2}\left(2\sqrt{6}+5\right)&=3\times 2\sqrt{2\times 6}+3\times 5\sqrt{2}\\&=6\sqrt{12}+15\sqrt{2}\\&=6\times 2\sqrt{3}+15\sqrt{2}\\&=12\sqrt{3}+15\sqrt{2}\end{aligned}$$

2.
$$\begin{aligned}\left(2\sqrt{5}-\sqrt{2}\right)\left(\sqrt{15}+7\right)&=2\sqrt{5\times 15}+2\times 7\sqrt{5}-\sqrt{2\times 15}-7\sqrt{2}\\&=2\sqrt{75}+14\sqrt{5}-\sqrt{30}-7\sqrt{2}\\&=2\times 5\sqrt{3}+14\sqrt{5}-\sqrt{30}-7\sqrt{2}\\&=10\sqrt{3}+14\sqrt{5}-\sqrt{30}-7\sqrt{2}\end{aligned}$$

3.
$$\begin{aligned}\left(\sqrt{7}+\sqrt{5}\right)\left(\sqrt{7}-\sqrt{5}\right)&=\sqrt{7\times 7}-\sqrt{7\times 5}+\sqrt{5\times 7}-\sqrt{5\times 5}\\&=\sqrt{49}-\sqrt{35}+\sqrt{35}-\sqrt{25}\\&=7-5\\&=2\end{aligned}$$

Conjugate surds

The surd expressions $a\sqrt{m}+b\sqrt{n}$ and $a\sqrt{m}-b\sqrt{n}$ are called **conjugate pairs**. The terms are the same in each expression but the connecting sign is different.

When conjugate pairs are multiplied the result is a **rational number**.

$$\begin{aligned}(a\sqrt{m}+b\sqrt{n})(a\sqrt{m}-b\sqrt{n})&=a\times a\sqrt{m\times m}-ab\sqrt{mn}+ab\sqrt{mn}-b\times b\sqrt{nn}\\&=a^2m-b^2n;\text{ a rational number}\\&=(\text{first term})^2-(\text{second term})^2\end{aligned}$$

Example D

Q. Simplify:

1. $\left(\sqrt{3}+\sqrt{2}\right)\left(\sqrt{3}-\sqrt{2}\right)$
2. $\left(2\sqrt{7}+\sqrt{3}\right)\left(2\sqrt{7}-\sqrt{3}\right)$
3. $\left(3\sqrt{11}-5\right)\left(3\sqrt{11}+5\right)$

A. 1. Checking that the terms in each bracket are the same but that the connecting sign is different we can see that the expressions to be multiplied are a conjugate pair.

Hence $\left(\sqrt{3}+\sqrt{2}\right)\left(\sqrt{3}-\sqrt{2}\right)=$ *(first term)*$^2-$ *(second term)*^{2}

$$\begin{aligned}&=\left(\sqrt{3}\right)^2-\left(\sqrt{2}\right)^2\\&=3-2\\&=1\end{aligned}$$

2. $\left(2\sqrt{7}+\sqrt{3}\right)\left(2\sqrt{7}-\sqrt{3}\right)$: terms are conjugate pairs.

$$\begin{aligned}\left(2\sqrt{7}+\sqrt{3}\right)\left(2\sqrt{7}-\sqrt{3}\right) &= \left(2\sqrt{7}\right)^2-\left(\sqrt{3}\right)^2\\ &= 2\times 2\times\left(\sqrt{7}\right)^2-3\\ &= 4\times 7-3\\ &= 25\end{aligned}$$

3. $\left(3\sqrt{11}-5\right)\left(3\sqrt{11}+5\right)$: terms are conjugate pairs.

$$\begin{aligned}\left(3\sqrt{11}-5\right)\left(3\sqrt{11}+5\right) &= 3\times 3\times\left(\sqrt{11}\right)^2-5\times 5\\ &= 9\times 11-25\\ &= 99-25\\ &= 74\end{aligned}$$

Unit 11.1 Activity 5D: Simplifying and expanding surds

1. Simplify:

a. $\sqrt{5}\times\sqrt{5}$ **b.** $\sqrt{15}\times\sqrt{5}$ **c.** $\sqrt{18}\times\sqrt{3}$

d. $\sqrt{45}\times\sqrt{10}$ **e.** $2\sqrt{3}\times 3\sqrt{2}$ **f.** $5\sqrt{12}\times\sqrt{6}$

2. Expand and simplify where possible:

a. $\left(2+\sqrt{5}\right)^2$ **b.** $\left(3\sqrt{2}-2\sqrt{3}\right)^2$ **c.** $\left(3\sqrt{5}-\sqrt{2}\right)^2$

d. $\left(4\sqrt{2}+3\right)^2$ **e.** $\left(12-5\sqrt{3}\right)^2$

3. Expand and simplify where possible (Hint: Simplify surds first):

a. $\sqrt{3}\left(2\sqrt{3}-1\right)$ **b.** $\left(\sqrt{7}+\sqrt{3}\right)\left(\sqrt{7}+\sqrt{3}\right)$

c. $\left(2\sqrt{3}+3\sqrt{2}\right)\left(\sqrt{3}-\sqrt{2}\right)$ **d.** $\left(\sqrt{12}-\sqrt{8}\right)\left(\sqrt{3}+1\right)$

e. $\left(\sqrt{50}-\sqrt{8}\right)\left(\sqrt{2}-\sqrt{75}\right)$ **f.** $\left(\sqrt{81}-\sqrt{27}\right)\left(1+2\sqrt{3}\right)$

4. Write, in simplest surd form, the conjugate of:

a. $2\sqrt{3}-4$ **b.** $2+\sqrt{24}$

c. $5-\sqrt{45}$ **d.** $\sqrt{125}+\sqrt{64}$

5. Simplify each of the following, writing your answer as a rational number:

a. $\left(2+\sqrt{5}\right)\left(2-\sqrt{5}\right)$ **b.** $\left(\sqrt{7}+\sqrt{4}\right)\left(\sqrt{7}-\sqrt{4}\right)$

c. $\left(\sqrt{11}-\sqrt{7}\right)\left(\sqrt{11}+\sqrt{4}\right)$ **d.** $\left(3\sqrt{2}+\sqrt{5}\right)\left(3\sqrt{2}-\sqrt{5}\right)$

e. $\left(4\sqrt{2}-3\right)\left(4\sqrt{2}+3\right)$ **f.** $\left(3\sqrt{7}+5\sqrt{2}\right)\left(3\sqrt{7}-5\sqrt{2}\right)$

g. $\left(12-5\sqrt{2}\right)\left(12+5\sqrt{2}\right)$

Rationalising the denominator of fractional surds

Fractional surds, for example $\frac{\sqrt{3}}{2\sqrt{5}+\sqrt{7}}$, are much easier to work with if they do not have surds in their denominators. The process of changing a fractional surd, with surds in the denominator, to one with a rational number in the denominator is called **rationalisation**.

- To rationalise the denominator of a fraction with a simple surd in the denominator, for example $\frac{a}{\sqrt{b}}$, we multiply that fraction by 1 in the form of $\frac{\sqrt{b}}{\sqrt{b}}$. This does not change the value of the fraction because we are multiplying by 1.

  $\frac{a}{\sqrt{b}} \times \frac{\sqrt{b}}{\sqrt{b}} = \frac{a\sqrt{b}}{b}$; the resulting fraction has a rational denominator.

- To rationalise the denominator of a fraction such as $\frac{\sqrt{3}}{2\sqrt{5}+\sqrt{7}}$ we multiply the fraction by 1 in the form of $\frac{\textit{conjugate of the denominator}}{\textit{conjugate of the denominator}}$. Multiplying by 1 does not change the fraction.

  $$\frac{\sqrt{3}}{2\sqrt{5}+\sqrt{7}} = \frac{\sqrt{3}}{2\sqrt{5}+\sqrt{7}} \times \frac{2\sqrt{5}-\sqrt{7}}{2\sqrt{5}-\sqrt{7}} = \frac{\sqrt{3}\left(2\sqrt{5}-\sqrt{7}\right)}{\left(2\sqrt{5}\right)^2-\left(\sqrt{7}\right)^2} = \frac{\sqrt{3}\left(2\sqrt{5}-\sqrt{7}\right)}{20-7} = \frac{\sqrt{3}\left(2\sqrt{5}-\sqrt{7}\right)}{13}$$

  The denominator is rationalised.

Example E

Q. Write the following fractional surds with a rational denominator.

1. $\frac{\sqrt{18}}{\sqrt{8}}$ **2.** $\frac{5}{\sqrt{3}}$ **3.** $\frac{7\sqrt{3}}{2\sqrt{5}}$ **4.** $\frac{3}{\sqrt{5}+\sqrt{2}}$ **5.** $\frac{\sqrt{5}}{3\sqrt{5}-\sqrt{7}}$

A. **1.** $\frac{\sqrt{18}}{\sqrt{8}} = \frac{\sqrt{9\times2}}{\sqrt{4\times2}} = \frac{\sqrt{9}\times\sqrt{2}}{\sqrt{4}\times\sqrt{2}} = \frac{3\sqrt{2}}{2\sqrt{2}} = \frac{3}{2}$

2. $\frac{5}{\sqrt{3}} = \frac{5}{\sqrt{3}} \times \frac{\sqrt{3}}{\sqrt{3}} = \frac{5\times\sqrt{3}}{\sqrt{3}\times\sqrt{3}} = \frac{5\sqrt{3}}{3}$: the denominator is rational

3. $\frac{7\sqrt{3}}{2\sqrt{5}} = \frac{7\sqrt{3}}{2\sqrt{5}} \times \frac{\sqrt{5}}{\sqrt{5}} = \frac{7\sqrt{3}\times\sqrt{5}}{2\sqrt{5}\times\sqrt{5}} = \frac{7\sqrt{15}}{2\times5} = \frac{7\sqrt{15}}{10}$

4.
$$\frac{3}{\sqrt{5}+\sqrt{2}} \times \frac{\textit{conjugate of the denominator}}{\textit{conjugate of the denominator}} = \frac{3}{\sqrt{5}+\sqrt{2}} \times \frac{\sqrt{5}-\sqrt{2}}{\sqrt{5}-\sqrt{2}}$$
$$= \frac{3\left(\sqrt{5}-\sqrt{2}\right)}{\left(\sqrt{5}+\sqrt{2}\right)\left(\sqrt{5}-\sqrt{2}\right)}$$
$$= \frac{3\left(\sqrt{5}-\sqrt{2}\right)}{5-2}$$
$$= \frac{3\left(\sqrt{5}-\sqrt{2}\right)}{3}$$

5. $\frac{\sqrt{5}}{3\sqrt{5}-\sqrt{7}} \times \frac{\textit{conjugate of the denominator}}{\textit{conjugate of the denominator}} = \frac{\sqrt{5}}{3\sqrt{5}-\sqrt{7}} \times \frac{3\sqrt{5}+\sqrt{7}}{3\sqrt{5}+\sqrt{7}}$

$$= \frac{\sqrt{5}\left(3\sqrt{5}+\sqrt{7}\right)}{\left(3\sqrt{5}-\sqrt{7}\right)\left(3\sqrt{5}+\sqrt{7}\right)}$$

$$= \frac{3 \times \sqrt{5} \times \sqrt{5} + \sqrt{5} \times \sqrt{7}}{45-7}$$

$$= \frac{15+\sqrt{35}}{38}$$

Unit 11.1 Activity 5E: Simplest form

1. Write the following surd fractions in simplest form, with a rational denominator.

   a. $\frac{\sqrt{2}}{\sqrt{50}}$ **b.** $\frac{\sqrt{24}}{\sqrt{54}}$ **c.** $\frac{3\sqrt{40}}{\sqrt{250}}$ **d.** $\frac{\sqrt{14}}{\sqrt{2}}$ **e.** $\frac{\sqrt{30}}{2\sqrt{5}}$

2. Rationalise the denominators of the following fractions, expressing the answer in simplest form.

   a. $\frac{1}{\sqrt{5}}$ **b.** $\frac{1}{2\sqrt{2}}$ **c.** $\frac{3}{\sqrt{3}}$ **d.** $\frac{1}{\sqrt{24}}$ **e.** $\frac{5+\sqrt{8}}{\sqrt{2}}$

3. Express the following surd fractions with rational denominators.

   a. $\frac{1}{\sqrt{3}-\sqrt{2}}$ **b.** $\frac{\sqrt{5}}{\sqrt{5}+2}$ **c.** $\frac{2\sqrt{3}}{\sqrt{3}-1}$

   d. $\frac{\sqrt{7}-2}{\sqrt{7}+2}$ **e.** $\frac{2\sqrt{5}}{2\sqrt{5}-1}$ **f.** $\frac{3\sqrt{6}-2}{3\sqrt{6}+2}$

4. Express the following fractions with a rational denominator, in simplest surd form.

   a. $\frac{3+2\sqrt{2}}{3-2\sqrt{2}}$ **b.** $\frac{2\sqrt{3}-\sqrt{6}}{2\sqrt{3}-1}$ **c.** $\frac{5-\sqrt{12}}{2+\sqrt{18}}$ **d.** $\frac{2\sqrt{20}+1}{\sqrt{45}-1}$

   e. $\frac{3+\sqrt{8}}{2+\sqrt{2}}$ **f.** $\frac{2\sqrt{10}+1}{\sqrt{40}-1}$ **g.** $\frac{\sqrt{3}}{\sqrt{5}+\sqrt{3}}$ **h.** $\frac{4\sqrt{7}}{2\sqrt{3}+\sqrt{7}}$

   i. $\frac{2+\sqrt{6}}{\sqrt{11}-\sqrt{6}}$ **j.** $\frac{4\sqrt{5}-\sqrt{3}}{3\sqrt{5}-\sqrt{3}}$

Unit 11.1 Number and Application

Topic 6: Basic numeracy—indices

This Topic presents an introduction to indices in line with the General Mathematics syllabus, p. 14. It covers:

- Use of fractional and negative indices.
- Simplifying algebraic expressions with indices.
- Using calculators to simplify calculations.
- Solution of equations involving indices.

Introduction

Indices involve expressions of the form x^n, where x is the **base** and n is the **power** or **index** or **exponent**. The index notation x^n is a short form of expressing a series of multiplication of the same number. That is, $\underbrace{x \times x \times x \times \ldots \times x}_{\text{n times}}$ is written in short form as x^n.

Indices involve expressions of the form x^n, where x is the **base** and n is the **power** or **index** or **exponent**.

Some important properties of indices already learned in NCEA Level 1 Mathematics are:

$x^m \times x^n = x^{m+n}$	$\frac{x^m}{x^n} = x^{m-n}$ or $\frac{1}{x^{n-m}}$ $[x \neq 0]$
$(xy)^m = x^m y^m$	
$(x^m)^n = x^{mn}$	$\left(\frac{x}{y}\right)^m = \frac{x^m}{y}$
$x^0 = 1$ $[x \neq 0]$	

Negative indices

Using the division law for powers, $\frac{x^2}{x^5} = x^{2-5} = x^{-3}$. Simplifying $\frac{x^2}{x^5}$ by writing in full and cancelling gives $\frac{\cancel{x} \times \cancel{x}}{\cancel{x} \times \cancel{x} \times x \times x \times x} = \frac{1}{x^3}$. This example illustrates the rule for negative indices:

$$x^{-n} = \frac{1}{x^n} \quad \text{where } x \neq 0$$

Zero indices

The same law can also be used to illustrate the rule for zero indices. That is,

$$\frac{x^2}{x^2} = x^{2-2} = x^0$$

By simplifying through cancellation:

$$\frac{x^2}{x^2} = \frac{\cancel{x} \times \cancel{x}}{\cancel{x} \times \cancel{x}} = 1$$

By directly comparing the two results:

$x^0 = 1$ for all a except $a = 0$

Example A

Q. Use the properties of indices to calculate the following, giving the answers as fractions or whole numbers.

1. 2^{-3} **2.** $\frac{1}{3^{-2}}$ **3.** $\left(\frac{2}{5}\right)^3$ **4.** $\left(\frac{3}{7}\right)^{-2}$ **5.** $\frac{2^{-3}}{3^{-2}}$ **6.** $\frac{2^{-1}}{7}$

7. $2^5 \times 2^3$ **8.** $\frac{5^3}{5^2}$ **9.** 7^0 **10.** $2 + 5^0$ **11.** $2 + 3^0 + 2^{-1}$

A. **1.** $2^{-3} = \frac{1}{x^m}$ [since $x^{-m} = \frac{1}{x^m}$]

$= \frac{1}{8}$ [$2^3 = 8$]

2. $\frac{1}{3^{-2}} = 3^2$ [since $\frac{1}{x^m} = x^{-m}$]

$= 9$

3. $\left(\frac{2}{5}\right)^3 = \frac{2^3}{5^3}$ [since $\left(\frac{x}{y}\right)^m = \frac{x^m}{y^m}$]

$= \frac{8}{125}$

4. $\left(\frac{3}{7}\right)^{-2} = \left(\frac{7}{3}\right)^2$ [since $\left(\frac{x}{y}\right)^{-m} = \left(\frac{y}{x}\right)^m$]

$= \frac{7^2}{3^2}$ [since $\left(\frac{y}{x}\right)^m = \frac{y^m}{x^m}$]

$= \frac{49}{9}$

5. $\frac{2^{-3}}{3^{-2}} = \frac{3^2}{2^3}$ [since $\frac{x^{-n}}{y^{-m}} = \frac{y^m}{x^n}$]

$= \frac{9}{8}$

6. $\frac{2^{-1}}{7} = \frac{1}{7} \times 2^{-1}$ [as dividing by 7 is the same as multiplying by $\frac{1}{7}$]

$= \frac{1}{7} \times \frac{1}{2}$ [as $2^{-1} = \frac{1}{2^1} = \frac{1}{2}$]

$= \frac{1}{14}$ [multiplying fractions]

7. $2^5 \times 2^3 = 2^{5+3}$

$= 2^8$

$= 256$

8. $\frac{5^3}{5^2} = 5^{3-2}$

$= 5^1$

$= 5$

9. $7^0 = 1$

10. $2 + 5^0 = 2 + 1$

$= 3$

11. $2 + 3^0 + 2^{-1} = 2 + 1 + \frac{1}{2}$

$= 3\frac{1}{2}$

Unit 11.1 Activity 6A: Negative indices

1. Calculate the following, giving the answers to no more than 4 decimal places:

a. 2^{-4} **b.** 1.2^{-5} **c.** 0.37^{-2} **d.** $0.37^{-1} + 2.3$

e. $(2.3)^{-2} + 4$ **f.** $\frac{1}{3^{-2}}$ **g.** $\frac{2}{(2.4)^{-3}}$ **h.** $(-3.4)^{-2}$

i. $\frac{1}{0.56^{-1}} + \frac{2}{0.5^{-2}}$ **j.** $(3.4^{-1} - (0.9)^{-2})^{-3}$ **k.** $(2.3)^{-3} + (-2.3)^{-2}$

2. Use the properties of indices to evaluate the following, giving the answers as fractions or whole numbers.

a. 3^{-2} **b.** 4^{-3} **c.** 2^{-4} **d.** $\left(\frac{3}{4}\right)^{3}$

e. $\frac{2^3}{5^2}$ **f.** $\frac{1}{3^{-2}}$ **g.** $\frac{1}{4^{-2}}$ **h.** $\left(\frac{3}{5}\right)^{-2}$

i. $\left(\frac{4}{5}\right)^{-3}$ **j.** $\frac{2^{-2}}{3^{-3}}$ **k.** $\frac{3^{-3}}{5^{-2}}$ **l.** $\frac{3^{-1}}{8}$

m. $\frac{5^{-1}}{2}$ **n.** $7(5^{-1})$ **o.** 8×3^{-1} **p.** $2^{-1} \times 3^{-1}$

q. $3^{-1}\,5^{-1}$ **r.** $3^{-1}\,2^{-2}$ **s.** $\frac{5}{2^{-1}}$ **t.** $\frac{3}{2^{-2}}$

u. $\frac{3^{-2}}{4}$ **v.** $\frac{5^{-2}}{7}$ **w.** $3^{-1}\,6 + 3^{0}$ **x.** $3^{-2}\,6^{2} + 6^{0}$

y. $2^{-1} \div 7$ **z.** $3^{-1} \div 4^{-1}$

3. Calculate the following, giving the answers as fractions:

a. $2^{-1} \times 3$ **b.** $3^{-2} \times 4^{-1}$ **c.** $(2^{-2} \times 3^{-1})^{-2}$ **d.** $(3^{-1}2^{-3})^{2}$

e. $2^{-1} + 3^{-1}$ **f.** $3^{-2} + 6^{-1}$ **g.** $2^{-1}\,3 + 5^{-1}\,3^{2}$ **h.** $2^{-2} - 8^{-1}$

i. $4^{-2} \div 8^{-1}$ **j.** $\frac{3^{-1}}{2^{-1}} + \frac{4^{-1}}{3^{-1}}$ **k.** $\frac{3^{-2}}{2^{-2}} - \frac{2^{-1}}{3^{-2}}$ **l.** $(3^{1}\,2^{-1} - 2^{1}\,3^{-1})^{-1}$

m. $(2^{2}\,3)^{-1} + 2^{-1}\,3^{-2}$ **n.** $5(2^{3})^{-1} + 4^{-2}$ **o.** $7(2^{-1}) - 3(2^{-2})$ **p.** $5^{0} - 5^{-1}$

4. Write the following as fractions or integers with no powers:

a. 2×4^{-1} **b.** $2^{-1} \times 3^{-1}$ **c.** 4×2^{-1} **d.** $3^{-2} \times 5$

e. $\left(\frac{2}{3}\right)^{-1}$ **f.** $\left(1\frac{1}{2}\right)^{-1}$ **g.** $3\left(\frac{1}{3}\right)^{-1}$ **h.** $\frac{2}{3}\left(\frac{2}{3}\right)^{-2}$

i. $\frac{2^{-1}}{5} \times \left(\frac{1}{4}\right)^{-2}$

5. Write each of the following as the product of **primes** raised to powers:
eg, $28 = 2^{2} \times 7^{1}$, $\frac{3}{8} = 3 \times 2^{-3}$.

a. 12 **b.** 200 **c.** 45 **d.** $\frac{25}{16}$ **e.** $\frac{2}{3}$

f. $\frac{4}{27}$ **g.** $\frac{16}{81}$ **h.** $\frac{64}{81}$ **i.** $2\frac{1}{4}$

6. Write the following as powers of 2, 3 and 5: eg $\frac{1}{25} = \frac{1}{5^2} = 5^{-2}$

a. 16 **b.** 125 **c.** 81 **d.** $\frac{1}{16}$

e. 64 **f.** $\frac{1}{625}$ **g.** $\frac{1}{9}$ **h.** $\frac{1}{243}$

7. Evaluate each of the following:

a. $(-5)^2$ **b.** -5^2 **c.** $(-5)^{-2}$ **d.** $3^{-1} + 2^{-1}$

e. $3^{-2} \times 2^{-2}$ **f.** $7^{-1} - 7^0$ **g.** $(-1)^{19}$

8. Write the following as powers of 2 and 3: eg $2^n \times 8 = 2^n \times 2^8 = 2^{n+2}$, $\frac{4}{2^n} = \frac{2^2}{2^n} = 2^{2-n}$ and $27 \times 9 = 3^3 \times 3^3 = 3^{3-2} = 3^5$.

a. $\frac{2^c}{4}$ **b.** $\frac{2^m}{2^{-m}}$ **c.** 8×2^t **d.** $\left(2^{x+1}\right)^2$

e. $\left(2^{1-n}\right)^{-1}$ **f.** $\frac{2^{x+1}}{2^x}$ **g.** $\frac{4}{2^{1-x}}$ **h.** 9×3^x

i. $\frac{3^y}{27}$ **j.** 243^a **k.** $\frac{9^{n+1}}{3^{2n-1}}$ **l.** $\frac{9^a}{3^{1-a}}$

Simplifying algebraic expressions with indices

Algebraic expressions may need to be simplified so that indices have positive powers only.

Example B

Q. Write with positive indices: **1.** x^{-5} **2.** $8a^{-2}$ **3.** $\frac{2^{-1}a^{-3}}{5}$

A. **1.** $x^{-5} = \frac{1}{x^5}$ [since $x^{-r} = \frac{1}{x^r}$]

2. $8a^{-2} = 8 \times a^{-2}$

$= 8 \times \frac{1}{a^2}$ [since $a^{-2} = \frac{1}{a^2}$]

$= \frac{8}{a^2}$

3. $\frac{2^{-1}a^{-3}}{5} = 2^{-1} \times a^{-3} \times \frac{1}{5}$ [dividing by 5 is the same as multiplying by $\frac{1}{5}$]

$= \frac{1}{2} \times \frac{1}{a^3} \times \frac{1}{5}$ [since $2^{-1} = \frac{1}{2}$ and $a^{-3} = \frac{1}{a^3}$]

$= \frac{1}{10a^3}$

Example C

Q. Express $\frac{15x^{-3}y^2}{(3x^{-2}y^4)^2}$ without negative indices.

A. $\frac{15x^{-3}y^2}{(3x^{-2}y^4)^2} = \frac{15x^{-3}y^2}{3^2(x^{-2})^2(y^4)^2}$ [since $(xy)^n = x^ny^n$]

$$= \frac{15x^{-3}y^2}{9x^{-4}y^8} \qquad [\text{since } (x^n)^m = x^{nm}]$$

$$= \frac{5x^1y^{-6}}{3} \qquad [\frac{x^m}{x^n} = x^{m-n} \text{ and } \frac{15}{9} = \frac{5}{3}]$$

$$= \frac{5x}{3} \times y^{-6}$$

$$= \frac{5x}{3} \times \frac{1}{y^6} \qquad [\text{since } y^{-n} = \frac{1}{y^n}]$$

$$= \frac{5x}{3y^6}$$

Unit 11.1 Activity 6B: Simplifying expressions with indices

1. Write each of the following without negative indices and as simply as possible:

eg, $3x^{-1} = \frac{3}{x}$

a. x^{-3} **b.** y^{-2} **c.** a^{-5} **d.** b^{-4} **e.** z^{-7}

f. y^{-6} **g.** z^{-11} **h.** $(xy)^{-1}$ **i.** $(abc)^{-1}$ **j.** $(xy)^{-2}$

k. $(2x)^{-1}$ **l.** $(2x)^{-2}$ **m.** $(3a)^{-2}$ **n.** $(2x)^{-3}$ **o.** $3x^{-1}$

p. ab^{-3} **q.** $a^{-3}b^2$ **r.** $a^{-3}a^2$ **s.** $\frac{x^{-1}}{2}$ **t.** $\frac{x^{-3}}{4}$

u. $\frac{3x^{-2}}{5}$ **v.** $\frac{2a^{-3}}{7}$ **w.** $\frac{4a^{-2}}{9}$ **x.** $\frac{3x^{-2}}{5x^2}$ **y.** $\frac{4x^{-3}}{5x^2}$

2. Write the following using negative indices for the variable: eg, $\frac{3}{x^2} = 3x^{-2}$

a. $\frac{1}{x^6}$ **b.** $\frac{1}{y^5}$ **c.** $\frac{1}{z^2}$ **d.** $\frac{4}{x}$ **e.** $\frac{5}{y^2}$

f. $\frac{7}{z^5}$ **g.** $\frac{1}{2x^2}$ **h.** $\frac{1}{3a^4}$ **i.** $\frac{1}{5y^4}$ **j.** $\frac{2}{5x^2}$

k. $\frac{3}{4y^3}$ **l.** $\frac{3}{(2x)^2}$ **m.** $\frac{4}{(5y)^2}$ **n.** $\frac{x^2}{(2x)^3}$ **o.** $\frac{y^2}{(2x)^3}$

3. Write the following as simply as possible without using negative indices and as a single fraction:

a. $\frac{4x^{-2}}{8y^{-3}}$ **b.** $\frac{6a^{-3}}{9b^{-4}}$ **c.** $\frac{5x^{-2}}{25x^{-3}}$ **d.** $\frac{5^{-1}x^{-2}}{10^{-1}y^{-4}}$ **e.** $\frac{(2x)^2}{4x^{-3}y}$

f. $\frac{(4x^{-1})^2}{8x^{-1}y^{-3}}$ **g.** $\frac{(2x^{-2})^3y^{-2}}{x}$ **h.** $\frac{5^{-1}x^{-2}}{5x}$ **i.** $\frac{2^{-1}x^{-1}}{3y}$ **j.** $\frac{3^{-1}x^{-2}}{(3x)^2}$

k. $2^{-1}y^{-2}x$ **l.** $3^{-2}y^{-3}x^2$ **m.** $2(3^{-1}x^{-2})$ **n.** $4(5^{-1}x^{-3})y^2$

o. $x^{-1} + y^{-1}$ **p.** $x^{-1} + x^{-2}$ **q.** $x(2^{-1}y + y^{-1}x^{-1})$

Rational powers (fractional indices)

Fractional indices refer to powers that appear as fractions and are used to express roots of numbers. For example, the square root of a number $\sqrt{x}$ can be expressed as $x^{\frac{1}{2}}$ and the cube root of a number $\sqrt[3]{x}$ can be expressed as $x^{\frac{1}{3}}$.

Generally, the nth root of a number can be written as:

$$x^{\frac{1}{n}} = \sqrt[n]{x}$$

Proof: The following steps set out the proofs for $n = 2$ and $n = 3$ and can be applied to any other value of n.

1. The square root of a number can be symbolically expressed as $\sqrt[2]{x}$ or $\sqrt{x}$ and the index representing the square root of a number is obtained from:

$$x^{\frac{1}{2}} \times x^{\frac{1}{2}} = \left(x^{\frac{1}{2}}\right)^2 \qquad [\text{ since } (x^m)^n = x^{mn}]$$

$$= x^1$$

$$= x$$

$$\therefore x^{\frac{1}{2}} = \sqrt[2]{x} \text{ or } \sqrt{x}$$

2. The index representing the cube root of a number is obtained from:

$$x^{\frac{1}{3}} \times x^{\frac{1}{3}} \times x^{\frac{1}{3}} = \left(x^{\frac{1}{3}}\right)^3$$

$$= x^1$$

$$= x$$

$$\therefore x^{\frac{1}{3}} = \sqrt[3]{x}$$

Generalising further:

$$x^{\frac{m}{n}} = \sqrt[n]{x^m} = \left(\sqrt[n]{x}\right)^m$$

Example D

$$(-27)^{\frac{5}{3}} = \left[(-27)^{\frac{1}{3}}\right]^5 \qquad [\text{since } (x)^{\frac{m}{n}} = \left[(x)^{\frac{1}{n}}\right]^m]$$

$$= \left(\sqrt[3]{-27}\right)^5 \qquad [\text{since } (x)^{\frac{1}{n}} = \sqrt[n]{x}]$$

$$= (-3)^5$$

$$= -243$$

Note: When you raise a negative number to an even power the answer is positive, but if you raise a negative number to an odd power the answer is negative. Therefore, if you take an even root of a negative number, the result will be undefined.

When simplifying algebraic expressions with fractional powers, the answer is often expressed in the form ax^n.

Example E

Q. Write the following in simplest index form:

1. $\sqrt{x}$ 2. $3\sqrt[5]{x^2}$ 3. $\frac{2}{\sqrt[3]{x^2}}$

A. 1. $\sqrt{x} = \sqrt[2]{x}$ [from the definition of $\sqrt{\ }$]

$= x^{\frac{1}{2}}$ [since $\sqrt[y]{x} = x^{\frac{1}{y}}$]

2. $3\sqrt[5]{x^2} = 3 \times \sqrt[5]{x^2}$

$= 3x^{\frac{2}{5}}$ [since $\sqrt[n]{x^m} = x^{\frac{m}{n}}$]

3. $\frac{2}{\sqrt[3]{x^2}} = 2 \times \frac{1}{\sqrt[3]{x^2}}$

$= 2 \times \frac{1}{x^{\frac{2}{3}}}$ [since $\sqrt[n]{x^m} = x^{\frac{m}{n}}$]

$= 2 \times x^{-\frac{2}{3}}$ [since $\frac{1}{x^n} = x^{-n}$]

$= 2x^{\frac{-2}{3}}$

Unit 11.1 Activity 6C: Fractional indices

1. Express the following in index form as simply as possible. eg, $3\sqrt[4]{x^5} = 3x^{\frac{5}{4}}$

a. $\sqrt{a}$ **b.** $\sqrt{c}$ **c.** $\sqrt[2]{x}$ **d.** $\sqrt[5]{a}$ **e.** $\sqrt[6]{u}$

f. $\sqrt[5]{x^2}$ **g.** $\sqrt[7]{x^3}$ **h.** $\sqrt[8]{a^3}$ **i.** $\frac{1}{\sqrt{x}}$ **j.** $\frac{1}{\sqrt{y}}$

k. $\frac{1}{\sqrt[3]{a}}$ **l.** $\frac{1}{\sqrt[5]{x}}$ **m.** $\frac{1}{\sqrt[7]{x^2}}$ **n.** $\frac{1}{\sqrt[9]{x^5}}$ **o.** $\frac{1}{\sqrt[7]{a^3}}$

p. $\frac{1}{\sqrt[9]{a^8}}$ **q.** $4\sqrt[3]{a}$ **r.** $\frac{6}{\sqrt[3]{y}}$ **s.** $\frac{8}{\sqrt[3]{x^7}}$ **t.** $\frac{1}{7\sqrt[3]{x^2}}$

u. $\frac{2}{5\sqrt[4]{x^3}}$ **v.** $(\sqrt{x})^{-2}$ **w.** $\frac{\sqrt{x}}{\sqrt[3]{x}}$ **x.** $\frac{1}{\sqrt[3]{a}\ \sqrt[3]{a^4}}$ **y.** $\frac{\sqrt{x}}{\sqrt[3]{x}\ \sqrt[4]{x}}$

2. Express the following using radical (or root) signs: eg, $x^{\frac{2}{3}} = \sqrt[3]{x^2}$

a. $a^{\frac{3}{7}}$ **b.** $b^{\frac{3}{4}}$ **c.** $x^{\frac{2}{5}}$ **d.** $x^{\frac{-2}{3}}$ **e.** $y^{\frac{-3}{4}}$

3. Express the following in index form and as simply as possible:

a. $\sqrt{xy}$ **b.** $\sqrt{xy^2}$ **c.** $\sqrt{\frac{x^2}{y}}$

d. $\sqrt[3]{\frac{x^6}{y}}$ **e.** $\sqrt[5]{\frac{32x^5}{y^3}}$ **f.** $\sqrt[4]{x^{-5}y^4}$

g. $\sqrt{x} \times \sqrt[3]{a}$ **h.** $\frac{1}{\sqrt{a}} \times \sqrt{c}$ **i.** $\frac{1}{\sqrt[4]{b^{-2}}} \times \sqrt{b}$

4. Simplify fully, writing with no negative powers or radical signs:

a. $\left(16x^4\right)^{\frac{-3}{4}}$ **b.** $\left(27x^9\right)^{\frac{-2}{3}}$ **c.** $\left(32y^{10}\right)^{-\frac{2}{5}}$ **d.** $\left(\sqrt[5]{243A^5}\right)^3$

e. $\left(\sqrt[6]{64B^{12}}\right)^{-5}$ **f.** $\dfrac{1}{\left(\sqrt[3]{8x^6}\right)^2}$ **g.** $\dfrac{1}{\left(\sqrt[4]{16A^4B^4}\right)^{-2}}$ **h.** $\left(\sqrt[3]{125A^3B^6}\right)^{-2}$

i. $\dfrac{3}{\left(\sqrt{x}\right)^{-3}}$ **j.** $\dfrac{8}{\left(\sqrt{64x^{12}}\right)^{-\frac{5}{6}}}$

Using calculators to evaluate $(-x)^{\frac{m}{n}}$

Calculators which have exponent keys such as x^y or $\wedge$ can simplify calculations involving indices. However, many calculators do not accept a negative base (such as –27.5) raised to a fractional power (such as $\frac{2}{3}$) and if the calculation $(-27.5)^{\frac{2}{3}}$ is entered, an error message results. In such cases the following procedure is used.

Calculation of $(-x)^{\frac{m}{n}}$ where *m* and *n* are integers and *x* is positive

The index $\frac{m}{n}$ should be simplified as much as possible. Then, regardless of whether $\frac{m}{n}$ is negative or positive:

- If n is even, $(-x)^{\frac{m}{n}}$ cannot be calculated. [cannot take square root, 4th root, etc, of a negative number]
- If n is odd and m is even, $(-x)^{\frac{m}{n}} = (x)^{\frac{m}{n}}$ and the answer is positive.
- If n and m are odd, $(-x)^{\frac{m}{n}} = -\left[(x)^{\frac{m}{n}}\right]$ and the answer is negative.

Example F

$(-27.5)^{\frac{2}{3}} = (27.5)^{\frac{2}{3}}$ [since 2 is even and 3 is odd]

$= 9.11$ [using a calculator, to 2 decimal places]

Note: Be sure to put brackets around fractional exponents. On a calculator $(-27.5)^{\frac{2}{3}}$ would be calculated by pressing:

     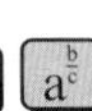   3) =

Unit 11.1 Activity 6D: Calculating with indices

Use a calculator to work out each of the following:

1. 2^4

2. 3^{-2}

3. 0.23^{-3}

4. $(-2)^5$

5. $\left(\frac{2}{3}\right)^{-2}$

6. $\left(1\frac{3}{4}\right)^{\frac{1}{2}}$

7. $(-3.47)^{\frac{3}{5}}$

8. $(-2.49)^{\frac{-1}{2}}$

9. $\left(\frac{12.37}{2.69}\right)^{\frac{14}{17}}$

10. $(3.52^4 - 2.15^3)^{\frac{3}{2}}$

11. $(17.63)^{\frac{1}{2}} - (3.165)^{\frac{1}{3}}$

12. $(15.86)^{\frac{-3}{7}} + (-4.39)^{\frac{2}{7}}$

13. $\frac{(11.45)^{0.467}}{(2.78)^{-3.1}}$

14. $\frac{(-17.45)^{\frac{13}{27}}}{(15.35)^{\frac{3}{4}}}$

15. $(3.2)^{\frac{1}{3}} \times (2.5)^{\frac{1}{2}} - (1.36)^{\frac{1}{4}}$

16. $(-3.2)^{\frac{2}{3}}$

17. $(56)^{\frac{2}{3}}$

18. $(63)^{\frac{-3}{5}}$

19. $(-4.57)^{\frac{-3}{4}}$

20. $(-31.2)^{\frac{-4}{5}}$

Solution of equations involving indices

Equations with indices often simplify to equations of the form $x^n = C$. By raising the expression on each side of the equation to the **reciprocal of the power of *x***, the solution can be found (the reciprocal of n is $\frac{1}{n}$).

Example G

Q. Solve $8x^4 - 3 = 2$, correct to 4 significant figures.

A. $8x^4 - 3 = 2$

$\therefore\ 8x^4 = 5$ [adding 5]

$x^4 = \frac{5}{8}$ [dividing by 8]

$\therefore\ (x^4)^{\frac{1}{4}} = \left(\frac{5}{8}\right)^{\frac{1}{4}}$ [taking **reciprocal powers**]

$\therefore\ x = 0.8891$ (4 sf)

The same technique works for fractional indices (the reciprocal of $\frac{m}{n}$ is $\frac{n}{m}$).

Example H

Q. Solve $(x)^{\frac{15}{23}} = 26.4$, correct to 4 significant figures.

A. $(x)^{\frac{15}{23}} = 26.4$

$\therefore \left[(x)^{\frac{15}{23}}\right]^{\frac{23}{15}} = (26.4)^{\frac{23}{15}}$ [raising both sides to the **reciprocal power** to isolate the variable]

$\therefore x = 151.3$ [using a calculator]

Example I

Q. Solve $(2A + 3)^{\frac{-4}{7}} = 8$, to 3 significant figures.

A. $(2A + 3)^{\frac{-4}{7}} = 8$

$\therefore \left[(2A + 3)^{\frac{-4}{7}}\right]^{\frac{-7}{4}} = (8)^{\frac{-7}{4}}$ [raising to the reciprocal power]

$\therefore 2A + 3 = (8)^{\frac{-7}{4}}$

$\therefore 2A = (8)^{\frac{-7}{4}} - 3$ [subtracting 3]

$\therefore A = \dfrac{(8)^{\frac{-7}{4}} - 3}{2}$ [dividing by 2]

$\therefore A = -1.49$

Sometimes *the unknown occurs as part of the index*. Equations of this nature are called **exponential** or **indicial** equations. When solving such equations, create same bases on both sides of the equation then equate the indices to one another and solve for the unknown. $2^x = 8$, $3^x = 10$ and $\dfrac{1}{10^x} = 100$ are examples of indicial equations.

Example J

Q. Solve the following exponential equations:

a. $2^x = 4$ **b.** $9^x = \dfrac{1}{27}$ **c.** $2 \times \dfrac{1}{4^x} = 32$ **d.** $5^{x+2} = \dfrac{1}{\sqrt{5}}$

A.

a. $2^x = 4$

$\therefore 2^x = 2^2$

$\therefore x = 2$

b. $9^x = \dfrac{1}{27}$

$\therefore 3^{2(x)} = \dfrac{1}{3^3}$

$\therefore 3^{2x} = 3^{-3}$

$\therefore 2x = -3$

c. $2 \times \dfrac{1}{4^x} = 32$

$\therefore \dfrac{1}{4^x} = 16$

$\therefore \dfrac{1}{(2^2)^x} = 2^4$

$\therefore 2^{-2x} = 2^4$

d. $5^{x+2} = \dfrac{1}{\sqrt{5}}$

$\therefore 5^{x+2} = \dfrac{1}{5^{\frac{1}{2}}}$

$\therefore 5^{x+2} = 5^{-\frac{1}{2}}$

$\therefore x + 2 = -\dfrac{1}{2}$

$\therefore x = \frac{-3}{2}$ $\qquad \therefore -2x = 4$ $\qquad \therefore x = -2\frac{1}{2}$

$= -\frac{3}{2}$ $\qquad \therefore x = -2$

Unit 11.1 Activity 6E: Equations with indices

1. Solve each of the following equations:

a. $x^3 = 2$ **b.** $x^5 = 243$ **c.** $x^{\frac{1}{4}} = 3$ **d.** $a^{\frac{1}{2}} = 5$

e. $a^{\frac{1}{5}} = \frac{2}{3}$ **f.** $y^{-2} = 25$ **g.** $x^{-4} = 256$ **h.** $y^{\frac{3}{4}} = 27$

i. $x^{\frac{2}{5}} = 4$ **j.** $\frac{y^{\frac{4}{3}}}{2} = 8$ **k.** $x^{\frac{-5}{3}} = 32$ **l.** $3y^{\frac{-7}{3}} = 384$

m. $2x^{\frac{1}{2}} - 1 = 3$ **n.** $4x^{\frac{1}{3}} + 1 = 9$ **o.** $2(x-1)^{\frac{1}{2}} = 4$ **p.** $\frac{x^{\frac{2}{3}}+1}{2} = 5$

q. $\frac{3}{x^{\frac{1}{3}}}$ **r.** $x^{\frac{-1}{2}} = 2$ **s.** $3y^{\frac{-3}{2}} - 1 = 80$ **t.** $\frac{z^{\frac{-2}{5}}}{3} - 1 = 2$

2. Solve the following equations:

a. $(x)^{\frac{2}{3}} = 16$ **b.** $(A)^{\frac{3}{7}} = 28$ **c.** $(A)^{\frac{-7}{3}} = 43$

d. $(p)^{1\frac{3}{4}} = 7$ **e.** $(p)^{\frac{4}{3}} = -2.3$ **f.** $(p)^{\frac{3}{5}} = -4$

g. $2(x)^{\frac{2}{3}} = 7$ **h.** $\left[(x)^{\frac{7}{9}} + 3\right]^{\frac{1}{2}} = 9$ **i.** $(2x-3)^{\frac{-3}{2}} = 5$

3. Solve the following indicial equations:

a. $2^x = 8$ **b.** $4^x = \frac{1}{2}$ **c.** $4^x = 32$ **d.** $2^x = 1$

e. $3^x = 0$ **f.** $3^{-x} = \frac{1}{9}$ **g.** $3^{2-x} = \frac{1}{27}$ **h.** $5^{x+1} = \frac{1}{25}$

i. $5^x = 125$ **j.** $10^x = 100\sqrt{10}$ **k.** $10^x = 0.001$ **l.** $5^{1-2x} = 1$

m. $49^x = \frac{1}{7}$ **n.** $4^x = \frac{1}{\sqrt{2}}$

4. Find the value for x for each of the following exponential equations:

a. $3 \times 2^x = 24$ **b.** $7 \times 2^x = 56$ **c.** $3 \times 2^{x+1} = 24$

d. $12 \times 3^x = \frac{4}{3}$ **e.** $9 \times 5^x = 225$ **f.** $5 \times \left(\frac{1}{2}\right)^x = 20$

5. The relationship connecting length, L, to cost, C, of a type of material is given by $C = 5L^{\frac{2}{3}} + 6$, where C is given in Kina and L in metres.
   a. What is the cost when 27 metres are purchased?
   b. What length can be purchased for K100?
6. The relationship between the volume, height and base radius of a shape is given by $V = KR^{\frac{5}{2}}H^{\frac{1}{2}}$, where R is base radius, H is height and K is a constant. When the height and base radius are both 2 the volume is 16:
   a. Find the volume when the radius is 16 and the height 49.
   b. Find the height when the volume is 128 and the radius is 4.
   c. Find the radius when the volume is 486 and the height is 3 125.
7. T and U are related by the formula $T\sqrt[5]{U^4} = 13.5$.
   a. Describe what happens to T as U gets larger.
   b. If the minimum value of T is 500, find the maximum value of U.

Unit 11.1 Number and Application

Topic 7: Basic numeracy—logarithms

Following the General Mathematics syllabus, p.14, this Topic covers the manipulation of algebraic expressions and solving equations:

- Use elementary properties of logarithms.
- Solve simple logarithmic equations.
- Solve problems in context.
- Choose algebraic techniques and strategies to solve problems.

Logarithmic functions

Any **function** of the type $y = a^x$ is called an **exponential function**. The inverse of this function is called the **logarithmic function to base *a*** and is written $\log_a x$.

Most scientific calculators have the logarithmic function to base 10 ($\log_{10} x$) and the exponential function, 10^x. On most calculators, $\log_{10}$ has a key labelled [log] and the exponential function, 10^x, is the inverse of the log key.

Example A

1. $\log_{10} 1\,000 = 3$ [since $1\,000 = 10^3$]
2. $\sqrt{\log_{10} 17} = \sqrt{1.2304489}$
   $= 1.11$ (2 dp)
3. $(10^{1.2} - 1)^{\frac{1}{3}} = (15.848932 - 1)^{\frac{1}{3}}$
   $= 2.46$ (2 dp)
4. $10^{\log_{10} 5} = 5$ [as '10^x' and 'log' are **inverse functions**]
5. $10^{2x} = 6$
   $\therefore\ 2x = \log_{10} 6$ [taking logs]
   $\therefore\ x = 0.39$ (2 dp)

Note: To evaluate powers of 10 press [shift] [10^x] or use the general power key [^] or [x^y].

[On some calculators [inv] replaces [shift]]

Unit 11.1 Activity 7A: Other logarithmic functions

Calculate to 2 decimal places using the [10^x] and [log] keys on your calculator:

1. $\log_{10} 125.7$
2. $10^{2.3}$
3. $10^{3.4} - \log_{10} 2347.5$
4. $\dfrac{10^{-2.3}}{\log_{10} 1.37}$
5. $10^{\log_{10} 3.678}$
6. $\log_{10} (10^{2.5})$

General properties of logarithmic functions

- If a is any positive number then $y = a^x$ and $y = \log_a x$ are **inverse functions**.

  Note: $\log_a x$ is the **logarithm** of x to the base a.

- The function $y = \log_a x$ is defined by:

  $y = \log_a x$ if and only if $x = a^y$

- $a^{\log_a x} = x$

  $a^{\log_a x} = x$ because a^x and $\log_a x$ are inverse functions of each other.

- $\log_a a^x = x$
- For all $a > 0$: $\log_a a = 1$ and $\log_a 1 = 0$
- $\log_a (xy) = \log_a x + \log_a y$

Proof: Let $u = \log_a x$ and $v = \log_a y$

$\therefore\ x = a^u$ and $y = a^v$ [from definition of log]

$\therefore\ xy = a^u a^v$ [multiplying x and y]

$\therefore\ xy = a^{u+v}$ [adding indices]

$\therefore\ u + v = \log_a xy$ [by definition of log]

$\therefore\ \log_a x + \log_a y = \log_a xy$, as required [since $u = \log_a x$, $v = \log_a y$]

- $\log_a\left(\dfrac{s}{t}\right) = \log_a s - \log_a t$

Proof: Let $x = \log_a s$ and $y = \log_a t$

$\therefore s = a^x$ and $t = a^y$

$\therefore \dfrac{s}{t} = \dfrac{a^x}{a^y}$ [divide by]

$\therefore \dfrac{s}{t} = a^{x-y}$ [subtract indices]

$\therefore x - y = \log_a\left(\dfrac{s}{t}\right)$

$\therefore \log_a s - \log_a t = \log_a\left(\dfrac{s}{t}\right)$ [since $x = \log_a s, v = \log_a t$]

- $\log_a x^n = n\log_a x$

Proof: $\log_a x^n = \log_a (\underbrace{x \times x \times x \times \dots x}_{n \text{ terms}})$

$= \underbrace{\log_a x + \log_a x + \dots \log_a x}_{n \text{ terms}}$

$= n\log_a x$

Worked examples

Example B

Write the equation $128 = 2^7$ in logarithmic form.

Solution

$128 = 2^7$

$\therefore\ 7 = \log_2 128$ [since $y = \log_a x$ when $x = a^y$]

$\therefore$ The required equation is $\log_2 128 = 7$.

Example C

Solve $4^x = 8$.

Solution

$$4^x = 8$$

$$\log_{10}(4^x) = \log_{10} 8 \quad \text{[taking logs of both sides]}$$

$$x\log_{10} 4 = \log_{10} 8 \quad \text{[since } \log A^n = n\log A\text{]}$$

$$x = \frac{\log_{10} 8}{\log_{10} 4} = 1.5 \quad \text{[dividing by } \log_{10} 4\text{]}$$

$$x = \frac{\log_{10} 2^3}{\log_{10} 2^2}$$

$$= \frac{3\log_{10} 2}{2\log_{10} 2}$$

$$= \frac{3}{2}$$

Example D

Solve the equation $5^{3t-1} = 8$.

Solution

$$5^{3t-1} = 8$$

$$\therefore\ \log_{10} 5^{3t-1} = \log_{10} 8 \quad \text{[taking logs of both sides]}$$

$$\therefore\ (3t-1)\log_{10} 5 = \log_{10} 8 \quad \text{[since } \log A^n = n\log A\text{]}$$

$$\therefore\ 3t - 1 = \frac{\log_{10} 8}{\log_{10} 5} \quad \text{[dividing by } \log_{10} 5\text{]}$$

$$\therefore\ 3t = \frac{\log_{10} 8}{\log_{10} 5} + 1 \quad \text{[adding 1]}$$

$$\therefore\ t = \left(\frac{\log_{10} 8}{\log_{10} 5} + 1\right) \div 3$$

$$= 0.764 \ \ (3\text{ dp})$$

Note: $\log_{10} 8$ and $\log_{10} 5$ are evaluated last to avoid rounding errors.

Example E

Find the value of $\log_{16} 15$.

Solution

$$
\begin{aligned}
\text{Let } x &= \log_{16} 15 \\
\therefore\ 16^x &= 15 && [\text{since } x = a^y \text{ when } y = \log_a x] \\
\therefore\ \log_{10} 16^x &= \log_{10} 15 && [\text{taking logs of both sides}] \\
\therefore\ x\log_{10} 16 &= \log_{10} 15 && [\text{since } \log A^n = n \log A] \\
\therefore\ x &= \frac{\log_{10} 15}{\log_{10} 16} \\
\therefore\ x &= 0.977 \text{ (3 dp)}
\end{aligned}
$$

Note: The **base changing rule** is: $\boxed{\log_a x = \dfrac{\log_b x}{\log_b a}}$

Here, $\log_{16} 15 = \dfrac{\log_{10} 15}{\log_{10} 16} = 0.977$.

Example F

Solve $\log_3 (2x + 1) = 5$.

Solution

$$
\begin{aligned}
\log_3 (2x + 1) &= 5 \\
\therefore\ 2x + 1 &= 3^5 && [x = a^y \text{ when } y = \log_a x] \\
\therefore\ 2x + 1 &= 243 && [\text{since } 3^5 = 243] \\
\therefore\ 2x &= 242 && [\text{subtracting } 1] \\
\therefore\ x &= 121 && [\text{dividing by } 2]
\end{aligned}
$$

Logarithm laws can be used to simplify expressions involving logarithms.

Example G

Express $2\log_a x - 3\log_a y + 4$ as a single logarithm.

Solution

$$
\begin{aligned}
2\log_a x - 3\log_a y + 4 &= \log_a x^2 - \log_a y^3 + 4 && [n\log u = \log u^n] \\
&= \log_a \frac{x^2}{y^3} + 4 && [\log u - \log v = \log \frac{u}{v}] \\
&= \log_a \frac{x^2}{y^3} + 4\log_a a && [\text{since } \log_a a = 1] \\
&= \log_a \frac{x^2}{y^3} + \log_a a^4 && [n\log u = \log u^n] \\
&= \log_a \frac{x^2 a^4}{y^3} && [\log u + \log v = \log uv]
\end{aligned}
$$

Unit 11.1 Activity 7B: Using logarithmic properties

1. Write the equivalent logarithmic statement for each of the following:

a. $2^3 = 8$ **b.** $3^4 = 81$ **c.** $5^2 = 25$ **d.** $6^3 = 216$ **e.** $x^a = b$

2. Write the equivalent exponential statement for the following:

a. $\log_3 9 = 2$ **b.** $\log_9 3 = \frac{1}{2}$ **c.** $\log_{25} 125 = 1.5$

d. $\log_{1000} 10 = \frac{1}{3}$ **e.** $\log_3 \frac{1}{3} = -1$

3. Solve the following for x (give answers as fractions or integers):

a. $2^x = 64$ **b.** $2^x = \frac{1}{2}$ **c.** $2^{x-1} = 32$ **d.** $2^x = 128$

e. $3^x = \frac{1}{3}$ **f.** $3^x = \frac{1}{27}$ **g.** $(3)^{\frac{1}{x}} = 9$ **h.** $125 = 5^x$

i. $9^x = 3$ **j.** $16^x = 2$ **k.** $25^x = 5$ **l.** $25^x = 125$

m. $25^x = 625$ **n.** $64^x = 32$ **o.** $49^x = \frac{1}{7}$ **p.** $121^x = 11^2$

q. $2^{1-x} = 4$ **r.** $\frac{1}{2^x} = 8$ **s.** $27^x = 243$ **t.** $81^x = 27$

4. Solve the following equations to 3 decimal places:

a. $10^x = 9$ **b.** $10^{-x} = 131$ **c.** $10^x = 323.3$

d. $2.5^x = 14.06$ **e.** $10^{3x} = 822.36$ **f.** $3^x = 4$

g. $2^{3x-1} = 9$ **h.** $3^x = 2^{x+1}$ **i.** $4^{2x} = 5^{x+2}$

j. $3.6^{4x+5} = 2.9^{2x+13}$ **k.** $500 = 200 \times 1.03^{2x}$

5. Calculate the following:

a. $\log_4 4$ **b.** $\log_3 3$ **c.** $\log_2 8$ **d.** $\log_9 3$

e. $\log_{16} 2$ **f.** $\log_{16} 32$ **g.** $\log_{625} 125$ **h.** $\log_{196} 14$

i. $\log_{32} 16$ **j.** $\log_{0.1} 10$

6. Use the laws of logarithms to express each of the following as a single logarithm:

a. $\log_a x + \log_a 2x$ **b.** $\log_a y - \log_a (2y)$

c. $3 \log_a x - \log_a (xy)$ **d.** $\log_a (x + 1) + \log_a (3x) - \log_a x$

e. $2 \log_b x + \log_b 2x - \log_b 3x$

f. $\frac{1}{2} \log_c 64 - \log_c 8 + \log_c 2$

g. $\log_a abc - \log_a b + \log_a c$

h. $\frac{1}{2} \log_a c - 2 \log_a c + \log_a c^3$

i. $\log_a \frac{pq}{r} - \log_a \frac{r}{q}$

j. $\frac{1}{4} \log_{10} 16 - 2 \log_{10} 2$

Unit 11.1 Number and Application

Topic 8: Units of measurement

Units of measurement is one sub-section within Unit 11.1 of the Grade 11 syllabus (see Syllabus p. 14). This topic deals with problem solving involving measurement of everyday objects and covers:

- Choosing and making measurements using appropriate equipment, scales and units.
- Conversion between units.
- Problems involving length, mass, capacity and time (including 24-hour clock time.
- Calculation and interpretation of rates.
- The precision (limits of accuracy) of a measurement.

Introduction

The **Imperial** system of units was defined by the British in 1824 and was adopted by countries in the British Empire. By the end of the 20th century most of these countries had converted to metric units.

Papua New Guinea declared the **International System of units (SI)** to be the legal units of **measurement** in 1980. It is illegal to use any units of measurement other than the SI units for trade purposes in PNG.

The International System of Units (SI) defines seven units of measure as a basic set from which all other SI units are derived. These SI **base units** and their physical quantities are:

- Metre for length.
- Kilogram for mass.
- Second for time.
- Ampere for electric current.
- Kelvin for temperature.
- Candela for luminous intensity.
- Mole for the amount of substance.

If the base unit of **length** is the metre then all the units for length, area and volume are defined in terms of this basic unit. Many other units, such as the litre, are not formally part of the SI, but are accepted for use with the SI.

The metric system

Appropriate units must be used when solving problems.

In the **metric system**, each quantity measured has a **basic unit** (gram, metre, litre), of which other units are multiples. These other units are identified by a **prefix** such as 'kilo', 'centi', and 'milli'. By using prefixes measurements can be expressed with suitably sized numbers, eg not 45 000 g but 45 kg.

The table shows commonly used metric prefixes.

Prefix	kilo, k	centi, c	milli, m
Means	10^3 1 000 one thousand	10^{-2} $\frac{1}{100}$ one hundredth	10^{-3} $\frac{1}{1\,000}$ one thousandth

Note: Less commonly used prefixes include hecto (h) meaning 100, deci (d) meaning $\frac{1}{10}$ and micro (μ) meaning $\frac{1}{1\,000\,000}$.

Example A

The kilometre, centimetre, and millimetre are all units *derived* from the basic unit, the metre.

1. The length and width of a classroom would probably be measured with the basic unit, the metre.
2. To measure a longer distance, such as the distance between Port Moresby and Brisbane, the kilometre is a more suitable unit to use.
3. For smaller lengths, such as the thickness of a telephone book, the unit millimetre might be preferred. Alternatively, the centimetre may be used.

Changing from one metric unit to another

To change from one unit to another, either **multipliers** or **divisors** are used. This **conversion** method is shown in the following diagram for the prefixes kilo, centi, and milli.

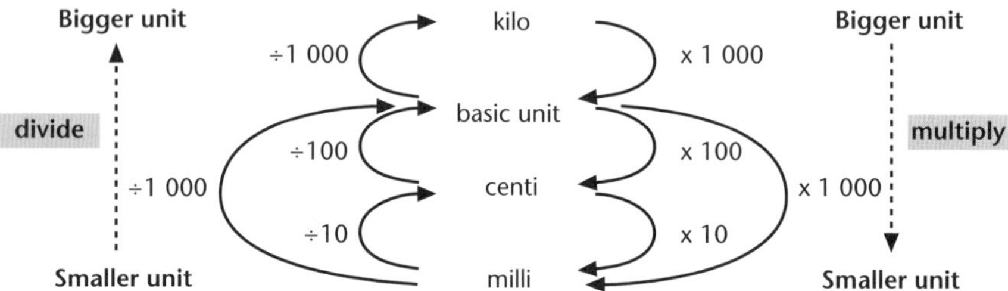

To convert from one unit to another follow the arrows, eg to change from the basic unit to kilo, divide by 1 000; to change from centi to milli, multiply by 10.

- To change from a bigger unit to a smaller unit (eg centi to milli), use multiplication, since there will be more of the smaller unit.
- To change from a smaller unit to a bigger unit, use division, since there will be fewer of the larger unit.

Example B

Q. Convert: **1.** 146 m to cm **2.** 13 400 mg to kg **3.** 1.75 km to cm

A. 1. m to cm is from a bigger unit to a smaller unit, so multiplication is used:

146 m = 146 x 100 cm [changing m to cm]
= 14 600 cm

2. mg to kg is from a smaller unit to a bigger unit, so division is used:

13 400 mg = 13 400 ÷ 1 000 g [changing mg to g]
= 13.4 g
= 13.4 ÷ 1 000 kg [changing g to kg]
= 0.0134 kg

3. 1.75 km = 1.75 x 1 000 m [changing km to m by multiplying]
= 1 750 x 100 cm [changing m to cm]
= 175 000 cm

Some important metric quantities

Length

The basic unit of length in the metric system is the **metre** (symbol **m**). Other commonly used units for length are **kilometre** (**km**), **centimetre** (**cm**), and **millimetre** (**mm**).

1 km = 1 000 m	100 cm = 1 m	1 000 mm = 1 m

It is useful to know approximately the width of your own handspan, the length of your foot or your 'pace' for estimating lengths in practical situations. Find out, for example, how far apart your hands need to be to approximate a metre.

Mass

The common units of **mass** in the metric system are the **gram** (symbol **g**) and **kilogram** (**kg**). Other units commonly used for mass are **milligram** (**mg**), and **tonne** (**t**).

1 kg = 1 000 g	1 000 mg = 1 g	1 tonne = 1 000 kg

Note: Mass is commonly (but incorrectly) referred to as weight. In this book mass will, on occasions, be referred to as weight.

Capacity

The **capacity** of a container is the amount of liquid (or other material) it will hold. The **litre** (**L**) and **millilitre** (**mL**) are the commonly used units of capacity in the metric system.

1 kL = 1 000 L	1 000 mL = 1 L

It is useful to know that a teaspoon holds 5 mL, a tablespoon holds 15 mL and a standard cup holds 250 mL. Also, 1 mL of water weighs 1 g and 1 L of water weighs 1 kg.

Capacity is commonly referred to as volume (see page 110).

Money

The basic unit of money in Papua New Guinea is the **Kina** (symbol **K**). The other unit is the **toea** (symbol **t**): K1 = 100 toea.

Note: 48 toea can be written 48t or K0.48. Do *not* write 48 toea as 0.48t.

Temperature

The temperature scale is measured in degrees, using the **Celsius**, or **centigrade scale** (**°C**). This scale is based on the freezing and boiling points of water.

Freezing point of water is 0°C. Boiling point of water is 100°C.

Time

Time is the only commonly used measurement that is not in the metric system.

60 seconds = 1 minute	60 minutes = 1 hour	24 hours = 1 day	
7 days = 1 week	2 weeks = 1 fortnight	365 days = 1 year	366 days = 1 leap year
1 century = 100 years			

You should also be familiar with the number of days in each month of the year.

There are two systems used to record time: the **12-hour clock** and the **24-hour clock**. It is essential, when writing time using the 12-hour clock, that either am or pm (whichever is appropriate) is written as well.

In the 24-hour clock, the hours in a day are numbered from 0–24, with 0 being midnight. Four digits are always used to express the time on a 24-hour clock.

- The first two digits represent the hours after midnight.
- The second two digits represent the minutes past the hour.

Example C

1. 3.00 am (in 12-hour time) is 0300 hours (in 24-hour time).
2. 1200 is midday or 12 noon. [12 hours after midnight]
3. 5.30 pm is 1730 hours. [1200 + 0530]
4. 1408 hours is 2.08 pm. [1408 – 1200 → 2 hours and 8 minutes after midday]
5. 1 minute after 1559 hours is written 1600 hours. [1559 + 0001]

When using 24-hour time, remember that the final two digits are minutes (60 minutes = 1 hour) and do not treat them as you would decimal (base 10) numbers.

Example D

Q. Elijah's plane leaves at 1015 and arrived at its destination at 2105. How long was the journey?

A. Number of minutes from 1015 to 1100 = 45 minutes

Hours and minutes from 1100 until 2105 = 10 hours 5 mins

Total time = 10 hours 50 mins [adding]

Note: The usual technique of finding the difference by subtracting the times (as decimal numbers) cannot be used for 24-hour times (2105 – 1015 = 1090).

Unit 11.1 Activity 8A: Units of measurement

1. From the list **i.– xx.** below, choose the measurement that best fits the following:

a. The distance between Goroka and Kavieng.

b. The height of a man.

c. The time for a good swimmer to swim 100 m.

d. The length of a pen.

e. The capacity of the petrol tank of a car.

f. The thickness of a match.

g. The length of a playing field.

h. The weight of a brick.

i. The temperature inside a refrigerator.

j. The diameter of a 10t coin.

k. The time for a car to travel 90 km.

l. The capacity of a teaspoon.

m. The weight of an average woman.

n. The height of Mt Wilhelm.

o. The temperature of a recently made cup of tea.

Choose from:

i.	2 mm	**ii.**	1 hour	**iii.**	5 mL	**iv.**	4509 m	**v.**	15 cm
vi.	60 g	**vii.**	1.8 m	**viii.**	2.3 cm	**ix.**	5° C	**x.**	717 km

xi. 50 L **xii.** 55 kg **xiii.** 10 sec **xiv.** 30° C **xv.** 300 m

xvi. 80°C **xvii.** 23 cm **xviii.** 2 kg **xix.** 1 min **xx.** 100 L

2. Change each of the following quantities to the unit in the brackets:

a. 8 m (cm) **b.** 4 cm (mm) **c.** 4 000 g (kg)

d. 3.5 m (mm) **e.** 0.3 km (cm) **f.** 600 mL (L)

g. 4.8 km (m) **h.** 8 300 kg (tonne) **i.** 0.37 g (mg)

j. 0.4 m (cm) **k.** 75 cL (mL) **l.** 945 mg (g)

m. 140 000 mm (km) **n.** 13 700 g (tonne) **o.** 45 000 mm (km)

p. 4.7 tonne (kg) **q.** 120 L (kL) **r.** 640 m (km)

s. 0.004 km (mm) **t.** 7.3 tonne (g) **u.** 186 cm (m)

v. 63 000 000 mg (kg) **w.** 3.8 kg (g) **x.** 9.37 L (mL)

3. a. A rope is 20 metres long. A piece 150 cm long is cut off it. What length (in metres) remains?

b. Mary has a handspan of approximately 20 cm. She finds that a table top is 12 handspans long. What is the approximate length of the table?

c. A child's building blocks are 35 mm cubes. How tall (in centimetres) is a tower of 11 blocks?

4. a. An average apple weighs 155 g and an average orange weighs 175 g. What is the approximate weight in kg of a bag of 3 apples and 5 oranges?

b. A plastic container of identical buttons weighs 150 g. If the container weighs 30 g, and there are 30 buttons in it, how much does each button weigh?

c. If an egg weighs 53 g and a carton containing a dozen eggs weighs 776 g, how much does the empty carton weigh?

5. a. Ravi entered a race, which began at 9.45 am. She finished 3 hours and 37 minutes later. At what time did she finish?

b. Express 1412 hours using the 12-hour clock.

c. Express half-past-four in the morning, using the 24-hour clock.

d. On March 5, Pania calculates that there are 30 days until her birthday. When is her birthday?

e. A parade starts at 1130 hours and finishes at 1315 hours. How long did the parade last?

6. a. An aeroplane, due to arrive at 4.10 pm, is 35 minutes early. When does it arrive?

b. A traveller leaves Lae at 9.05 am and reaches Kainantu at 11.10 am. She stops for refreshments then starts again at 11.47 am. She reaches Goroka at 3.14 pm. How long did it take her to get from Lae to Goroka, excluding stopping time at Kainantu?

c. A traveller left Port Moresby on Tuesday at 1425 and arrived at her destination the following Wednesday at 0835. How much time did her journey take?

d. How many seconds are there in the month of March?

e. A flight leaves Auckland, New Zealand at 2130 for Port Moresby. The flight takes 4 hours and 55 minutes. In Port Moresby, time is 3 hours behind Auckland time. What time will it be in Port Moresby when the plane arrives (local PNG time)?

Scales and dials

Measurements are usually made by using various tools and instruments such as measuring tapes, stop watches or meters. On many such instruments taking a measurement means reading a **scale** or **dial**.

A scale, in this context, is shown by marks at set intervals, such as those on a ruler or protractor, the speedometer of a car or bathroom scales. Small divisions between markings on a scale indicate intermediate values that can be read with reasonable accuracy.

Example E

Q. The scale indicates the number of kilograms an object weighs. The pointer stops as shown. What weight is the object?

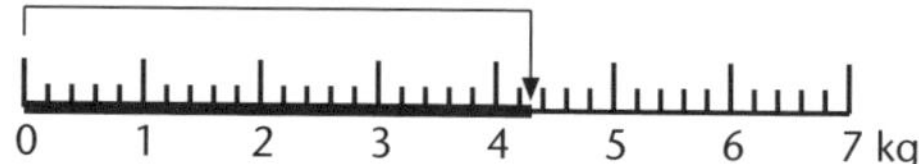

A. Between each kilogram marking are 5 equal divisions. Each division therefore represents $\frac{1}{5}$ kg = 0.2 kg.
The pointer has stopped halfway between the first and second division, which is $0.2 + \frac{1}{2} \times 0.2 = 0.3$ kg beyond 4 kg. The weight is 4.3 kg.

Circular dials, speedometers and scales on measuring jugs are other examples of linear scales, which are read in a similar way (be aware that some dials are read anticlockwise).

If a jug has sloping sides the scale may not be linear, ie the gaps between divisions may not be equal. Even if gaps between divisions on a scale are of unequal size, the divisions marked on the scale indicate measures of equal size. For example, in the scale shown below, the interval between 0 and 1 has been divided into five parts of unequal length; however, A is 0.2, B is 0.4, C is 0.6 and D is 0.8.

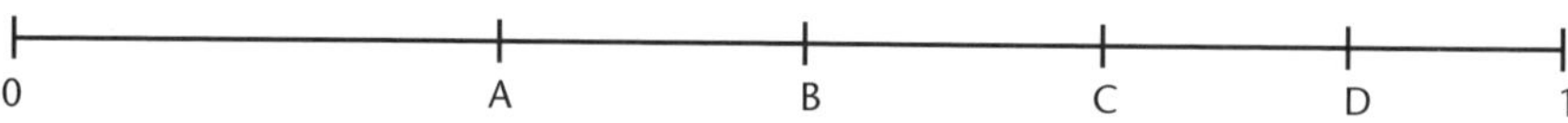

Example F

Q. A measuring cup has a capacity of 300 mL. A point labelled B is halfway between the base of the cup and the 100 mL mark. Which of the three positions A, B or C shown is the most likely to be the 50 mL mark?

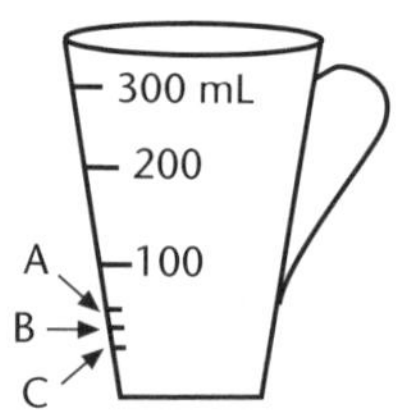

A. Position A is the most likely to be the 50 mL mark.

[the capacity up to level B is smaller than the capacity between level B and the 100 mL mark]

Unit 11.8 Activity 8B: Reading measuring instruments

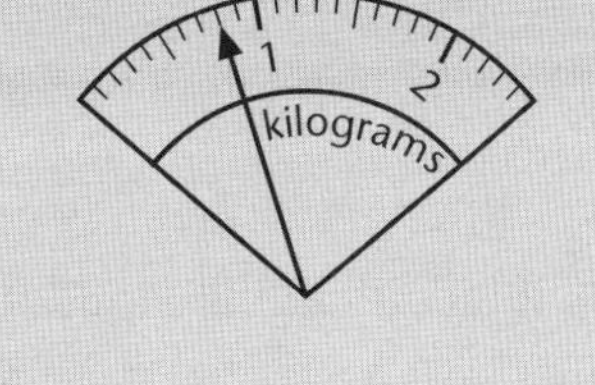

1. Flour is weighed for a recipe (see diagram).

a. Give the weight in kilograms.

b. How many grams did the flour weigh?

c. Copy the diagram and draw an arrow to show a weight of 1.75 kg.

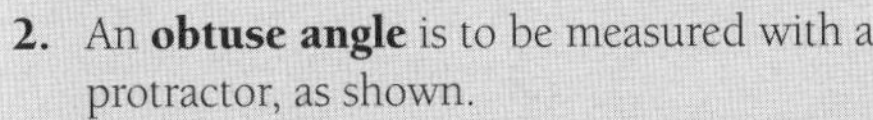

2. An **obtuse angle** is to be measured with a protractor, as shown.

a. How large is the angle to the nearest degree?

b. Two angles are supplementary if they add up to 180°. What size is the supplement of the angle shown?

3. For the following scales, give the reading or measurement shown by each arrow. Give units with each answer.

a.

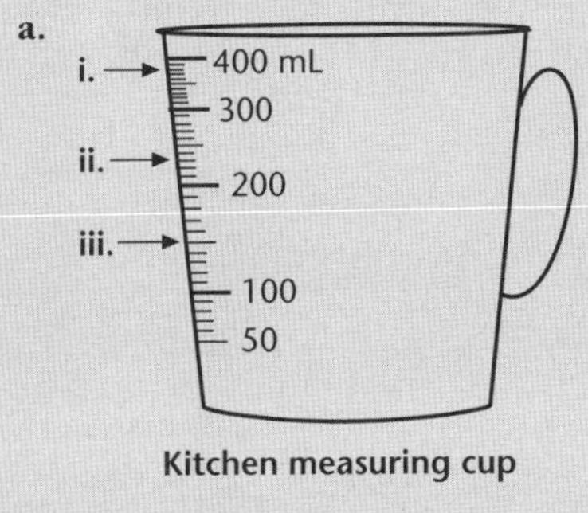

Kitchen measuring cup

b.

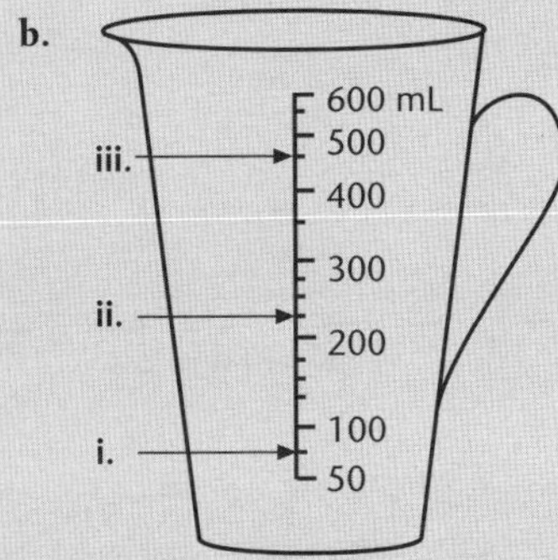

Kitchen measuring jug

c.

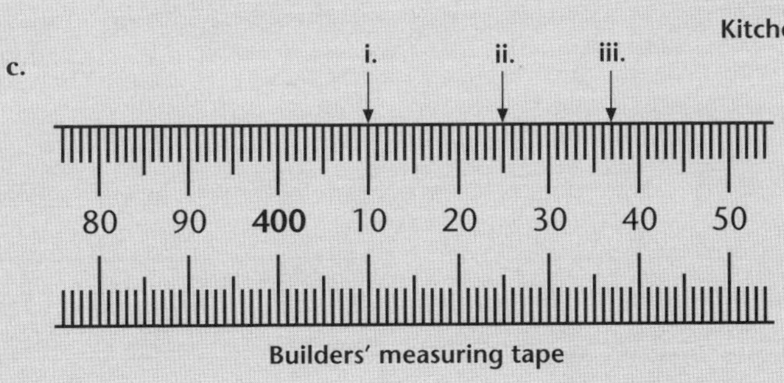

Builders' measuring tape

d.

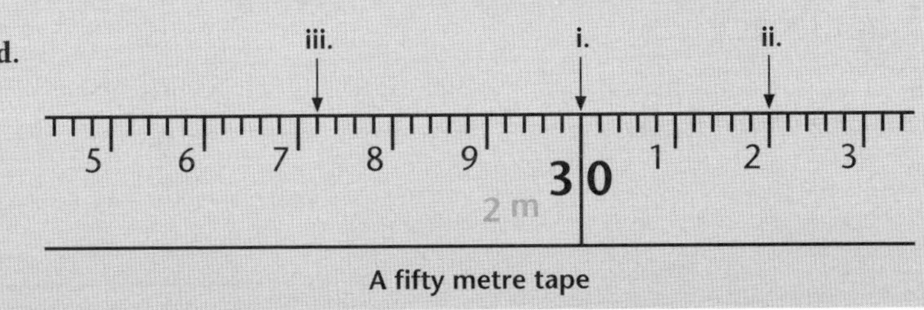

A fifty metre tape

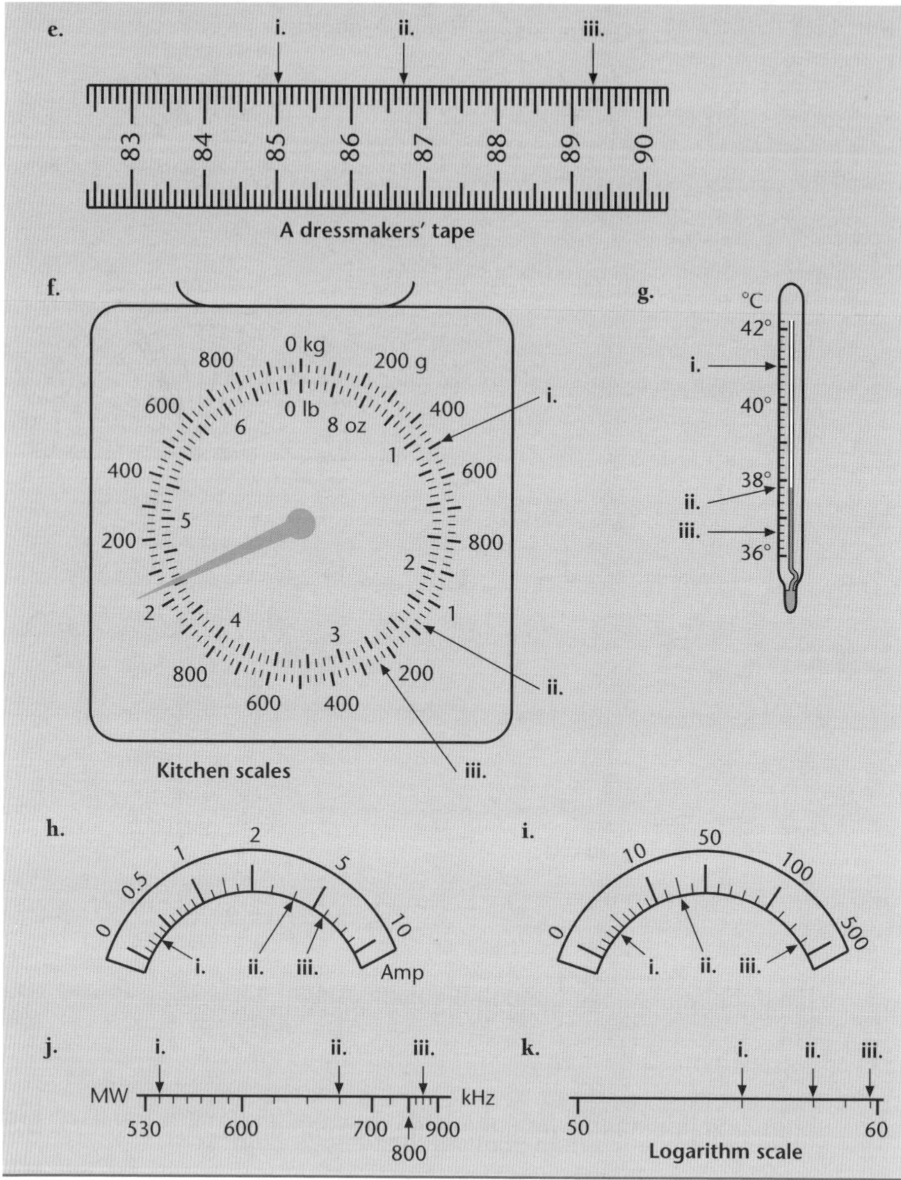

Limits of accuracy

When a quantity is measured, the measurement is never exact. It is only accurate to *half of the place value of the last* ***significant figure***.

Example G

1. If the length of a pen is recorded as being 14.4 cm, its actual length lies between 14.35 cm and 14.45 cm.
   [any measurement in this range of values rounds to 14.4 cm (3 sf)]

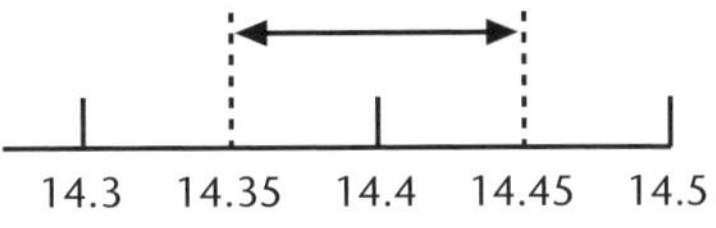

2. The time taken for a car to travel down a street was written down as 12 s. The actual time taken lies between 11.5 s and 12.5 s.

The **limits of accuracy** are often expressed as an **inequality**. In Example G, the actual length of the pen, x, recorded as being 14.4 cm, lies between 14.35 cm and 14.45 cm. This can be written as $14.35 \leq x < 14.45$ (note that the upper value 14.45 rounds to 14.5 cm, so is excluded from the set of values x can take).

- The number 14.35 is the **lower limit** for the measurement 14.4 cm.
- The number 14.45 cm is the **upper limit** for the measurement 14.4 cm.

To calculate the limits of accuracy for any measurement, proceed as follows:

> **Lower limit:** reduce the last significant figure by 1 then 'add a 5' in the next place right (often the end of the number), adding a decimal point if required.

Thus 14.4 becomes 14.35, and 12 becomes 11.5. Note that 2.50 becomes 2.495.

> **Upper limit:** 'add a 5' in the place to the right of the last significant figure (often the end of the number), adding a decimal point if required.

Thus 14.4 becomes 14.45, and 12 becomes 12.5. Note that 4 700 becomes 4 750.

Example H

Q. The side length of a square is given as 11.5 cm. Between what limits does its **perimeter** lie?

A. The lower limit of accuracy of the measurement is 11.45 cm. The upper limit of accuracy of the measurement is 11.55 cm. Therefore the perimeter lies between 4 x 11.45 = 45.8 cm and 4 x 11.55 = 46.2 cm, ie 45.8 cm ≤ Perimeter < 46.2 cm.

Unit 11.1 Activity 8C: Limits of accuracy

1. Write the limits of accuracy for each of the following measurements:

a. 6 m **b.** 4.3 cm **c.** 5.24 m **d.** 1.8 cm **e.** 24 mm

f. 758 km **g.** 1.04 L **h.** 0.6 mL **i.** 7.0 mg **j.** 9.00 m

2. A stopwatch shows a swimmer's time for a 25 m freestyle event to be 15.42 seconds. What are the limits of accuracy for this time?

3. The length of a laboratory bench is measured and found to be 3.40 m long. Give the limits of accuracy for this measurement.

4. The dimensions of a room are given as 3.2 m by 2.1 m. Find the limits of accuracy for the perimeter of the room.

5. The capacity of a pool is given as 48 000 L. Between what limits does the capacity lie?

6. The old record for a running race was given as 12.4 seconds. At a recent event, and using an electronic timer, a runner was timed as running the event in 12.41 seconds. Is the runner necessarily slower in this event than the record-holder?

Conversion of units

Many countries, although having officially converted to metric (SI) units, have had trouble convincing their population to work in these units. This is particularly evident in Britain and the United States of America where there are often references to the Imperial system of units.

Conversion tables

Length

In the Imperial system of units:

- Small lengths are measured in inches (in). For example, the length of my index finger is 3.5 inches.
- Medium lengths are measured in feet (ft) or yards (yd). For example, the length of my desk is two feet, the width across our classroom is 6 yards.
- Large distances are measured in miles. For example, the distance between Lae and Madang is 133 miles.

The following tables can be used for conversion from Imperial units to Metric units and vice versa.

Imperial		Metric
1 inch [in]		2.54 cm
1 foot [ft]	12 in	0.3048 m
1 yard [yd]	3 ft	0.9144 m
1 mile	1760 yd	1.6093 km
1 nautical mile	2025.4 yd	1.853 km

Metric		Imperial
1 millimetre [mm]		0.03937 in
1 centimetre [cm]	10 mm	0.3937 in
1 metre [m]	100 cm	1.0936 yd
1 kilometre [km]	1000 m	0.6214 mile

Area

Imperial units for **area** are square inches (in^2), square feet (ft^2), square yards (yd^2), square miles ($mile^2$) and acres where 1 acre = 4840 yd^2.

Metric		Imperial
1 sq cm [cm^2]	100 mm^2	0.1550 in^2
1 sq m [m^2]	10,000 cm^2	1.1960 yd^2

Metric		Imperial
1 hectare [ha]	10,000 m^2	2.4711 acres
1 sq km [km_2]	100 ha	0.3861 square miles

Imperial		Metric
1 sq inch [in^2]		6.4516 cm^2
1 sq foot [sq ft]	144 in^2	0.0929 m^2
1 sq yd [yd^2]	9 sq ft	0.8361 m^2
1 acre	4840 yd^2	4046.9 m^2
1 sq mile [$mile^2$]	640 acres	640 acres

Volume

Imperial units for volume are cubic inches (in^3), cubic feet (ft^3), and cubic yards (yd^3). For liquids the units are the fluid ounce (fl oz), the pint, where 1 pint = 20 fl ounces, and the gallon, where 1 gallon = 8 pints.

Metric		Imperial
1 cubic cm [cm^2]		0.0610 in^3
1 cubic decimetre [dm^3]	1,000 cm^3	0.0353 ft^3
1 cubic metre [m^3]	1,000 dm^3	1.3080 yd^3
1 litre [l]	1 dm^3	1.76 pt

Imperial		Metric
1 cubic inch [in^3]		16.387 cm^3
1 cubic foot [ft^3]	1,728 in^3	0.0283 m^3
1 fluid ounce [fl oz]		28.413 ml
1 pint [pt]	20 fl oz	0.5683 l
1 gallon [gal]	8 pt	4.5461 l

Example I

Q. Convert

a. 5 metres to feet

b. 10 miles to kilometres

c. 3 hectares to acres

d. 15 square feet to square metres

e. 2 litres to pints

f. 20 gallons to litres

A. **a.** 5 (metres) × 1.0936 (yards) × 3 (feet) = 16.404 feet

b. 10 (miles) ×1.6093 (km) = 16.093 km

c. 3 (hectares) × 2.4711 (acres) = 7.4133 acres

d. 15 (sq ft) × 0.0929 (m^2) = 1.3935 m_2

e. 2 (litres) × 1.76 (pints) = 3.52 pints

f. 20 (gallon) × 4.5461 (litres) = 90.922 litres

Unit 11.1 Activity 8D: Conversion of units

1. Convert the following lengths:

a. 3 metres to yards **b.** 200 miles to kilometres **c.** 40 inches to metres

d. 5 kilometres to yards **e.** 1 nautical mile to metres **f.** 10 centimetres to inches

2. Convert the following areas:

a. 1000 m^2 to square yards **b.** 2000 ha to square miles **c.** 1 km^2 to acres

d. 2000 square yards to hectares **e.** 3 square inches to mm^2 **f.** 0.35 acres to m^2

3. Convert the following volumes:

a. 4 m^3 to cubic feet **b.** 204 litres to cubic feet **c.** 60 fluid ounces to litres

d. 2 pints to cm^3 **e.** 350 cm^3 to pints **f.** 0.75 litres to cubic inches

Unit 11.1 Activity 8E: Multiple choice—measurement

1. 2.563 kilometres, in centimetres, is:

A. 256.3 **B.** 2563 **C.** 25630 **D.** 256300

2. One cubic centimetre is equivalent to:

A. 10 mm^3 **B.** 100 mm^3 **C.** 1000 mm^3 **D.** 1 000 000 mm^3

3. A rectangle has a length of 1.2 metres and a width of 80 centimetres. The area of this rectangle, in square metres, is:

A. 0.96 **B.** 9.6 **C.** 96 **D.** 9600

4. The limits of accuracy for a measurement of 2.54 m are:

A. 2.53 and 2.55 **B.** 2.545 and 2.546

C. 2.535 and 2.545 **D.** 2.50 and 2.60

5. The interval below is divided into 5 parts of unequal length but they indicate measures of equal size.

The measurement indicated by the arrow on the scale is closest to:

A. 83.25 **B.** 86.4 **C.** 87.0 **D.** 87.25

6. Jak leaves at 22:15 on a flight that takes 3 hours and 55 minutes. His destination is in a time zone that is 2 hours ahead of his departure time zone. The local time when Jak arrives at his destination is:

A. 1:10 **B.** 2:10 **C.** 3:10 **D.** 4:10

7. 0.9144 metres is equivalent to 36 inches in Imperial units. 15 inches, in centimetres, is equivalent to:

A. $\frac{15}{36} \times 0.9144 \times 100$ **B.** $\frac{36}{15} \times 0.9144 \times 100$

C. $\frac{15}{0.9144} \times 36 \times 100$ **D.** $\frac{15 \times 36}{0.9144 \times 100}$

Unit 11.1 Number and Application

Topic 9: Ratio and proportion

Ratio and proportion is a sub-section within Unit 11.1 of the Grade 11 syllabus (see Syllabus p. 14). This Topic introduces ratio and covers:

- Applying scales on a map with actual lengths on the ground.
- Solving problems on direct and inverse variation.

Ratios

A **ratio** is a *comparison* of two or more *like* quantities.

A ratio can be expressed in different ways. The different quantities can be compared using colons (:) between the quantities, eg 3 : 5 or 1 : 2 : 4. Two quantities can also be compared by writing their ratio as a fraction.

Example A

If a bag contains 30 black and 40 white marbles, then the ratio of black marbles to white marbles can be expressed as 30 : 40. This can also be expressed as the fraction $\frac{30}{40}$.

Simplifying ratios

Since a ratio can be expressed as a fraction, it can be **simplified** like a fraction by dividing by a **common factor**. Thus, in Example P, the ratio of the black marbles to white marbles can be written more simply as 3 : 4 or $\frac{3}{4}$. A ratio is in its **simplest form** when all **terms** are *whole numbers with no common factors*.

Example B

1. $80 : 40 = 2 : 1$	[dividing both terms by 40]
2. $1\frac{1}{2} : 2\frac{1}{4} = 6 : 9$	[multiplying by 4 to remove fractions]
$= 2 : 3$	[dividing by 3 to simplify]
3. $0.5 : 1.5 : 2.25 = 50 : 150 : 225$	[multiplying each term by 100 to remove decimal points]
$= 2 : 6 : 9$	[dividing by 25]

Note: **1.** A ratio should *not* include fractions or decimal points in any term.

2. A ratio does *not* have a unit.

3. Ratio pairs may be simplified using the fraction key on a calculator.

Quantities with different units must be expressed in the same unit before they can be compared as a ratio.

Example C

16 mm to 4 cm = 16 mm to 40 mm	[changing 4 cm to 40 mm so that both terms have the same unit]
= 16 : 40	
= 2 : 5	[dividing each term by 8]

Calculations with ratios

Calculations with ratios can be done by treating ratios in a similar way to equivalent fractions.

Example D

Q. A supermarket ordered lamb chops and chickens for Christmas sales in the ratio 3 : 5. If 60 chickens arrived in one order, how many lamb chops arrived in the same order?

A. Let the number of lamb chops in the order be x.

Then, lamb chops : chickens = 3 : 5 = x : 60

$$\therefore \frac{3}{5} = \frac{x}{60} \quad \text{[expressing the ratios as fractions]}$$

$$\frac{3}{5} \times \frac{60}{1} = x \quad \text{[multiplying by 60]}$$

$$x = 36$$

∴ there were 36 lamb chops in the order.

Sharing quantities in a given ratio

To share a quantity in a given ratio:

- *Find the total number of parts* by adding all numbers in the ratio.
- *Find the value of each part* by dividing the quantity by this total.
- Multiply each number in the ratio by the value of each part.

Example E

Q. Divide 16 oranges between two people in the ratio 3 : 5.

A. The ratio 3 : 5 gives 8 parts, since 8 = 3 + 5.

Therefore each part is $\frac{16}{8}$ = 2 oranges.

One share is 3 x 2 = 6 oranges and the other is 5 x 2 = 10 oranges.

So, 16 oranges divided in the ratio 3 : 5 gives 6 oranges to the first person and 10 oranges to the second person.

This process works no matter how many numbers are in the ratio.

Example F

Q. Divide K36 in the ratio 2 : 3 : 5.

A. Number of parts = 2 + 3 + 5 = 10 Value of each part = K36 ÷ 10 = K3.60

K36 in the ratio 2 : 3 : 5 divides into: 2 x 3.60 : 3 x 3.60 : 5 x 3.60

= 7.20 : 10.80 : 18.00 [simplifying]

The K36 divides into K7.20, K10.80, and K18.00.

Note: The final figures must always add up to the number or quantity being divided. In Example E, 6 oranges and 10 oranges totalled 16 oranges; and in Example F, K7.20 + K10.80 + K18.00 = K36.00.

Unit 11.1 Activity 9A: Ratios

1. Simplify each of the following ratios:

a. 8 : 16 **b.** 5 : 15 **c.** 20 : 12 **d.** 0.5 : 4

e. $\frac{1}{2} : \frac{1}{3}$ **f.** K2 to 50t **g.** 1 cm to 1 mm **h.** 20 min to 1 hr

i. 0.4 kg to 20 g **j.** 4 m to 1 km **k.** 10 mL to 5 L **l.** $4\frac{1}{5} : \frac{3}{10}$

2. **a.** Divide 40 in the ratio 3 : 5. **b.** Divide 56 in the ratio 5 : 2.

c. Divide 90 in the ratio 1 : 4 : 5. **d.** Divide K8.00 in the ratio 2 : 3.

e. Divide K32.40 in the ratio 5 : 7. **f.** Divide 180 in the ratio 2 : 3 : 5.

3. Mr Arua and Ms Tai bought a truck for K8 000. Mr Arua paid K5 000. They then hired the truck out for K300.

a. How much did Ms Tai pay for her share of the truck?

b. What should be Ms Tai's share of the $300?

4. Mr Bik invested K50 000 to buy a K70 000 company in partnership with Mr Liklik, who paid the balance.

a. How much did Mr Liklik pay?

b. What was the ratio of the investments, Mr Liklik's to Mr Bik's?

c. A profit of $175 000 was shared between the two in the same ratio as their investments. How much did each get?

5. A fruit concentrate is sold in 250 mL packs. The instructions are 'Add water to make one litre of ready-to-drink fruit juice'.

a. What is the ratio of concentrate to water in the made-up juice?

b. How many litres of water are required to dilute 2 L of concentrate?

c. Find the volume of water in 6 L of made-up juice.

d. If one litre of made-up juice fills 5 glasses, how many packs of concentrate are needed to make the juice to fill 80 glasses?

6. A nut mixture has peanuts, almonds and cashews in the ratio 5 : 3 : 1.

a. If Helen buys 2.7 kg of nuts, what weight of almonds will she have?

b. How much nut mixture should Helen buy is she wants to ensure she has 750 g of almonds in the mix?

Map scales

Ratio scales

A ratio scale is often found on a **map**, usually at the top or the bottom of the map.

For example, if the ratio scale is 1 : 1 000 000, then this means that 1 unit, measured on the map, represents 1 000 000 units in actual distance.

Simple scales

A simple scale is a diagram on the map showing a map measurement and stating the actual distance that this represents.

For example, if the simple scale is,

where each division on the scale is exactly 1 cm long, then 5 centimetres on the map represents 10 kilometres in actual distance (or 1 cm on the map represents 2 km in actual distance)

Example G

Q. Below is a map of the National Capital District

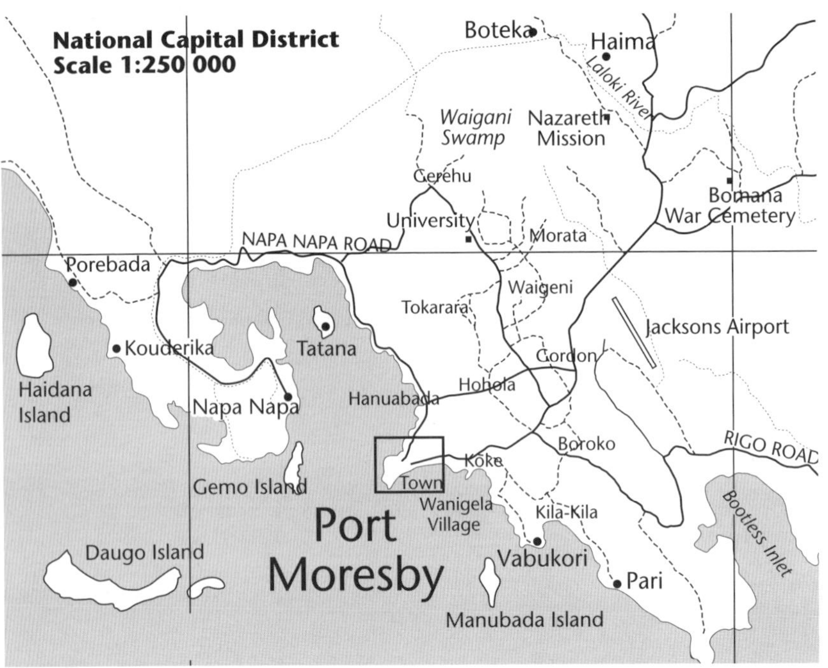

a. Use the map and the ratio scale to find the actual straight line distance between Haima and Pari.

b. Draw a simple scale for this map.

A. a. The straight line distance between Haima and Pari on the map is measured as 7 centimetres.

Substituting in the ratio:

Map distance : Actual distance = 1 : 250 000

7: Actual distance = 1 : 250 000 (centimetres)

$$\frac{\textit{Actual distance}}{7} = \frac{250\ 000}{1}$$

$$\textit{Actual distance} = 7 \times 250\ 000$$
$$= 1750\ 000 \text{ centimetres}$$
$$= 17\ 500 \text{ metres}$$
$$= 17.5 \text{ kilometres}$$

b. Map distance : Actual distance = 1 : 250 000

So 1 cm on the map is equivalent to 250 000 cm actual distance.

250 000 cm = 2500 metres

= 2.5 kilometres

1 centimetre on the map is equivalent to 2.5 kilometres in actual distance so 4 cm is equivalent to 10 km.

Simple scale:

Example H

Q. Two towns that are known to be 20 kilometres apart measure 5 cm apart on a map. What is the ratio scale of the map?

A. Map distance : Actual distance = 5 cm : 20 km

The quantities have different units so we need to convert the larger unit, kilometres, to the smaller unit, centimetres:

Map distance : Actual distance = 5 cm : 20 km

= 5 cm : 20 000 metres

= 5 cm : 2 000 000 cm

= 5 : 2 000 000 (Units are now the same)

= 1 : 400 000 (Dividing by 5 for simplest form)

Multiply kilometres by 1000 to convert to metres

The ratio scale of the map is 1 : 400 000

Example I

Q. Convert the following simple scale to a ratio scale.

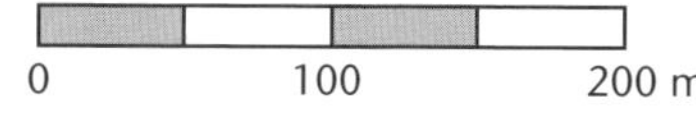

A. The marks on the scale are 1 cm apart.

This simple scale tells us that 4 centimetres on the map represent 200 metres in actual distance (or 1 centimetre represents 50 metres)

To convert this to a ratio scale:

Map distance : Actual distance = 4 cm : 200 metres

= 4 : 20 000 centimetres

= 1 : 5000 (Dividing by 4 for simplest form)

The ratio scale is 1 : 5000

Unit 11.1 Activity 9B: Map scales

1.

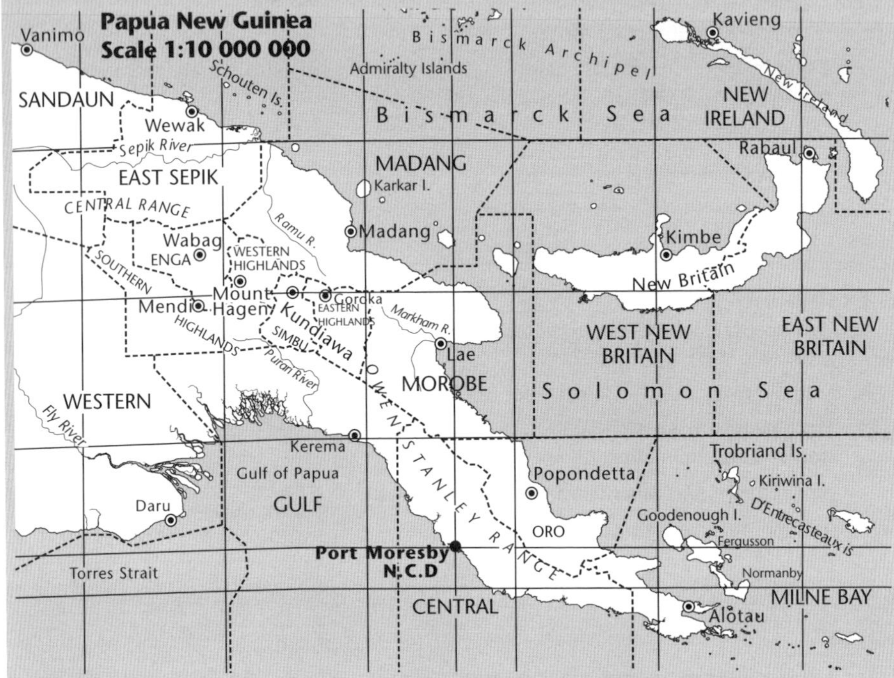

Papua New Guinea
Scale 1:10 000 000

a. The ratio scale for the map above is 1 : 10 000 000. Convert this so that you can say 'One centimetre on the map represents .... kilometres'.

b. Construct a simple scale for this map.

c. Measure the straight line distance on the map, to the nearest millimetre, between

i. Wewak and Goroka

ii. Popondetta and Mt Hagen

d. Convert the map distances from part **c**. to actual distances in kilometres.

2. Convert the following simple scales to ratio scales:

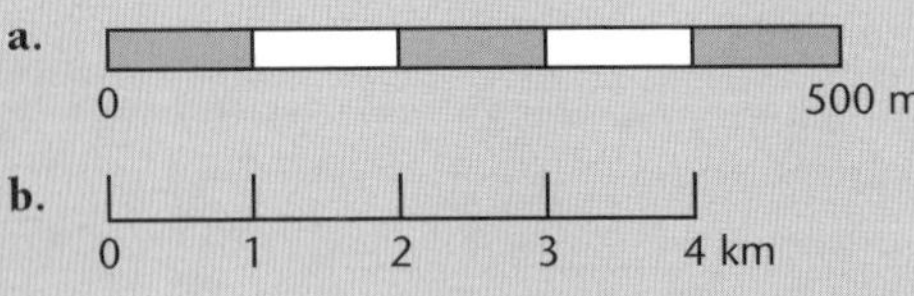

c.

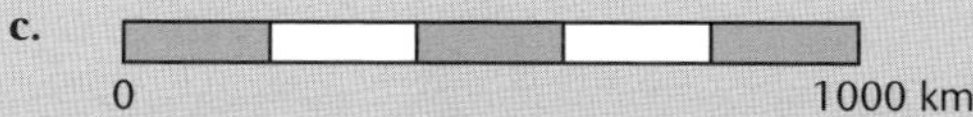

3. Draw a simple scale for each of the following ratio scales:

a. 1 : 2000

b. 1 : 150 000

c. 1 : 25 000

4. The map of Rabaul (below) has a ratio scale of 1 : 13 000.

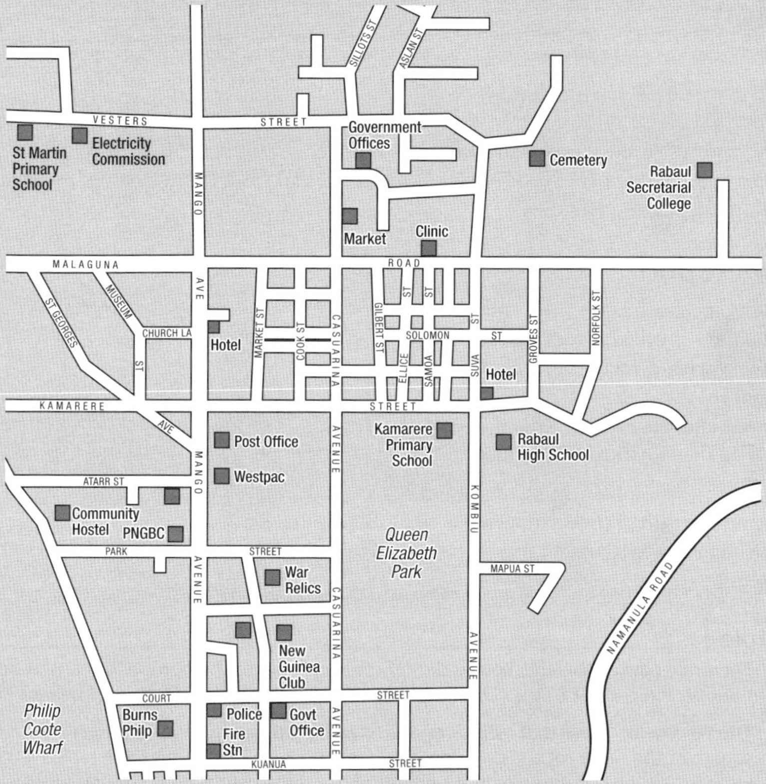

a. Use the ratio scale to find the actual straight line distance between:

i. The Police Station and the Post Office.

ii. The Community Hostel and the Market.

b. The straight line distance between two locations in Rabaul is 871 metres. How far apart are the two locations on the map?

c. Construct a simple scale for this map.

5. Two locations that are known to be 77 m apart are drawn on a map which has a ratio scale of 1 : 500. What is the distance between the locations on the map?

6. The straight line distance between two cities is known to be 560 km. If the simple scale on a map is:

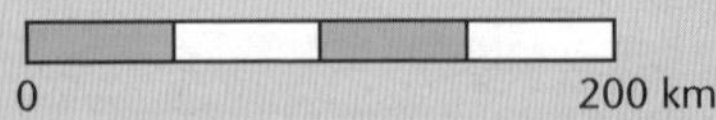

How far apart are the two cities on the map?

7. On a map two cities are 7.7 cm apart and the actual distance between the two cities is known to be 385km. What is the ratio scale of the map?

Variation

Variation describes the situation where a change in one quantity causes a change in another quantity. For example, the more cans of fish that are bought at the supermarket, the more money you pay.

The amount paid is said to be **proportional** to the number of cans bought. If the total cost paid is C kina and the number of cans bought is n then we can write $C \propto n$ where the $\propto$ symbol means 'is proportional to' or 'varies as'.

Direct variation

If one quantity increases as the other quantity increases then this is called **direct variation**. If the quantity y **varies directly** as (or **is directly proportional to**) the quantity $\boldsymbol{x}$, then we can write $y \propto x$ and hence $y = kx$ where $\boldsymbol{k}$ is a constant, called the **constant of variation**.

If there is direct variation between two quantities, x and y, then the **graph** of the relationship between the quantities is a straight line with gradient k.

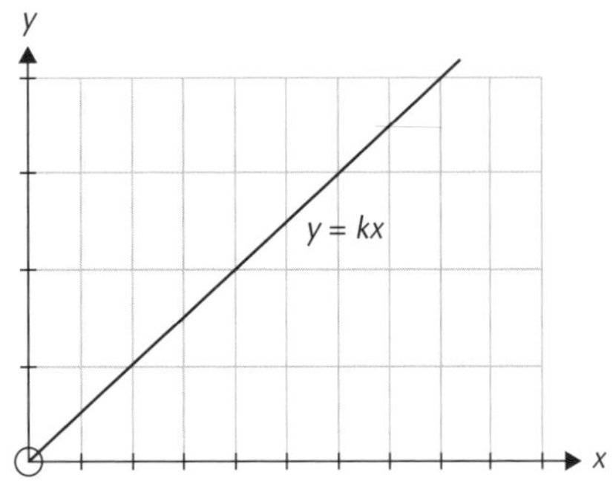

Example J

Q. 1. The price paid at the checkout is directly proportional to the number of cans of beans bought. If 6 cans of beans cost K3.78 then what is the cost of 15 cans of beans?

2. The volume of water that **cylinders** of a fixed height can hold varies directly as the square of the radius of the cylinder. If the volume is 3770 mL when the radius is 10 cm what is the volume of water in a cylinder of the same height when the radius is 6 cm?

A. 1. If b is the number of cans bought and P kina is the price paid at the checkout then we can say $P \propto b$ and $P = kb$ where k is a constant.

We know that when $b = 6$ then $P = 3.78$; substituting these values into the equation $P = kb$ will enable us to find a value for k.

$P = kb$

$3.78 = k \times 6$: divide both sides of the equation by 6

$k = 0.63$ (this is price of one can of beans)

so $P = 0.63b$

Substitute $b = 15$

$P = 0.63 \times 15 = 9.45$ You will pay K9.45 at the checkout for 15 cans of beans.

2. If V mL is the volume of water the cylinder can hold and *r* cm is the radius of the cylinder $V \propto r^2$ then and $V = kr^2$ where *k* is a constant.

Substituting V = 3770 and r = 10 into $V = kr^2$ to find k :

$$3770 = k \times 10^2$$
$$\frac{3770}{10^2} = k$$
$$k = 37.70$$

So $V = 37.70r^2$

Substituting $r = 6$

$$V = 37.70 \times 6^2$$
$$= 1357.2$$: the volume of the cylinder will be 1357.2 mL

Unit 11.1 Activity 9C: Variation

1. If y varies directly as *x*:

a. Write a variation statement and an equation using the constant *k*.

b. If $y = 34$ when $x = 1.7$ find the value of *k*.

c. Find the value of *y* when $x = 3.2$.

d. Find the value of *x* when $y = 56$.

2. If *p* is directly proportional to *q* and $p = 1.25$ when $q = 10.5$, find:

a. *p* when $q = 63$

b. *q* when $p = 30$

3. If $a \propto b^2$ and $a = 30$ when $b = 4$, find:

a. *a* when $b = 12$

b. *b* when $a = 187.5$

4. Which of the following pairs would you expect to be directly proportional?

a. petrol consumption and distance travelled at a constant speed

b. speed of travel and distance covered

c. speed of travel and time taken for a journey

5. The exchange rate on a particular day is 1 Papua New Guinea kina (PGK) = 0.421362 Australian dollars (AUD).

a. Write down a rule connecting Australian dollars, *A*, and Papua New Guinea kina, *P*, on that day.

b. How many Australian dollars, to the nearest **cent**, would you buy with K80?

c. How many Papua New Guinea kina, to the nearest toea, would you buy with $100 Australian?

6. Hooke's law states that the extension (**E**) of an elastic body, such as a spring, is directly proportional to the force (***f***), such as a weight, acting on it.

When a weight of 2 kilograms is hung on a spring the extension is 5.2 centimetres. Find:

a. the extension of the spring when a weight of 5 kg is hung on the spring.

b. the weight required to extend the spring by 10 cm.

7. The value of a garnet (a type of gemstone) varies directly as the square of its weight. A garnet that weighs 4 carat is valued at K244. Find the value of a garnet that weighs 5 carat.

8. The **surface area** of a sphere, A, varies directly as the square of its radius, r. $A \propto r^2$. If a sphere of radius 6 cm has a surface area of approximately 452 cm^2, find the approximate radius (to the nearest mm) of a sphere of surface area 1000 cm^2.

9. When a stone falls freely, the time taken to hit the ground varies directly as the square root of the distance fallen. If it takes 4 seconds to fall 78.4 m, find how long it would take for a stone to fall 500 m.

10. The distance to the visible horizon at sea varies directly as the square root of the height of the observer's eye above sea level. If an observer's eye level is 5.4 metres then the distance to the visible horizon is 9 km. How far can an observer see to the horizon if his eye is at a height of 10 metres above sea level?

11. The period, T seconds, is the time that a pendulum takes to swing through one oscillation. The period is directly proportional to the square root of the length of the pendulum, l cm. $T \propto \sqrt{l}$.

If a pendulum 12 cm long has a period of 0.7 seconds, find:

a. the period of a pendulum that is 8 cm long

b. the length of a pendulum that has a period of 1 second

Inverse variation

- If there is a decrease in one quantity corresponding to an increase in the other quantity then this is called inverse variation.
- For inverse variation $y \propto \frac{1}{x}$.
- If $y \propto \frac{1}{x}$ then $y = \frac{k}{x}$ where k is a constant (the constant of variation).
- A test for inverse variation is to find the product of the variables, xy, and show that it is a constant: $xy = k$.
- $y \propto \frac{1}{x}$ can be referred to as 'y varies inversely as x', 'y is inversely proportional to x' or 'y is proportional to $\frac{1}{x}$'
- If y varies inversely as x then the graph of y versus x has the shape shown at right.
- For inverse variation neither of the variables can be zero.

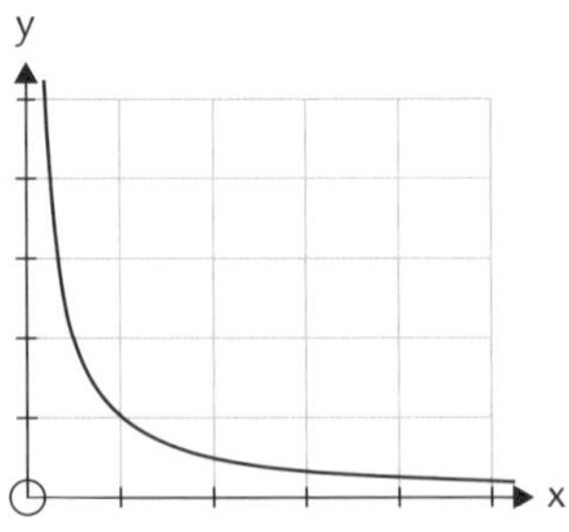

Example J

Q. **1.** If y is inversely proportional to x and $y = 5$ when $x = 8$, find:

a. y when $x = 5$

b. x when $y = 4$

2. The amount of light that enters a camera lens depends on the area of the opening of the shutter. The relationship between the f- number on the camera settings and the area, A mm^2, of the opening of the shutter is shown in the table below.

f	2	4	8	16
A	480	120	30	7.5

a. Show that A and f^2 are inversely proportional

b. Find the value of A when $f = 2.8$; an actual setting on a manually operated camera.

A. **1.** If y is inversely proportional to x then $y \propto \frac{1}{x}$ and $y = \frac{k}{x}$ where k is a constant.

a. If $y = \frac{k}{x}$ then $k = xy$. Substituting $y = 5$ and $x = 8$ gives $k = 5 \times 8 = 40$.

So $xy = 40$ is the general rule for this question.

Substituting $x = 5$ in $xy = 40 : 5 \times y = 40$ giving $y = 8$

b. Substituting $y = 4$ in $xy = 40 : x \times 4 = 40$ giving $x = 10$

2. **a.** We need to show that $A \propto \frac{1}{f^2}$ which means that $A = \frac{k}{f^2}$ or $Af^2 = k$ a constant.

Extending the table:

f	2	4	8	16
A	480	120	30	7.5
f^2	4	16	64	256
Af^2	1920	1920	1920	1920

$Af^2 = 1920$, a constant, so we can say that A and f^2 are inversely proportional.

b. Substituting $f = 2.8$ in $Af^2 = 1920$:

$Af^2 = 1920$

$$A = \frac{1920}{2.8^2} = 244.9 \text{ mm}^2$$

Unit 11.1 Activity 9D: Inverse variation

1. If y varies inversely as x and $y = 2.2$ when $x = 15$, find:

a. y when $x = 3$

b. x when $y = 5$

2. The variables in the table are inversely proportional.

i. Find the value of the constant of proportionality.

ii. Find the missing values.

a.

x	2	3	8	
y	60	40		10

b.

p	100		25	
q	0.75	1.5	3	10

3. Boyle's law states that 'for a given mass of gas at constant temperature, the volume, V, is inversely proportional to the pressure, P'.

a. Write an equation involving the variables V and P and a constant of proportionality, k.

b. A gas occupies 2 litres if it is at 740 mm of mercury pressure. Find the value of k.

c. What volume will the gas occupy if it is at 500 mm of mercury pressure?

4. The time taken, t, for a journey of fixed distance is inversely proportional to the average speed of travel, s. If the journey takes 2 hours 15 minutes travelling at an average speed of 50 km/hour, how long will it take if the average speed is 27 km/hr? Give your answer in hours and minutes.

5. On a guitar the frequency of vibration of a string, measured in Hertz (Hz), or vibrations per second, determines the note. The frequency of vibration, f, of a guitar string is inversely proportional to the length, l, of the string. The length of a string is shortened by pressing the string behind a fret. If the note A, which has a frequency of vibration of 440 Hz, is produced by a string of length 63 cm, find:

a. the length of a string that will produce the note C, which has a frequency of vibration of 523.2 Hz

b. the length of a string that will produce the note G, which has a frequency of vibration of 784 Hz

6. For a light with a constant output, the illumination, I units, at a distance d m from the light source is inversely proportional to d^2, ie. $I \propto \frac{1}{d^2}$.

a. If a light source has an illumination of 850 units at a distance of 1 metre, what will the illumination be at a distance of from the source?

i. 2 metres

ii. 5 metres

b. At what distance from the light source will the illumination be 68 units?

Unit 11.1 Number and Application

Topic 10: Basic algebra—revision

Basic algebra is a sub-section within Unit 11.1 of the Grade 11 syllabus (see Syllabus p. 14). This Topic deals with using straightforward algebraic methods and solving equations. It covers:

- Simplifying algebraic expressions involving exponents such as $(2x)^3$ and $\frac{12a^5}{8a^2}$.
- Manipulating and simplifying expressions such as $\frac{x}{4} + \frac{x}{3}$.

Algebraic terms

In **algebra**, letters such as x or a are used to represent unknown quantities or numbers. These letters are often called **pronumerals** or **variables**.

Algebraic terms are formed by combining numbers and variables. A **term** can be

- A **constant** such as 5 or $\frac{-1}{2}$.
- A **variable** such as y or A.
- A variable raised to a **power** or **exponent** eg y^3.
- Any product or quotient of constants and variables with exponents, eg $4r^2$ or $\frac{ab}{c}$.

In an algebraic term, the **coefficient** of the variable is the factor multiplying the variable, eg in the term $4x$, the coefficient of x is 4.

Algebraic expressions

Numerical expressions, such as $3 + \frac{4}{5}$ or $1 - 2 \times 3^2$, are formed using numbers along with the **operations** of addition, subtraction, multiplication and division etc. **Algebraic expressions** are formed in a similar way by adding and subtracting algebraic terms.

Example A

1. The **sum** of 8 and 4 is written 8 + 4. Similarly,
   - The sum of 8 and x is written $8 + x$.
   - The sum of y and x is written $y + x$.
2. 7 more than the **product** of 5 and 6 is written 5 x 6 + 7. Similarly,
   - 7 more than the product of 5 and a is written 5 x a + 7 or $5a + 7$.
   - c more than the product of a and b is written $ab + c$.
3. Nine less than the **square** of 7 is written $7^2 - 9$. Similarly,
   - 9 less than the square of x is written $x^2 - 9$.
   - y less than the square of x is written $x^2 - y$.

Simplifying algebraic expressions

Numerical expressions are simplified in the usual way using number rules, for example $1 - 2 \times 3^2 = 1 - 2 \times 9 = -17$. Some algebraic expressions may also be simplified, using the laws of algebra.

Since letters are representing numbers, the rules for arithmetic calculations hold true for algebraic calculations. These include:

Algebraic Rule	Numeric Example
$a + b = b + a$	$5 + 4 = 4 + 5$
$ab = ba$	$2 \times 3 = 3 \times 2$
$(a + b) + c = a + (b + c)$	$(3 + 4) + 5 = 3 + (4 + 5)$
$(ab)c = a(bc)$	$(3 \times 5) \times 6 = 3 \times (5 \times 6)$
$a(b + c) = ab + ac$	$7(3 + 2) = 7 \times 3 + 7 \times 2$
$a + 0 = 0 + a = a$	$2 + 0 = 0 + 2 = 2$
$a \times 1 = 1 \times a = a$	$5 \times 1 = 1 \times 5 = 5$
$a \times -1 = -1 \times a = -a$	$3 \times -1 = -1 \times 3 = -3$

Order of operations

Whether a problem is written in words or as a **mathematical sentence**, punctuation symbols are as important in mathematics as they are in English. In mathematics, the main punctuation symbols are **brackets**.

Example B

The messages 'No, cost too much' and 'No cost too much' have entirely different meanings, yet they differ by just one comma.

Likewise, the mathematical expressions $2 + 3 \times 4$ and $(2 + 3) \times 4$ give different answers:

$2 + 3 \times 4 = 2 + 12 = 14$ and $(2 + 3) \times 4 = 5 \times 4 = 20$.

The mnemonics **BEMA** or BEDMAS (brackets; exponents; multiplication/division; addition/subtraction) gives the convention for the correct order of operations.

Take care to include brackets in algebraic or numeric expressions wherever necessary to indicate the required order of operations.

Example C

Q. Write each of the following using a mathematical sentence:

1. 6 times the sum of 3 and 5.
2. 20 divided by the sum of p and q.
3. The product of 4, and another number that is increased by 5.
4. The square root of the difference between a and b (where $a > b$).

A.
1. $6(3 + 5)$ [the addition of 3 and 5 is done *before* multiplication by 6]
2. $\dfrac{20}{p + q}$ [the bracket around $p + q$ is understood, and need not be shown]
3. $4(x + 5)$ [x is the number, and it is increased by 5 before multiplication by 4]
4. $\sqrt{a - b}$ [the bracket around $a - b$ is understood, and need not be shown]

Unit 11.1 Activity 10A: Algebraic expressions

Write the following as algebraic expressions.

1. The sum of 9 and a.

2. b decreased by 4.

3. The product of 5 and c.

4. 6 more than x.

5. 7 less than y.

6. The product of 4 and the sum of x and y.

7. A number divided by 6.

8. The sum of x and y, divided by 10.

9. The difference between 12 and y (where $y < 12$).

10. x is increased by 4, then tripled.

11. y is decreased by 7, then halved.

12. The square root of the sum of x and 10.

13. The sum of two consecutive whole numbers.

14. The sum of a and 10 divided by the product of 7 and b.

15. A number which is 3 less than twice the value of y.

16. The cube of two less than p.

17. Jane is x years old. What was her age 5 years ago?

18. A trailer weighs t kg. It is loaded with 3 logs, each weighing l kg. Write down the weight of the loaded trailer.

19. The cost of x 10t stamps plus y 15t stamps.

20. I am a years old and my brother is b years old. Write down the sum of our ages in 2 years' time.

Like terms

Like terms have the same variables to the same powers. In **unlike terms** the variables or exponents (or both) are different.

Example D

1. $3a$, $4a$, and $-5a$ are *like* terms because they have the same variable, a, and the same exponent, 1 (ie $a = a^1$).
2. $3a$ and $4a^2$ are *unlike* terms because, although they have the same variable, a, they have different exponents (1 and 2 respectively).
3. $3ab$ and $4ba$ are like terms, but $3a$ and $4b$ are unlike terms.
   [the order of the variables (ab or ba) does not affect whether terms are/are not like terms]

Addition and subtraction of algebraic terms

Algebraic expressions containing *like* terms *can* be simplified by addition and subtraction. *Unlike* terms *cannot* be simplified by addition or subtraction. Gather like terms together and combine them by adding/subtracting their coefficients. (Numbers are combined with each other in the usual way.)

Example E

The following expressions contain like terms and can be simplified:

1. $3a + 4a = (3 + 4)a$ [adding coefficients]
   $= 7a$
2. $8 - 3x - 2 + x = 8 - 2 - 3x + x$ [gathering like terms]
   $= 6 + (-3 + 1)x$
   $= 6 - 2x$ [adding coefficients]
3. $x^2 + 3x + 4x^2 - 7x = x^2 + 4x^2 + 3x - 7x$ [gathering like terms]
   $= (1 + 4)x^2 + (3 - 7)x$ [adding/subtracting coefficients of like terms]
   $= 5x^2 + -4x$, or (better) $5x^2 - 4x$
4. $4a - 3b + 2ab - a + 2b = 4a - a - 3b + 2b + 2ab$ [gathering like terms]
   $= 3a - b + 2ab$

Note: In part **3**, the terms $5x^2$ and $-4x$ are unlike terms (unequal exponents) and so the expression $5x^2 - 4x$ will not simplify further. Similarly, in part **4**, $3a$, $-b$, and $2ab$ are all unlike terms.

Multiplication and division of algebraic terms

Algebraic expressions involving *like* or *unlike terms* can be multiplied together eg $a \times b \times c = abc$. Exponents are used to simplify answers eg $a \times a$ is written as a^2. Note that multiplication signs are not included in the final answer.

Example F

Q. Simplify **1.** $3a \times 2b$ **2.** $4w \times 3y$ **3.** $8ab \times 12ac$

A. 1. $3a \times 2b = 3 \times a \times 2 \times b$
$= 3 \times 2 \times a \times b$
$= 6 \times a \times b$
$= 6ab$

2. $4w \times 3y = 4 \times 3 \times w \times y$
$= 12wy$

3. $8ab \times 12ac = 8 \times a \times b \times 12 \times a \times c$
$= 96a^2bc$ [$a \times a = a^2$]

Algebraic division is set out in fraction form and common factors are 'cancelled', using the fact that, if $x \neq 0$ then $\frac{x}{x} = \frac{1}{1} = 1$.

Example G

The following expressions involve division of unlike terms:

1. $\frac{8ab}{4a} = \frac{{}^{2}\cancel{8} \times \cancel{a}^{1} \times b}{{}_{1}\cancel{4} \times \cancel{a}_{1}}$ [dividing numerator and denominator by 4 and a]
   $= 2b$

2. $\frac{24abc}{16ac} = \frac{\overset{3}{\cancel{24}}\ \overset{1}{\cancel{a}}\ b\ \overset{1}{\cancel{c}}}{\underset{2}{\cancel{16}}\ \underset{1}{\cancel{a}}\ \underset{1}{\cancel{c}}}$ [dividing numerator and denominator by 8, a and c]
   $= \frac{3b}{2}$

Unit 11.1 Activity 10B: Combining algebraic terms

For questions 1–20, simplify the expressions where possible:

1. $5a + 3a$ **2.** $7a - a$ **3.** $3 + x + x$

4. $2x + 3 + x^2 - x - 2$ **5.** $x^2 - x + 5 + 3x^2 - 2x$ **6.** $6x^2y - 3xy^2 - 2yx^2$

7. $7x + 3x$ **8.** $5a - 2a$ **9.** $4 + x + x + x$

10. $5x - 2 + x^2 + 3x + 4$ **11.** $x^2 + 2x - 3 + 2x^2 - 3x$

12. $2x^2 - 4x + 7 - x^2 - 5x + 2$ **13.** $8a \times 3b$ **14.** $4a \times 3ab$

15. $4p \times 3q \times 2r$ **16.** $5a \times 2ab^2$ **17.** $3x^6 + x^4 - 2x^4$

18. $\frac{9xy}{3y}$ **19.** $\frac{12abc}{8a}$ **20.** $\frac{6x}{9xy}$

Algebraic fractions

Algebraic fractions are fractions which have variables (letters) as well as numbers in them, eg $\frac{x}{3}$, $\frac{5a}{3b}$, $\frac{x^2 + 5}{x - 2}$, $\frac{1}{a - 7}$, $\frac{p + q}{4}$.
Since the letters stand for numbers, algebraic fractions obey the same rules as 'ordinary' fractions do when used in calculations.

Simplifying algebraic fractions

Algebraic fractions, like ordinary fractions, are usually written in their simplest form. To simplify an algebraic fraction, divide the numerator and denominator by the same factor (ie cancel common factors). The **laws of indices** are often required.

Example H

Q. Simplify: **1.** $\frac{12b^3}{30b}$ **2.** $\frac{5x^4y^2}{15x^3y^3}$

A. 1. $\frac{12b^3}{30b} = \frac{2b^3}{5b}$ [dividing numerator and denominator by 6]

$= \frac{2b^2}{5}$ [using index laws, $\frac{b^3}{b} = b^{3-1} = b^2$]

2. $\frac{5x^4y^2}{15x^3y^3} = \frac{\overset{1}{\cancel{5}}\overset{x}{\cancel{x^4}}\cancel{y^2}}{\underset{3}{\cancel{15}}\cancel{x^3}\underset{y}{\cancel{y^3}}}$ [dividing numerator and denominator by 5 and using index laws for division]

$= \frac{x}{3y}$

Addition and subtraction of algebraic fractions

To add/subtract fractions when the denominators are the *same*.

- The numerators are added together (or subtracted).
- The denominator stays the same.

If the denominators are *different*, the fractions must first be changed to **equivalent fractions** with the *same* denominators before the rule is applied.

Example I

Q. Simplify: **1.** $\frac{a}{4} + \frac{3}{4}$ **2.** $\frac{b}{2} + \frac{b}{3}$ **3.** $\frac{3c}{4} - \frac{c}{8}$ **4.** $\frac{1}{x} + \frac{1}{y}$

A. **1.** $\frac{a}{4} + \frac{3}{4} = \frac{a + 3}{4}$ [denominators the same, add numerators together]

2. $\frac{b}{2} + \frac{b}{3} = \frac{b \times 3}{2 \times 3} + \frac{b \times 2}{3 \times 2}$ [**common denominator** is the lowest common multiple of the denominators 2 and 3, which is 6]

$= \frac{3b + 2b}{6}$

$= \frac{5b}{6}$ [adding numerators]

3. $\frac{3c}{4} - \frac{c}{8} = \frac{3c \times 2}{4 \times 2} - \frac{c}{8}$ [LCM of 4 and 8 is 8]

$= \frac{6c - c}{8}$

$= \frac{5c}{8}$

4. $\frac{1}{x} + \frac{1}{y} = \frac{1 \times y}{x \times y} + \frac{1 \times x}{y \times x}$

$= \frac{y + x}{xy}$

Take care with signs when expanding brackets in algebraic fractions.

Example J

$\frac{x + 1}{2} - \frac{2x + 1}{3} = \frac{3(x + 1)}{3 \times 2} - \frac{2(2x + 1)}{2 \times 3}$ [making equivalent fractions with denominator 6]

$= \frac{3x + 3 - 4x - 2}{6}$ [expanding brackets]

$= \frac{1 - x}{6}$ [simplifying]

Multiplication of algebraic fractions

To multiply algebraic fractions, all the numerators are multiplied together and all the denominators are multiplied together. Often the process can be made easier by cancelling first.

Example K

Q. Simplify: **1.** $\frac{a}{2} \times \frac{a}{3}$ **2.** $\frac{a^2}{b^3} \times \frac{b^2}{a}$ **3.** $\frac{3a^2}{4bc} \times \frac{8c^3}{9a}$

A. 1. $\frac{a}{2} \times \frac{a}{3} = \frac{a \times a}{2 \times 3}$

$= \frac{a^2}{6}$

2. $\frac{a^2}{b^3} \times \frac{b^2}{a} = \frac{a^2 b^2}{ab^3}$

$= \frac{a}{b}$ [using the laws of indices]

Alternatively, $\frac{\cancel{a^2}^{\,a}}{{}_{b}\cancel{b^3}} \times \frac{\cancel{b^2}}{\cancel{a}} = \frac{a}{b}$ [cancelling first]

3. $\frac{3a^2}{4bc} \times \frac{8c^3}{9a} = \frac{{}^{2}\cancel{24}\,\cancel{a^2}^{\,a}\,\cancel{c^3}^{\,c^2}}{{}_{3}\cancel{36}\,\cancel{a}\,b\,\cancel{c}}$ [multiplying numerators and multiplying denominators then simplifying by cancelling and index laws]

$= \frac{2ac^2}{3b}$

Division of algebraic fractions

To divide by an algebraic fraction, multiply by its **reciprocal** (the reciprocal of $\frac{a}{b}$ is $\frac{b}{a}$).

Example L

Q. Simplify: **1.** $\frac{a}{2} \div \frac{a}{3}$ **2.** $\frac{4a^2}{b} \div \frac{8ab}{c}$

A. 1. $\frac{a}{2} \div \frac{a}{3} = \frac{a}{2} \times \frac{3}{a}$ [changing '$\div \frac{a}{3}$' to '$\times \frac{3}{a}$']

$= \frac{3}{2}$ [cancelling a's]

2. $\frac{4a^2}{b} \div \frac{8ab}{c} = \frac{4a^2}{b} \times \frac{c}{8ab}$ [$\frac{c}{8ab}$ is the reciprocal of $\frac{8ab}{c}$]

$= \frac{{}^{1}\cancel{4}\,\cancel{a^2}^{\,a}\,c}{{}_{2}\cancel{8}\,\cancel{a}\,b^2}$ [multiplying numerators and denominators then cancelling]

$= \frac{ac}{2b^2}$ [using laws of indices to simplify]

Order of operations in algebraic fraction expressions

Be sure to follow the correct order of operations (BEMA or BEDMAS) (see p. 5) when simplifying algebraic fraction expressions with more than one operation.

Example M

Q. Simplify $2x - \frac{x}{2} \times \frac{1}{3}$

A. $2x - \frac{x}{2} \times \frac{1}{3} = 2x - \frac{x}{6}$ [multiplying $\frac{x}{2} \times \frac{1}{3}$ first]

$= \frac{12x - x}{6}$ [subtracting]

$= \frac{11x}{6}$

More complicated algebraic fractions

When there is more than one term in either the denominator or numerator, simplifying is done by factorising first, then cancelling. This is discussed in Topic 11.

Unit 11.1 Activity 10C: Algebraic fractions

1. Simplify each of the following:

a. $\frac{x^{10}}{x^2}$ b. $\frac{3y^{12}}{y^4}$ c. $\frac{16a^6}{4a^3}$ d. $\frac{20b^2}{5b}$ e. $\frac{36x^2y}{12xy}$

f. $\frac{(3a^3)^2}{3a^2}$ g. $\frac{15a^4b^2}{30ab}$ h. $\frac{24x^3y^2}{16xy^3}$ i. $\frac{(2a^3b^2)^2}{10a^4b^3}$ j. $\frac{8x^6y^2z}{12xy^2z^3}$

2. Write each of the following as a single fraction:

a. $\frac{x}{3} + \frac{1}{3}$ b. $\frac{y}{9} - \frac{2}{9}$ c. $\frac{x}{3} + \frac{x}{4}$ d. $\frac{3a}{5} - \frac{a}{2}$

e. $\frac{4k}{5} - \frac{k}{3}$ f. $\frac{5y}{6} - \frac{2y}{3}$ g. $\frac{3}{x+1} + \frac{4}{x+1}$ h. $\frac{x}{2} + \frac{1}{x}$

i. $\frac{4}{a} + \frac{3}{b}$ j. $\frac{1}{3x} + \frac{1}{x}$ k. $\frac{1}{x^2} + \frac{2}{x}$ l. $\frac{1}{y^2} - \frac{2}{3y}$

3. Simplify each of the following:

a. $\frac{a}{3} \times \frac{a}{4}$ b. $\frac{a}{5} \times \frac{10}{a}$ c. $\frac{3a}{b} \times \frac{b}{3a}$ d. $\frac{x}{2} \times \frac{x}{5}$

e. $\frac{a}{5} \div \frac{a}{2}$ f. $\frac{a}{3} \div \frac{a}{4}$ g. $\frac{a}{3b} \times \frac{b}{3a}$ h. $\frac{4a^2}{3} \div \frac{15}{2a}$

i. $\frac{6x^2}{5} \times \frac{15}{4x}$ j. $\frac{12a^2b}{15ab} \times \frac{5ab^2}{4b}$ k. $\frac{6xy^2}{5x} \div \frac{4y}{3xy}$ l. $\frac{3xy^3}{4y} \div \frac{9x^2}{8xy}$

4. Simplify each of the following:

a. $\frac{x+1}{4} - \frac{x-1}{5}$ b. $\frac{x+1}{4} + \frac{x-1}{5}$ c. $\frac{x+1}{4} - \frac{x-1}{2}$

d. $\frac{x+1}{7} + \frac{x-1}{2}$ e. $\frac{x+2}{3} - \frac{3x}{4}$ f. $\frac{1}{4x} + \frac{1}{3}$

5. Use the correct order of operations to simplify the following.

a. $\frac{a}{2} + \frac{a}{3} \times \frac{1}{4}$ b. $\frac{3}{x} - \frac{2}{3} \div \frac{x}{6}$ c. $\left(\frac{a}{4} - \frac{a}{8}\right) \times \frac{a}{3}$

d. $\frac{y}{3} - \frac{y}{5} + \frac{y}{10}$ e. $\frac{2w}{3} \times \frac{1}{4} + \frac{w}{6}$ f. $\frac{5a^4}{12b} \div \frac{3}{4b} \times \frac{6}{a^2}$

Unit 11.1 Number and Application

Topic 11: Basic algebra—expanding and factorising

Basic algebra is a sub-section within Unit 11.1 of the Grade 11 syllabus (see Syllabus p. 14). This Topic deals with using straightforward algebraic methods and solving equations. It covers:

- Factorising and expanding.
- Manipulating and simplifying expressions such as $\frac{x^2 - 4}{x - 2}$.

Bracketed expressions

Brackets around an algebraic expression mean that the expression in the brackets is regarded as one **term**.

Example A

1. The expression $x + 4$ consists of two terms, 'x' and '4', but the bracketed expression $(x + 4)$ is considered to be only one term, '$x + 4$'.
2. An expression such as $3(x + 4)$ means 3 lots of $(x + 4)$, thus

$3(x + 4) = (x + 4) + (x + 4) + (x + 4)$

$= 3 \times x + 3 \times 4$ [3 lots of x plus 3 lots of 4].

$= 3x + 12$ [simplifying]

Expanding brackets

The process of removing brackets from an algebraic expression is called **expanding**.

- The number in front of the brackets (or sometimes behind) multiplies every term inside the bracket.
- The answer is then **simplified** (where possible) by combining like terms.

Example B

Q. Expand each of the following expressions and simplify where possible:

1. $2(x + 3)$ 2. $y(y - 2)$

A. 1. $2(x + 3) = 2 \times x + 2 \times 3$ [x and +3 are *each* multiplied by 2]

$= 2x + 6$

2. $y(y - 2) = y \times y + y \times -2$ [y and –2 are each multiplied by y]

$= y^2 - 2y$

By considering areas in the diagram alongside, a general rule for expanding brackets (called the distributive law) can be seen:

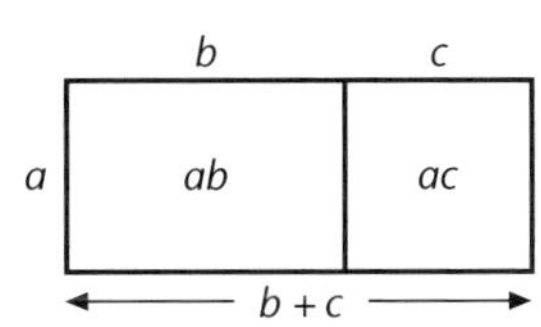

$$a(b + c) = ab + ac$$

These rules also apply to expanding brackets with more than two terms.

Example C

1. $-3(a + 2b - c) = -3 \times a + -3 \times 2b + -3 \times -c$ [each term is multiplied by –3]

   $= -3a - 6b + 3c$

   Note: The product of two negatives is a positive, so $-3 \times -c = +3c$.

2. $(x + y - 2)y = y(x + y - 2)$ [swapping y to front of bracket since $a \times b = b \times a$]

   $= yx + y^2 - 2y$ [removing the bracket]

   $= xy + y^2 - 2y$ [since $yx = xy$]

3. $-3x^2(x - xy + 4y) = -3x^2 \times x + -3x^2 \times -xy + -3x^2 \times 4y$

   $= -3x^3 + 3x^3y - 12x^2y$ [simplifying]

Take care to follow the correct order of operations (BEMA or BEDMAS) when removing brackets.

Example D

1. $4 + 2(y - 5) = 4 + 2 \times y + 2 \times -5$ [y and –5 are each multiplied by +2]

   $= 4 + 2y - 10$

   $= 2y - 6$ [collecting like terms]

   Note: The order of operations (BEMA) applies when removing brackets. Therefore, in $4 + 2(y - 5)$, the brackets are expanded *before* the addition of 4.

2. $3 - (4 - x) = 3 + -1 \times 4 + -1 \times -x$ [$-(a + b)$ means $-1 \times (a + b)$]

   $= 3 - 4 + x$

   $= -1 + x$ or $x - 1$

Unit 11.1 Activity 11A: Expanding brackets

Expand each of the following expressions and simplify where possible:

1. $3(x + 1)$
2. $4(x + 2)$
3. $a(b - 2)$
4. $x(x - 3)$
5. $-4(x - 5)$
6. $2x(x^2 + 3x - 4)$
7. $5 - 3(2x + 1)$
8. $8 - (x - 3)$
9. $4 + 2(x + 3)$
10. $4(2x + 3y) + 3(x - 2y)$
11. $3(4a + 5b) - 2(a - 3b)$
12. $3(x - 2y) - (x - 5y)$
13. $x(x + 3) + 2(x - 1)$
14. $p(3p - q) - q(p - 2q)$
15. $x(x - 5) - 2x(x - 2)$
16. $2x(x - 3) - x^2$
17. $3x^2(1 - 2x - x^2)$
18. $-2xy(x - y - 3z)$
19. A square room of side length x is made 3 m longer (its width stays the same). What is the new area of the room?

20. Lucy has Ky. She spends K10.
    a. How much money does Lucy have now?
    b. Lucy's brother says, 'I've got three times as much money as you have now.' How much money does Lucy's brother have?

Multiplying pairs of brackets

When two bracketed expressions are multiplied together, each term in the first bracket multiplies the terms in the second bracket.

Example F

$(3x + 2)(2x - 5) = 3x \times 2x + 3x \times -5 + 2 \times 2x + 2 \times -5$ [applying FOIL method]

$= 6x^2 - 15x + 4x - 10$

$= 6x^2 - 11x - 10$

Some important expansions

Take care when squaring a bracket. $(x - 3)^2$ means $(x - 3)(x - 3)$. It *does not* mean $x^2 - 3^2$ or $x^2 + (-3)^2$.

Example G

Q. Expand $(x - 3)^2$.

A. $(x - 3)^2 = (x - 3)(x - 3)$ [$(x - 3)^2$ means $(x - 3) \times (x - 3)$]

$= x^2 - 3x - 3x + 9$ [expanding using FOIL method]

$= x^2 - 6x + 9$ [adding like terms]

The expressions $(a + b)^2$ and $(a - b)^2$ are known as **perfect squares** and their expansions are:

$$(a + b)^2 = a^2 + 2ab + b^2$$
$$(a - b)^2 = a^2 - 2ab + b^2$$

For example, $(x + 5)^2 = x^2 + 2 \times x \times 5 + 5^2 = x^2 + 10x + 25$ and
$(3x - 4y)^2 = (3x)^2 - 2 \times 3x \times 4y + (4y)^2 = 9x^2 - 24xy + 16y^2$

Another important expansion gives the **difference between two squares**.

$$(a + b)(a - b) = a^2 - b^2$$

For example, $(x + 2)(x - 2) = x^2 - 2^2 = x^2 - 4$.

Unit 11.1 Activity 11B: Further expansions

Expand each of the following expressions and simplify where possible:

1. $(x + 4)(x + 5)$
2. $(x - 3)(x + 2)$
3. $(x + 4)(x - 5)$
4. $(x - 5)(x + 3)$
5. $(x - 7)(x - 4)$
6. $(x - 5)(x - 6)$
7. $(2x + 3)(x + 4)$
8. $(4x + 3)(3x - 4)$
9. $(2x - 3)(x - 6)$
10. $(3x + 2)(3 - x)$
11. $(2x + 7)(5 - x)$
12. $(2x - 5)(3x - 4)$
13. $(3x + 1)(x + 3)$
14. $(5x - 2)(3x + 4)$
15. $(3 - 2x)(4 + 3x)$
16. $(x + 6)^2$
17. $(x + 4)^2$
18. $(x - 7)^2$
19. $(x - 10)(x + 10)$
20. $(4x + 1)(4x - 1)$
21. $(2x - 5)(2x + 5)$
22. $(2x + 3)^2$
23. $(3x - 5)^2$
24. $(x + 2)(x^2 + x + 1)$
25. $(x - 3)(x^2 + 2x - 4)$
26. $(2x - 5)(x^2 - 2x - 3)$

27. Pythagoras' Theorem is a relationship involving the three sides of a right-angled triangle. For the triangle ABC, shown alongside, $a^2 = b^2 + c^2$.

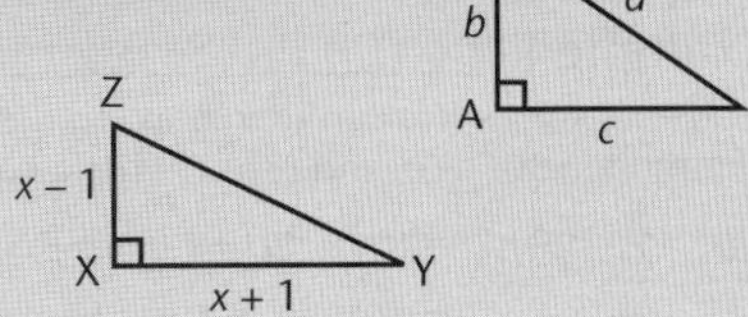

Use Pythagoras' Theorem to find the length of the hypotenuse (ie side YZ) in terms of x.

28. The diagram shows a plan of a garden. Find an expanded expression for the area of the garden in terms of x.

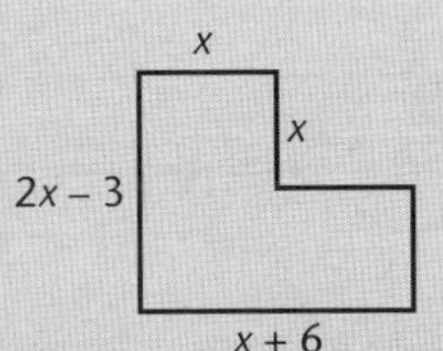

29. A square of side length x cm is cut from the centre of a square piece of card whose side length is 5 cm longer than the side length of the square that is cut out. Find an expression for the area of the card that remains.

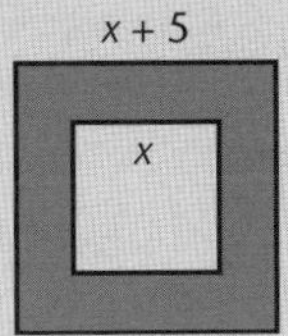

30. A cube has side length $(x + 2)$. Find an expanded expression for its volume $(x + 2)^3$.

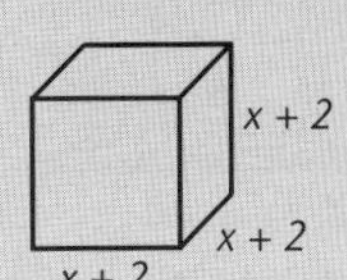

Factorising

Factorising means writing an expression as a **product** of its **factors**. To do this the rules for expanding are used in reverse.

$$ab + ac = a(b + c)$$

Thus, the expression $ab + ac$ is expressed as a product of the two factors a and $(b + c)$.

To factorise an expression, the first step is taking out the **highest common factor** (HCF) of the terms in the expression.

Example H

Q. Factorise: **1.** $8x + 12y$ **2.** $3x + xy$ **3.** $x^2y^3 - xy^2$

A. **1.** In $8x + 12y$, the terms $8x$ and $12y$ have a highest common factor of 4.

$\therefore 8x + 12y = 4(2x + 3y)$ [since $8x = 4 \times 2x$ and $12y = 4 \times 3y$]

Note: 2 is also a common factor of $8x$ and $12y$, but it is not the *highest* common factor.

2. In $3x + xy$, the terms $3x$ and xy have a common factor of x.

$\therefore 3x + xy = x(3 + y)$ [since $3x = x \times 3$ and $xy = x \times y$]

3. In $x^2y^3 - xy^2$, the terms x^2y^3 and xy^2 have a highest common factor of xy^2.

$\therefore x^2y^3 - xy^2 = xy^2(xy - 1)$ [since $x^2y^3 = xy^2 \times xy$ and $xy^2 = xy^2 \times 1$]

Note: The HCF of two powers of the same variable is the lower power of the variable. For example, the HCF of x^7 and x^4 is x^4.

Some expressions may not have a common factor, but *groups of terms* within the expression may have.

Example I

If $pq - 3q + 3p - 9$ is considered as $(pq - 3q)$ plus $(3p - 9)$, it can be seen that $(pq - 3q)$ has a common factor of q, and $(3p - 9)$ has a common factor of 3. This gives:

$pq - 3q + 3p - 9$	$= (pq - 3q) + (3p - 9)$	[grouping terms]
	$= q(p - 3) + 3(p - 3)$	[factorising]
	$= (p - 3)(q + 3)$	[since $(p - 3)$ is a common factor]

Unit 11.1 Activity 11C: Basic factorising

Factorise each of the following completely by taking out the highest common factor:

1. $4a + 12$	**2.** $7x - 14$	**3.** $5y + 25$
4. $9x + 15y$	**5.** $xy - 3x$	**6.** $y^3 + 5y$
7. $a^2 + ab$	**8.** $x^2 - 7x$	**9.** $6x + 8y$
10. $ab - a^2$	**11.** $3p^2 - 6p$	**12.** $4x + 14y$
13. $3x^3 + 6x^2 - 3x$	**14.** $8x + 12y - 20z$	**15.** $15x + 10y - 25$
16. $8x^2y^3 + 12xy^4 - 10xy^2$	**17.** $8x - 24y + 20$	**18.** $3a^2b + 5a^3b^2 + ab$
19. $4x^2 + 8x - 10$	**20.** $2y - 6y^2 + y^3$	**21.** $3p^2q - 9pq^2$
22. $p^4 - 3p^2$	**23.** $4a^2 - 18a^3$	**24.** $a^2b^3 + a^3b^2$
25. $2x(x + 3) + 5(x + 3)$	**26.** $6x(x - 2) - 5(x - 2)$	**27.** $ab + xb + ac + xc$
28. $3x + 6y + ax + 2ay$	**29.** $5x + 15y - px - 3py$	**30.** $ax - bx + ay - by$

Factorising quadratic expressions of the form $x^2 + bx + c$

Quadratic expressions are expressions of the form $ax^2 + bx + c$.

When the **coefficient** of x^2 is 1 (ie $a = 1$), the quadratic expression is written as $x^2 + bx + c$. Such expressions are factorised by finding two numbers which

- *Multiply* together to give c.
- *Add* together to give b.

If both b and c are positive, then both the numbers required for the factorisation will be positive.

Example J

Q. Factorise: **1.** $x^2 + 7x + 12$ **2.** $x^2 + 8x + 12$

A. 1. To factorise $x^2 + 7x + 12$, find two numbers which multiply to give 12 and add to give 7.

The numbers are 3 and 4, because 3 x 4 = 12 and 3 + 4 = 7.

$\therefore x^2 + 7x + 12 = (x + 3)(x + 4)$

2. To factorise $x^2 + 8x + 12$, find two numbers which multiply to give 12 and add to give 8.

The numbers are 2 and 6, because 2 x 6 = 12 and 2 + 6 = 8.

$\therefore x^2 + 8x + 12 = (x + 2)(x + 6)$

Note: Factorisations may be checked by **expanding** the answer. For example, in part **2**,

$(x + 2)(x + 6) = x^2 + 6x + 2x + 12 = x^2 + 8x + 12$ as required [using FOIL method]

If c is positive and b is negative in the expression $x^2 + bx + c$ then both numbers in the factorisation will be negative.

Example K

Q. Factorise $x^2 - 11x + 30$.

A. The two numbers required to factorise $x^2 - 11x + 30$ must multiply to give 30 and add to give –11.

The numbers are –5 and –6, because –5 x –6 = 30 and –5 + –6 = –11.

$\therefore x^2 - 11x + 30 = (x - 5)(x - 6)$

If c is negative in the expression $x^2 + bx + c$ then the two numbers used for the factorisation will have *different signs*, regardless of whether b is negative or positive. If b is positive, the bigger of the two numbers will also be positive and the second number will be negative.

Example L

Q. Factorise $x^2 + 2x - 48$.

A. The numbers required to factorise $x^2 + 2x - 48$ are +8 and –6, because –6 x 8 = –48 and –6 + 8 = +2.

$\therefore x^2 + 2x - 48 = (x + 8)(x - 6)$

Note: +8 and –6 are chosen so that the bigger number is positive. This gives a positive b as required.

If both c and b are negative in the expression $x^2 + bx + c$ then the two numbers will have *different signs* and the bigger number will be negative.

Example M

Q. Factorise $x^2 - 5x - 24$.

A. The numbers –8 and +3 are used to factorise $x^2 - 5x - 24$, because –8 x 3 = –24 and –8 + 3 = –5

$\therefore x^2 - 5x - 4 = (x - 8)(x + 3)$

Factorising the difference of two squares

Quadratic expressions of the type $x^2 - c^2$ are called the **difference of two squares**. These expressions factorise to give $(x + c)(x - c)$.

Example N

Q. Factorise: **1.** $x^2 - 36$ **2.** $4p^2 - 9q^2$

A. **1.** $x^2 - 36 = (x + 6)(x - 6)$ [since $\sqrt{36} = 6$]

2. $4p^2 - 9q^2 = (2p + 3q)(2p - 3q)$ [$\sqrt{4p^2} = 2p$, $\sqrt{9q^2} = 3q$]

When the coefficient of x^2 is a number other than 1, the method of factorising becomes more involved. In cases where a common *factor can be removed*, the problem still involves factorising a quadratic expression of the form $x^2 + bx + c$.

Example O

Q. Factorise: **1.** $2x^2 + 8x - 24$ **2.** $x^3 + 3x^2 + 2x$

A. **1.** $2x^2 + 8x - 24 = 2(x^2 + 4x - 12)$ [2 is a common factor]

$= 2(x + 6)(x - 2)$ [factorising $x^2 + 4x - 12$]

2. $x^3 + 3x^2 + 2x = x(x^2 + 3x + 2)$ [x is a common factor]

$= x(x + 2)(x + 1)$ [factorising $x^2 + 3x + 2$]

Factorising quadratic expressions of the form $ax^2 + bx + c$

A procedure for factorising expressions of the form $ax^2 + bx + c$ is illustrated in the following example.

Example P

Q. Factorise $3x^2 + 11x - 20$.

A. **a.** Multiply the coefficient of x^2 (3) by the constant (–20) to get –60.

b. Find the factors of –60 that add to the coefficient of x (11).

The pairs of numbers that multiply together to give 60 are:
(1,60), (2,30), (3,20), (4,15), (5,12), (6,10).

Thus the factors of –60 that add to 11 are –4 and 15 since $-4 \times 15 = -60$ and $15 - 4 = 11$.

c. Rewrite the term in x as two terms using these numbers, ie $11x = 15x - 4x$.

d. This results in the expression $3x^2 + 15x - 4x - 20$ to factorise:

$3x^2 + 15x - 4x - 20 = 3x(x + 5) - 4(x + 5)$ [factorise in pairs]

$= (x + 5)(3x - 4)$ [$x + 5$ is a common factor]

Example Q

Q. Factorise $4x^2 - 5x - 6$.

A. $4 \times -6 = -24$. Factor pairs of 24 are (1, 24), (2, 12), (3, 8), (4, 6).

The required numbers are 3 and –8 since $3 - 8 = -5$ and $3 \times -8 = -24$.

$\therefore\ 4x^2 - 5x - 6 = 4x^2 - 8x + 3x - 6$ [rewriting $-5x$ as $-8x + 3x$]

$= 4x(x - 2) + 3(x - 2)$ [factorising in pairs]

$= (x - 2)(4x + 3)$ [$x - 2$ is a common factor]

Unit 11.1 Activity 11D: Factorising quadratic expressions

Factorise each of the following completely:

1. $x^2 + 6x + 8$	**2.** $x^2 + 10x + 25$	**3.** $x^2 - 8x + 16$	**4.** $x^2 + 2x - 15$
5. $x^2 + 5x - 24$	**6.** $x^2 - 5x - 14$	**7.** $x^2 + 9x + 20$	**8.** $x^2 - 10x + 25$
9. $x^2 + 7x + 10$	**10.** $x^2 - 4x + 3$	**11.** $x^2 - 11x + 28$	**12.** $x^2 + x - 20$
13. $x^2 + 4x - 21$	**14.** $x^2 + 9x + 18$	**15.** $x^2 - 12x + 36$	**16.** $x^2 - 9$
17. $x^2 - 49$	**18.** $x^2 - 64$	**19.** $25x^2 - 16$	**20.** $4x^2 - 81$
21. $5x^2 - 20$	**22.** $3x^2 + 12x - 63$	**23.** $2x^2 - 2x - 4$	**24.** $4x^2 - 16x + 12$
25. $2x^2 + 5x - 3$	**26.** $2x^2 + 9x + 10$	**27.** $3x^2 - 7x + 2$	**28.** $5x^2 - 13x - 6$
29. $3x^2 + 4x - 7$	**30.** $6x^2 + x - 12$		

Simplifying rational expressions by factorising

Rational expressions are simplified by cancelling common factors. When a numerator or denominator of an algebraic fraction has more than one term, then factorisation must be done before cancelling.

Example R

Q. Simplify: **1.** $\dfrac{5a + 6a}{a}$ **2.** $\dfrac{x^2 + 5x + 6}{x + 2}$

A. **1.** $\dfrac{5a + 6a}{a} = \dfrac{11a}{a}$ [simplifying the numerator]

$= 11$ [cancelling by dividing top and bottom by a]

2. $\dfrac{x^2 + 5x + 6}{x + 2} = \dfrac{(x + 3)(x + 2)}{(x + 2)}$ [factorising the numerator and placing brackets around the denominator]

$= x + 3$ [cancelling by dividing top and bottom by $(x + 2)$]

Sometimes both the numerator and denominator can be factorised, as shown below.

Example S

Q. Factorise $\frac{x^2-9}{2x+6}$.

A. $\frac{x^2-9}{2x+6}=\frac{(x+3)(x-3)}{2(x+3)}$ [factorising numerator and denominator]

$=\frac{x-3}{2}$ [dividing top and bottom by $(x+3)$]

Unit 11.1 Activity 11E: Simplifying rational expressions by factorising

Simplify each of the following rational expressions.

1. $\frac{6a+3a}{a}$
2. $\frac{10b-3b}{b}$
3. $\frac{4x+12}{4}$
4. $\frac{3x+15}{x+5}$
5. $\frac{x^2+4x}{x+4}$
6. $\frac{x^2+8x+12}{x+2}$
7. $\frac{x^2+x-12}{x-3}$
8. $\frac{x^2+3x-4}{x^2-3x+2}$
9. $\frac{x^2-5x-24}{x^2+5x+6}$
10. $\frac{x^2+6x+8}{x^2-4}$
11. $\frac{x^2-x-6}{x^2-9}$
12. $\frac{x^2-x-20}{2x^2-50}$

Unit 11.1 Activity 11F: Multiple choice—factorising quadratic expressions

1. In simplest form the expression $4ab + 3a + 3b + 5a \times -2b$ is equal to:

   A. $3a + 3b - 6ab$

   B. $-5ab$

   C. $4ab + 8a + b$

   D. $ab + 3a + 3b$

2. Jacob is x years old and his sister is two years younger. In three years time the sum of the ages of Jacob and his sister will be:

   A. $2x + 6$

   B. $2x + 4$

   C. $x + 4$

   D. $2x + 8$

3. $\frac{2a}{3}+\frac{3}{2}=$

   A. $\frac{2a+3}{6}$

   B. $\frac{2a+3}{5}$

   C. $\frac{2a+9}{3}$

   D. $\frac{4a+9}{6}$

Example E

1. An expression such as $(x + 3)(x + 4)$ means $(x + 3)$ lots of $(x + 4)$, therefore

$$\begin{aligned}(x + 3)(x + 4) &= x(x + 4) + 3(x + 4) && [x \text{ lots of } (x + 4) \text{ plus 3 lots of } (x + 4)]\\ &= x^2 + 4x + 3x + 12 && [\text{expanding}]\\ &= x^2 + 7x + 12 && [\text{simplifying}]\end{aligned}$$

2. $$\begin{aligned}(x - 2)(x + 5) &= x(x + 5) - 2(x + 5) && [\text{multiplying } (x + 5) \text{ by } x \text{ and } -2]\\ &= x^2 + 5x - 2x - 10\\ &= x^2 + 3x - 10\end{aligned}$$

   Note: The second bracket is multiplied by -2, *not* 2, because the second term in the first bracket is -2.

3. $$\begin{aligned}(a + 2)(b + c - 3) &= a(b + c - 3) + 2(b + c - 3)\\ &= ab + ac - 3a + 2b + 2c - 6\end{aligned}$$

By considering areas in the diagram alongside a general rule for expanding pairs of brackets can be seen:

$$\begin{aligned}(a + b)(c + d) &= a(c + d) + b(c + d)\\ &= ac + ad + bc + bd\end{aligned}$$

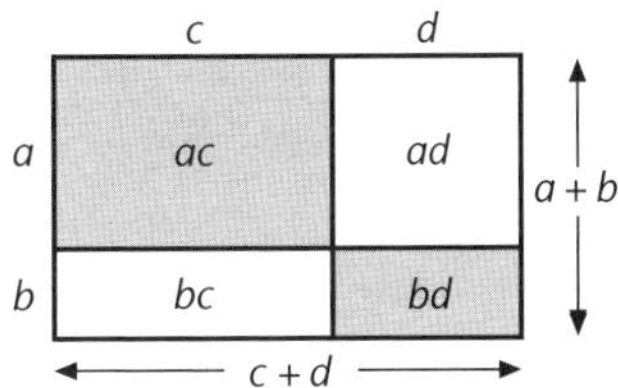

The mnemonic '**FOIL**' is a useful way to remember how to expand two pairs of brackets, each containing two terms.

F **F**irst terms in each bracket are multiplied together. $(a + b)(c + d) = ac$ ($a \times c$)

O **O**utside terms are multiplied together. $(a + b)(c + d) = ad$ ($a \times d$)

I **I**nside terms are multiplied together. $(a + b)(c + d) = bc$ ($b \times c$)

L **L**ast terms in each bracket are multiplied together. $(a + b)(c + d) = bd$ ($b \times d$)

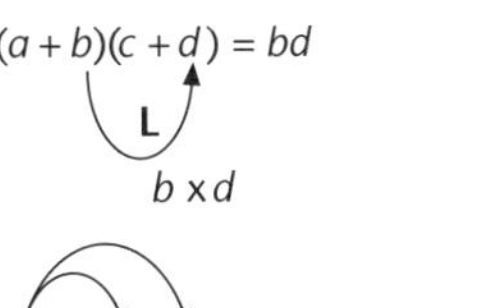

Thus the **FOIL** method gives $(a + b)(c + d) = ac + ad + bc + bd$

4. $\frac{x}{2} + \frac{2}{x} =$

   A. 2

   B. $\frac{4x}{2 + x}$

   C. $\frac{x^2 + 4}{2x}$

   D. $\frac{x + 2}{x}$

5. $(2x - 3)(x + 1)$ expanded and simplified is equal to:

   A. $2x^2 - 3$

   B. $2x^2 + 5x - 3$

   C. $2x^2 - 3x + 1$

   D. $2x^2 - x - 3$

6. $(x - 1)^2 =$

   A. $x^2 - 1$

   B. $x^2 + 1$

   C. $x^2 - 2x + 1$

   D. $x^2 - 2x - 1$

7. An expression that represents the shaded area, at right, is:

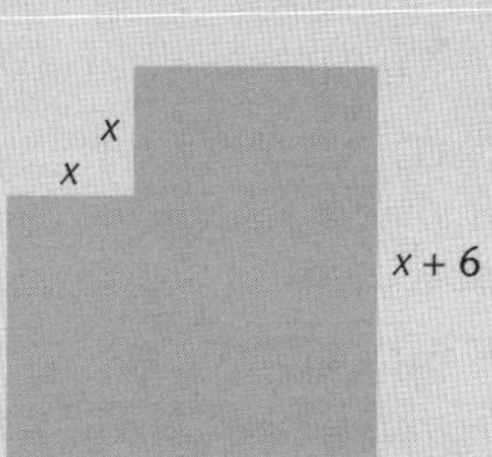

   A. $11x + 30$

   B. $x^2 + 10x + 30$

   C. $2x^2 + 11x + 30$

   D. $4x + 22$

8. When factorised, the expression $ab + ax + bc + cx$ is equal to:

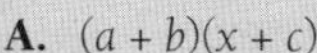

   A. $(a + b)(x + c)$

   B. $(b + x)(a + c)$

   C. $(a + x)(b + c)$

   D. $2(a + b + c + x)$

9. When factorised, the expression $x^2 + 3x - 4$ is equal to:

   A. $(x + 4)(x - 1)$

   B. $(x + 4)(x + 1)$

   C. $(x - 4)(x + 1)$

   D. $(x - 4)(x - 1)$

10. When fully factorised, the expression $4x^2 - 64$ is equal to:

   A. $(2x - 8)(2x + 8)$

   B. $4(x + 4)(x - 4)$

C. $4(x-1)^2$

D. $4(x-8)(x+8)$

11. When factorised, $3x^2 - 16x - 12$ is equal to:

A. $(3x+4)(x-3)$

B. $(3x-12)(x-1)$

C. $(3x+6)(x-2)$

D. $(3x+2)(x-6)$

12. When factorised, $4x^2 + 4x - 15$ is equal to:

A. $(2x-5)(2x+3)$

B. $(4x+5)(x-3)$

C. $(2x+5)(2x-3)$

D. $(4x+15)(x-1)$

13. The rational expression $\frac{x^2+3x-10}{x^2-4}$, when simplified, is equal to:

A. $\frac{x+5}{x+2}$

B. $\frac{x-5}{x+2}$

C. $\frac{x-5}{x-2}$

D. $\frac{x+5}{x-2}$

14. The rational expression $\frac{2x^2+x-1}{x^2-1}$, when simplified, is equal to:

A. $\frac{2x+1}{x+1}$

B. $\frac{2x-1}{x-1}$

C. $\frac{2x-1}{x+1}$

D. $\frac{2x+1}{x-1}$

Unit 11.1 Number and Application

Topic 12: Basic algebra—solving quadratic equations

Basic algebra is a sub-section within Unit 11.1 of the Grade 11 syllabus (see Syllabus p. 14). This Topic deals with using straightforward algebraic methods and solving. It covers:

- Solving factorised equations such as $(x - 1)(x + 3) = 0$.
- Solving simple quadratic equations such as $x^2 + 30x = 400$ and interpreting the results.
- Modelling by forming and solving appropriate equations.
- Interpretation in context.

Introduction

The general form of a quadratic equation in x is

$$ax^2 + bx + c = 0$$

In a **quadratic equation**, the highest power of the variable is 2.

For example: $x^2 - 4x + 1 = 0$, $y^2 = y + 1$, and $3a^2 = 4$ are all quadratic equations.

Solving quadratic equations

The following method for solving quadratic equations depends on an important property of zero:

If two real numbers multiply to make zero, then one or the other (or both) *of the numbers must be zero.*

This result is expressed using symbols as follows:

$$\text{If } ab = 0 \text{ then either } a = 0 \text{ or } b = 0 \text{ (or both)}$$

This result makes the solving of quadratic equations straightforward if the quadratic is given in factored form (and is equal to zero).

Example A

Q. Solve the quadratic equations: **1.** $(x + 4)(x - 2) = 0$ **2.** $3x(2x + 3) = 0$

A. **1.** If $(x + 4)(x - 2) = 0$ then

either $(x + 4) = 0$ or $(x - 2) = 0$ [setting each factor in turn to zero]

$x = -4$ or $x = 2$ [solving both equations for x]

2. $3x(2x + 3) = 0$

$\therefore 3x = 0$ or $(2x + 3) = 0$ [setting each factor in turn to zero]

$\therefore x = 0$ [dividing by 3] or $2x = -3$ [subtracting 3]

$\therefore x = 0$ or $x = -\frac{3}{2}$ [dividing second equation by 2]

If quadratic equations are not given in factored form then the quadratic must be factorised first.

Example B

Q. Solve: **1.** $x^2 + 5x - 6 = 0$ **2.** $2y^2 - 18 = 0$

A. 1.

$$\begin{aligned} x^2 + 5x - 6 &= 0 \\ (x + 6)(x - 1) &= 0 \quad \text{[factorising]} \\ \therefore x + 6 &= 0 \quad \text{or} \quad x - 1 = 0 \\ \therefore x &= -6 \quad \text{or} \quad x = 1 \end{aligned}$$

2.

$$\begin{aligned} 2y^2 - 18 &= 0 \\ 2(y^2 - 9) &= 0 \quad \text{[factorising]} \\ 2(y + 3)(y - 3) &= 0 \quad \text{[factorising]} \\ y + 3 &= 0 \quad \text{or} \quad y - 3 = 0 \\ \therefore y &= -3 \quad \text{or} \quad y = 3 \end{aligned}$$

If necessary, the quadratic equation must be **rearranged** before factorising, so that every term of the equation is on one side of the equals sign, leaving zero on the other side.

Example C

Q. Solve: **1.** $x^2 = 4x + 12$ **2.** $(x + 2)(x - 3) = 50$

A. 1. To solve $x^2 = 4x + 12$, the equation must be rearranged and then factorised.

$$\begin{aligned} x^2 &= 4x + 12 \\ x^2 - 4x - 12 &= 0 \quad \text{[subtracting 4x and 12 from both sides]} \\ \therefore (x - 6)(x + 2) &= 0 \quad \text{[factorising]} \\ x - 6 &= 0 \quad \text{or} \quad x + 2 = 0 \\ \therefore x &= 6 \quad \text{or} \quad x = -2 \end{aligned}$$

2.

$$\begin{aligned} (x + 2)(x - 3) &= 50 \\ x^2 - 3x + 2x - 6 &= 50 \quad \text{[expanding LHS]} \\ x^2 - x - 6 &= 50 \quad \text{[simplifying]} \\ x^2 - x - 56 &= 0 \quad \text{[subtracting 50]} \\ (x - 8)(x + 7) &= 0 \quad \text{[factorising]} \\ x - 8 &= 0 \quad \text{or} \quad x + 7 = 0 \quad \text{[solving]} \\ x &= 8 \quad \text{or} \quad x = -7 \end{aligned}$$

Note: Although the quadratic in part **2** was factorised, the RHS was not zero, so the quadratic could not be solved using the techniques described previously. The quadratic needed to be expanded and the RHS made equal to zero before re-factorising and solving.

The solutions to a quadratic equation can be checked by **substituting** back into the original equation. (This can sometimes be done mentally.)

Example D

Q. Check that the solutions to $x^2 + 5x - 6 = 0$ are $x = -6$ or $x = 1$ (as found in Example B, part **1**).

A.
$$x = -6 \Rightarrow (-6)^2 + 5 \times -6 - 6 = 0 \quad \text{[substituting } x = -6 \text{ into } x^2 + 5x - 6]$$
$$36 - 30 - 6 = 0$$
$$36 - 36 = 0 \text{ which is true.}$$
$$x = 1 \Rightarrow 1^2 + 5 \times 1 - 6 = 0 \quad \text{[substituting } x = 1 \text{ into } x^2 + 5x - 6]$$
$$1 + 5 - 6 = 0$$
$$6 - 6 = 0 \text{ which is true.}$$

∴ both solutions, $x = -6$ and $x = 1$, are correct.

Once it is recognised that an equation (in x say) is **quadratic** (the equation has terms in x^2 and/or x and/or constants) it is important to remember the correct strategy for solving such equations:

- **Z**ero: the right-hand side of the equation must equal zero.
- **F**actorise: express the quadratic as a product of its factors.
- **S**olve: set each factor in turn to zero and solve for the variable.

A useful reminder of this strategy is the mnemonic **ZFS**.

Example E

Q. Solve $(x - 1)(x + 3) = 0$.

A. A quadratic equation such as $(x - 1)(x + 3) = 0$ has already been made equal to zero (**Z**) and factorised (**F**) so that all that remains is to solve (**S**) the equation by setting each factor to zero.

Thus the solution is $x = 1$ or $x = -3$ [solving $x - 1 = 0$ and $x + 3 = 0$]

Note: A common error in solving equations such as $(x - 1)(x + 3) = 0$ is to expand the left-hand side of the equation, then attempt to isolate the x (as if the equation were linear).

Unit 11.1 Activity 12A: Solving quadratic equations

1. Solve the following quadratic equations given in factored form.

a. $(x + 2)(x + 3) = 0$ **b.** $(x - 2)(x - 4) = 0$ **c.** $(x + 1)(x - 3) = 0$
d. $(x - 5)(x + 2) = 0$ **e.** $x(x - 4) = 0$ **f.** $x(x + 5) = 0$
g. $(2x - 3)(x + 6) = 0$ **h.** $(2x + 1)(x - 1) = 0$ **i.** $(5x + 3)(2x + 5) = 0$
j. $(3 - 4x)(2x - 9) = 0$ **k.** $(3x - 4)(x - 5) = 0$ **l.** $(3x - 2)(3x + 2) = 0$

2. Factorise and solve the following quadratic equations.

a. $x^2 + 3x = 0$ **b.** $2x^2 - 10x = 0$ **c.** $x^2 + x - 6 = 0$
d. $x^2 + 5x + 4 = 0$ **e.** $x^2 - 3x - 40 = 0$ **f.** $x^2 - 6x + 9 = 0$
g. $x^2 - 8x + 12 = 0$ **h.** $x^2 + 3x - 10 = 0$ **i.** $x^2 - 4 = 0$

3. Solve the following quadratic equations.

a. $x^2 = 2x + 15$ **b.** $x^2 + 3x = 18$ **c.** $x^2 + 3 = 4x$
d. $x^2 = 6x$ **e.** $2x^2 = 5x$ **f.** $x^2 = 49$
g. $x^2 - 2x = 8$ **h.** $3x^2 = 12$ **i.** $2x^2 + x = 6$

4. Solve the following quadratic equations to find n.

a. $(n + 1)(n - 1) = 63$ **b.** $(n - 8)(n + 2) = 11$ **c.** $(n + 2)(n - 2) = 140$

d. $(2n + 3)(2n - 3) = 91$ **e.** $n(2n + 1) = 36$ **f.** $(3n - 1)(2n + 3) = 35$

Solving word problems with quadratic equations

When solving **word problems**, the solutions must be interpreted correctly so that they are meaningful within the given context. For example, if the unknown value, x, is a side length, then only positive values for x make sense. In word problems at the Merit level, the practical application is described and the quadratic equation to be solved is given.

Example F

Q. An aircraft dropped a box of emergency supplies to some villagers. The height of the box above the ground, h, at time, t, is given by the formula:

$h = 520 - 5t^2$, where h is in metres and t is in seconds.

To find how long the box takes to fall to the ground, the equation $520 - 5t^2 = 0$ must be solved. Find how long it takes for the box to fall to the ground.

A.

$520 - 5t^2 = 0$ [at ground level, $h = 0$]

$t^2 - 144 = 0$ [dividing by –5 and rearranging]

$(t + 12)(t - 12) = 0$ [factorising]

$t = -12$ or 12 [solving, giving both solutions]

The time taken is 12 seconds [ignore the negative solution, since time taken is positive].

At the 'very high achievement' level, the context of the problem will be described but the forming and solving of the quadratic equation will be left to the student.

Example G

Q. In an art shop, canvases are sold by the **square centimetre**. Two canvases in the shop have the same price. One canvas is a rectangle four times as long as it is wide. The other is a square canvas of side length 18 cm longer than the width of the rectangular canvas. Find the dimensions of the square canvas.

A. Let x be the width of the rectangular canvas. Thus, the length of this canvas is $4x$ and the side length of the square canvas is $(x + 18)$.

4x

x

x + 18

x + 18

The area of the rectangular canvas is $x.4x = 4x^2$ and the area of the square canvas is $(x + 18)^2$. Equating areas gives the following quadratic to be solved.

$(x + 18)^2 = 4x^2$

$x^2 + 36x + 324 = 4x^2$ [expanding]

$3x^2 - 36x - 324 = 0$ [collecting terms on one side]

$x^2 - 12x - 108 = 0$ [dividing by 3]

$(x + 6)(x - 18) = 0$ [factorising]

$x = -6$ or $x = 18$

Only $x = 18$ applies, since x is a length and therefore can't be negative.

The side length of the square is 18 + 18 = 36 cm.

Unit 11.1 Activity 12B: Solving word problems with quadratic equations

1. Sam is the middle child of a 3-child family. His brother is 2 years older than Sam and his sister is 3 years younger than Sam. The product of their (Sam's brother and sister) ages is 176. The equation $(x + 2)(x - 3) = 176$ can be used to find the ages of the 3 children.

  a. What does 'x' represent?

  b. Solve the equation to find the ages of the 3 children.

2. In a village, the equation $L = -n^2 + 12n + 40$ modelled the number of pigs born each day for a period of time in the village. L = number of pigs born and n = day number.

To find n when 67 pigs were born in a day, the equation $-n^2 + 12n + 40 = 67$ needs to be solved. Solve this equation to find when 67 pigs were born.

3. A field is 20 m longer than it is wide. The area of the field is 384 m^2. In order to find the dimensions of the field, the quadratic equation $x(x + 20) = 384$ must be solved.

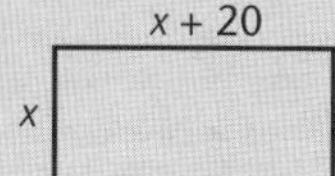

Solve this equation to find the length and width of the field.

4. A square room has side length x m. The room is made 2 m wider and 4 m longer so that its area is now 24 m^2.

Solve the equation $(x + 2)(x + 4) = 24$ to find the original dimensions of the room.

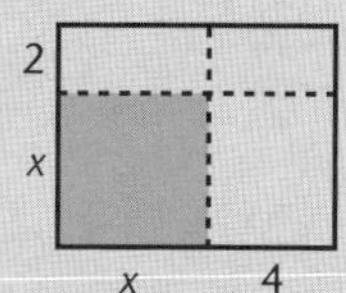

5. William is playing a series of computer games in which he has to capture aliens. William scores one point for each alien captured. So far William has a total score of 45 points. At the end of his next game, William notices that if he squared and doubled the number of points he got for that game then he would get the same number of points as his new total score.

Solve the equation $2x^2 = x + 45$ to find how many aliens William captured in his latest game.

6. A box for soap has a square base and is 6 cm deep. Its top is open. The area of cardboard used for the box is 256 cm^2. The surface area of the outside of the box is $A = a^2 + 24a$, where a is the side length of the base of the box in cm.

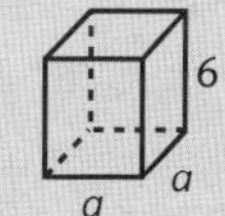

Solve the equation $a^2 + 24a = 256$ to calculate the side length of the base of the box.

7. Alice and Felix think of two numbers. The numbers are 10 apart and their product is 56. Solve the equation $x(x - 10) = 56$ to work out the numbers Alice and Felix thought of (there are two different possible pairs of numbers).

8. Two square pieces of cardboard are required as backing for a photo. The smaller square has side length 4 cm less than the side length of the larger square.

If the total area of cardboard required is 400 cm^2, find the dimensions of the larger square.

9. The product of two numbers is 360. If one number is 16 less than double the other number, find the two possible number pairs.

10. Three boys played a game in which they won points for skill. At the endof the game the winner has 6 points less than double the middle player's score. The loser has 4 points less than the middle player's score. The product of the winner's and loser's scores is equal to the square of the middle player's score. Find the three scores.

Unit 11.1 Activity 12C: Multiple choice—solving quadratic equations

1. Which one of the following is not a quadratic equation?

   A. $2x^2 + 3x = 6x - 1$ **B.** $x^2 + 5x + 6 = 0$

   C. $(x + 2)(x - 5) = x^2$ **D.** $x^2 = 5x + 24$

2. The solution(s) of the equation $x(x - 4) = 0$ are:

   A. $x = 4$ **B.** $x = -4$

   C. $x = 2$ or $x = -2$ **D.** $x = 0$ or $x = 4$

3. The solutions of the equation $x^2 + 2x = 3$ are:

   A. $x = 1$ or $x = 3$ **B.** $x = 1$ or $x = -3$

   C. $x = -2$ or $x = 3$ **D.** $x = 2$ or $x = -3$

4. Which one of the following equations has only one solution?

   A. $x^2 + 9 = 6x$ **B.** $x^2 - 3x = 0$

   C. $x^2 = 16$ **D.** $x^2 + 4x + 16 = 0$

5. The solutions of the equation $x^2 + 2x - 8$ are:

   A. $x = 2$ or $x = 4$ **B.** $x = -2$ or $x = 4$

   C. $x = -2$ or $x = -4$ **D.** $x = 2$ or $x = -4$

6. Which one of the following equations has the solutions $x = -1$ or $x = 2$?

   A. $(x - 1)(x + 2) = 0$ **B.** $x^2 = x + 2$

   C. $x(x + 1) = 2$ **D.** $x^2 - 1 = 1$

7. The solutions to the equation $(3x - 1)(2x + 5) = 0$ are:

   A. $x = 1$ or $x = -5$ **B.** $x = 3$ or $x = -2.5$

   C. $x = \frac{1}{3}$ or $x = -2.5$ **D.** $x = -\frac{1}{3}$ or $x = 2.5$

8. The solutions of the equation $(x + 4)(x - 10) + 24 = 0$ are:

   A. $x = -4$ or $x = 10$ **B.** $x = 20$ or $x = 34$

   C. $x = 8$ or $x = -2$ **D.** $x = -6$ or $x = 2.4$

Unit 11.1 Number and Application

Topic 13: Basic algebra—graphs of quadratic functions

Basic algebra is a sub-section within Unit 11.1 of the Grade 11 syllabus (see Syllabus p. 14). This Topic deals with sketching and interpreting graphs. It covers:

- Sketch and interpret features of quadratic graphs.
- Determining and applying an appropriate model for a situation involving graphs.
- Writing equations from a graph to solve a problem (a combination of only two different types of transformation is expected, eg $y = 2x^2 + 3$ or $y = (x - 2)^2 + 1$).
- Drawing a graph to find the solution to a problem.

The parabola

The equation of a **quadratic function** can be written in the form

$$y = ax^2 + bx + c, \text{ where } a, b, c \text{ are constants, } a \neq 0$$

The following are all examples of equations of quadratic functions:

$y = 3x^2 - 2x + 4$ $\quad y = x^2 + 2x$ $\quad y = -x^2 + 3$ $\quad y = -2x^2$ $\quad y = (x - 3)^2$

The graph of a quadratic function is called a **parabola**. The parabola is a smooth curve, with one **axis of symmetry** and a **turning point** called the **vertex**.

The graph of $y = x^2$

The simplest quadratic function has the equation $y = x^2$. The graph is drawn by setting up a table of values and **plotting** points.

A table of values and graph for $y = x^2$ are shown.

x	–4	–3	–2	–1	0	1	2	3	4
$y = x^2$	16	9	4	1	0	1	4	9	16

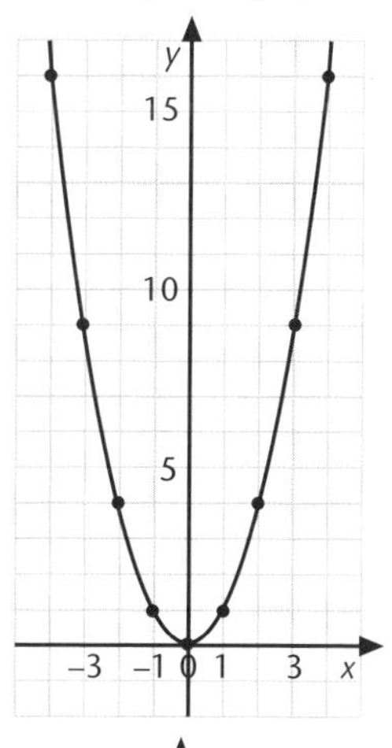

Note: **1.** The graph is symmetrical about the y-axis, ie the line $x = 0$.

2. There is a vertex at (0, 0) giving a minimum value for y of 0 when $x = 0$.

3. The points are joined with a smooth curve. Notice how the curve 'flattens out' at the vertex; it is *not* a sharp point.

To see more clearly the behaviour of y between $x = -1$ and $x = 1$ consider the following table of values.

x	–1	$-\frac{3}{4}$	$-\frac{1}{2}$	$-\frac{1}{4}$	0	$\frac{1}{4}$	$\frac{1}{2}$	$\frac{3}{4}$	1
$y = x^2$	1	$\frac{9}{16}$	$\frac{1}{4}$	$\frac{1}{16}$	0	$\frac{1}{16}$	$\frac{1}{4}$	$\frac{9}{16}$	1

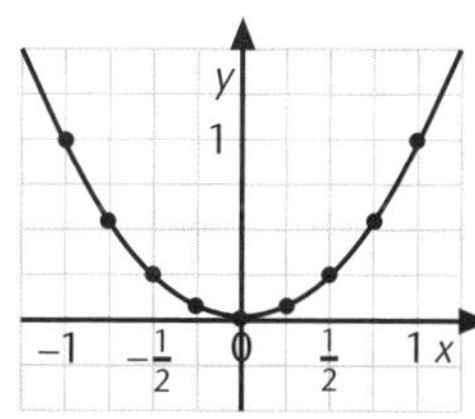

The 'magnification' of the part of the graph near the turning point shows the 'flattening out' of the curve much more clearly.

Transformations of the curve $y = x^2$

Changes to the basic equation of a parabola, $y = x^2$, cause various **transformations** of the curve. The general equation of a **transformed** parabola is

$$y = a(x - h)^2 + k \text{ where } a, h, k \text{ are constants}$$

Four transformations are now discussed. Their graphs are drawn by plotting points.

Graphs of the form $y = ax^2$

By changing the value of a in the equation $y = ax^2$, the steepness of the graph changes. The larger the size of a, the steeper the graph.

Example A

The graphs of $y = 2x^2$ and $y = \frac{1}{2}x^2$ are drawn below and compared with the graph of $y = x^2$.

x	−2	−1	0	1	2
x^2	4	1	0	1	4
$2x^2$	8	2	0	2	8
$\frac{1}{2}x^2$	2	$\frac{1}{2}$	0	$\frac{1}{2}$	2

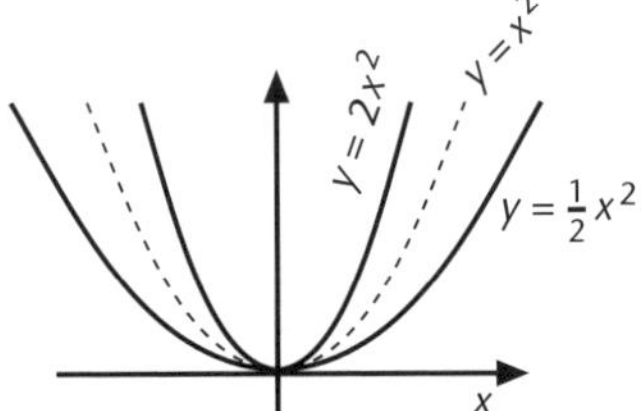

Note:
1. The graph of $y = ax^2$ is steeper than the graph of $y = x^2$ if $a > 1$. For example above, the graph of $y = 2x^2$ lies above the graph of $y = x^2$.
2. The graph of $y = ax^2$ is shallower than the graph of $y = x^2$ if $0 < a < 1$. For example above, the graph of $y = \frac{1}{2}x^2$ lies below the graph of $y = x^2$.
3. For each of these graphs, the vertex is (0, 0) and the axis of symmetry is the y-axis.

If a is negative, the parabola is **inverted** (upside-down). This gives a **maximum** value at the vertex, *not* a **minimum**. (Axis of symmetry remains the y-axis.)

Example B

The graphs of $y = -2x^2$ and $y = -\frac{1}{2}x^2$ are shown and compared with the graph of $y = -x^2$:
[Tabled values for x^2, $2x^2$, $\frac{1}{2}x^2$ (in Example A) are multiplied by −1.]

The graphs of $y = -2x^2$, $y = x^2$ and $y = -\frac{1}{2}x^2$ are *inverted* compared with the graphs of $y = 2x^2$, $y = x^2$ and $y = \frac{1}{2}x^2$.

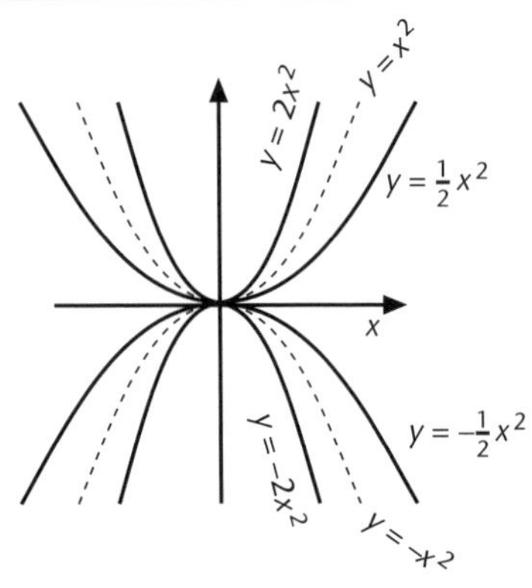

Graphs of the form $y = x^2 + k$

The graph of $y = x^2 + k$ is the graph of $y = x^2$ **transformed** by the **translation** $\begin{pmatrix} 0 \\ k \end{pmatrix}$.

In other words, it is the graph of $y = x^2$ shifted vertically k units.

- If k is positive, the shift is up.
- If k is negative, the shift is down.

Example C

By setting up a table of values, the graph of $y = x^2 + 2$ can be drawn and compared with the graph of $y = x^2$.

x	–3	–2	–1	0	1	2	3
x^2	9	4	1	0	1	4	9
$y = x^2 + 2$	11	6	3	2	3	6	11

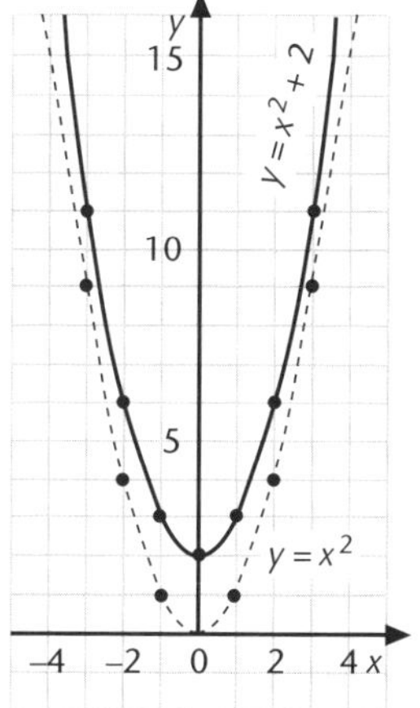

Note: The table and graph show that:

1. The graph of $y = x^2 + 2$ results from moving the graph of $y = x^2$ *up* 2 units.
2. The graph of $y = x^2 + 2$ is symmetrical about the y-axis.
3. The vertex is at (0, 2) and the minimum value for y is 2.

Graphs of the form $y = (x - h)^2$

The graph of $y = (x - h)^2$ is the graph of $y = x^2$ transformed by the translation $\begin{pmatrix} h \\ 0 \end{pmatrix}$ ie the graph of $y = x^2$ moved sideways h units.

- The axis of symmetry is $x = h$ and the vertex is $(h, 0)$.
- The **y-intercept** is $(0, h^2)$ and the **x-intercept** is $(h, 0)$.

Example D

Drawing up a table and plotting the graph of $y = (x - 2)^2$, and then comparing it with $y = x^2$, gives:

x	–1	0	1	2	3	4	5
$x - 2$	–3	–2	–1	0	1	2	3
$y = (x - 2)^2$	9	4	1	0	1	4	9

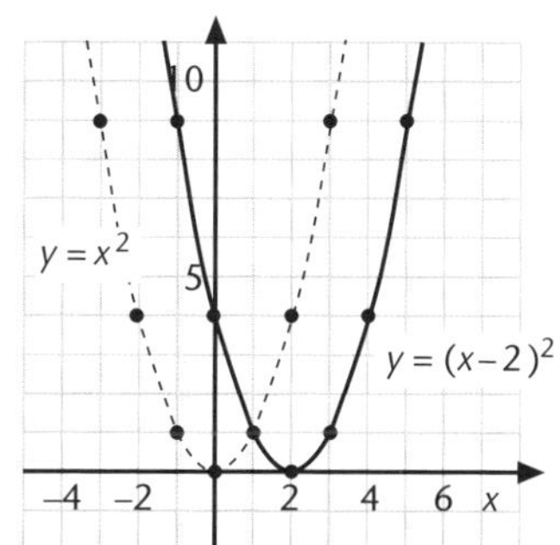

Note:
1. The axis of symmetry is $x = 2$ and the vertex is (2, 0).
2. The y-intercept is (0, 4) and the x-intercept is (2, 0).
3. The graph of $y = x^2$ has been shifted 2 units to the right.

Graphs of the form $y = (x - h)^2 + k$

The graph of $y = (x - h)^2 + k$ is the graph of $y = x^2$ transformed by the translation $\begin{pmatrix} h \\ k \end{pmatrix}$, ie the graph of $y = x^2$ is shifted horizontally h units and vertically k units.

- The axis of symmetry is the line $x = h$.
- The vertex is (h, k), so the minimum value of y is k.
- The y-intercept is $(0, h^2 + k)$.

Example E

Q. Draw the graph of $y = (x + 3)^2 - 2$.

A. Drawing up a table and plotting the points gives:

x	–6	–5	–4	–3	–2	–1	0
$x + 3$	–3	–2	–1	0	1	2	3
$(x + 3)^2$	9	4	1	0	1	4	9
$y = (x + 3)^2 - 2$	7	2	–1	–2	–1	2	7

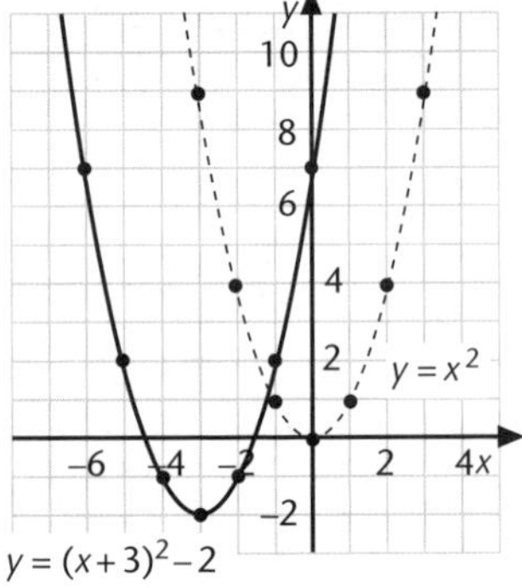

Note:
1. The axis if symmetry is the line $x = -3$.
2. The vertex is $(-3, -2)$, which is a **minimum turning point**.
3. The y-intercept is $(0, 7)$.

Sketching quadratic graphs by transformations

If the equation of a parabola is given in transformation form, $y = a(x - h)^2 + k$, its graph can be sketched by considering the various transformations involved.

Example F

1. The graph of $y = (x + 3)^2 - 2$ was drawn in Example E by plotting points. However, this graph can be drawn directly from the equation. The graph of $y = x^2$ is transformed as follows:

   $y = \underbrace{(x + 3)^2}_{\text{graph is moved 3 units left}} \; \underbrace{- 2}_{\text{graph is moved 2 units down}}$

   To find the y-intercept, substitute $x = 0$ in the equation.

   The graph is as in Example E.

2. The graph of $y = -(x + 3)^2 - 2$ is the graph in part **1** inverted (the negative outside the bracket causes this).

   When $x = 0$, $y = -(0 + 3)^2 = -11$

   The graph is as shown.

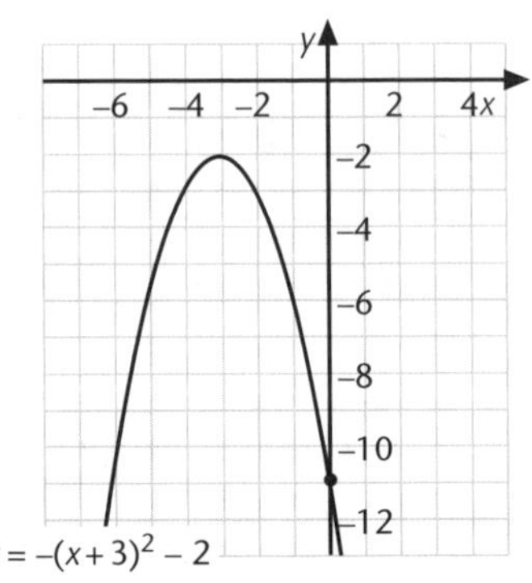

If the coefficient of x^2 is not ±1, the steepness of the graph changes.

Example G

Q. Sketch the graph of $y = \frac{1}{2}(x - 4)^2 - 3$.

A. The graph of $y = x^2$ is transformed as follows:

$y = \underbrace{\tfrac{1}{2}}_{\text{shallower}} \; \underbrace{(x-4)^2}_{\text{4 units right}} \; \underbrace{-\,3}_{\text{down 3}}$

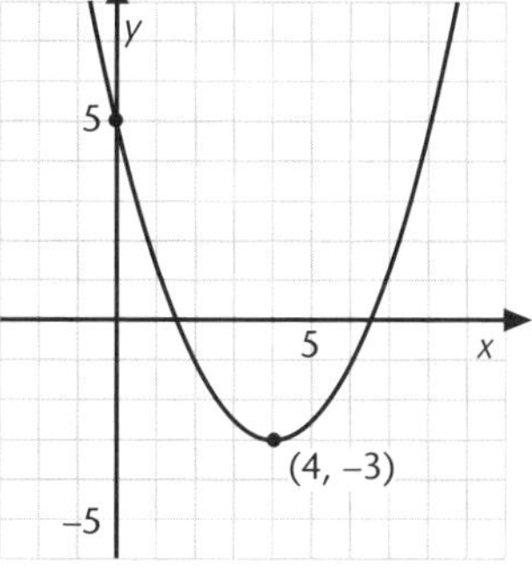

The vertex is at (4, –3) and the y-intercept is

$y = \frac{1}{2}(0 - 4)^2 - 3$ [substituting $x = 0$]

$= 5$

The graph is as shown.

Note: 1. The basic shape of the parabola is the same as that of $y = \frac{1}{2}x^2$, ie from the vertex travel out ±1 and up $\frac{1}{2}$ to reach a point on the curve. Another point is out ±2 up 2 from the vertex, etc.

2. The exact positions of the x-intercepts are not given. (If required, substitute $y = 0$ in the equation of the curve and solve the resulting equation.)

Unit 11.1 Activity 13A: Quadratic graphs

1. Complete the tables of corresponding values of x and y which fit the equations.

a. $y = x^2 - 2$

x	–2	–1	0	1	2
x^2	4				
$y = x^2 - 2$	2				

b. $y = (x - 2)^2$

x	0	1	2	3	4
$x - 2$					2
$y = (x - 2)^2$					4

c. $y = (x + 3)^2$

x	–5	–4	–3	–2	–1
$x + 3$	–2				2
$y = (x + 3)^2$	4				

d. $y = 2 - x^2$

x	–2	–1	0	1	2
x^2					
$y = 2 - x^2$					–2

e. $y = x^2 + 3$

x	–2	–1	0	1	2
x^2					4
$y = x^2 + 3$					7

f. $y = 2x^2$

x	–2	–1	0	1	2
x^2					4
$y = 2x^2$					8

2. Match the graphs below with the equations from Question 1.

a.
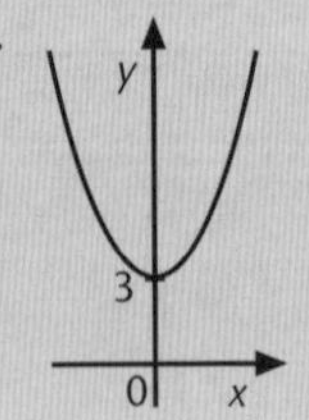

b.
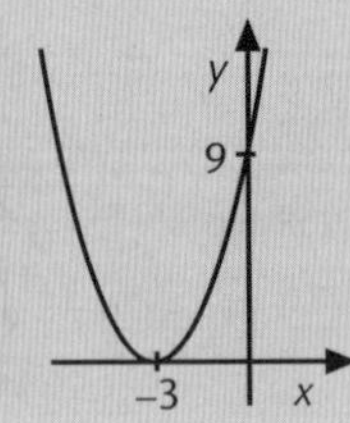

c.
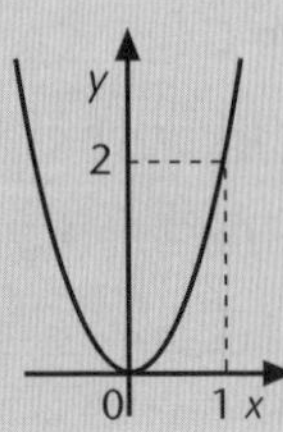

d.
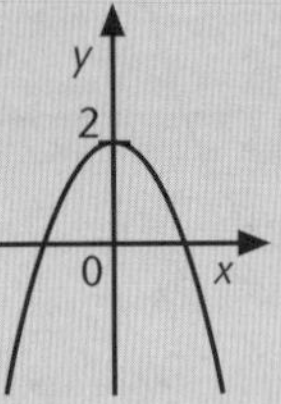

e.
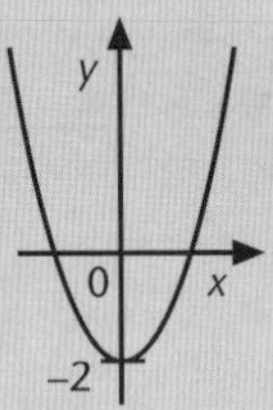

f.
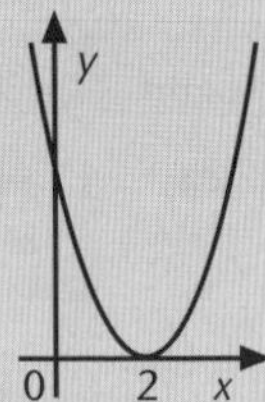

3. Sketch the graphs of the following equations, labelling carefully all intercepts (where possible) and the vertex for each graph.

a. $y = x^2 + 1$ **b.** $y = x^2 - 4$ **c.** $y = (x - 1)^2$

d. $y = (x + 1)^2$ **e.** $y = -2x^2$ **f.** $y = 4 - x^2$

4. Sketch the graphs of the following functions, labelling carefully all intercepts (where possible) and the vertex for each graph.

a. $y = (x + 1)^2 - 4$ **b.** $y = (x - 2)^2 + 3$ **c.** $y = (x + 2)^2 + 1$

d. $y = 2(x - 1)^2 - 3$ **e.** $y = -(x + 4)^2 + 1$ **f.** $y = -\frac{1}{2}x^2 - 2$

Sketching graphs of quadratic functions in factored form

When the equation of a quadratic function is given in factored form, eg $y = (x + 1)(x - 2)$, a different approach is used for drawing the graph. The steps are as follows:

1. Find the *x-intercepts* and the *y-intercept*.
2. Find the *axis of symmetry*.
3. Find the *vertex*.
4. Sketch the parabola.

Example H

Q. Sketch the graph of $y = (x - 4)(x + 2)$.

A. • Intercepts

For the *y*-intercept substitute $x = 0$

$y = (0 - 4)(0 + 2) = -4 \times 2 = -8$ $\quad\therefore$ *y*-intercept is $(0, -8)$

For the *x*-intercepts substitute $y = 0$

$(x - 4)(x + 2) = 0 \Rightarrow x = 4$ or -2 $\quad\therefore$ *x*-intercepts are $(4, 0)$ and $(-2, 0)$

- The axis of symmetry lies midway between the x-intercepts, so the axis of symmetry is:

  $x = \dfrac{4 + -2}{2}$ [averaging the values of the x-intercepts]

  $x = 1$ [shown by dotted line on graph]

- The vertex lies on the axis of symmetry ie on the line $x = 1$, and will have a **y-coordinate** of:

  $y = (1 - 4)(1 + 2)$ [substituting $x = 1$ in function]
  $= -3 \times 3$
  $= -9$

  $\therefore$ vertex is $(1, -9)$

The graph is as shown.

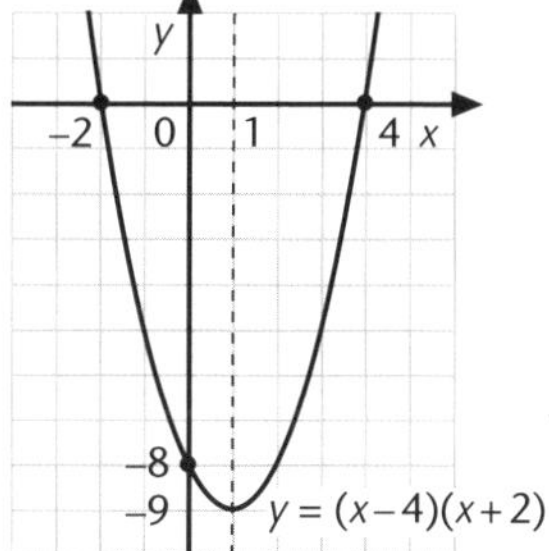

If the equation of the quadratic function is not in factored form, the quadratic will need to be factorised first.

Example I

Q. Sketch the graph of $y = x^2 + x - 6$

A. $y = (x + 3)(x - 2)$ [factorising]

- x-intercepts are -3 and 2

  y-intercept is $3 \times -2 = -6$

- Axis of symmetry is $x = \dfrac{-3 + 2}{2} = -\dfrac{1}{2}$
- Vertex is $(-\frac{1}{2}, -6.25)$

  [substituting $x = -\frac{1}{2}$ in equation]

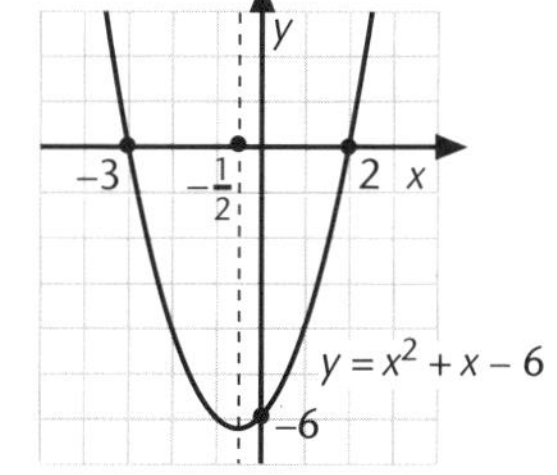

If the term in x^2 is negative then the graph will be inverted (upside down).

Example J

Q. Sketch the graph of $y = (2 - x)(4 + x)$

A. $y = (2 - x)(4 + x)$

- x-intercepts are 2 and -4

  y-intercept is $2 \times 4 = 8$

- Axis of symmetry is $x = \dfrac{-4 + 2}{2} = -1$
- Vertex is $(-1, 9)$

  [substituting $x = -1$ in equation]

Note: Expansion of $(2 - x)(4 + x)$ gives $-x^2 - 2x + 8$ so graph is 'upside down.' [coefficient of x^2 is negative]

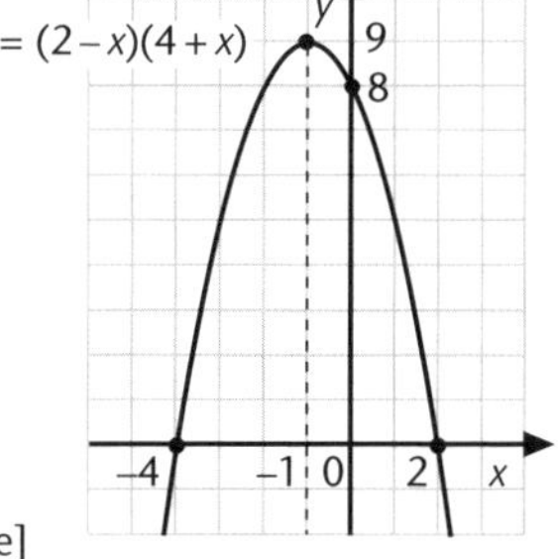

Writing equations of parabolas

By examining features of the graph of a parabola, such as intercepts, vertex, etc, the equation of a parabola can be written down.

Transformation method

If the vertex and one other point are known, the transformation method works well. The equation will be of the form

$$y = a(x - h)^2 + k$$

where h is the sideways translation, k is the vertical translation and a is a factor indicating the change in steepness.

Example K

Q. Write down the equations of the parabolas shown.

1.

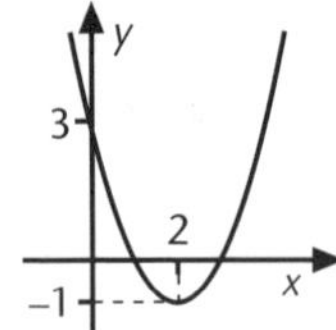

2.

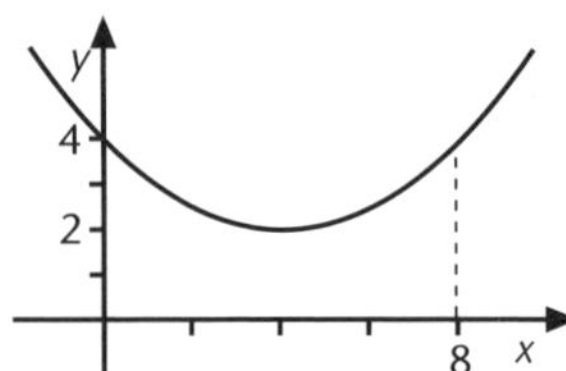

A. 1. The parabola has been translated by the **vector** $\begin{pmatrix} 2 \\ -1 \end{pmatrix}$ so its equation is of the form

$y = a(x - 2)^2 - 1$ [general transformation equation with $h = 2$, $k = -1$]

Substituting the point (0, 3) gives
$$3 = a(0 - 2)^2 - 1$$
$$3 = 4a - 1$$
$$4 = 4a$$
$$a = 1$$

So the equation of the parabola is $y = (x - 2)^2 - 1$.

2. The axis of symmetry of the parabola is $x = 4$. The minimum value is 2.

$\therefore$ the parabola $y = x^2$ has been translated by the vector $\begin{pmatrix} 4 \\ 2 \end{pmatrix}$ and so its equation is

$y = a(x - 4)^2 + 2$ [general transformation equation with $h = 4$, $k = 2$]

Substituting the point (0, 4) gives
$$4 = a(0 - 4)^2 + 2$$
$$4 = 16a + 2 \quad \text{[simplifying]}$$
$$16a = 2$$
$$a = \frac{2}{16} = \frac{1}{8}$$

So the equation of the parabola is $y = \frac{1}{8}(x - 4)^2 + 2$.

Factorised form method

If the x-intercepts are known, or can be worked out, then the quadratic function can be written down in factored form. The equation of the parabola will be of the form

$$y = a(x - b)(x - c)$$

where b and c are the x-intercepts and a is a factor affecting the steepness of the curve. Another known point is then substituted to evaluate a.

Example L

Q. Write down the equations of the parabolas shown.

1. **2.**

A. **1.** The x-intercepts are 1 and 9. Thus, the equation is of the form $y = a(x - 1)(x - 9)$.

The axis of symmetry of the graph is $x = 5$ [since $\frac{1 + 9}{2} = 5$]

From the graph $y = -32$, when $x = 5$:

$$-32 = a(5 - 1)(5 - 9) \quad \text{[substituting } x = 5, y = -32 \text{ in } y = a(x - 1)(x - 9)\text{]}$$
$$-32 = a \times 4 \times -4$$
$$-32 = -16a$$
$$a = 2$$

$\therefore$ the equation is $y = 2(x - 1)(x - 9)$

The equation could also be written as $y = 2x^2 - 20x + 18$ after expanding.

2. Only one intercept is given, the x-intercept, 15; the maximum point is (9, 36). This means the axis of symmetry is $x = 9$ and the other x-intercept is 3 (by symmetry, $15 - 9 = 6$, $9 - 6 = 3$). So, the equation will be of the form:

$y = -a(x - 3)(x - 15)$ [$-a$ since the parabola is 'upside down']

When $x = 9$, $y = 36$:

$$\therefore 36 = -a(9 - 3)(9 - 15) \quad \text{[substituting]}$$
$$36 = 36a \quad \text{[simplifying]}$$
$$a = 1$$

$\therefore$ the equation will be $y = -(x - 3)(x - 15)$ (or $y = -x^2 + 18x - 45$ in expanded form)

Unit 11.1 Activity 13B: Quadratic graphs

1. Sketch the graphs whose equations are given.

a. $y = (x + 2)(x - 4)$ **b.** $y = (x - 5)(x - 1)$ **c.** $y = x(x - 3)$

d. $y = (x + 3)(x + 1)$ **e.** $y = (x - 4)(x - 2)$ **f.** $y = (3 - x)(1 + x)$

g. $y = x^2 + 6x + 5$ **h.** $y = x^2 + 4x - 12$ **i.** $y = 5x - x^2$

2. The sketch shows the graph of $y = x^2 + 2x - 8 = (x + 4)(x - 2)$.

CD is the axis of symmetry.

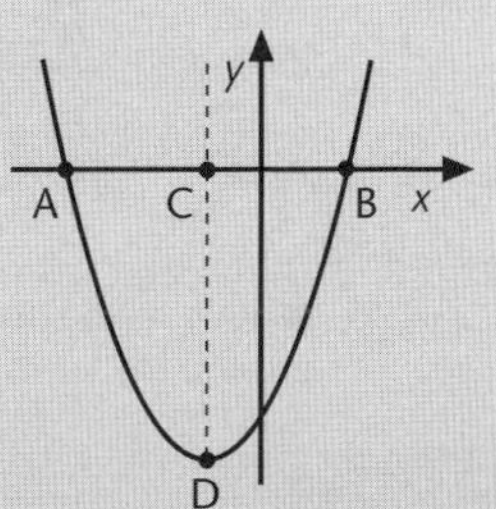

a. Write the coordinates of:

i. A **ii.** B **iii.** C **iv.** D

b. What is the least possible value of $x^2 + 2x - 8$?

c. Write down the equation of the parabola in the form $y = a(x - h)^2 + k$.

3. Find the equations of the following parabolas (*not* drawn to scale):

a.

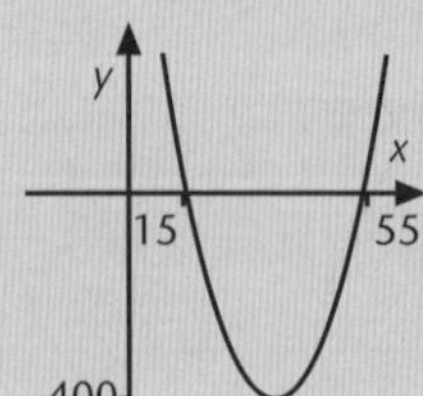

b.

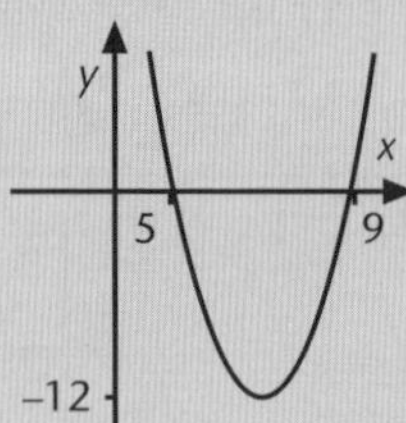

c.

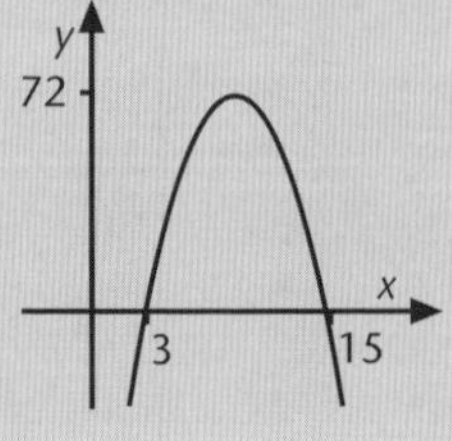

d.

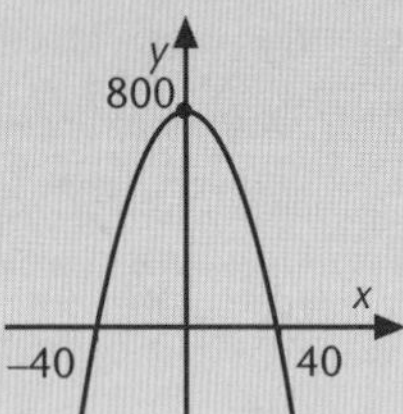

e.

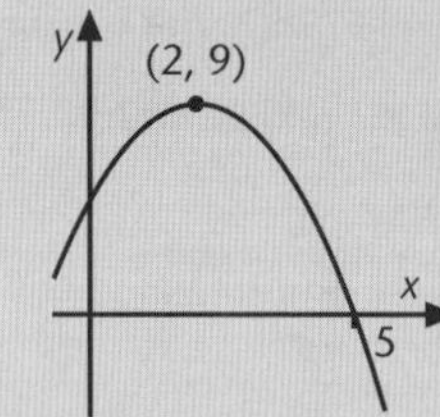

f.

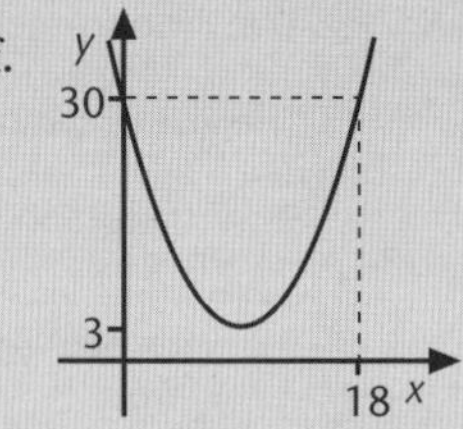

Using quadratic graphs to solve problems

Quadratic functions and their graphs can be used to model many different situations and to solve many problems in a practical context.

Example M

Q. Johnny's birthday present is a rocket launcher which projects a small toy in a parabolic arc.

The equation of the path of the toy is

$$h = \frac{(4 - d)(d + 2)}{5}$$

h (height)
d metres
Johnny
marker
landing point

where d is the distance in metres from a marker and h is the height in metres of the toy above the ground.

1. How far away from Johnny does the toy land?
2. What is the maximum height the toy reaches?
3. What is the height of the toy as it passes over the marker?

A. **1.** When the toy lands, $h = 0$ [height of toy above ground is 0 metres]

Solving $\frac{(4 - d)(d + 2)}{5} = 0$

gives $(4 - d)(d + 2) = 0$ [multiplying by 5]

$d = 4$ or -2

Relative to the marker, Johnny is at position –2 and the toy lands 6 m away at position 4.

h
–2
4
d

2. Toy reaches maximum height halfway between Johnny and the landing point, ie at

$d = \frac{-2 + 4}{2} = 1$

When $d = 1$ $\quad h = \frac{(4 - 1)(1 + 2)}{5}$ [substituting $d = 1$ in equation of curve]

$= 1.8$ m

The maximum height reached is 1.8 m.

3. As the toy passes over the marker, $d = 0$

$\therefore h = \frac{(4 - 0)(0 + 2)}{5} = 1.6$ m

The properties of the parabola can be used to find maximum or minimum values of quantities modelled by quadratic functions.

Example N

Q. Sue has 100 metres of fencing wire and wishes to build a rectangular garden. What are the dimensions of the garden that will give the rectangle the largest area?

A. Let the length of the rectangle be x and the width y.

x

y

Hence $x + y + x + y = 100$ [perimeter = 100 m]

$2x + 2y = 100$

$x + y = 50$ [dividing by 2]

$y = 50 - x$ [rearranging]

The area, A, of the garden is $A = xy$

$= x(50 - x)$ [substituting for y]

$= 50x - x^2$

From the table alongside, the graph of $A = 50x - x^2$ can be drawn.

x	0	5	10	15	20	25	30	35	40	45	50
A	0	225	400	525	600	625	600	525	400	225	0

The graph of $A = 50x - x^2$ shows that the maximum area is 625 m^2 when $x = 25$ m, and hence $y = 50 - x = 25$, ie a square garden of side 25 metres gives the maximum area.

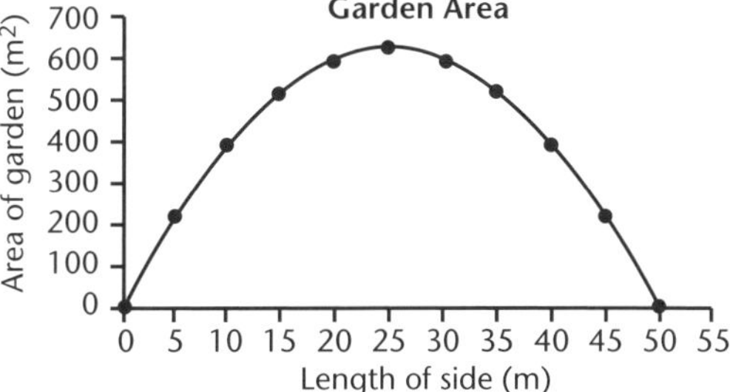

Note: The graph of $A = 50x - x^2$ can also be drawn by factorising first to give A = x(50 – x), then using factored form techniques (x-intercepts at (0, 0) and (50, 0), vertex at $x = 25$, $A = 625$ etc).

The equation of a curve may need to be found in order to solve a problem.

Example O

Q. A large trench is made so that its **cross-section** is in the shape of a parabola which is 3 m across. A pole inserted in the trench one metre out from one side of the trench shows a depth of 3 m. What is the depth of the trench at its deepest point?

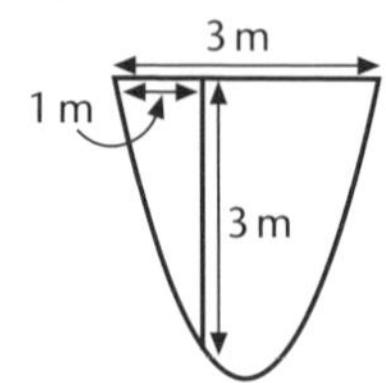

A. The equation of the parabola drawn on the axes shown is

$y = k(x + 1)(x - 2)$

[since $x = -1$ and $x = 2$ are the x-intercepts]

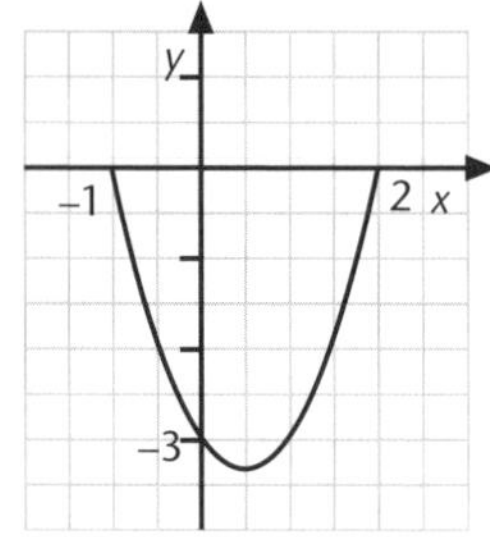

Since the point $(0, -3)$ lies on the curve, it follows that

$-3 = k(0 + 1)(0 - 2)$ [substituting $x = 0$, $y = -3$ in equation of curve]

$-3 = -2k$ [simplifying]

$k = \dfrac{-3}{-2}$ [dividing by –2]

$= 1.5$

The equation of the parabola is $y = 1.5(x + 1)(x - 2)$.

The trench is deepest halfway between the intercepts, ie at $x = \dfrac{-1 + 2}{2} = 0.5$

At $x = 0.5$, $y = 1.5(0.5 + 1)(0.5 - 2)$ [substituting $x = 0.5$ in equation]

$= -3.375$

The trench is 3.375 m deep at its deepest point.

Unit 11.1 Activity 13C: Using quadratic graphs to solve problems

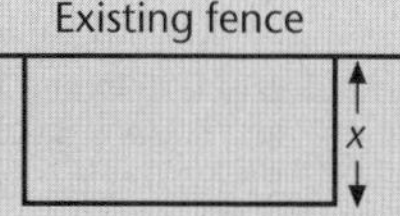

1. A villager has 24 m of fencing and builds a rectangular garden using an existing fence as one side. The area, A m^2, of the garden is given by $A = 24x - 2x^2$, where x is the width of the garden.

a. Copy and complete the following table:

x	0	1	2	3	4	5	6	7	8	9	10	11	12
A	0	22	40	54									

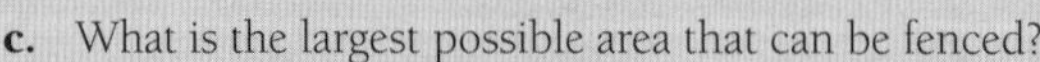

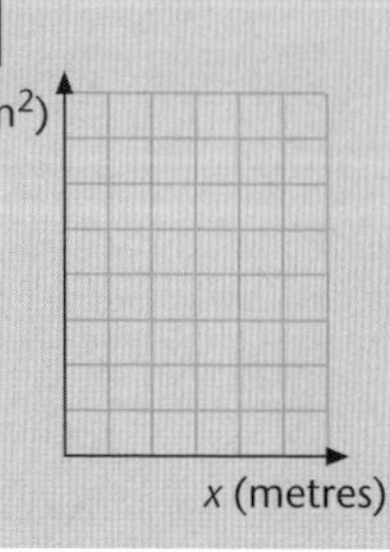

b. Draw the graph of $A = 24x - 2x^2$, using axes as shown.

c. What is the largest possible area that can be fenced?

d. Why is $x \geq 0$?

e. Use your graph to find the width of the garden when its area is 60 m^2.

2. Hosea finds that the depth of water in a small drain in his village is related to the distance across the drain by the quadratic equation

$d = (w - 10)(w - 70)$

where d is the depth, in millimetres, of the water in the drain and w is the distance, in centimetres, measured from a post on one of its banks.

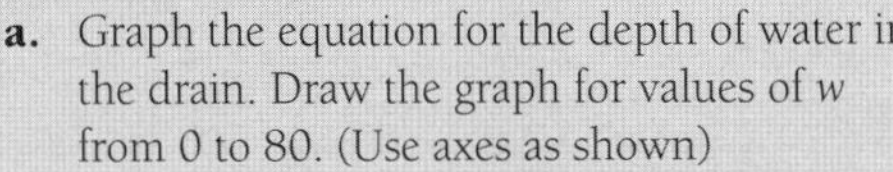

a. Graph the equation for the depth of water in the drain. Draw the graph for values of w from 0 to 80. (Use axes as shown)

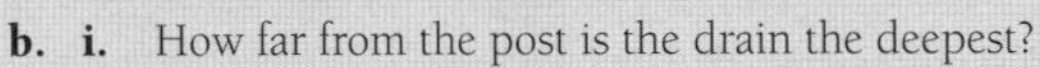

b. i. How far from the post is the drain the deepest?
ii. What is the depth at this distance?

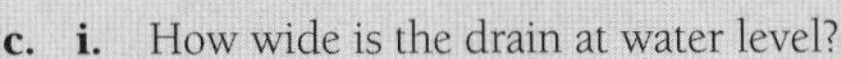

c. i. How wide is the drain at water level?
ii. How is this shown on the graph?

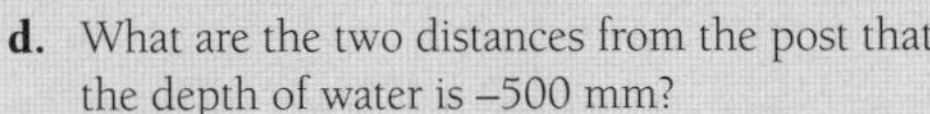

d. What are the two distances from the post that the depth of water is –500 mm?

e. For what values of w does the quadratic equation apply to the drain depth? Explain your answer.

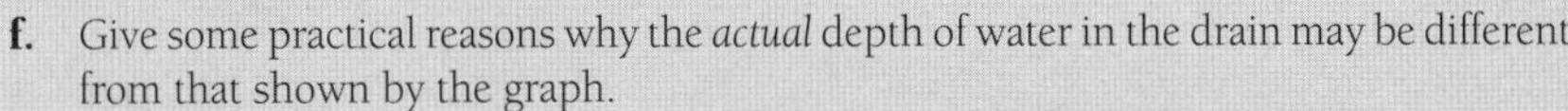

f. Give some practical reasons why the *actual* depth of water in the drain may be different from that shown by the graph.

3. In Gabriel's village there is an electrical wire suspended between two poles across a gully. The poles are on the edges of the gully and the wire is at the same height on both sides of the gully. The curve of the wire can be modelled by the equation

$h = (d - 6)^2 + 3$

where h is the height, in metres, of the wire above the floor of the gully and d is the distance, in metres, from the left-hand post.

a. Draw the graph of $h = (d - 6)^2 + 3$ for the values of distance, d, from 0 to 12.

b. What is the minimum height that the wire is above the floor of the gully?

c. i. How far apart are the two poles?
ii. How is this shown on the graph?

4. Judy has a piece of string hung across a wall in her room that has her birthday cards on it. The string takes on a shape which is approximately that of a parabola. The ends of the string are 2 m above the floor of each end, and the string is 1.5 metres above the floor at the lowest point. The ends of the string are 4 metres apart.

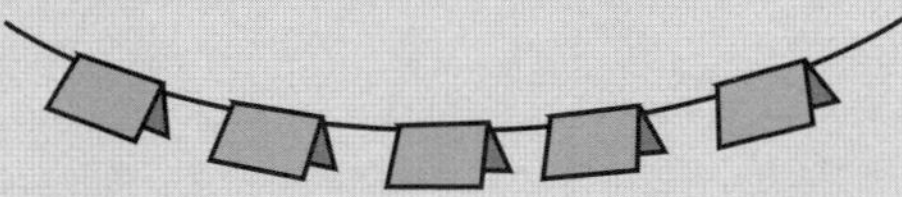

a. Find an equation that represents the shape of the string and draw its graph.

b. Judy is 1.65 m tall. She stands under the string and 1.2 m to the right of the lowest point of the string. Will her head touch the string?

5. Josie has a stack of yams built into a small pit near her village. She measures the height, h cm, from the floor of the pit, every metre from the left-hand end.

   Josie's data are shown in the table below. A graph of these data is shown

Distance d (m)	0	1	2	3	4	5	6	7	8	9	10
Height h (cm)	10	60	110	160	210	260	310	360	320	200	0

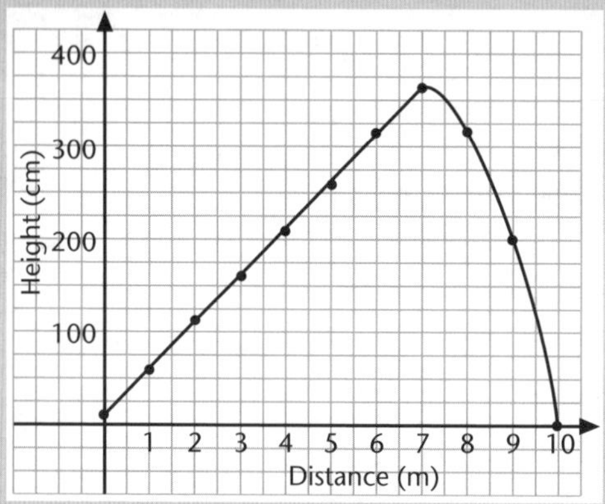

   a. Write the equation for h in terms of d that models the **first seven metres** of the stack.

   b. Write the equation for h in terms of d that models the **last three metres** of the stack. (Assume (7, 360) is the turning point of this quadratic graph).

6. The path of a bouncing ball was recorded by multiple flash photography.

   The flash operated every 0.1 seconds and the height of the ball from the floor was measured. The measurements of the path of the first bounce are shown in the table below.

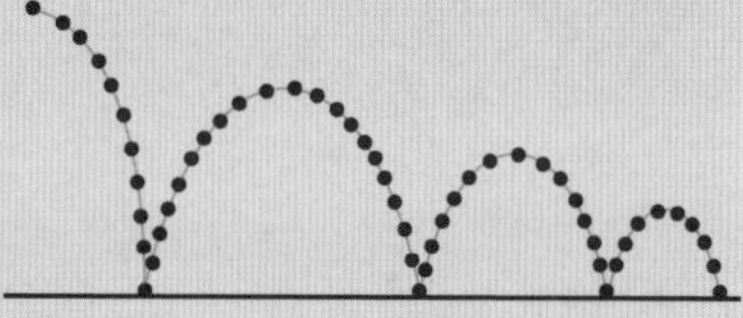

Time (secs)	1.0	1.1	1.2	1.3	1.4	1.5	1.6	1.7	1.8	1.9	2.0
Height (m)	0	0.45	0.80	1.05	1.20	1.25	1.20	1.05	0.80	0.45	0

   a. Graph the results, joining the points with a smooth curve.

   b. What was the maximum height reached by the ball on its first set of bounces?

   c. Write an equation for the curve.

   d. How high would the ball have been above the floor after 1.25 seconds?

7. A gutter for the roof of a shed is to have a rectangular cross-section and to be made from a long strip of sheet metal 30 cm wide. The sides are to be turned up at right angles to the base and the sides are to be of equal height.

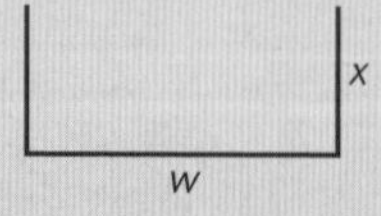

   For what value of x is the cross-sectional area the greatest?

   Hint: Express w in terms of x then sketch the graph of the quadratic function which models the area of the cross-section.

Unit 11.1 Number and Application
Topic 14: Basic algebra—simultaneous equations

In the Grade 11 syllabus (see Syllabus p.14), the basic algebra sub-section requires students to solve simultaneous equations using elimination, substitution and graphical methods. This topic covers:

- Solving pairs of simultaneous linear equations.
- Informal methods such as 'guess and check' for solving simultaneous equations.
- Solving problems involving modelling by forming and solving appropriate equations.

Introduction

Simultaneous equations are two or more equations which are true for the *same values* of their unknowns or variables. In this chapter, pairs of simultaneous equations with *two unknowns* are solved.

Solving simultaneous equations requires finding the value of each of the unknowns so that *both* equations are true.

- 'Guess and check' methods take a numerical approach, and may offer a quick solution if the answers are whole numbers.
- Algebraic techniques usually offer a more direct path to the answer.

The solutions to problems involving simultaneous equations should be checked by substituting the values in both equations. (This can often be carried out mentally.)

A graphical calculator, such as the *fx*-975, will find solutions to a pair of simultaneous **linear** equations immediately, providing the equations are entered in a a certain form.

Guess and check method

Some simultaneous equations can be solved using repeated substitutions, improving these estimates as you proceed.

Example A

Q. Two numbers a and b are four apart in size. If I triple the smaller number and double the bigger number, the sum of these new numbers is 73. This gives the following pair of simultaneous equations to solve:

$$b = a + 4$$
$$3a + 2b = 73$$

where a is the smaller number. Solve these equations to find the values of a and b.

A. A starting value of $a = 10$ is chosen.

Setting out estimates for a and b in a table:

a	$b = a + 4$	$3a + 2b$	Compare answer with 73
10	14	30 + 28 = 58	Numbers too small
16	20	48 + 40 = 88	Numbers too large
14	18	42 + 36 = 78	Numbers a little too large
13	17	39 + 34 = 73	Numbers correct

The numbers are 13 and 17.

Algebraic solutions to simultaneous equations

Two commonly used methods for solving simultaneous equations follow.

Substitution method

The **substitution method** is used when one of the unknowns is the **subject** of one (or both) equation(s). This expression is then substituted for the unknown in the other equation.

Example B

Q. Solve the following pairs of simultaneous equations using the substitution method.

1. $x + 2y = 8$ and $y = 3x - 10$ **2.** $x = y - 4$ and $2y - x = 6$

A. 1.

$x + 2y = 8$ ... (1)

$y = 3x - 10$ ... (2) [labelling the equations]

Substituting $3x - 10$ for y in (1) gives:

$x + 2(3x - 10) = 8$ [writing $3x - 10$ in place of y]

$x + 6x - 20 = 8$ [expanding]

$7x - 20 = 8$ [simplifying]

$7x = 28$ [adding 20]

$x = 4$ [dividing by 7]

Substituting $x = 4$ in (2) gives $y = 3 \times 4 - 10$

$= 2$

The solution is $x = 4$ and $y = 2$.

To check the answer $x = 4$ and $y = 2$, substitute into equation (1) to give: $4 + 2 \times 2 = 8$ which is true, so the solutions are correct.

2.

$x = y - 4$ ... (1)

$2y - x = 6$ ... (2) [labelling the equations]

Substituting $x = y - 4$ in (2) gives:

$2y - (y - 4) = 6$ [replacing x with $y - 4$]

$2y - y + 4 = 6$ [expanding bracket]

$y + 4 = 6$ [simplifying]

$y = 2$ [subtracting 4]

Substituting $y = 2$ in (1) gives $x = 2 - 4$

$= -2$

The solution is $x = -2$, $y = 2$

Elimination method

The **elimination method** is best used when both unknowns in both equations are on the same side of the equals sign. (This is the form required by graphical calculators.)

In the elimination method, the two equations are added together (or subtracted from each other), so that one of the two variables is eliminated. This leaves one equation to solve, in which there is only one unknown.

Example C

Q. Solve the simultaneous equations $2x + y = 8$ and $x - y = 1$ using the elimination method.

A. The equations $2x + y = 8$ and $x - y = 1$ have 'y' in one equation and '$-y$' in the other. If the equations are *added* together, the two y's will be eliminated, because $y + -y = 0$.

	$2x + y = 8$	... (1)	
	$x - y = 1$	... (2)	[labelling the equations]
Adding (1) and (2)	$3x + 0 = 9$	[eliminating the y's]	
	$x = 3$	[dividing by 3]	

Substituting $x = 3$ in (1) gives $2 \times 3 + y = 8$

$y = 2$ [subtracting 6]

The solution is $x = 3$ and $y = 2$.

Check: Substituting $x = 3$ and $y = 2$ in (2) gives $3 - 2 = 1$, which is true.

Sometimes one (or both) of the equations must be 'scaled up' first (ie multiplied through by a constant) so that addition or subtraction will eliminate one of the two variables.

Example D

Q. Solve the equations $2x + 3y = 14$ and $3x + y = 10.5$ using elimination.

A. If the equation $3x + y = 10.5$ is multiplied through by 3, an equation with the term $3y$ results. One equation can then be subtracted from the other to eliminate the term in y.

	$2x + 3y = 14$	... (1)	
	$3x + y = 10.5$	... (2)	[labelling the equations]
3 x (2) gives	$9x + 3y = 31.5$	... (3)	[labelling the new equation]
(3) – (1) gives	$7x = 17.5$		[subtracting, (3) – (1)]
	$x = 2.5$		[dividing by 7]

Substituting $x = 2.5$ in (1) gives:

$2 \times 2.5 + 3y = 14$

$3y = 9$ [subtracting 5]

$y = 3$ [dividing by 3]

The solution is (2.5, 3).

Check: Substituting $x = 2.5$ and $y = 3$ in (2) gives $7.5 + 3 = 10.5$, which is true.

Note: **1.** In the elimination method, the two equations to be subtracted or added together are best written one above the other (with variables lined up) before doing the addition or subtraction.

2. It is usually best to use the two *original* equations for finding the value of the second unknown and for the check.

The better **algebraic method** for solving simultaneous equations depends on the form in which the equations are given, as the following example shows. The substitution method is used in the first example, while the elimination method is used in the second example.

Example E

Q. Solve the simultaneous equations **1.** $y = x + 4$, $y = 2x - 1$ **2.** $4x + 3y = 1$, $5x - 2y = 30$

A. **1.** $y = x + 4$ ... (1)

$y = 2x - 1$ ... (2)

Substitute for y in (1) $2x - 1 = x + 4$

$x - 1 = 4$ [subtracting x]

$x = 5$

Substitute for x in (1) $y = 5 + 4 = 9$

Solution is (5, 9)

2. $4x + 3y = 1$ ... (1)

$5x - 2y = 30$ ... (2)

Multiply (1) by 2 $8x + 6y = 2$ ... (3)

Multiply (2) by 3 $15x - 6y = 90$ ... (4)

Add (3) and (4) $23x = 92$

$\therefore x = 4$ [dividing by 23]

Substitute $x = 4$ in (1) $16 + 3y = 1$

$3y = -15$

$y = -5$

Solution is $x = 4$, $y = -5$.

Graphical solution to simultaneous linear equations

The graphical solution to a pair of **linear equations** is *the point of intersection of the graphs of the two lines*. This is the point that satisfies both equations at the same time, or simultaneously.

Example F

Q. Find the simultaneous solution to the linear equations $y = 4x - 5$ and $y = -2x + 7$

A. The line $y = 4x - 5$ goes through the point (0, –5) and, substituting $x = 1$, the point (1, –1).

The line $y = -2x + 7$ goes through the point (0, 7) and, substituting $x = 1$, the point (1, 5).

Graphing both lines on a set of axes:

Accurate graphing of the linear equations reveals the simultaneous solution to be $x = 2$ and $y = 3$

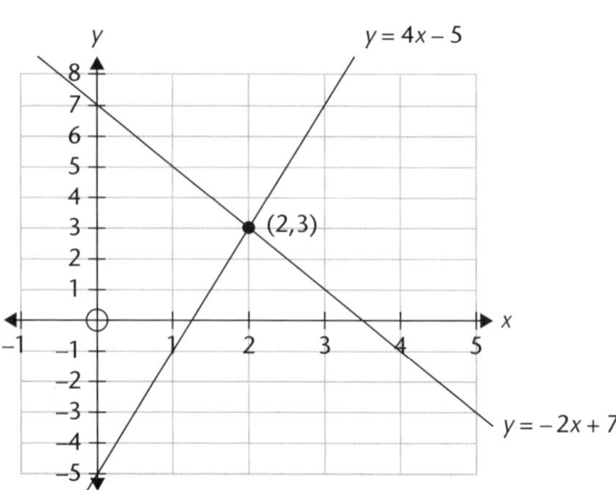

Unit 11.1 Activity 14A: Simultaneous equations

1. Solve the following pairs of simultaneous equations using 'guess and check' techniques.

a. $y = x + 5$, $y + 2x = 47$ **b.** $y = x + 4$, $3x - y = 18$ **c.** $y = 2x + 3$, $x + 3y = 58$ **d.** $y = 2x + 3$, $y - x = 15$

2. Solve the following pairs of simultaneous equations by elimination.

a. $2x + y = 11$, $x + y = 7$ **b.** $4x - 3y = 8$, $2x + 3y = 22$ **c.** $x + y = 8$, $3x + y = 14$ **d.** $2x + y = 11$, $x - 2y = 13$

e. $2x + y = 6$, $x - 2y = \frac{1}{2}$ **f.** $3x - 2y = 13$, $5x + y = 0$ **g.** $2x + 3y = 23$, $5x - 2y = 10$ **h.** $2x - 3y = 10$, $3x + 5y = 15$

3. Solve the following pairs of simultaneous equations by substitution.

a. $x + y = 7$, $y = 2x - 2$ **b.** $y = 2x + 3$, $x + 2y = 16$ **c.** $y = 2x - 5$, $3y - 2x = 5$ **d.** $y = x + 5$, $x - 2y = -7$

e. $2y + x = 14$, $x = y + 5$ **f.** $y = 2x + 3$, $2x + 3y = 5$ **g.** $y = x - 3$, $2y = 3x - 8$ **h.** $x = 9 - y$, $y - 2x = 12$

4. Solve the following pairs of simultaneous equations using any appropriate technique.

a. $3x - 4y = 6$, $2x + 3y = 4$ **b.** $3x - 4y = 4$, $y = x - 1$ **c.** $3x + 2y = 7$, $4x + 3y = 9$ **d.** $1.1x + 3.2y = 22$, $1.5x - 4y = -16$

5. On a set of axes, accurately graph each of the following pairs of linear equations and find the simultaneous solution.

a. $y = 1 - 2x$, $y = 10 + x$ **b.** $y = 3x$, $y = 8x - 6$
c. $y = 2x + 7$, $y + 4x = 16$ **d.** $c = 30n$, $c = 126 + 12n$

Solving problems using simultaneous equations

You may be required to form and solve a set of simultaneous equations in order to find the solution to a problem.

Example G

Q. A trip to a rugby game is arranged. The total number of children and adults going on the trip is 20. Each adult's ticket is K25 and each child's ticket is K15. The total price of the tickets is K370. How many adults and children went to the game?

A. Let a = number of adults and c = number of children.

$a + c = 20$ ... (1) [20 people go altogether]

Adults' tickets cost a total of $a \times 25 = 25a$ kina

Children's tickets cost a total of $c \times 15 = 15c$ kina

$25a + 15c = 370$ ... (2) [total cost is K370]

$15a + 15c = 300$ ... (3) [multiplying equation (1) by 15]

Subtracting (3) from (2) gives

$10a = 70$

$a = 7$ [dividing by 10]

Substituting a = 7 in (1) gives $7 + c = 20$ and $c = 13$.

So 7 adults and 13 children went to the rugby game.

Unit 11.1 Activity 14B: Word problems with simultaneous equations

1. Linus bought some takeaway burgers for his family. The hamburgers cost 80 toea less than the cheeseburgers. The total cost for three hamburgers and four cheeseburgers was K28.40. Solve the following pair of simultaneous equations to find the price of a cheeseburger.

$$h = c - 0.80$$
$$3h + 4c = 28.40$$

2. Greg spends ten hours each weekend at his two part-time jobs. His gardening job pays K8 per hour and his shop job pays K9 per hour. Altogether Greg earns K86.50 per weekend. Solve the following pair of simultaneous equations to find how long Greg spends gardening each weekend.

$$g + s = 10$$
$$8g + 9s = 86.50$$

3. A company employed two consultants, Artemius and Brown, for a total of 50 hours. Artemius charged K120 per hour and Brown K90 per hour. The company paid a total of K5 460 for the two consultants. Solve the following pair of simultaneous equations to find how many hours each consultant worked:

$$120\text{A} + 90\text{B} = 5\,460$$
$$\text{A} + \text{B} = 50$$

4. Vavine bought some vegetable plants for her garden. She bought beans at K2.95 a box and aibika at K1.45 a box. Vavine bought 3 more boxes of aibika than she bought boxes of beans, and she spent a total of K30.75.

Solve the simultaneous equations $295x + 145y = 3\,075$ and $y = x + 3$ to find the number of boxes of beans (x) and the number of boxes of aibika (y) that Vavine bought.

5. A shop offered a special on socks and handkerchiefs for Father's Day. For K50 you could buy 7 pairs of socks and 3 boxes of handkerchiefs or for K24 you could buy 3 pair of socks and 2 boxes of handkerchiefs.

a. Form a pair of simultaneous equations for this information.

b. Solve your equations to find the cost of a pair of socks.

6. Tom decides to sell flowers and vegetables at the market, so he makes two rectangular gardens. The flower garden is 3 metres longer than it is wide. The length of the vegetable garden is 4 metres more than double its width. The combined perimeters of the two gardens is 126 metres.

Tom decides to fence both gardens to keep out his pigs. The fencing for the flower garden costs K4 per metre and the fencing for the vegetable garden costs K5 per metre. The total fencing cost for the two gardens is K584.

Find the dimensions of each garden.

Unit 11.1 Number and Application

Topic 15: Basic algebra—inequalities

This Topic covers the last bullet point listed under 'Basic algebra' (see p. 14 of the Grade 11 syllabus):

- Solve inequality and plot on number line or plane.

Inequalities

Solving linear inequations

A linear **equation** has the form $mx + c = 0$ where m and c are constants.

A linear **inequation** has a similar form but the equals sign (=) is replaced by one of the inequality signs:

- $>$, greater than.
- $\geq$, greater than or equal to.
- $<$, less than.
- $\leq$, less than or equal to.

The process for solving linear inequations is the same as that for solving linear equations except for the following rule:

The inequality sign is reversed when multiplying or dividing an inequation by a negative number.

An inequality can be graphed on a number line or written in **interval notation**.

For example : $x \geq 3.5$.

On a number line:

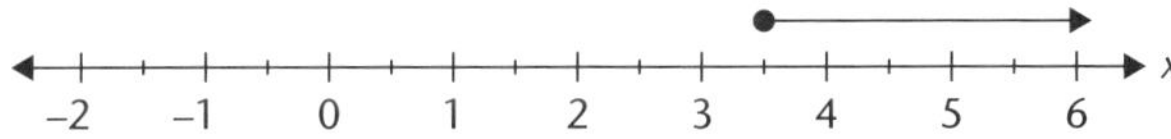

● means the value is included and ○ means the value is not included

This inequality can be expressed in **interval notation** as $[3.5, \infty)$ where the square bracket, [, means the value is included and the round bracket,), means the value is not included.

Note: Round brackets are always used with ∞ and $-\infty$.

Example A

Q. For each of the following inequations:

1. $3x + 2 \leq 14$
2. $4x - 3 > 6x - 7$

   i. Solve for x.

   ii. Show the solutions on a number line.

   iii. Write the answer in interval notation.

A. **1.** **i.** $3x + 2 \leq 14$: subtract 2 from both sides

$3x \leq 12$: divide both sides by 3

$x \leq 4$

ii. $x \leq 4$

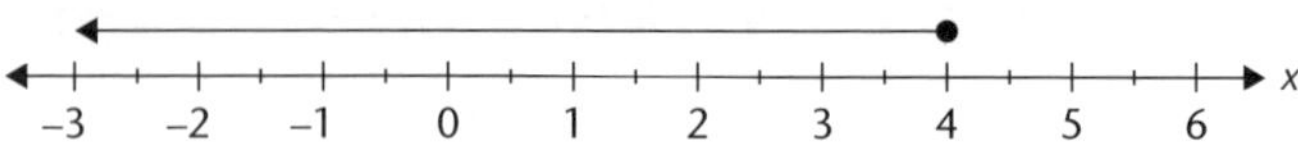

iii. $(-\infty, 4]$

2. i. $4x - 3 > 6x - 7$: subtract $6x$ from both sides

$-2x - 3 > -7$: add 3 to both sides

$-2x > -4$: divide both sides by -2; we need to reverse the inequality sign

$x < 2$

ii. $x < 2$

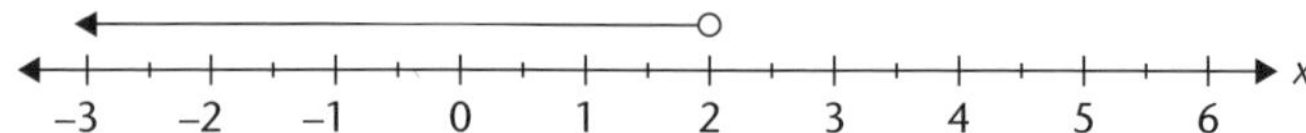

iii. $(-\infty, 2)$

Unit 11.1 Activity 15A: Inequations

1. For each of the following inequations:

i. Solve for x.

ii. Show the solutions on a number line.

iii. Write the answer in interval notation.

a. $5x \leq 10$ **b.** $-7 > 6x + 5$ **c.** $9 \geq 5 - x$

d. $\frac{x}{7} \geq \frac{3}{14}$ **e.** $\frac{x}{3} + 4 > 1.5$ **f.** $15 - 2x < 8$

2. Solve for x, giving your answer in interval notation.

a. $3x + 5 > 0$ **b.** $-3x \geq 9$ **c.** $\frac{x}{6} \leq \frac{7}{2}$ **d.** $5 - 6x \leq -7$

e. $\frac{2x+1}{2} \geq 5$ **f.** $\frac{x-3}{-2} < 7$ **g.** $\frac{4-x}{5} \geq -2$ **h.** $\frac{3-2x}{2} \leq -1$

3. Solve for a.

a. $3a - 2 > a - 5$ **b.** $2a - 3 > 5a - 7$ **c.** $5 - 2a \geq a + 4$

d. $7 - 3a \leq 5 - a$ **e.** $3(a-1) > a + 2$ **f.** $3(a + 2) > 4 - a$

g. $4(a + 1) < 3(2 + a)$ **h.** $3 + 2(a - 5) \leq 5 - 3(a + 1)$

4. Solve for b.

a. $5 + \frac{b+3}{2} > 1$ **b.** $3 - \frac{3-b}{4} \leq -1$ **c.** $\frac{b}{2} - \frac{b}{6} < 4$

d. $\frac{b+2}{3} + \frac{b-3}{4} \leq 1$ **e.** $\frac{1-b}{2} + \frac{b+2}{3} \leq 1$ **f.** $\frac{b+1}{3} - \frac{b}{6} \geq \frac{2b-3}{2}$

Regions

On a set of Cartesian **axes** (the **Cartesian plane**) a **linear equation** defines a set of points that are in a straight line. A **linear inequation** defines a **region** of the Cartesian plane.

Example B

Q. 1. For each of the following, graph the region defined by the inequation

a. $x \geq 2$ **b.** $x + y > 3$

c. $2x + 3y \geq 6$

2. Find the equation of the boundary line and hence state the inequality that defines the following shaded region.

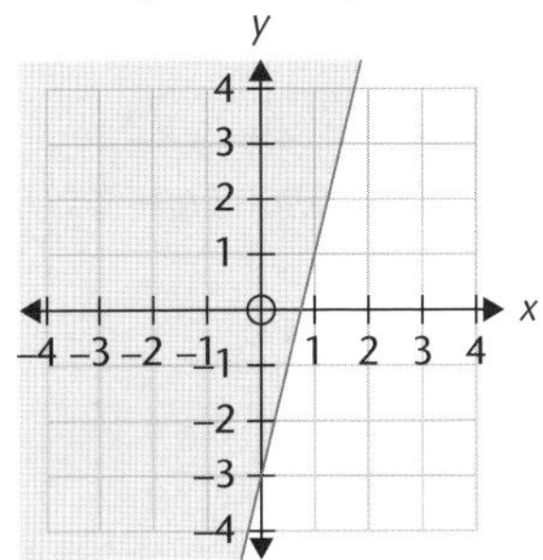

A. 1. a. We need to graph all the points that have an **x-coordinate** that is **greater than or equal to 2**.

The points that have an x-coordinate **equal to 2** will be the points on the straight line $x = 2$ and this line forms a boundary for the region.

The points that have an x-coordinate greater than 2 will be all the points to the right of the line $x = 2$, shaded opposite.

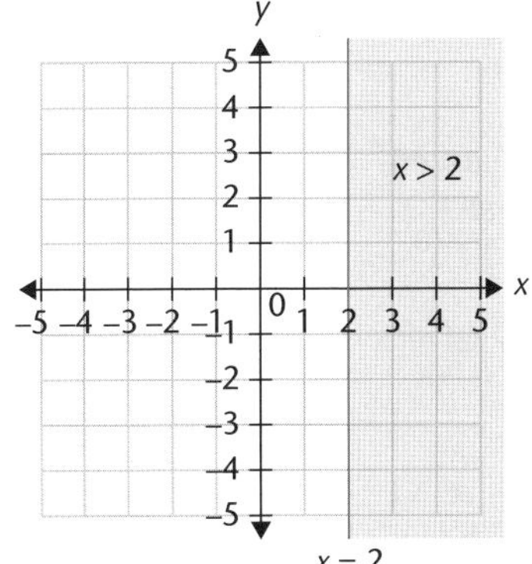

The region defined by $x \geq 2$ includes all points on the line $x = 2$ and the shaded region $x > 2$.

b. Two points on the line $x + y = 3$ are (0, 3) and (3, 0).

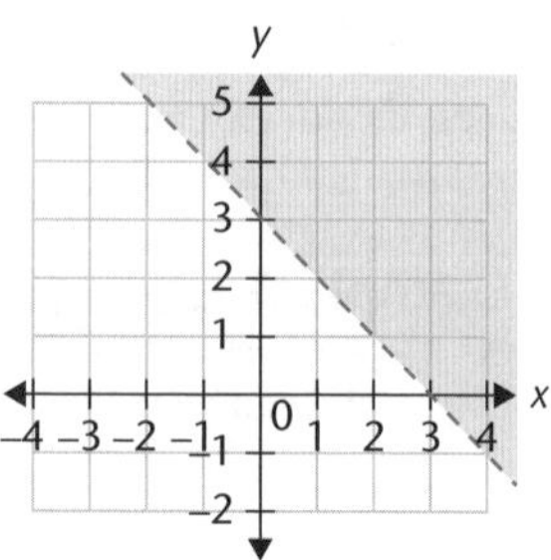

The straight line $x + y = 3$ can be drawn through these two points.

For the region $x + y > 3$ the points on the line $x + y = 3$ are **not** included so we draw this line as a dotted line.

Testing that the **origin** is in the region: substituting (0, 0) into $x + y > 3$, gives $0 + 0 > 3$ which is **not** true. The region is the side of the line not containing the origin.

c. To graph the inequality $2x + 3y \le 6$ we first graph the equality part as the boundary; the straight line $2x + 3y = 6$.

The line $2x + 3y = 6$ has intercepts with the axes $x = 3$ and $y = 2$.

One side of the line $2x + 3y = 6$ will contain all the points where $2x + 3y < 6$ and the other .

$2x + 3y > 6$.

To find which side of this line is the 'less than' side we choose a point that is not on the line and test the inequation with the x and y values.

For example: Choosing (1, 4) as the point.

Testing: substituting in $2x + 3y$ gives $2 \times 1 + 3 \times 4 = 14$. Is this less than 6?

Note. So the point (1, 4) is not on the side of the line that includes all the points $2x + 3y < 6$.

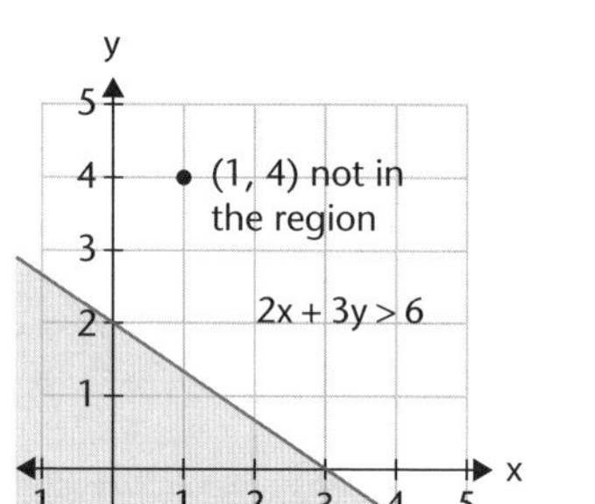

The side required will be the other side of the line.

The region $2x + 3y \le 6$ is the shaded region and the line.

Note: If the boundary line of an inequation does not go through the origin then it is usually convenient to choose the point (0, 0) as the testing point.

2. Looking at the boundary line of the shaded region we can see that it goes through the points (0, –3) and (1, 1). This means that the

line has a gradient of $\dfrac{1-(-3)}{1-0} = \dfrac{4}{1} = 4$ and

the y-intercept is –3 so the equation of the line is $y = 4x - 3$.

The **origin** (0, 0) is in the region so testing $y \le 4x - 3$ gives $0 \le 4 \times 0 - 3$ or $0 \le -3$ which is **not true**. So the region must be $y \ge 4x - 3$.

Unit 11.1 Activity 15B: Regions

1. Sketch the region defined by each of the following inequalities:

a. $x \geq 0$ **b.** $y < -2$

c. $x + y \leq 5$ **d.** $2x + y \geq 6$

e. $4x - 3y \leq 24$ **f.** $y \geq x$

g. $y \leq 5x - 3$

2. Find the equation of the boundary line and hence state the inequality that defines each of the following shaded regions:

a.

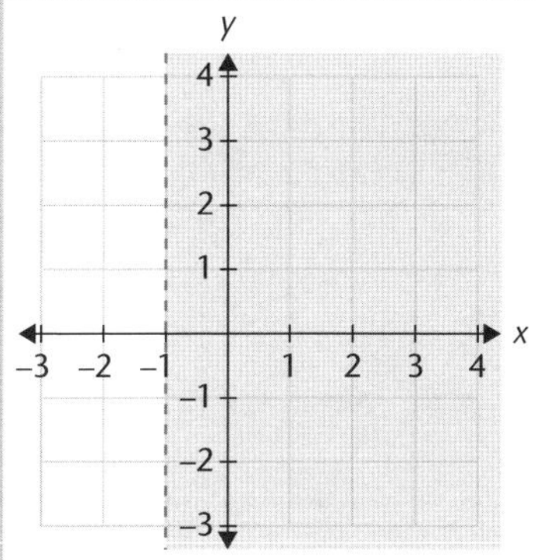

b.

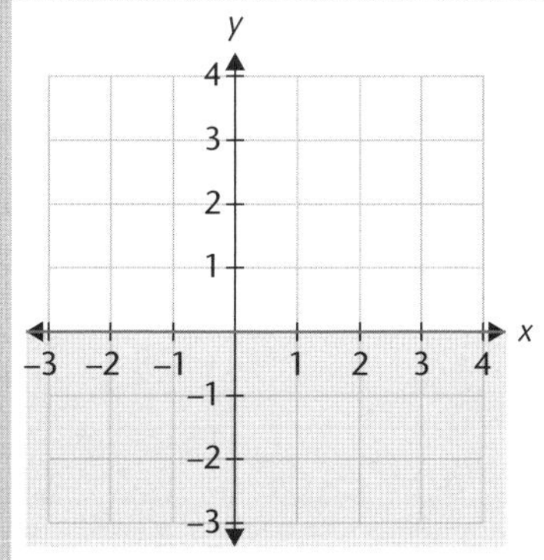

c.

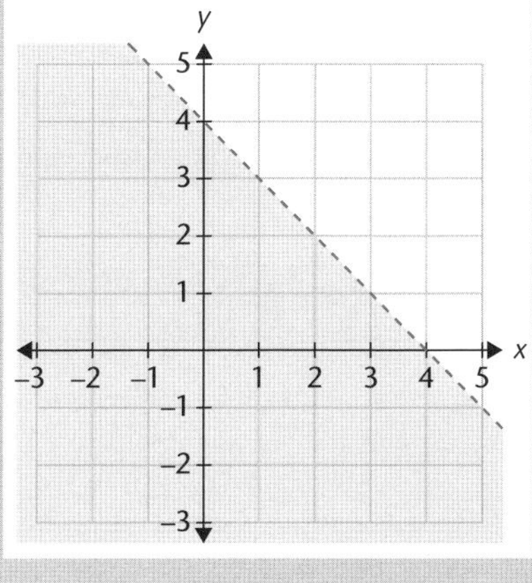

d.

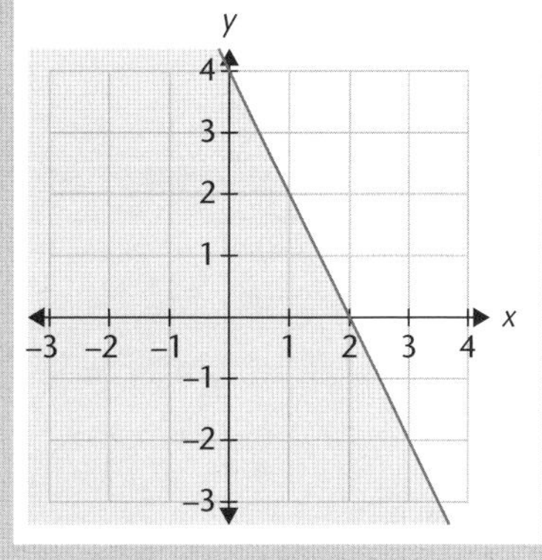

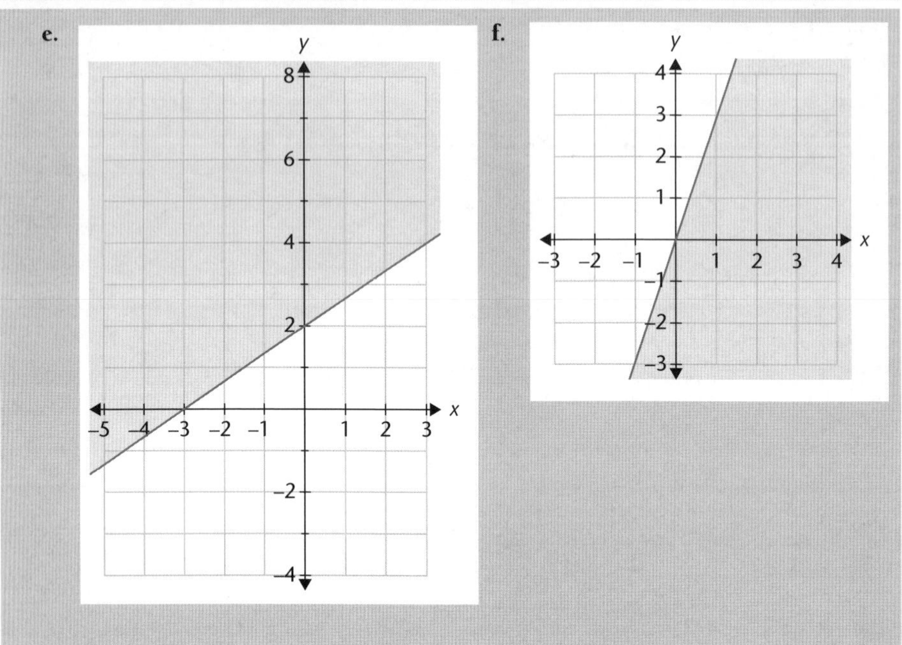
e.
y
8
6
4
2
x
–5 –4 –3 –2 –1 1 2 3
–2
–4
f.
y
4
3
2
1
x
–3 –2 –1 1 2 3 4
–1
–2
–3

Unit 11.2 Graphs and Functions

Topic 1: Algebraic expressions

This Topic and the next aim to cover the content listed under the heading 'Algebraic expressions' on p. 15 of the Syllabus. Topic 1 covers:

- Forming algebraic expressions from worded problems.
- Simplifying and expanding algebraic expressions.
- Factorising.
- The difference of two squares.
- Simplifying quotients.

Forming algebraic expressions from worded problems

Problems given in words (**word problems**) need to be changed to mathematical equations to be solved. This process is called **mathematical modelling**. The general approach is illustrated by the following example.

Example A

Shane is taller than Jason by 2.4 cm. Jason is taller than Ian by 1.3 cm. The three heights total 452 cm. What are their heights?

Solution

Let Ian's height be h

$\therefore$ Jason's height is $h + 1.3$ [Jason is 1.3 cm taller than Ian]

Shane's height is $h + 1.3 + 2.4 = h + 3.7$ [Shane is 2.4 cm taller than Jason]

Since their combined height is 452 cm

$h + h + 1.3 + h + 3.7 = 452$

$\therefore 3h + 5 = 452$ [simplifying]

$\therefore h = 149$ [subtracting 5 and dividing by 3]

Thus Ian is 149 cm tall; Jason's height is 149 + 1.3 = 150.3 cm and Shane's height is 149 + 1.3 + 2.4 = 152.7 cm.

Unit 11.2 Activity 1A: Solving word problems

For each of the following, find: **a.** an equation; **b.** a solution using your equation or any other method.

1. Notebooks cost K1 more than pens. Nadia purchases 5 pens and 6 notebooks. Her total expenditure is K32.40. How much does each pen cost?

2. Tom has some money. Arona and Dave each have K1 more than Tom, and Paul, Dan and Luke each have twice as much as Tom. Together, the six boys have K15.50. How much does each have?

3. A man has three children who each want a calculator for Christmas. He buys the two older children the same model and the youngest child gets a basic model costing K20 less. What is the value of each type of calculator, if the man spends a total of K86.50 on the calculators?

4. The perimeter of a room is 50 m. The length exceeds the width by 3.3 m. Find the length of the room.

5. Four consecutive natural numbers have a sum of 110. What is the smallest of these numbers?

6. A girl purchases a number of items each costing K17.35. Her bank account was K1000.00 before she made the purchase. Afterwards it was K687.70. How many items did she purchase?

7. Tina and Jane purchase raffle tickets and agree to share the prize money in the ratio of how much each paid. They win K117.00 and Jane, who put in K2.00, gets K26.00. What was Tina's share of the cost of the raffle tickets?

8. The distance from B to C is $\frac{7}{11}$ of the distance from A to B. The distance from A to B exceeds that from B to C by 20 km. How far is it from A to B?

9. A man purchases a number of tools at K5 each. Unfortunately, five do not work, but by selling those that do at K12 each he is able to make K45 profit. How many tools did the man purchase initially?

10. All prices are to rise by 15% as a result of a new tax. A shopkeeper had increased the cost of a particular item by K5.00 before the new tax came into effect. As a result of the second price rise, this item now costs K13.80. What did it cost originally?

11. Inflation is running at a certain percentage. After one year, an item which cost \$36.00 costs K40.86. What is the inflation rate?

12. The denominator of a fraction is 44 more than the numerator. The fraction, when simplified, is $\frac{13}{17}$. Find the numerator.

13. My son is 31 years younger than I am. In one year's time my son's age will be one quarter of my current age. How old am I?

14. One girl has four times as much money as her friend. She gives her friend K12.00, and as a result they now have the same amount of money. How much did each have originally?

15. A husband and his wife decide to invest in a business. The first time they do this, he invests K3 000.00 and she invests K5 000.00. The next time they invest, they each contribute the same amount. After the two investments are made, the ratio of the husband's investment to his wife's is 13:17. How much did each invest the second time?

Expanding

Brackets are removed from and inserted into algebraic expressions using the **distributive law**. The distributive law states that if a, b and c are numbers, then:

$$\begin{aligned} a(b + c) &= a \times b + a \times c \\ &= ab + ac \end{aligned}$$

Expanding an expression means removing brackets by multiplying, using the distributive law. Where possible, the expression is then simplified.

Example B

1. $5x(3 - 2x) = 5x \times 3 - 5x \times 2x$ [expanding by using the distributive law]

$= 15x - 10x^2$

2. $2x^2(3x^3 + 2x) = 2x^2 . 3x^3 + 2x^2 . 2x$ [using the distributive law]

$= 6x^5 + 4x^3$

3. $(x - 4)(2x + 3)(4x + 5) = (x - 4)(8x^2 + 10x + 12x + 15)$ [expanding second pair of brackets]

$= (x - 4)(8x^2 + 22x + 15)$ [simplifying]

$= 8x^3 + 22x^2 + 15x - 32x^2 - 88x - 60$ [multiplying each term in the second bracket by x then by -4]

$= 8x^3 - 10x^2 - 73x - 60$ [simplifying]

Note: $(2x + 3)$ $(4x + 5)$ can be expanded quickly by multiplying every possible pair from the brackets (indicated by arrows) then simplifying.

$$\begin{aligned} (2x + 3)(4x + 5) &= 8x^2 + 10x + 12x + 15 \\ &= 8x^2 + 22x + 15 \end{aligned}$$

4. $(2x - 1)(x + 3)^2 = (2x - 1)(x + 3)(x + 3)$ [$(x + 3)^2$ means $(x + 3)(x + 3)$]

$= (2x - 1)(x^2 + 3x + 3x + 9)$ [expanding $(x + 3)(x + 3)$]

$= (2x + (-1))(x^2 + 6x + 9)$ [simplifying, writing $(2x - 1)$ as $(2x + (-1))$]

$= 2x^3 + 12x^2 + 18x + (-x^2) + (-6x) + (-9)$ [expanding brackets]

$= 2x^3 + 11x^2 + 12x - 9$ [simplifying]

It is useful to be able to expand squares involving brackets using a formula (rather than writing the expansion out in full, as in Example A. **4.** above).

$$(A \pm B)^2 = A^2 \pm 2AB + B^2$$

These formulae can be verified by expanding (A + B)(A + B) and (A – B)(A – B).

Example C

1. $(x + 5)^2 = x^2 + 2 \times 5 \times x + 5^2$ [using the formula $(A + B)^2 = A^2 + 2AB + B^2$]

   $= x^2 + 10x + 25$

2. $(7x - 2y)^2 = (7x)^2 - 2 \times 7x \times 2y + (2y)^2$ [using the formula $(A - B)^2 = A^2 - 2AB + B^2$]

   $= 49x^2 - 28xy + 4y^2$

Unit 11.2 Activity 1B: Expanding

Expand and simplify these expressions:

1. $5(3x + 1)$ **2.** $5x(2x + 3)$ **3.** $(x + y)(2x + y)$
4. $(2x + 1)(3x + 4)$ **5.** $(3A - 2)(4A + 5)$ **6.** $(2x + 7)(x - 3)$
7. $(6P - 8)(2Q + 3)$ **8.** $(2A + B)(3A - B)$ **9.** $(x - y)(3x - y + 2)$
10. $(x + 2y)(3x + 2y - 3)$ **11.** $(2x - y + 4)(x - y + 2)$ **12.** $(x + 1)^2$
13. $(3x - 4y)^2$ **14.** $(P + R - 2)^2$ **15.** $(A + B)^3$
16. $(x + 2)^2(x - 2)^2$ **17.** $((x + 1) - (x - 1))^3$ **18.** $(3x + 4)(x - 5)^2$
19. $(3A^2 + 2BA + B^2)(4A^2 - 3AB + 5B^2)$ **20.** $(x^2 - 2x + 3)^2$
21. $(2x + 3y)^3$ **22.** $(1 - x)(2 - x)(3 - x)$ **23.** $(2x - 5)(x - 6)(3x + 4)$
24. $6x(3x + 5)(8 - x)$ **25.** $-2x(5x - 4)(3x - 2)$

Example D

Expand and simplify $4(2x + y - 2) - 3(x - y - 3)$.

Solution

$4(2x + y - 2) - 3(x - y - 3) = 8x + 4y - 8 - 3(x - y - 3)$ [expanding first bracket]

$= 8x + 4y - 8 - 3x + 3y + 9$ [expanding second bracket]

$= 5x + 7y + 1$ [simplifying]

Unit 11.2 Activity 1C: Expanding and simplifying

Expand and simplify the following expressions as much as possible:

1. $5(x + 1) + 2(x + 3)$ **2.** $4(A + B) + 3(A - B)$
3. $6(x + y) - 3y$ **4.** $6(x + 2y + 3) + 2(x - 2y - 2)$
5. $7(2x + 3y) - 2(x + y)$ **6.** $5(A - 3B) - 2(B + A)$
7. $x^2 + 3x + 2(x - 1)$ **8.** $x(2x + 3) - 3(x + 4)$
9. $3x^2 + 4x + 5 - x(x - 2)$ **10.** $5x - (3x - 7)$
11. $7(x + 1)^2 - (2x + 3)^2(2x^2 - 3x + 2)$ **12.** $(2A + 1)^3 - 2(A^3 - 2A^2 + 3A + 1)$
13. $(2A + 3)^3 - (2A - 1)^3$ **14.** $x^2(x + 1) - (x - 1)^3$

Factorising

A **factor** of a term is a number or expression which divides that term without remainder. **Factorising** an algebraic expression means writing the expression as a product of factors. These factors will be simpler algebraic expressions or numbers. To factorise, first look for **common factors** (factors shared by all terms in the expression).

Example E

1. $6x - 12$ can be written $6 \times x - 6 \times 2$ [6 is a common factor]

   Thus $6x - 12$ can be factorised to give $6(x - 2)$.

2. $2A^2 + 7A = A \times 2A + A \times 7$ [A is a common factor]

   $= A(2A + 7)$

3. $5L^3m^2n^2 - 10L^2m^3n^2 + 15L^2m^2n^3 = 5L^2m^2n^2(L - 2m + 3n)$ [$5L^2m^2n^2$ is a common factor]

When an expression has no common factor, first try factorising groups of terms separately.

Example F

Factorise $AP - AQ + 2Q - 2P$

Solution

$AP - AQ + 2Q - 2P = A(P - Q) + 2(Q - P)$ [factorising $AP - AQ$ and $2Q - 2P$ separately]

$= A(P - Q) - 2(P - Q)$ [$-(P - Q) = Q - P$]

$= (A - 2)(P - Q)$ [$(P - Q)$ is a common factor]

Factorising quadratics

An algebraic expression of the form $ax^2 + bx + c$, where a, b and c are numbers, is called a **quadratic expression** or just a **quadratic**.

In the simplest case, when the coefficient of x^2 is 1, the quadratic simplifies to $x^2 + bx + c$. Such expressions are factorised by finding two numbers which *multiply together to give c and add together to give b*. Begin by considering the factor pairs of c.

Example G

1. To factorise $x^2 + 7x + 12$, first consider the factor pairs of 12. These are 1×12, 2×6 and 3×4. The required numbers are 3 and 4, since $3 + 4 = 7$. Thus

   $x^2 + 7x + 12 = (x + 4)(x + 3)$.

2. $A^2 - 7A - 18 = (A - 9)(A + 2)$ [since $-9 \times 2 = -18$ and $-9 + 2 = -7$]

If a is not 1, the quadratic to be factorised is $ax^2 + bx + c$ (where $a \neq 1$). This can be factorised by a trial-and-error technique, in which different pairs of factors of a and c are tried.

There is, however, a more systematic method which is illustrated in the following examples.

Example H

Factorise $4B^2 + 7B - 15$

Solution

Step 1: Multiply the **coefficient** of B^2 (which is 4) and the constant term (–15) together:

$4 \times (-15) = -60$

Step 2: Now look for a pair of factors of –60 whose sum is equal to the coefficient of B (which is 7). The numbers are –5 and 12.

Using these two numbers, the middle term is expressed as the sum of two terms.

$4B^2 + 7B - 15 = 4B^2 - 5B + 12B - 15$ [since $-5B + 12B = 7B$]

$= B(4B - 5) + 3(4B - 5)$ [factorising pairs of terms]

$= (B + 3)(4B - 5)$ [since $(4B - 5)$ is a common factor]

Example I

Factorise $6x^2 - 13x - 28$

Solution

Step 1: Find the product of the coefficient of x^2 and the constant term:
$6 \times (-28) = -168$

Step 2: Express the term in x as the sum of two terms (whose coefficients have this product).

Find two numbers whose product is –168 and whose sum is –13.

By considering factor pairs of –168, these numbers are –21 and 8.
[since $-21 \times 8 = -168$ and $-21 + 8 = -13$]

The factorisation can now be carried out as follows:
$6x^2 - 13x - 28 = 6x^2 - 21x + 8x - 28$ [since $-21x + 8x = -13x$]

$= 3x(2x - 7) + 4(2x - 7)$ [factorising pairs of terms]

$= (3x + 4)(2x - 7)$ is the required factorisation.

The difference of two squares

The **difference of two squares**, $x^2 - y^2$, has a particularly important factorisation which is shown below. (This formula can be confirmed by expanding the right-hand side.)

$$x^2 - y^2 = (x - y)(x + y)$$

Expressions which can be written as the difference of two squares are easily factorised using this formula.

Example J

1. $49x^2 - 36B^2 = (7x)^2 - (6B)^2$ [writing $49x^2$ and $36B^2$ as squares]

$= (7x - 6B)(7x + 6B)$ [factorising using formula]

2. $128 - 18a^4 = 2(64 - 9a^4)$ [2 is a common factor]

$= 2((8)^2 - (3a^2)^2)$ [writing bracket as difference of squares]

$= 2(8 + 3a^2)(8 - 3a^2)$ [factorising using formula]

Unit 11.2 Activity 1D: Factorising

Factorise the following (using brackets):

1. $ax - ab$
2. $ab - a^2$
3. $11x^2 - x$
4. $4ax + 8ay$
5. $x^2 + 7x + 12$
6. $x^2 + 13x + 36$
7. $P^2 + 12P + 27$
8. $p^2 - 3pq + 2q^2$
9. $R^2 - 6RP + 8P^2$
10. $x^2 - xy - 42y^2$
11. $p^2 + 4pq - 5q^2$
12. $p^2 - 3pq - 10q^2$
13. $R^2 + 10RS + 24S^2$
14. $x^2 - 49$
15. $P^2 - 36P$
16. $L^2 - 5L + 6$
17. $M^2 + 16M + 60$
18. $M^2 - 7M - 8$
19. $25M^2 - 100$
20. $2x^2 + 3x + 1$
21. $x^2 - 6xb + 5b^2$
22. $2x^2 + 7x + 3$
23. $4P^2 + 21P + 5$
24. $3z^2 + 4z - 7$
25. $4P^2 + 5P - 21$
26. $12x^2 - 11x + 2$
27. $L^2P^2 - 2LP - 24$
28. $14 + 11x - 15x^2$
29. $7p^2 + 6pq - q^2$
30. $8L^2 - 32x^2$
31. $27Px^2 - 3Py^2$
32. $50 - 2x^2$
33. $3AP^2 - 4AP + A$
34. $9x^2 + x^7$
35. $18xy - 21x^2 - 14x$
36. $8p^3q^3 + 16p^4q^3 + 24p^5q^4$
37. $p^6q^4r^4 - 9x^{10}y^{12}$
38. $x^5 - b^2x^3 + b^3x^2 - b^5$
39. $A(C - D) + B(D - C)$
40. $x^2 + 2x + 1 + ax + a$

Check by 1

Often a student would like to know whether their expansion or factorisation is correct. A very useful technique is to substitute 1 into both the original expression and the final answer. If the results are different, the expansion (or factorisation) is incorrect.

Example K

A student expands $(x^2 + 1)(2x^2 + x + 1) - (3 - 2x + x^2 - 2x^3)$ and gets an answer $2x^4 + x^3 + 2x^2 - x + 2$. Is this correct?

Solution

Substituting $x = 1$ into the original expression gives

$$(1 + 1)(2 + 1 + 1) - (3 - 2 + 1 - 2) = 2 \times 4 - 0$$
$$= 8$$

Substituting $x = 1$ into the student's answer gives,

$$2 + 1 + 2 - 1 + 2 = 6$$

Since the results are different, the student's expression is incorrect.

Note 1: If the results of the substitutions are the same this does not guarantee that the expansion is correct (although it is likely to be).
For example, the correct expansion in the example is, $2x^4 + 3x^3 + 2x^2 + 3x - 2$.
Substituting $x = 1$ gives $2 + 3 + 2 + 3 - 2 = 8$ as expected.
However, substituting $x = 1$ into an incorrect expansion, such as
$3x^4 + 2x^3 + 3x^2 + 2x - 2$ also gives $3 + 2 + 3 + 2 - 2 = 8$, even though it is wrong.

Note 2: Substitution by 1 is particularly easy to evaluate, hence its use.
Substitution by other numbers, such as 2 or –3, could also be used but the resulting calculations would take longer and be more error-prone.

Simplifying quotients

Quotients are expressions which have a *denominator* and a *numerator*. A quotient can be simplified by factorising the numerator and denominator, and cancelling any common factors.

Example L

$$\frac{x^2-8x+12}{x^2-36}=\frac{(x-6)(x-2)}{(x-6)(x+6)}$$ [factorising numerator and denominator]

$$=\frac{x-2}{x+6}$$ [cancelling the common factor, $(x-6)$]

Note: Cancelling can only be done if the numerator and denominator are both expressed as products which have common factors.

Unit 11.2 Activity 1E: Simplifying quotients

Simplify the following expressions as much as possible:

1. $\frac{x^2-4x-5}{(x+1)^2}$ **2.** $\frac{x^2+4x+4}{x^2+5x+6}$ **3.** $\frac{x^2+5x}{x^2+6x+5}$

4. $\frac{x^2-49}{x^2-5x-14}$ **5.** $\frac{(x+1)^2}{(x+1)^3}$ **6.** $\frac{x^2+9x+20}{x^2-16}$

7. $\frac{p^2+11p+28}{p^2+7p+12}$ **8.** $\frac{5p+15}{p^2+4p+3}$ **9.** $\frac{a^2-ab}{ab}$

10. $\frac{3y^2-27}{12y^2+36y}$ **11.** $\frac{2y^2+3y+1}{y^2+2y+1}$ **12.** $\frac{a^2-(b-c)^2}{b^2-(a-c)^2}$

13. $\frac{a^3+a^2}{(a+1)^2}$ **14.** $\frac{(x-y)^3}{x^2z-2xyz+y^2z}$ **15.** $\frac{x^2+ax+bx+ab}{x^2+2ax+a^2}$

16. A student factorises $6x^3+29x^2-7x-10$ and gets $(x+5)(2x+1)(3x-2)$. Use substitution of $x=1$ to comment on whether or not this factorisation is correct.

17. A student expands $(x+2y)^3+(x-y)^3$ and gets $2x^3+6x^2y+6xy^2+3y^3$. Use substitution of $x=1$ and $y=1$ to see whether this expansion is likely to be correct.

Unit 11.2 Graphs and Functions
Topic 2: Simplifying rational expressions

Topic 2 continues the coverage of the content listed under the heading 'Algebraic Expressions' on p. 15 of the Syllabus. It covers:

- Simplifying rational expressions.
- Factorising expressions using quadratics.

Rational expressions are sometimes called **algebraic fractions**. They are fractions involving variables (letters) as well as numbers. The procedures used to simplify rational expressions are similar to those used with fractions involving only numbers.

Multiplication

Numerators are multiplied together and **denominators** are multiplied together.

Example A

Express $\frac{x}{y} \times \frac{a}{c}$ as a single fraction.

Solution

$\frac{x}{y} \times \frac{a}{c} = \frac{xa}{yc}$ [numerators then denominators are multiplied together]

Answers are simplified by cancelling any common factors

Example B

Simplify $\frac{x}{y} \times \frac{a}{c} \times \frac{z^2}{x^3}$.

Solution

$$\frac{x}{y} \times \frac{y}{z} \times \frac{z^2}{x^3} = \frac{xyz^2}{x^3yz}$$

$$= \frac{x.y.z.z}{x.x^2.y.z}$$ [writing $z^2 = z \,.\, z$, $x^3 = x \,.\, x^2$]

$$= \frac{z}{x^2}$$ [cancelling common factors x, y and z]

Some expressions are best factorised *before* any multiplication is done, so that common factors can be identified.

Example C

$$\frac{x^2 + 3x + 2}{(x + 2)^2} \times \frac{5}{x + 1} = \frac{(x+1)(x+2)}{(x+2)(x+2)} \times \frac{5}{(x + 1)}$$ [factorising]

$$= \frac{5}{x + 2}$$ [cancelling $(x + 1)$ and $(x + 2)$]

Unit 11.2 Activity 2A: Multiplication of rational expressions

1. Simplify each of the following:

a. $\frac{3}{4} \times \frac{a}{b}$ **b.** $\frac{5a}{b} \times \frac{c}{d}$ **c.** $\left(\frac{3}{4} \times \frac{a}{d}\right) \times \frac{c}{b}$

d. $\frac{a}{b} \times \frac{a}{c}$ **e.** $\frac{3}{a} \times \frac{ab}{6}$ **f.** $\frac{pq}{r} \times \frac{3r}{q}$

g. $\frac{5a^2}{3} \times \frac{9}{a}$ **h.** $\left(\frac{5}{a} \times \frac{a^2}{10}\right) \times \frac{1}{3}$ **i.** $\frac{6k}{3q} \times \frac{qk}{3}$

j. $3\frac{1}{4} \times \frac{b}{2}$ **k.** $\frac{x^2}{y} \times \frac{y^3}{x^4} \times \frac{y}{x}$ **l.** $\frac{2x^2}{y} \times \frac{xy^2}{6x^4}$

2. Simplify each of the following (factorise first):

a. $\frac{x+1}{x+3} \times \frac{x^2+3x}{x^2+2x+1}$ **b.** $\frac{x}{x^2-1} \times \frac{x^2+4x+3}{x^2+3x}$

c. $\frac{(x+3)^2}{x^2+5x+6} \times \frac{x^2-9x+14}{x^2-4x-21}$ **d.** $\frac{4x+8}{x^2+11x-28} \times \frac{x^2+9x+14}{x^2+4x+4}$

Division

Division means multiplication by the **reciprocal** of the **divisor** (the reciprocal of $\frac{a}{b}$ is $\frac{b}{a}$).

Example D

Simplify $\frac{x^2}{3} \div \frac{x}{9}$

Solution

$\frac{x^2}{3} \div \frac{x}{9} = \frac{x^2}{3} \times \frac{9}{x}$ [multiplying by $\frac{9}{x}$ which is the reciprocal of $\frac{x}{9}$]

$= \frac{9x^2}{3x}$ [multiplying numerators then denominators]

$= 3x$ [cancelling common factors 3, x]

Remember that expressions must be in factored form before common factors can be cancelled.

Example E

$\frac{x^2-y^2}{x^2+2xy+y^2} \div \frac{ax-ay}{2x+2y} = \frac{x^2-y^2}{x^2+2xy+y^2} \times \frac{2x+2y}{ax-ay}$ [multiplication by reciprocal]

$= \frac{(x-y)(x+y)}{(x+y)(x+y)} \times \frac{2(x+y)}{a(x-y)}$ [factorising]

$= \frac{2}{a}$ [cancelling]

Unit 11.2 Activity 2B: Division of rational expressions

1. Simplify each of the following:

a. $\frac{2}{3} \div \frac{a}{c}$ **b.** $\frac{ab}{c} \div \frac{d}{e}$ **c.** $\frac{x^2}{y} \div \frac{y}{3}$

d. $\left(\frac{2}{5} \times \frac{a}{3}\right) \div \frac{2}{3}$ **e.** $\left(\frac{a}{c} \times \frac{5}{3}\right) \div \left(\frac{a}{b} \times \frac{5}{3}\right)$ **f.** $\frac{ab}{3} \div \frac{a^2}{9}$

g. $\frac{4a^2b}{3} \div \frac{2ab}{3}$

2. Simplify each expression to a single fraction.

a. $\frac{2}{3} \div \frac{5}{a+1}$ **b.** $\frac{5}{7} \div \frac{x}{x^2-1}$ **c.** $\frac{b-1}{3} \div \frac{3}{4}$

d. $\frac{2}{3} \div \left(\frac{y}{x+y}\right)$ **e.** $\frac{3x-1}{2} \div \frac{2}{3}$ **f.** $\frac{3x}{x+1} \div 2\frac{1}{2}$

3. Simplify each expression (factorise first where possible).

a. $\frac{5x-10}{x+3} \div \frac{x-2}{x+3}$ **b.** $\frac{x+4}{x-1} \div \frac{x+4}{x+2}$ **c.** $\frac{x^2+x}{x+2} \div \frac{x}{x^2-4}$

d. $\frac{x+3}{x+7} \div \frac{x^2+4x+3}{x^2+10x+21}$ **e.** $\frac{x^2+2x+1}{x^2+3x+2} \div \frac{x^2+6x+5}{x^2-4}$

Addition and subtraction

Addition and subtraction of fractions involves changing the fractions to equivalent fractions with a **common denominator** (choose the lowest common multiple of the denominators).

Example F

$\frac{x}{3} + \frac{y}{4} = \frac{4x}{12} + \frac{3y}{12}$ [changing both fractions to equivalent fractions with common denominator 12]

$= \frac{4x + 3y}{12}$ [expressing as a single fraction]

Example G

Express $\frac{a}{x^2} - \frac{b}{ax^3} - \frac{1}{a^2}$ as a single fraction.

Solution

$\frac{a}{x^2} - \frac{b}{ax^3} - \frac{1}{a^2} = \frac{a^3x}{a^2x^3} - \frac{ab}{a^2x^3} - \frac{x^3}{a^2x^3}$ [changing to equivalent fractions with common denominator a^2x^3]

$= \frac{a^3x - ab - x^3}{a^2x^3}$ [expressing as a single fraction]

Use brackets to avoid errors with negative signs.

Example H

$$\frac{a+3}{2}-\frac{a-3}{3}=\frac{3a+9}{6}-\frac{2a-6}{6}$$ [changing to a common denominator of 6]

$$=\frac{(3a+9)-(2a-6)}{6}$$ [using brackets to avoid errors]

$$=\frac{3a+9-2a+6}{6}$$ [removing brackets (note 6 is positive)]

$$=\frac{a+15}{6}$$ [simplifying]

Factorise denominators first (where possible) so that the lowest common multiple of the denominators can be seen more easily.

Example I

$$\frac{3}{x+y}-\frac{2}{x-y}+\frac{1}{x^2-y^2}$$

$$=\frac{3}{x+y}-\frac{2}{x-y}+\frac{1}{(x-y)(x+y)}$$ [factorising the denominator]

$$=\frac{3(x-y)}{(x+y)(x-y)}-\frac{2(x+y)}{(x+y)(x-y)}+\frac{1}{(x+y)(x-y)}$$ [changing to a common denominator]

$$=\frac{3(x-y)-2(x+y)+1}{(x+y)(x-y)}$$

$$=\frac{3x-3y-2x-2y+1}{(x+y)(x-y)}$$ [removing brackets]

$$=\frac{x-5y+1}{(x+y)(x-y)}$$ [simplifying]

Unit 11.2 Activity 2C: Addition and subtraction of rational expressions

1. Express each of the following as a single fraction in simplest form:

a. $\frac{a}{b}+\frac{b}{3}$ **b.** $\frac{a}{8}+\frac{a}{4}$ **c.** $\frac{a}{b}+\frac{1}{b}$ **d.** $\frac{a}{b}+\frac{1}{4}$

e. $\frac{a}{3}-\frac{b}{4}$ **f.** $\frac{a}{c}-\frac{b}{a}$ **g.** $\frac{1}{a}+\frac{2}{a^2}$ **h.** $\frac{2}{3}\times\frac{a}{b}+\frac{1}{b}$

i. $2\frac{1}{2}+\frac{b}{4}$ **j.** $3p+\frac{2q}{7}$

2. Simplify each of these expressions to a single fraction in simplest form:

a. $\frac{a+3}{2}+\frac{a}{2}$ **b.** $\frac{2b+3a}{3}+\frac{a+b}{3}$ **c.** $\frac{a+b}{2}+\frac{2a+7b}{3}$

d. $a+\frac{a+4}{3}$ **e.** $\frac{5b+11}{2}-\frac{1}{3}$ **f.** $\frac{4a+1}{a}+\frac{1}{2}$

3. Simplify these expressions to a single fraction in simplest form:

a. $\frac{a+1}{3}-\frac{a-1}{4}$ **b.** $\frac{2b+3}{3}-\frac{3-b}{5}$ **c.** $\frac{p-q}{4}-\frac{p+q}{3}$

d. $\frac{3p}{4}-\frac{p-2q}{5}$ **e.** $\frac{a}{3}-\frac{4-3a}{5}$ **f.** $\frac{7-p}{2}-\frac{p}{3}$

g. $\frac{18}{5}-\frac{2-p}{2}$ **h.** $\frac{3+r}{4}-\frac{2-r}{3}$ **i.** $\frac{6p-q}{3}-q$

4. Simplify each of these expressions to a single fraction in simplest form:

a. $\frac{1}{x+1}+\frac{1}{x+2}$ **b.** $\frac{2}{p+3}+\frac{1}{p+1}$ **c.** $\frac{2}{p+1}+\frac{1}{p-3}$

d. $\frac{3}{R+4}+\frac{2}{R-2}$ **e.** $\frac{3}{R}-\frac{1}{R+3}$ **f.** $\frac{1}{A+3}-\frac{1}{R+2}$

g. $\frac{4}{A-3}-\frac{1}{A-1}$ **h.** $\frac{A+2}{A+1}+\frac{A}{A-1}$ **i.** $\frac{x}{2(x+1)}-\frac{1}{3(x+1)}$

j. $\frac{y}{3(x+y)}-\frac{x}{4(x-y)}$ **k.** $\frac{2}{(x+1)(x+2)}+\frac{3}{(x+2)(x+3)}$

l. $\frac{4}{x(x+1)}-\frac{3}{(x+1)(x+2)}$ **m.** $\frac{1}{x-y+2}+\frac{2}{x+y-3}$

n. $\frac{3}{x(x+y-2)}-\frac{2}{x(x-y+2)}$ **o.** $\frac{4}{x(x-y)}-\frac{3}{y(x-y)}$

5. Express each of the following as single fractions, simplifying where possible:

a. $\frac{2ab}{10a^2b^3}\times\frac{25a^3b}{15ab^2}$ **b.** $\frac{6xyz^2}{15x^2z}\times\frac{10y^3z^3}{9xy^2}$

c. $\frac{9a^2c}{a-c}\times\frac{2a^2-2c^2}{6ac^3}$ **d.** $\frac{x^2+4x+4}{x^2-9}\times\frac{x^2+5x+6}{x(x+2)^2}$

e. $\frac{u^2-36}{x^2+5x+4}\div\frac{u^2+5u-6}{x^2-1}$ **f.** $\frac{2a^2-98}{a^2-25}\times\frac{a^2-3a-10}{3a-21}\div\frac{a^2+5a-14}{2a+10}$

g. $\frac{2}{x-y}+\frac{1}{2x-2y}$ **h.** $\frac{2}{x+1}+\frac{3}{x-1}$

i. $\frac{2}{x-2y}-\frac{1}{x+2y}$ **j.** $\frac{1}{x+y}+\frac{y}{x^2-y^2}$

k. $\frac{1}{x+1}+\frac{1}{x+2}-\frac{3}{x^2+3x+2}$ **l.** $\frac{1}{a-4}-\frac{2}{a-5}+\frac{1}{a-6}$

m. $\frac{1}{x^2-1}+\frac{1}{x^2+2x-3}+\frac{2}{x^2+4x+3}$ **n.** $\frac{a(a-1)(a-2)}{4}+\frac{a(a-1)(a-2)(a-3)}{3}$

Unit 11.2 Activity 2D: Simplifying algebraic expressions—multiple choice

1. When simplified $\frac{p}{q} + \frac{q}{5}$ is equal to:

A. $\frac{p+q}{q+5}$ **B.** $\frac{p}{5}$ **C.** $\frac{5p+q^2}{5q}$ **D.** $\frac{5p+q}{5}$

2. When simplified $\frac{3a-1}{2} + \frac{4}{3}$ is equal to:

A. $\frac{9a+7}{6}$ **B.** $\frac{2(3a-1)}{3}$ **C.** $\frac{3a+3}{5}$ **D.** $\frac{9a+5}{6}$

3. When simplified $\frac{z^3-z^2}{(z-1)^2}$ is equal to:

A. z^2 **B.** $\frac{z}{z-1}$ **C.** $\frac{z(z+1)}{z-1}$ **D.** $\frac{z^2}{z-1}$

4. When simplified $\frac{x}{x^2-1} \times \frac{x-1}{x^2+x}$ is equal to:

A. $\frac{1}{(x+1)^2}$ **B.** $\frac{1}{x(x+1)}$ **C.** $\frac{1}{x(x-1)}$ **D.** $\frac{1}{x^2-1}$

5. When simplified $\frac{x^2-4}{2x} \div \frac{2x+4}{x^2-2x}$ is equal to:

A. $(x+2)^2$ **B.** $\frac{(x-2)^2}{4}$ **C.** $\frac{4(x+2)}{x}$ **D.** $\frac{(x-2)^2}{4x}$

6. When $\frac{2b}{a(a+b)} - \frac{3a}{b(a+b)}$ is simplified the denominator will be:

A. $2b - 3a$ **B.** $ab(a+b)^2$ **C.** $ab(a+b)$ **D.** $a^2 - b^2$

7. When simplified $\frac{xyz^2}{x(x-1)^2} \times \frac{2x^2+x-3}{yz}$ is equal to:

A. $\frac{z(2x+3)}{(x-1)^2}$ **B.** $\frac{z(2x-3)}{x-1}$ **C.** $\frac{z^2(2x+3)}{x-1}$ **D.** $\frac{z(2x+3)}{x-1}$

8. The lowest common denominator of $\frac{2x(x-4)}{(x+3)(x-1)}$ and $\frac{2}{x(x^2-9)}$ is:

A. $x(x-1)(x^2-9)$ **B.** $x(x+3)(x-1)$ **C.** $(x+3)(x-3)(x+1)$ **D.** 2

9. $\frac{2}{a+b} - \frac{3}{a-b} + \frac{5}{b^2-a^2}$ is equal to:

A. $\frac{5-a-5b}{b^2-a^2}$ **B.** $\frac{5+a+5b}{b^2-a^2}$ **C.** $\frac{-30}{b^4-a^4}$ **D.** $\frac{5+a+5b}{(a-b)^2(b-a)^2}$

Unit 11.2 Graphs and Functions

Topic 3: Formulae

In this Topic we continue the coverage of 'Algebraic Expressions':

- Manipulating algebraic expressions.
- Solving equations.
- Choosing algebraic techniques and strategies to solve problems.

Introduction

The content of this Topic provides a revision of work on formulae, mathematical modelling, writing formulae, changing the subject of a formula, and using formulae.

Mathematical modelling

Mathematical modelling involves expressing a relationship as a **formula** in which mathematical symbols and letters are used to represent **variables**. The ability to do this accurately and use the resulting formula is important in many situations.

Example A

Three quantities used in accounting are assets (A), proprietorship (P), and liabilities (L). The relationship between these quantities is that assets are equal to the sum of proprietorship and liabilities. In mathematical form, the relationship can be expressed: $A = P + L$.

Note: A, P and L are variables, and the equation, $A = P + L$ is a formula for A in terms of the two variables, P and L.

Example B

The volume of a **cone** is one third of the product of its height and the area of its base. Express this relationship in mathematical form.

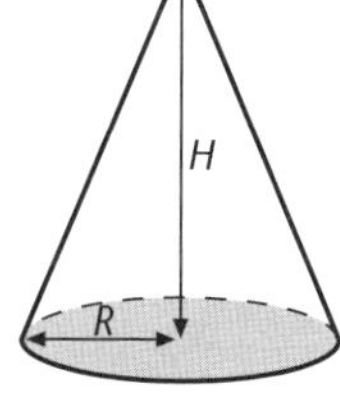

Solution

Let the volume of the cone be V, the **radius** of the base be R, and the height be H.

The base area is πR^2 [formula for area of circle]

$\therefore$ the volume is $V = \frac{1}{3}\pi R^2 H$ [from information given]

Example C

A **trapezium** is a **quadrilateral** with a pair of parallel sides.

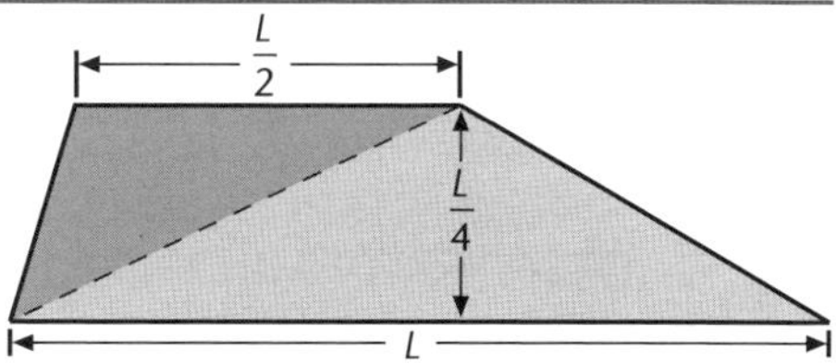

Find an expression for the area of the trapezium shown where one parallel side is one half the length of the other parallel side, and the distance between the two parallel sides is equal to one quarter of the length of the longer parallel side. Express the area in terms of the length of the longer parallel side.

Solution

Let the length of the longer parallel side be L and the area be A. The area of the trapezium is the sum of the areas of the two shaded triangles (as shown on the diagram, where the shorter parallel side is $\frac{L}{2}$ and the height is $\frac{L}{4}$).

Area of the smaller triangle is $\frac{1}{2} \times \frac{L}{2} \times \frac{L}{4} = \frac{L^2}{16}$ [area is half × base × height]

Area of the larger triangle is $\frac{1}{2} \times L \times \frac{L}{4} = \frac{L^2}{8}$

∴ area of the trapezium is $A = \frac{L^2}{8} + \frac{L^2}{16} = \frac{3L^2}{16}$ [adding fractions]

Unit 11.2 Activity 3A: Writing formulae

1. A pair of socks costs KD. Write an expression for the cost of the socks in toea.

2. a. Express the time of $3x$ minutes in hours.
   b. Express this same time in seconds.

3. Peter weighs x kilograms and y grams.
   a. Express Peter's weight in grams.
   b. Express his weight in kilograms.

4. A rectangle has length L metres and width W centimetres. Write an expression for:
   a. the area in cm^2.
   b. the area in m^2.
   c. the perimeter in mm.

5. A square has area $4L^2$ in square metres. Write an expression for:
   a. the perimeter in cm.
   b. the perimeter in mm.

6. A right-angled triangle has base b and height a. Both have the same unit.
   a. What is the area of the triangle?
   b. What is the length of the **hypotenuse** (the third side)?

7. a. Find an expression for the volume of the cylinder shown.
   b. Find an expression for the total surface area of the cylinder.

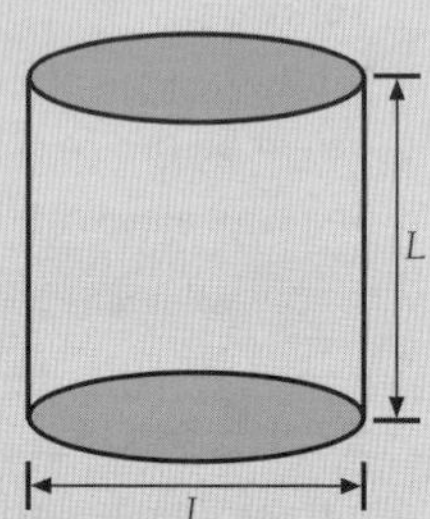

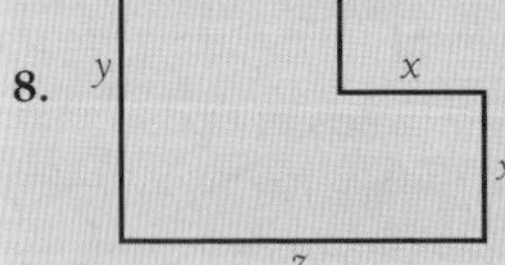

8. a. Find an expression for the perimeter of the shape shown.
   b. Find a formula for the area of this shape.

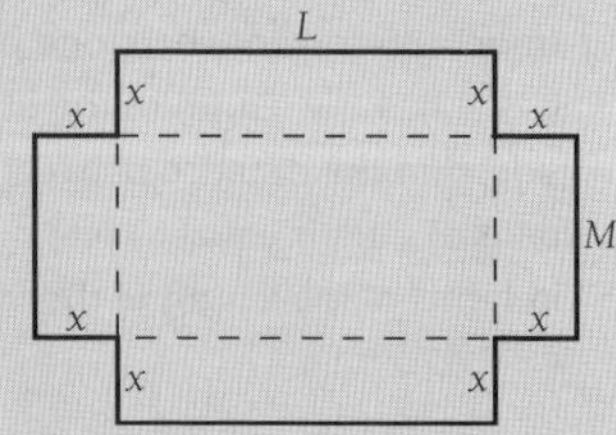

9. a. What is the perimeter of the shape shown?
   b. What is the volume of the box which forms when the shape is folded along the dotted lines?

10. The length of the closed box shown is $4L$. The height and width of the box are both half the length.

a. Write down an expression for the volume of the box.

b. Write down an expression for the surface area of the box.

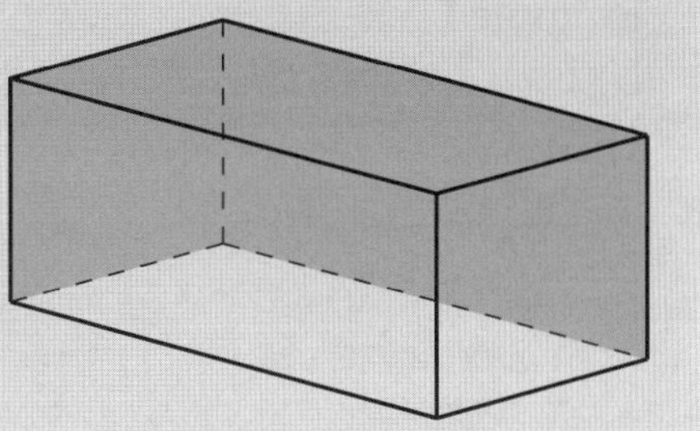

Changing the subject of a formula

The **subject** of a formula is a variable which appears *alone* on one side (usually the left side) of the equals sign. The subject is expressed in terms of the other variables. **Changing the subject** of a formula means **rearranging** the formula (using **inverse operations**) so that another variable becomes the subject.

Example D

Make x the subject of $y = 2x - 6$.

Solution

$y = 2x - 6$

$\therefore y + 6 = 2x$ [adding 6 to both sides]

$\therefore 2x = y + 6$ [swapping sides]

$\therefore x = \dfrac{y+6}{2}$ [dividing both sides by 2]

Sometimes the variable to be made subject appears more than once in the formula. You will need to collect terms in this variable on one side and then factorise.

Example E

Make T the subject of $\dfrac{AT + D}{T} = E$.

Solution

$\dfrac{AT + D}{T} = E$

$\therefore AT + D = ET$ [multiplying both sides by T]

$\therefore ET = AT + D$ [swapping sides]

$\therefore ET - AT = D$ [collecting terms in T on one side]

$\therefore T(E - A) = D$ [factorising $ET - AT$]

$\therefore T = \dfrac{D}{E - A}$ [dividing by $E - A$ (providing $E - A \neq 0$)]

Note: $D = 0$ if $E - A = 0$ [substituting $E - A = 0$ in $T(E - A) = D$]

Squaring and taking the square root are inverses. Take care with signs when taking a square root (if $x^2 = y$ then $x = \pm\sqrt{y}$).

Example F

Make M the subject of $\frac{AM^2 - B}{D} = M^2 + E$ when $A \neq D$.

Solution

$$\frac{AM^2 - B}{D} = M^2 + E$$

$\therefore AM^2 - B = DM^2 + ED$ [multiplying by D]

$\therefore AM^2 = DM^2 + ED + B$ [adding B]

$\therefore AM^2 - DM^2 = ED + B$ [subtracting DM^2]

$\therefore M^2(A - D) = ED + B$ [factorising $AM^2 - DM^2$]

$\therefore M^2 = \frac{ED + B}{A - D}$ [dividing by $A - D$ (where $A \neq D$)]

$\therefore M = \pm\sqrt{\frac{ED + B}{A - D}}$ [taking the positive or negative square roots]

Note: Once again these are restrictions on the values the variables can take. Division by zero is undefined, so $A - D \neq 0$. Also taking the square root of a negative quantity is not defined for real numbers, so $\frac{ED + B}{A - D} \geq 0$.

Unit 11.2 Activity 3B: Changing the subject of formulae

Make M the subject of each of the following formulae:

1. $3M = T$
2. $AM = T$
3. $\frac{M}{P} = T$
4. $\frac{M}{P} = P$
5. $\frac{M}{P} = A + B$
6. $\frac{2M}{P} = 5$
7. $\frac{5M}{T} = T$
8. $M + B = A$
9. $M - B = A$
10. $M - B = B$
11. $M - B - B^2 = B^2 + 2B$
12. $3M - B^2 = 2B^2$
13. $\frac{M}{AB} = BC$
14. $\frac{M}{B} - 3 = C$
15. $\frac{M}{A} = \frac{C}{B}$
16. $M^2 = A$
17. $CM^2 = A + B$
18. $\frac{AM - D}{B} = B$
19. $\frac{AM - DM}{B} = B$
20. $\frac{M}{A} + \frac{M}{B} = C$
21. $\frac{M - A}{M + D} = E$
22. $\sqrt{M} = C$
23. $\frac{\sqrt{M}}{C} = D$
24. $\frac{2A}{3}\sqrt{\frac{M}{G}} = L$
25. $\frac{A}{C}\sqrt{\frac{G}{M}} = X$
26. $\frac{3M^2 - D}{M} = AM$
27. $\frac{A\sqrt{M} - D}{B} = \sqrt{M}$
28. $\frac{A}{\sqrt{M}} + \frac{C}{\sqrt{M}} = E$
29. $y = \frac{AM - 5}{BM - 4}$
30. $y = \frac{3M^3 - E}{4M^3 - F}$

Using formulae

Formulae need to be used accurately. Using formulae often means **substituting** numbers for some of the variables and solving the equation to find the value of an unknown variable.

Example G

From Example B, the volume of a cone is $\frac{1}{3}\pi R^2 H$,

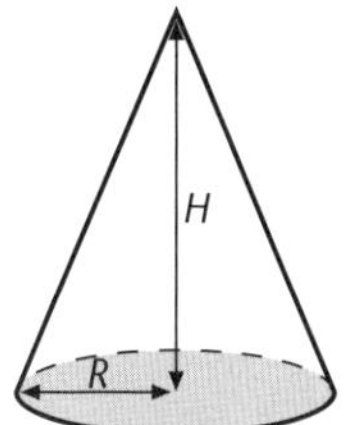

where R is the radius of the base and H is the height. (In this example, π is taken to be 3.142.)

1. Find the volume when $R = 2.152$ cm and $H = 4.365$ cm.
2. Find the radius of the base when the volume is 63.25 cm^2 and the height is 5.230 cm.

Solutions

1. $V = \frac{1}{3}\pi R^2 H$

$= \frac{1}{3} \times 3.142 \times (2.152)^2 \times 4.365$ [substituting into $V = \frac{1}{3}\pi R^2 H$]

$= 21.17$ cm^3 (2 dp) [rounding answer]

2. $V = \frac{1}{3}\pi R^2 H$

$\therefore 63.25 = \frac{1}{3} \times 3.142 \times R^2 \times 5.230$ [substituting into $V = \frac{1}{3}\pi R^2 H$]

$\therefore R^2 = \frac{3 \times 63.25}{3.142 \times 5.230}$ [solving for R^2]

$R = \sqrt{\frac{3 \times 63.25}{3.142 \times 5.230}}$ [taking the positive square root since R is a length]

$= 3.40$ cm (2 dp) [rounding answer]

Note: *Do not* round off calculations until the very end of the calculation. Rounding as you progress through a calculation introduces errors at each stage of the calculation, which accumulate over a number of steps producing significant errors in the final answer.

Unit 11.2 Activity 3C: Using formulae

1. The formula for the area of the walls of a room is $A = 2h(l + w)$ where h is the height of the room, l is the length and w the width.

- **a.** Find the area if $h = 2.3$ m, $l = 5.3$ m, $w = 4.5$ m.
- **b.** Find the area if $h = 3.41$ m, $l = 4.6$ m, $w = 5.83$ m.
- **c.** Find the height if the area if 60.42 m^2, the length is 4.86 m and the width is 3.92 m.
- **d.** Find the width if the area is 73.4 m^2, the height is 4.2 m and the length is 6.8 m.
- **e.** Find the length if the area is 165.6 m^2, the height is 5.3 m and the width is 4.8 m.

2. The formula for the volume of a cylinder is $V = \pi R^2 h$, where R is the radius of the cross-section, h is the height and π is taken to be 3.14.

- **a.** Find the volume if $R = 4.6$ cm, $h = 12.3$ cm.
- **b.** Find the volume if $R = 2.67$ cm, $h = 4.76$ cm.
- **c.** Find the height if the volume is 283 cm^3 and the radius is 5.12 cm.
- **d.** Find the radius if the volume is 645 cm^3 and the height is 6.32 cm.
- **e.** Find the radius if the volume is 1367 cm^3 and the height is 15.32 cm.

3. The formula for the simple interest earned by investing KP at R% per year for R years is $I = \frac{PRT}{100}$. Use a graphical calculator or spreadsheet to answer the following questions.

   a. Find the principal invested if the rate of simple interest is $12\frac{1}{2}$%, the time is 3 years and the interest earned is K1 031.25.

   b. Find the annual rate of interest earned if K4 700 earns K998.75 simple interest after 2.5 years.

   c. Patrick borrows K7 500 at $6\frac{1}{4}$ % simple interest per year for a number of years. If he pays a total of K1 875 interest, how long was the loan?

Unit 11.2 Graphs and Functions

Topic 4: Polynomials, factor theorem and remainder theorem

In this Topic we continue the coverage of 'Algebraic Expressions' following the Syllabus, p. 15. In particular, the Topic covers:

- Polynomials.
- Factorising polynomials.
- The factor theorem.
- The remainder theorem.
- Factorising polynomials using the factor and remainder theorem.

Introduction

A **polynomial** is an expression, in x, of the form

$P(x) = a_n x^n + a_{n-1} x^{n-1} + a_{n-2} x^{n-2} \ldots \ldots + a_1 x + a_0$

Properties of polynomials:

- All powers of x are positive whole numbers or zero.
- $a_n, \ a_{n-1}, \ldots\ldots a_1, \ a_0$ are real numbers, called **co-efficients**.
- The number a_0 is called the **constant term**.
- The highest power of x in the polynomial determines the **degree of the polynomial**.
- The term with the highest power of x is called the **leading term**.
- All polynomials are functions, ie they have one-to-one correspondence.
- If a value, b, is substituted for x in the expression then this is denoted by $P(b)$.

Special polynomials

Polynomials of degree

- One are called **linear** functions. eg $P(x) = 2x - 3$.
- Two are called **quadratic** functions eg. $P(x) = 3x^2 - 4x + 1$.
- Three are called **cubic** functions eg. $P(x) = 3x^2 + x^2 - 2x + 1$.
- Four are called **quartic** functions eg. $P(x) = x^4 - 1$.

Example A

Q. Which of the following expressions are polynomials?

$x^2 + \frac{1}{x^2}$; $3 - x$; $5x^3 + 2x + 4\sqrt{x}$; $7 + 7x^7$; $x^2 + x^3 - 5x^5 - x + 20$

A. $x^2 + \frac{1}{x^2}$ is not a polynomial because the term $\frac{1}{x^2} = x^{-2}$ is not a positive power of x.

$3 - x$ is a polynomial as all powers of x are zero or positive whole numbers $3 - x = 3x^0 - x^1$.

$5x^3 + 2x + 4\sqrt{x}$ is not a polynomial because $\sqrt{x} = x^{\frac{1}{2}}$ which is not a whole number.

$7 + 7x^7 = 7x^0 + 7x^2$ is a polynomial as all terms are zero or positive whole numbers.

$x^2 + x^3 - 5x^5 - x^1 + 20x^0$ is a polynomial as all terms are zero or positive whole numbers.

Example B

Q. For each of the following:

i. State the degree of the polynomial and write down the leading term.
ii. State the coefficient of the x^3 term.

a. $x^2 + 3x^3$
b. $5 - x$
c. $5x^3 + 2x + 4$
d. $7 + 7x^7$
e. $x^2 + x^3 - 5x^5 - x + 20$

A. **a.** **i.** The highest power of x is 3 so the degree is three. The leading term is $3x^3$.
ii. The number in front of the x^3 term is 3 so this is the coefficient.
b. **i.** The highest power of x is 1 so the degree is one. The leading term is $-x$.
ii. There is no x^3 term so 0 is the coefficient.
c. **i.** The highest power of x is 3 so the degree is three. The leading term is $5x^3$.
ii. The number in front of the x^3 term is 5 so this is the coefficient.
d. **i.** The highest power of x is 7 so the degree is seven. The leading term is $7x^7$.
ii. There is no x^3 term so 0 is the coefficient.
e. **i.** The highest power of x is 5 so the degree is five. The leading term is $-5x^3$.
ii. The number in front of the x^3 term is 1 so this is the coefficient.

Example C

Q. For the polynomial $P(x) = 2x^3 - 3x^2 + x - 5$ find:

a. $P(2)$
b. $P(-1)$
c. $P(u)$

A. **a.** $P(2) = 2 \times 2^3 - 3 \times 2^2 + 2 - 5 = 2 \times 8 - 3 \times 4 + 2 - 5 = 1$
b. $P(-1) = 2 \times (-1)^3 - 3 \times (-1)^2 + (-1) - 5 = -2 - 3 - 1 - 5 = -11$
c. $P(u) = 2u^3 - 3u^2 + u - 5$

Example D

Q. If the polynomials $P(x) = 2x^2 - 4x$ and $Q(x) = 1 + x - x^2$ find, in simplest form:

a. $P(x) + Q(x)$
b. $2P(x) - Q(x)$
c. $P(x) \times Q(x)$

A. **a.** $P(x) + Q(x) = 2x^2 - 4x + (1 + x - x^2) = (2x^2 - x^2) + (-4x + x) + 1 = x^2 - 3x + 1$
b. $2P(x) - Q(x) = 2(2x^2 - 4x) - (1 + x - x^2) = 4x^2 - 8x - 1 - x + x^2 = 5x^2 - 9x - 1$
c.
$$
\begin{aligned}
P(x) \times Q(x) &= (2x^2 - 4x)(1 + x - x^2) \\
&= 2x^2 \times 1 + 2x^2 \times x + 2x^2 \times -x^2 - 4x \times 1 + -4x \times x + -4x \times -x^2 \\
&= 2x^2 + 2x^3 - 2x^4 - 4x - 4x^2 + 4x^3 \\
&= -2x^4 + 6x^3 - 2x^2 - 4x
\end{aligned}
$$

Unit 11.2 Activity 4A: Polynomials

1. Which of the following expressions are not polynomials? State the reason that they are not polynomials.
   $x^4 + 3x^2 - 2x + 24$; $\frac{2}{x+2}$; $3\sqrt{x^2-1}$; $9-\frac{1}{3}x^2$; $(2x-3)^2$; $\frac{x^2+3x-4}{3}$; $x^2 + x + x^{-1}$
2. For each of the following polynomials:
   i. State the degree of the polynomial and write down the leading term.
   ii. State the coefficient of the term in brackets.
   a. $2x^4 + 3x^3 + 6x + 1.2$ (x)
   b. $25 - x - 14x^3$ (x^3)
   c. $15x^3 + 3x^2 - 2x + 8$ (constant term)
   d. $7x^2 + 7x^4 + 6x^3 - 2x^2 + 3$ (x^2)
   e. $4x^2 + 2x^3 - 7x^5 + x + 22$ (x)
3. For the polynomial $P(x) = x^3 - 4x^2 - 7x + 10$ find:
   a. $P(2)$ **b.** $P(-4)$ **c.** $P(5)$ **d.** $P(3) - P(-1)$ **e.** $P(m)$
4. For the polynomials $P(x) = 2x - 3$, $Q(x) = x^2 + 4x - 3$ and $R(x) = x^3 + 2x^2 - 4x + 6$ find:
   a. $2P(x) + 2Q(x)$ **b.** $R(x) - Q(x)$ **c.** $P(x) \times Q(x)$
   d. $[P(x)]^2$ **e.** $R(2) + Q(2) - P(2)$

Factorising polynomials

It is useful to be able to factorise polynomials as this is part of the process of solving equations.

You have already learnt how to factorise linear and quadratic polynomials. Factorising polynomials of degree three or more requires techniques that rely on two theorems, the remainder theorem and the factor theorem, and the technique of re-expression of polynomials.

To illustrate the remainder theorem we will look at *dividing a polynomial by a linear polynomial*.

Example E

Q. Divide $2x + 3$ by $x - 1$

A. $\frac{2x+3}{x-1} = \frac{2(x-1)+5}{x-1}$: creating a term in $(x - 1)$ using the leading term and then adjusting the constant.

$= \frac{2(x-1)}{x-1} + \frac{5}{x-1}$: separating the two terms

$= 2 + \frac{5}{x-1}$: cancelling the factors $(x + 1)$

We can say that when $2x + 1$ is divided by $x - 1$ the answer is 2 plus a remainder of 5.

Example F

Q. Divide $x^2 + 3x + 3$ by $x + 2$

A. $\frac{x^2+3x+3}{x+2} = \frac{x(x+2)+x+3}{x+2}$: making the factor $(x + 2)$ from the x^2 term.

$= \frac{x(x+2)}{x+2} + \frac{x+3}{x+2}$: separating the two terms

$= \frac{x(x+2)}{x+2} + \frac{(x+2)+1}{x+2}$: making the factor $(x + 2)$ from the x term.

$= \frac{x(x+2)}{x+2} + \frac{(x+2)}{x+2} + \frac{1}{x+2}$: separating the remaining terms

$= x+1+\frac{1}{x+1}$: cancelling the factors $(x + 2)$

We can say that when $x^2 + 3x + 3$ is divided by $x + 2$ the answer is $x + 1$plus a remainder of 1.

Example G

Q. Divide $2x^2 + x - 15$ by $x + 3$

A. $\frac{2x^2+x-15}{x+3} = \frac{2x(x+3)-5x-15}{x+3}$

$= \frac{2x(x+3)}{x+3} - \frac{5x+15}{x+3}$

$= \frac{2x(x+3)}{x+3} - \frac{5(x+3)}{x+3}$

$= 2x-5$

We can say that when $2x^2 + x - 15$ is divided by $x + 3$ the answer is $2x - 5$ plus a remainder of 0.

In fact in this case we have found the factors of $2x^2 + x - 15$

If

$$\frac{2x^2+x-15}{x+3} = 2x-5$$

then $2x^2 + x - 15 = (x + 3)(2x - 5)$

The remainder theorem

The remainder theorem states:

> If a polynomial $P(x)$ is divided by a linear polynomial $(x - a)$ then the remainder will be $P(a)$

Confirming that the remainder theorem works for the examples above:

Example H

The remainder theorem states:

If $P(x) = 2x + 3$ is divided by $x - 1$ then the remainder will be given by $P(1)$.

$P(1) = 2 \times 1 + 3 = 5$; the same value that we found using division.

Example I

The remainder theorem states:

If $P(x) = x^2 + 3x + 3$ is divided by $x + 2$ then the remainder will be $P(-2)$

$P(-2) = (-2)^2 + 3 \times (-2) + 3 = 4 - 6 + 3 = 1$; the same value that we found using division.

Example J

The remainder theorem states:

If $P(x) = 2x^2 + x - 15$ is divided by $x + 3$ then the remainder will be $P(-3)$

$P(-3) = 2 \times (-3)^2 + (-3) - 15 = 18 - 3 - 15 = 0$; the same value that we found using division.

If the **remainder is zero** when a polynomial $P(x)$ is divided by a linear polynomial $(x - a)$ then we have found a factor, $(x - a)$, of the polynomial and this is the basis of the Factor Theorem.

The factor theorem

If, for a polynomial $P(x)$, $P(a) = 0$ then $(x - a)$ is a factor of $P(x)$

For example:

Consider the cubic polynomial $P(x) = x^3 - 4x^2 - 7x + 10$.

To find a factor of this polynomial we could try any number but it is wise to try a factor of the constant term, 10. This can be any of 1, −1, 2, −2, 5, −5, 10, −10.

Trying $P(1)$ which is the easiest to calculate.

$P(1) = 1^3 - 4 \times 1^2 - 7 \times 1 + 10 = 0$

As there is zero remainder then the factor theorem states that $(x - 1)$ will be a factor of $P(x)$.

There are likely to be other factors of 10 that will give a remainder of zero but having found one factor of our polynomial we can use the **process of re-expression** to find the others:

$x^3 - 4x^2 - 7x + 10 = x^2(x - 1) - 3x^2 - 7x + 10$: creating a factor of $(x - 1)$ with the x^3 term and then adjusting the x^2 term.

$= x^2(x - 1) - 3x(x - 1) - 10x + 10$: creating a factor of $(x - 1)$ with the x^2 term and then adjusting the x term.

$= x^2(x - 1) - 3x(x - 1) - 10(x - 1)$: creating a factor of $(x - 1)$ with the x term and then adjusting the constant term. **Note:** there is zero remainder.

$= (x - 1)(x^2 - 3x - 10)$: taking $(x - 1)$ out as a common factor.

$= (x - 1)(x - 5)(x + 2)$: factorising the quadratic factor to two linear factors.

Unit 11.2 Activity 4B: Polynomials and the remainder theorem

1. For the polynomial $P(x) = 2x^3 - 4x^2 - 4x - 6$:

a. Use the remainder theorem to find the remainder when $P(x)$ is divided by:

i. $x - 1$ **ii.** $x + 2$ **iii.** $x - 3$

b. State a factor of $P(x)$.

2. For the polynomial $P(x) = x^3 + 4x^2 + x + 4$:

a. Use the remainder theorem to find the remainder when $P(x)$ is divided by:

i. $x + 1$ **ii.** $x - 2$ **iii.** $x + 4$

b. State a factor of $P(x)$.

3. Use the remainder theorem for each of the following polynomials to find a factor. State the factor. Hint: Substitute factors of the constant term until you find one that results in zero.

a. $P(x) = x^3 - 2x^2 + x - 2$ **b.** $F(x) = -x^3 - 4x^2 + 3x + 12$

c. $Q(x) = x^3 + 9x^2 + 9x + 8$ **d.** $R(x) = 2x^3 - 3x^2 - 23x + 12$

Factorising polynomials

You have already seen techniques for factorising linear and quadratic polynomials in previous chapters. In this exercise we will use the remainder and factor theorem and re-expression of polynomials to factorise polynomials of degree three or more.

Example K

Q. For the polynomial $P(x) = x^3 + 3x^2 - 4x - 12$:

i. Use the remainder and factor theorem to find a factor of $P(x)$.

ii. Use the process of re-expression to factorise $P(x)$.

iii. Factorise the polynomial completely.

A. i. Looking at factors of the constant term (–12) we can choose from 1, –1, 12, –12, 2, –2, 6, –6, 3, –3, 4, –4.

Try $P(1)$

If $P(1) = 0$ the remainder theorem states that there is zero remainder and the factor theorem states that $(x - 1)$ is a factor of $P(x)$.

$P(1) = 1^3 + 3 \times 1^2 - 4 \times 1 - 12 = 1 + 3 - 4 - 12 = -12$ So $x - 1$ is not a factor.

Try $P(2) = 2^3 + 3 \times 2^2 - 4 \times 2 - 12 = 8 + 12 - 8 - 12 = 0$ So $x - 2$ is a factor.

Note: Other factors of the constant term (–12) will also give a remainder of 0.

ii. $P(x) = x^3 + 3x^2 - 4x - 12$

$P(x) = x^2(x - 2) + 5x^2 - 4x - 12$: creating a term in $(x - 2)$ from the leading term and adjusting the x^2 term.

$P(x) = x^2(x - 2) + 5x(x - 2) + 6x - 12$: creating a factor of $(x - 2)$ with the x^2 term and then adjusting the x term.

$P(x) = x^2(x - 2) + 5x(x - 2) + 6(x - 2)$: creating a factor of $(x - 2)$ with the x^2 term and then adjusting the x term. (Zero remainder)

$P(x) = (x - 2)(x^2 + 5x + 6)$: taking $(x - 2)$ out as a common factor.

iii. $P(x) = (x - 2)(x + 2)(x + 3)$: factorising the remaining quadratic.

$P(x)$ is completely factorised.

With practice, the process of re-expression can be completed in one line.

Example L

Q. Factorise completely the quartic polynomial $F(x) = x^4 - x^3 - 5x^2 - x - 6$.

A. Trying factors of –6:

$F(1) = 1^4 - 1^3 - 5 \times 1^2 - 1 - 6 = 1 - 1 - 5 - 1 - 6 \neq 0$

$F(2) = 2^4 - 2^3 - 5 \times 2^2 - 2 - 6 = 16 - 8 - 20 - 2 - 6 \neq 0$

$F(-1) = (-1)^4 - (-1)^3 - 5 \times (-1)^2 - (-1) - 6 = 1 + 1 - 5 + 1 - 6 = -8 \neq 0$
$F(-2) = (-2)^4 - (-2)^3 - 5 \times (-2)^2 - (-2) - 6 = 16 + 8 - 20 + 2 - 6 = 0$
So $(x + 2)$ is a factor of $F(x)$.
Re-expressing $F(x)$ in terms of $(x + 2)$:

$$\begin{aligned} F(x) &= x^3(x + 2) - 3x^3 - 5x^2 - x - 6 \\ &= x^3(x + 2) - 3x^2(x + 2) + x^2 - x - 6 \\ &= x^3(x + 2) - 3x^2(x + 2) + x(x + 2) - 3x - 6 \\ &= x^3(x + 2) - 3x^2(x + 2) + x(x + 2) - 3(x + 2) \\ &= (x + 2)(x^3 - 3x^2 + x - 3) \end{aligned}$$

We now have the task of factorising a cubic $P(x) = x^3 - 3x^2 + x - 3$
Using the remainder theorem with factors of -3.
$P(3) = 3^3 - 3 \times 3^2 + 3 - 3 = 27 - 27 + 3 - 3 = 0$
So $(x - 3)$ is a factor of $P(x)$
Re-expressing $P(x)$ in terms of $(x - 3)$:

$$\begin{aligned} P(x) &= x^2(x - 3) + x - 3 \\ &= x^2(x - 3) + (x - 3) \\ &= (x - 3)(x^2 + 1) \end{aligned}$$

The quadratic $x^2 + 1$ cannot be factorised further.
So $F(x)$ completely factorised is $F(x) = (x + 2)(x - 3)(x^2 + 1)$.

Unit 11.2 Activity 4C: Factorising polynomials

1. Copy and complete the following re-expressions:

a. $x^3 - 2x^2 - 5x + 6 = \cdots(x-1) - x(\ldots\ldots) - \ldots(x-1)$
$= (x-1)(x^2 - \ldots - \cdots)$

b. $x^3 - 3x - 2 = x^2(x+1) - \cdots(x+1) - 2(\ldots\ldots)$
$= (x+1)(x^2 - \ldots - 2)$

c. $2x^3 + x^2 - 2x - 1 = 2x^2(x+1) - (\ldots\ldots) - (x+1)$
$= (x+1)(2x^2 - \ldots - \cdots)$

2. If, for the polynomial $P(x) = x^3 + x^2 - 17x + 15$, $P(1) = 0$:

a. Write down a factor of $P(x)$.
b. Re-express $P(x)$ in terms of this factor.
c. Completely factorise $P(x)$.

3. If, for the polynomial $P(x) = x^3 - x^2 - 10x - 8$, $P(4) = 0$:

a. Write down a factor of $P(x)$.
b. Re-express $P(x)$ in terms of this factor.
c. Completely factorise $P(x)$.

4. For the polynomial $P(x) = x^3 + 2x^2 - 20x + 24$,

a. Use the remainder and factor theorems to find a factor of $P(x)$.

b. Re-express $P(x)$ in terms of this factor.

c. Completely factorise $P(x)$.

5. For the polynomial $P(x) = x^3 - 3x^2 - 9x + 27$,

a. Use the remainder and factor theorems to find a factor of $P(x)$.

b. Re-express $P(x)$ in terms of this factor.

c. Completely factorise $P(x)$.

6. Factorise completely each of the following polynomials:

a. $x^3 - 2x^2 - 11x + 12$

b. $x^3 + x^2 - 4x - 4$

c. $x^3 + 7x^2 - x - 7$

d. $x^3 - 12x - 16$

e. $x^3 + 6x^2 + 11x + 6$

f. $x^3 - x^2 - 4$

g. $3x^3 + 17x^2 - 171x + 55$

h. $2x^3 - 5x^2 - 23x - 10$

i. $6x^3 - 25x^2 + 3x + 4$

j. $x^4 - 1$

k. $x^4 - 9x^3 + x - 9$

l. $x^4 - 2x^3 - 21x^2 + x + 20$

Unit 11.2 Graphs and Functions

Topic 5: Working with indices

In Topic 5 we continue with the manipulation of algebraic expressions and solving equations:

- Using fractional and negative indices.

Introduction

Indices involve expressions of the form x^n, where x is the **base** and n is the **power** or **index** or **exponent**.

Some important properties of indices already learned in NCEA Level 1 Mathematics are:

$x^m \times x^n = x^{m+n}$	$\frac{x^m}{x^n} = x^{m-n}$ $[x \neq 0]$ or $\frac{1}{x^{n-m}}$
$(xy)^m = x^m y^m$	
$(x^m)^n = x^{mn}$	$\left(\frac{x}{y}\right)^m = \frac{x^m}{y^m}$
$x^0 = 1$ $[x \neq 0]$	

Negative indices

Using the division law for powers, $\frac{x^2}{x^5} = x^{2-5} = x^{-3}$. Simplifying $\frac{x^2}{x^5}$ by writing in full and cancelling gives $\frac{\cancel{x} \times \cancel{x}}{\cancel{x} \times \cancel{x} \times x \times x \times x} = \frac{1}{x^3}$. This example illustrates the rule for negative indices:

$$x^{-n} = \frac{1}{x^n} \quad \text{where } x \neq 0$$

In the following example, expressions with negative indices are evaluated by calculator.

Example A

Calculate the following:

1. 3^{-3} **2.** $\frac{1}{2^{-4}}$ **3.** $(-2)^{-2}$ **4.** $\frac{2^{-1}}{4}$

Solution

By using the fraction button $a\frac{b}{c}$ on the *Casio fx-82* calculator, answers can be converted to fraction form.

1. Using the most recent *Casio fx-82* calculator:

   3^{-3} is calculated by:

   The answer is 0.037037. [to 6 decimal places]

   By pressing the fraction key $a\frac{b}{c}$ the answer is converted to the fraction $\frac{1}{27}$.

2. $\dfrac{1}{2^{-4}}$ is calculated by:

The answer is 16.

Similarly:

3. $(-2)^{-2} = 0.25$

or $\dfrac{1}{4}$ [pressing a b/c]

4. $\dfrac{2^{-1}}{4} = 0.125$

or $\dfrac{1}{8}$ [pressing a b/c]

Note: Older versions of this calculator have x^y as the power key.

Using the division law for indices the following two rules also hold for negative indices:

$\left(\dfrac{x}{y}\right)^{-m} = \left(\dfrac{y}{x}\right)^{m}$	$\dfrac{x^{-n}}{y^{-m}} = \dfrac{y^m}{x^n}$

Example B

Use the properties of indices to calculate the following, giving the answers as fractions or whole numbers.

1. 2^{-3} 2. $\dfrac{1}{3^{-2}}$ 3. $\left(\dfrac{2}{5}\right)^3$ 4. $\left(\dfrac{3}{7}\right)^{-2}$ 5. $\dfrac{2^{-3}}{3^{-2}}$ 6. $\dfrac{2^{-1}}{7}$

Solution

1. $2^{-3} = \dfrac{1}{2^3}$ [since $x^{-m} = \dfrac{1}{x^m}$]

$= \dfrac{1}{8}$ [$2^3 = 8$]

2. $\dfrac{1}{3^{-2}} = 3^2$ [since $\dfrac{1}{x^m} = x^{-m}$]

$= 9$

3. $\left(\dfrac{2}{5}\right)^3 = \dfrac{2^3}{5^3}$ [since $\left(\dfrac{x}{y}\right)^m = \dfrac{x^m}{y^m}$]

$= \dfrac{8}{125}$

4. $\left(\dfrac{3}{7}\right)^{-2} = \left(\dfrac{7}{3}\right)^2$ [since $\left(\dfrac{x}{y}\right)^{-m} = \left(\dfrac{y}{x}\right)^m$]

$= \dfrac{7^2}{3^2}$ [since $\left(\dfrac{y}{x}\right)^m = \dfrac{y^m}{x^m}$]

$= \dfrac{49}{9}$

5. $\dfrac{2^{-3}}{3^{-2}} = \dfrac{3^2}{2^3}$ [since $\dfrac{x^{-n}}{y^{-m}} = \dfrac{y^m}{x^n}$]

$= \dfrac{9}{8}$

6. $\frac{2^{-1}}{7} = \frac{1}{7} \times 2^{-1}$ [as dividing by 7 is the same as multiplying by $\frac{1}{7}$]

$= \frac{1}{7} \times \frac{1}{2}$ [as $2^{-1} = \frac{1}{2^1} = \frac{1}{2}$]

$= \frac{1}{14}$ [multiplying fractions]

Unit 11.2 Activity 5A: Working with indices

1. Calculate the following, giving the answers to no more than 4 decimal places:

a. 2^{-4} **b.** 1.2^{-5} **c.** 0.37^{-2} **d.** $0.37^{-1} + 2.3$

e. $(2.3)^{-2} + 4$ **f.** $\frac{1}{3^{-2}}$ **g.** $\frac{2}{(2.4)^{-3}}$ **h.** $(-3.4)^{-2}$

i. $\frac{1}{0.56^{-1}} + \frac{2}{0.5^{-2}}$ **j.** $(3.4^{-1} - (0.9)^{-2})^{-3}$ **k.** $(2.3)^{-3} + (-2.3)^{-2}$

2. Use the properties of indices to evaluate the following, giving the answers as fractions or whole numbers.

a. 3^{-2} **b.** 4^{-3} **c.** 2^{-4} **d.** $\left(\frac{3}{4}\right)^3$

e. $\frac{2^3}{5^2}$ **f.** $\frac{1}{3^{-2}}$ **g.** $\frac{1}{4^{-2}}$ **h.** $\left(\frac{3}{5}\right)^{-2}$

i. $\left(\frac{4}{5}\right)^{-3}$ **j.** $\frac{2^{-2}}{3^{-3}}$ **k.** $\frac{3^{-3}}{5^{-2}}$ **l.** $\frac{3^{-1}}{8}$

m. $\frac{5^{-1}}{2}$ **n.** $7(5^{-1})$ **o.** 8×3^{-1} **p.** $2^{-1} \times 3^{-1}$

q. $3^{-1}\,5^{-1}$ **r.** $3^{-1}\,2^{-2}$ **s.** $\frac{5}{2^{-1}}$ **t.** $\frac{3}{2^{-2}}$

u. $\frac{3^{-2}}{4}$ **v.** $\frac{5^{-2}}{7}$ **w.** $3^{-1}\,6$ **x.** $3^{-2}\,6^2$

y. $2^{-1} \div 7$ **z.** $3^{-1} \div 4^{-1}$

3. Calculate the following, giving the answers as fractions:

a. $2^{-1} \times 3$ **b.** $3^{-2} \times 4^{-1}$ **c.** $(2^{-2} \times 3^{-1})^{-2}$ **d.** $(3^{-1}2^{-3})^2$

e. $2^{-1} + 3^{-1}$ **f.** $3^{-2} + 6^{-1}$ **g.** $2^{-1}\,3 + 5^{-1}\,3^2$ **h.** $2^{-2} - 8^{-1}$

i. $4^{-2} \div 8^{-1}$ **j.** $\frac{3^{-1}}{2^{-1}} + \frac{4^{-1}}{3^{-1}}$ **k.** $\frac{3^{-2}}{2^{-2}} - \frac{2^{-1}}{3^{-2}}$ **l.** $(3^1\,2^{-1} - 2^1\,3^{-1})^{-1}$

m. $(2^2\,3)^{-1} + 2^{-1}\,3^{-2}$ **n.** $5(2^3)^{-1} + 4^{-2}$ **o.** $7(2^{-1}) - 3(2^{-2})$

4. Write the following as fractions or integers with no powers:

a. 2×4^{-1} **b.** $2^{-1} \times 3^{-1}$ **c.** 4×2^{-1} **d.** $3^{-2} \times 5$

e. $\left(\frac{2}{3}\right)^{-1}$ **f.** $\left(1\frac{1}{2}\right)^{-1}$ **g.** $3\left(\frac{1}{3}\right)^{-1}$ **h.** $\frac{2}{3}\left(\frac{2}{3}\right)^{-2}$

i. $\frac{2^{-1}}{5} \times \left(\frac{1}{4}\right)^{-2}$

5. Write each of the following as the product of **primes** raised to powers:
eg, $28 = 2^2 \times 7^1$, $\frac{3}{8} = 3 \times 2^{-3}$.

a. 12 **b.** 200 **c.** 45 **d.** $\frac{25}{16}$ **e.** $\frac{2}{3}$

f. $\frac{4}{27}$ **g.** $\frac{16}{81}$ **h.** $\frac{64}{81}$ **i.** $2\frac{1}{4}$

Simplifying algebraic expressions with indices

Algebraic expressions may need to be simplified so that indices have positive powers only.

Example C

Write with positive indices: **1.** x^{-5} **2.** $8a^{-2}$ **3.** $\frac{2^{-1}a^{-3}}{5}$

Solution

1. $x^{-5} = \frac{1}{x^5}$ [since $x^{-r} = \frac{1}{x^r}$]

2. $8a^{-2} = 8 \times a^{-2}$

$= 8 \times \frac{1}{a^2}$ [since $a^{-2} = \frac{1}{a^2}$]

$= \frac{8}{a^2}$

3. $\frac{2^{-1}a^{-3}}{5} = 2^{-1} \times a^{-3} \times \frac{1}{5}$ [dividing by 5 is the same as multiplying by $\frac{1}{5}$]

$= \frac{1}{2} \times \frac{1}{a^3} \times \frac{1}{5}$ [since $2^{-1} = \frac{1}{2}$ and $a^{-3} = \frac{1}{a^3}$]

$= \frac{1}{10a^3}$

Example D

Express $\frac{15x^{-3}y^2}{(3x^{-2}y^4)^2}$ without negative indices.

Solution

$\frac{15x^{-3}y^2}{(3x^{-2}y^4)^2} = \frac{15x^{-3}y^2}{3^2(x^{-2})^2(y^4)^2}$ [since $(xy)^n = x^ny^n$]

$= \frac{15x^{-3}y^2}{9x^{-4}y^8}$ [since $(x^n)^m = x^{nm}$]

$= \frac{5x^1y^{-6}}{3}$ [$\frac{x^m}{x^n} = x^{m-n}$ and $\frac{15}{9} = \frac{5}{3}$]

$= \frac{5x}{3} \times y^{-6}$

$= \frac{5x}{3} \times \frac{1}{y^6}$ [since $y^{-n} = \frac{1}{y^n}$]

$= \frac{5x}{3y^6}$

Unit 11.2 Activity 5B: Simplifying expressions with indices

1. Write each of the following without negative indices and as simply as possible:

eg, $3x^{-1} = \frac{3}{x}$

a. x^{-3} **b.** y^{-2} **c.** a^{-5} **d.** b^{-4} **e.** z^{-7}

f. y^{-6} **g.** z^{-11} **h.** $(xy)^{-1}$ **i.** $(abc)^{-1}$ **j.** $(xy)^{-2}$

k. $(2x)^{-1}$ **l.** $(2x)^{-2}$ **m.** $(3a)^{-2}$ **n.** $(2x)^{-3}$ **o.** $3x^{-1}$

p. ab^{-3} **q.** $a^{-3}b^{2}$ **r.** $a^{-3}a^{2}$ **s.** $\frac{x^{-1}}{2}$ **t.** $\frac{x^{-3}}{4}$

u. $\frac{3x^{-2}}{5}$ **v.** $\frac{2a^{-3}}{7}$ **w.** $\frac{4a^{-2}}{9}$ **x.** $\frac{3x^{-2}}{5x^{2}}$ **y.** $\frac{4x^{-3}}{5x^{2}}$

2. Write the following using negative indices for the variable: eg, $\frac{3}{x^2} = 3x^{-2}$

a. $\frac{1}{x^6}$ **b.** $\frac{1}{y^5}$ **c.** $\frac{1}{z^2}$ **d.** $\frac{4}{x}$ **e.** $\frac{5}{y^2}$

f. $\frac{7}{z^5}$ **g.** $\frac{1}{2x^2}$ **h.** $\frac{1}{3a^4}$ **i.** $\frac{1}{5y^4}$ **j.** $\frac{2}{5x^2}$

k. $\frac{3}{4y^3}$ **l.** $\frac{3}{(2x)^2}$ **m.** $\frac{4}{(5y)^2}$ **n.** $\frac{x^2}{(2x)^3}$ **o.** $\frac{y^2}{(2x)^3}$

3. Write the following as simply as possible without using negative indices and as a single fraction:

a. $\frac{4x^{-2}}{8y^{-3}}$ **b.** $\frac{6a^{-3}}{9b^{-4}}$ **c.** $\frac{5x^{-2}}{25x^{-3}}$ **d.** $\frac{5^{-1}x^{-2}}{10^{-1}y^{-4}}$ **e.** $\frac{(2x)^2}{4x^{-3}y}$

f. $\frac{(4x^{-1})^2}{8x^{-1}y^{-3}}$ **g.** $\frac{(2x^{-2})^3 y^{-2}}{x}$ **h.** $\frac{5^{-1}x^{-2}}{5x}$ **i.** $\frac{2^{-1}x^{-1}}{3y}$ **j.** $\frac{3^{-1}x^{-2}}{(3x)^2}$

k. $2^{-1}y^{-2}x$ **l.** $3^{-2}y^{-3}x^{2}$ **m.** $2(3^{-1}x^{-2})$ **n.** $4(5^{-1}x^{-3})y^{2}$

o. $x^{-1} + y^{-1}$ **p.** $x^{-1} + x^{-2}$ **q.** $x(2^{-1}y + y^{-1}x^{-1})$

Rational powers (fractional indices)

Fractional powers are used to express the roots of numbers. For example $x^{\frac{1}{2}}$ is $\sqrt{x}$, the **square root** of x. $x^{\frac{1}{3}}$ is $\sqrt[3]{x}$, the **cube root** of x, and so on.

Generally, the yth root of x can be written:

$$x^{\frac{1}{y}} = \sqrt[y]{x}$$

Proof: The following proof for $y = 3$ can be applied to any value of y.

$$\left[x^{\frac{1}{3}}\right]^3 = x^1 \qquad \text{[since } (x^m)^n = x^{mn}\text{]}$$

$$\therefore\ x^{\frac{1}{3}} = \sqrt[3]{x} \qquad \text{[taking cube roots]}$$

Generalising further:

$$x^{\frac{m}{n}} = \sqrt[n]{x^m} = \left(\sqrt[n]{x}\right)^m$$

Example E

$(-27)^{\frac{5}{3}} = \left[(-27)^{\frac{1}{3}}\right]^5$ [since $(x)^{\frac{m}{n}} = \left[(x)^{\frac{1}{n}}\right]^m$]

$= \left(\sqrt[3]{-27}\right)^5$ [since $(x)^{\frac{1}{n}} = \sqrt[n]{x}$]

$= (-3)^5$

$= -243$

When simplifying algebraic expressions with fractional powers, the answer is often expressed in the form ax^n.

Example F

Write the following in simplest index form:

1. $\sqrt{x}$
2. $3\sqrt[5]{x^2}$
3. $\dfrac{2}{\sqrt[3]{x^2}}$

Solution

1. $\sqrt{x} = \sqrt[2]{x}$ [from the definition of $\sqrt{\ }$]

$= x^{\frac{1}{2}}$ [since $\sqrt[y]{x} = x^{\frac{1}{y}}$]

2. $3\sqrt[5]{x^2} = 3 \times \sqrt[5]{x^2}$

$= 3x^{\frac{2}{5}}$ [since $\sqrt[n]{x^m} = x^{\frac{m}{n}}$]

3. $\dfrac{2}{\sqrt[3]{x^2}} = 2 \times \dfrac{1}{\sqrt[3]{x^2}}$

$= 2 \times \dfrac{1}{x^{\frac{2}{3}}}$ [since $\sqrt[n]{x^m} = x^{\frac{m}{n}}$]

$= 2 \times x^{-\frac{2}{3}}$ [since $\dfrac{1}{x^n} = x^{-n}$]

$= 2x^{\frac{-2}{3}}$

Unit 11.2 Activity 5C: Fractional indices

1. Express the following in index form as simply as possible. eg, $3\sqrt[4]{x^5} = 3x^{\frac{5}{4}}$

a. $\sqrt{a}$ b. $\sqrt{c}$ c. $\sqrt[3]{x}$ d. $\sqrt[5]{a}$ e. $\sqrt[6]{u}$

f. $\sqrt[5]{x^2}$ g. $\sqrt[7]{x^3}$ h. $\sqrt[8]{a^3}$ i. $\dfrac{1}{\sqrt{x}}$ j. $\dfrac{1}{\sqrt{y}}$

k. $\dfrac{1}{\sqrt[3]{a}}$ l. $\dfrac{1}{\sqrt[5]{x}}$ m. $\dfrac{1}{\sqrt[7]{x^2}}$ n. $\dfrac{1}{\sqrt[9]{x^5}}$ o. $\dfrac{1}{\sqrt[7]{a^3}}$

p. $\dfrac{1}{\sqrt[9]{a^8}}$ q. $4\sqrt[3]{a}$ r. $\dfrac{6}{\sqrt[3]{y}}$ s. $\dfrac{8}{\sqrt[3]{x^7}}$ t. $\dfrac{1}{7\sqrt[3]{x^2}}$

u. $\dfrac{2}{5\sqrt[4]{x^3}}$ **v.** $\left(\sqrt{x}\right)^{-2}$ **w.** $\dfrac{\sqrt{x}}{\sqrt[3]{x}}$ **x.** $\dfrac{1}{\sqrt[3]{a}\,\sqrt[3]{a^4}}$ **y.** $\dfrac{\sqrt{x}}{\sqrt[3]{x}\,\sqrt[4]{x}}$

2. Express the following using radical (or root) signs: eg, $x^{\frac{2}{3}} = \sqrt[3]{x^2}$

a. $a^{\frac{3}{7}}$ **b.** $b^{\frac{3}{4}}$ **c.** $x^{\frac{2}{5}}$ **d.** $x^{\frac{-2}{3}}$ **e.** $y^{\frac{-3}{4}}$

3. Express the following in index form and as simply as possible:

a. $\sqrt{xy}$ **b.** $\sqrt{xy^2}$ **c.** $\sqrt{\dfrac{x^2}{y}}$

d. $\sqrt[3]{\dfrac{x^6}{y}}$ **e.** $\sqrt[5]{\dfrac{32x^5}{y^3}}$ **f.** $\sqrt[4]{x^{-5}y^4}$

g. $\sqrt{x} \times \sqrt[3]{a}$ **h.** $\dfrac{1}{\sqrt{a}} \times \sqrt{c}$ **i.** $\dfrac{1}{\sqrt[4]{b^{-2}}} \times \sqrt{b}$

4. Simplify fully, writing with no negative powers or radical signs:

a. $\left(16x^4\right)^{-\frac{3}{4}}$ **b.** $\left(27x^9\right)^{\frac{-2}{3}}$ **c.** $\left(32y^{10}\right)^{-\frac{2}{5}}$ **d.** $\left(\sqrt[5]{243A^5}\right)^3$

e. $\left(\sqrt[6]{64B^{12}}\right)^{-5}$ **f.** $\dfrac{1}{\left(\sqrt[3]{8x^6}\right)^2}$ **g.** $\dfrac{1}{\left(\sqrt[4]{16A^4B^4}\right)^{-2}}$ **h.** $\left(\sqrt[3]{125A^3B^6}\right)^{-2}$

i. $\dfrac{3}{\left(\sqrt{x}\right)^{-3}}$ **j.** $\dfrac{8}{\left(\sqrt{64x^{12}}\right)^{-\frac{5}{6}}}$

Using calculators to evaluate $(-x)^{\frac{m}{n}}$

Calculators which have exponent keys such as $\boxed{x^y}$ or $\boxed{\wedge}$ can simplify calculations involving indices. However, many calculators do not accept a negative base (such as –27.5) raised to a fractional power (such as $\frac{2}{3}$) and if the calculation $(-27.5)^{\frac{2}{3}}$ is entered, an error message results. In such cases the following procedure is used.

Calculation of $(-x)^{\frac{m}{n}}$ where *m* and *n* are integers and *x* is positive

The index $\frac{m}{n}$ should be simplified as much as possible. Then, regardless of whether $\frac{m}{n}$ is negative or positive:

- If n is even, $(-x)^{\frac{m}{n}}$ cannot be calculated. [cannot take square root, 4th root, etc, of a negative number]
- If n is odd and m is even, $(-x)^{\frac{m}{n}} = (x)^{\frac{m}{n}}$ and the answer is positive.
- If n and m are odd, $(-x)^{\frac{m}{n}} = -\left[(x)^{\frac{m}{n}}\right]$ and the answer is negative.

Example G

$(-27.5)^{\frac{2}{3}} = (27.5)^{\frac{2}{3}}$ [since 2 is even and 3 is odd]

$= 9.11$ [using a calculator, to 2 decimal places]

Note: Be sure to put brackets around fractional exponents. On a calculator $(-27.5)^{\frac{2}{3}}$ would be calculated by pressing:

[2] [7] [·] [5] [^] [(] [2] [a b/c] [3] [)] [=]

Unit 11.2 Activity 5D: Calculating with indices

Use a calculator to work out each of the following:

1. 2^4

2. 3^{-2}

3. 0.23^{-3}

4. $(-2)^5$

5. $\left(\frac{2}{3}\right)^{-2}$

6. $\left(1\frac{3}{4}\right)^{\frac{1}{2}}$

7. $(-3.47)^{\frac{3}{5}}$

8. $(-2.49)^{\frac{-1}{2}}$

9. $\left(\frac{12.37}{2.69}\right)^{\frac{14}{17}}$

10. $(3.52^4 - 2.15^3)^{\frac{3}{2}}$

11. $(17.63)^{\frac{1}{2}} - (3.165)^{\frac{1}{3}}$

12. $(15.86)^{\frac{-3}{7}} + (-4.39)^{\frac{2}{7}}$

13. $\dfrac{(11.45)^{0.467}}{(2.78)^{-3.1}}$

14. $\dfrac{(-17.45)^{\frac{13}{27}}}{(15.35)^{\frac{3}{4}}}$

15. $(3.2)^{\frac{1}{3}} \times (2.5)^{\frac{1}{2}} - (1.36)^{\frac{1}{4}}$

16. $(-3.2)^{\frac{2}{3}}$

17. $(56)^{\frac{2}{3}}$

18. $(63)^{\frac{-3}{5}}$

19. $(-4.57)^{\frac{-3}{4}}$

20. $(-31.2)^{\frac{-4}{5}}$

Solution of equations involving indices

Equations with indices often simplify to equations of the form $x^n = C$. By raising the expression on each side of the equation to the **reciprocal of the power of x**, the solution can be found (the reciprocal of n is $\frac{1}{n}$).

Example H

Solve $8x^4 - 3 = 2$, correct to 4 significant figures.

Solution

$8x^4 - 3 = 2$

$\therefore\ 8x^4 = 5$ [adding 5]

$x^4 = \frac{5}{8}$ [dividing by 8]

$\therefore\ (x^4)^{\frac{1}{4}} = \left(\frac{5}{8}\right)^{\frac{1}{4}}$ [taking reciprocal powers]

$\therefore\ x = 0.8891$ (4 sf)

The same technique works for fractional indices (the reciprocal of $\frac{m}{n}$ is $\frac{n}{m}$).

Example I

Solve $(x)^{\frac{15}{23}} = 26.4$, correct to 4 significant figures.

Solution

$$(x)^{\frac{15}{23}} = 26.4$$

$$\therefore \left[(x)^{\frac{15}{23}}\right]^{\frac{23}{15}} = (26.4)^{\frac{23}{15}}$$ [raising both sides to the **reciprocal power** to isolate the variable]

$$\therefore x = 151.3$$ [using a calculator]

Example J

Solve $(2A + 3)^{\frac{-4}{7}} = 8$, to 3 significant figures.

Solution

$$(2A + 3)^{\frac{-4}{7}} = 8$$

$$\therefore \left[(2A + 3)^{\frac{-4}{7}}\right]^{\frac{-7}{4}} = (8)^{\frac{-7}{4}}$$ [raising to the reciprocal power]

$$\therefore 2A + 3 = (8)^{\frac{-7}{4}}$$

$$\therefore 2A = (8)^{\frac{-7}{4}} - 3$$ [subtracting 3]

$$\therefore A = \frac{(8)^{\frac{-7}{4}} - 3}{2}$$ [dividing by 2]

$$\therefore A = -1.49$$

Unit 11.2 Activity 5E: Equations with indices

1. Solve each of the following equations:

a. $x^3 = 2$ **b.** $x^5 = 243$ **c.** $x^{\frac{1}{4}} = 3$ **d.** $a^{\frac{1}{2}} = 5$

e. $a^{\frac{1}{5}} = \frac{2}{3}$ **f.** $y^{-2} = 25$ **g.** $x^{-4} = 256$ **h.** $y^{\frac{3}{4}} = 27$

i. $x^{\frac{2}{5}} = 4$ **j.** $\frac{y^{\frac{4}{3}}}{2} = 8$ **k.** $x^{\frac{-5}{3}} = 32$ **l.** $3y^{\frac{-7}{3}} = 384$

m. $2x^{\frac{1}{2}} - 1 = 3$ **n.** $4x^{\frac{1}{3}} + 1 = 9$ **o.** $2(x - 1)^{\frac{1}{2}} = 4$ **p.** $\frac{x^{\frac{2}{3}} + 1}{2} = 5$

q. $\frac{3}{x^{\frac{1}{3}}}$ **r.** $x^{\frac{-1}{2}} = 2$ **s.** $3y^{\frac{-3}{2}} - 1 = 80$ **t.** $\frac{z^{\frac{-2}{5}}}{3} - 1 = 2$

2. Solve the following equations:

a. $(x)^{\frac{2}{3}} = 16$ b. $(A)^{\frac{3}{7}} = 28$ c. $(A)^{\frac{-7}{3}} = 43$

d. $(p)^{1\frac{3}{4}} = 7$ e. $(p)^{\frac{4}{3}} = -2.3$ f. $(p)^{\frac{3}{5}} = -4$

g. $2(x)^{\frac{2}{3}} = 7$ h. $\left[(x)^{\frac{7}{9}} + 3\right]^{\frac{1}{2}} = 9$ i. $(2x-3)^{\frac{-3}{2}} = 5$

3. The relationship connecting length, L, to cost, C, of a type of material is given by $C = 5L^{\frac{2}{3}} + 6$, where C is given in dollars and L in metres.
   a. What is the cost when 27 metres are purchased?
   b. What length can be purchased for \$100?

4. The relationship between the volume, height and base radius of a shape is given by $V = KR^{\frac{5}{2}}H^{\frac{1}{2}}$, where R is base radius, H is height and K is a constant. When the height and base radius are both 2 the volume is 16:
   a. Find the volume when the radius is 16 and the height 49.
   b. Find the height when the volume is 128 and the radius is 4.
   c. Find the radius when the volume is 486 and the height is 3 125.

5. T and U are related by the formula $T\sqrt[5]{U^4} = 13.5$.
   a. Describe what happens to T as U gets larger.
   b. If the minimum value of T is 500, find the maximum value of U.

Unit 11.2 Graphs and Functions

Topic 6: Working with logarithms

In Topic 6 we continue with the manipulation of algebraic expressions and solving equations:

- Using logarithms.

General properties of logarithms

If a is any positive number then:

- $y = a^x$ and $y = \log_a x$ are equivalent statements.
- $\log_a a = 1$ since $a^1 = a$.
- $\log_a 1 = 0$ since $a^0 = 1$.

Laws of logarithms

The following log laws are derived from the **laws of indices**.

1. $\log_a (xy) = \log_a x + \log_a y$

2. $\frac{\log_a x}{y} = \log_a x - \log_a y$

3. $\log_a x^n = n\log_a x$

Solving simple logarithmic equations

Simple logarithmic equations can be solved by changing the logarithmic equation to an indice equation.

Example A

Q. Solve $\log_2 x = 4$

A. Writing the equivalent indice equation:
If $\log_2 x = 4$ then $2^4 = x$ so $x = 16$

Example B

Q. Solve $\log_x 125 = 3$

A. Writing the equivalent indice equation:
If $\log_x 125 = 3$ then $x^3 = 125$.
We know that $5^3 = 125$ so $x^3 = 5^3$
If the indices are the same then the bases must be equal so $x = 5$.

Example C

Q. Solve $\log_3 81 = 2x + 1$

A. Writing the equivalent indice equation:
If $\log_3 81 = 2x + 1$ then $3^{2x+1} = 81$.

We know that $81 = 3^4$ so $3^{2x+1} = 3^4$.

Both sides of the equation have the same base so the indices are equal:

$$\begin{aligned} 2x+1 &= 4 \\ 2x &= 3 \\ x &= \frac{2}{3} \end{aligned}$$

Logarithm laws can be used to simplify expressions involving logarithms.

Example D

Express $2\log_a x - 3\log_a y + 4$ as a single logarithm.

Solution

$$\begin{aligned} 2\log_a x - 3\log_a y + 4 &= \log_a x^2 - \log_a y^3 + 4 && [n\log u = \log u^n] \\ &= \log_a \frac{x^2}{y^3} + 4 && [\log u - \log v = \log \frac{u}{v}] \\ &= \log_a \frac{x^2}{y^3} + 4\log_a a && [\text{since } \log_a a = 1] \\ &= \log_a \frac{x^2}{y^3} + \log_a a^4 && [n\log u = \log u^n] \\ &= \log_a \frac{x^2 a^4}{y^3} && [\log u + \log v = \log uv] \end{aligned}$$

Unit 11.2 Activity 6A: Further logarithmic problems

1. The population p of a species of creatures after t years is given by $\frac{6(3)^t - 2}{2(3)^t + 5}$ where p is measured in millions.
   a. What is the initial population?
   b. What is the population after 3 years?
   c. After how many years is the population equal to 2 million?
   d. What size does the population get close to as the years pass?

2. The number of cassowaries in a bush after t years is $N = 2\,000 - 1\,000(2)^{-t}$.
   a. How many cassowaries are there in the bush when records are first collected?
   b. How many cassowaries are there in the bush $3\frac{1}{2}$ years later?
   c. What is the maximum number of cassowaries which can live in the bush?
   d. At what time is the population of cassowaries equal to 1 500?

3. The value of a computer is given by the equation $V = 4\,500(0.3)^{0.25t}$, where:
   V = value of computer in dollars, and
   t = time in years since the computer was purchased.
   a. How much was the computer worth initially?
   b. How much was the computer worth after three years?
   c. When will the computer be worth $1 500?
   d. How long is it before the computer loses half its value?

4. The cost of a hectare of land is believed to be $A(2.7)^{kt}$ in kina, where t is the number of years, and A and k are constants. After 3 years the value of the land is K112 486 and after 7 years it is K131 593.

a. Find the values of A and k.

b. Find the value of the land after 10 years.

c. When will the land have a value of K1 000 000?

5. The number of particles of ash deposited up a chimney is given by $\log_{10} N = A\log_{10} h + \log_{10} k$, where N is the number of particles, h is the height of the chimney, and A and k are constants. At a height of 1 metre, the number of particles deposited is 1 000. At a height of 5 metres the number of particles deposited is 25 000. Find the height of a chimney if 100 000 particles are deposited.

6. Solve the following equations for x:

a. $\log_2 x = 6$ **b.** $\log_{3x} x = 4$ **c.** $\log_4 x = \frac{1}{2}$

d. $\log_6 x = 3$ **e.** $\log_x 3 = 2$ **f.** $\log_x 4 = \frac{1}{3}$

g. $\log_x 9 = \frac{1}{2}$ **h.** $\log_x 3 = 1$ **i.** $\log_x 16 = 4$

j. $\log_x 216 = \frac{3}{2}$ **k.** $\log_2 32 = x$ **l.** $\log_3 27 = x + 2$

m. $\log_{\frac{1}{4}} 64 = x$ **n.** $\log_5 25 = 3x + 1$ **o.** $\log_8 4 = x$

7. Express each of the following as a single logarithm:

a. $\log x + \log y + \log z$ **b.** $2\log x + \log y$

c. $\log x + \log y + \log y$ **d.** $\log x + \log y - \log z$

e. $\log a + 3\log b - 2\log c$ **f.** $\frac{1}{2}\log p - \frac{3}{2}\log q$

g. $\log xy + \log z$ **h.** $\log xyz - \log x - \log z$

i. $\log x^2y^3 - \log x + \log y$ **j.** $\log (\frac{xy}{z}) + \log (\frac{z}{x})$

k. $2\log (\frac{x}{y}) - \log x^2$ **l.** $\log (\frac{1}{x}) + \log x + 2\log y$

m. $\frac{1}{2}\log x - \log \sqrt{x} + \log xy$ **n.** $\log \frac{x}{yz} + \log \frac{1}{z} - \log \frac{x}{y}$

o. $2\log xy - \log \frac{x}{y}$ **p.** $2\log_a x + 5$

8. Express the following as logarithms of single numbers:

a. $\log 12 + \log 5$

b. $\log 60 - \log 5$

c. $2\log 5 + \log 3$

d. $\frac{1}{2}\log 9 + \frac{2}{3}\log 8$

e. $\frac{3}{2}\log 4 + \frac{1}{4}\log 81$

f. $\frac{1}{3}\log 27 + \frac{1}{5}\log 32 - \log 7$

g. $\frac{2}{5}\log 32 - \log 4$

h. $3\log 5 - \log 15$

i. $7\log 2 - \frac{1}{2}\log 16 + \log 3$

j. $\log 15 + \log 20 - \log 25$

9. Simplify to a single number, given as a fraction or integer without using a calculator:

eg, $\frac{\log 7}{\log 49} = \frac{\log 7}{\log 7^2} = \frac{\log 7}{2\log 7} = \frac{1}{2}$

a. $\frac{\log 8}{\log 2}$

b. $\frac{\log 25}{\log 125}$

c. $\frac{\log 36}{\log 216}$

d. $\frac{\log 625}{\log 125}$

e. $\frac{\log 64}{\log 32}$

f. $\frac{\log 196}{\log 14}$

g. $\frac{\log 64 + \log 32}{\log 32}$

h. $\frac{\log 7 + \log 7}{\log 49}$

i. $\frac{\log 32 - \log 8}{\log 4}$

j. $\frac{\log 625 + \log 25}{\log 125}$

10. Solve the index equations

a. $2^x \times 3^x = 10$

b. $5^{2x} \times 25^x = 100$

c. $8^x \times 4^{x+1} = 48$

d. $2^x \times 3^{x+1} = 73$

Unit 11.2 Graphs and Functions

Topic 7: Relations and functions

This Topic continues coverage of the content under the heading 'Graphs and Functions' as set out on p. 15 of the Syllabus:

- Solving and sketching linear and quadratic equations.
- Solving domain and range of functions.
- Sketching and writing quadratic, cubic and absolute value functions.
- Sketching rational, exponential and logarithmic functions.

Relations and functions

A relation is a set of **ordered pairs**.

If a relation is graphed on a set of Cartesian axes then the ordered pairs are in the form (x, y).

Example A

A relation is the set of ordered pairs: (2, 3), (2, 4), (3, 3), (3, 4).

When graphed, this relation is shown at right.

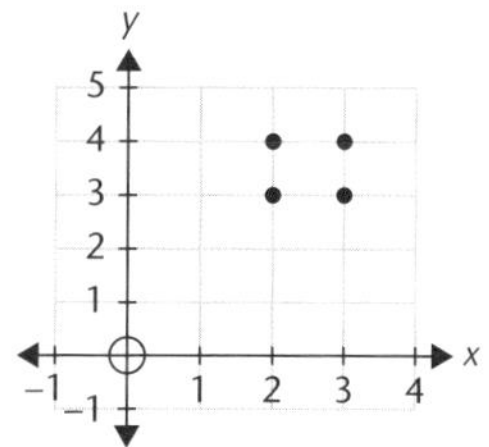

Example B

Some ordered pairs for the linear relation $y = 2x - 3$ are (1, –1) and (0, –3).

If we graph the whole set of ordered pairs for this relation then we will have the straight line graph shown at right.

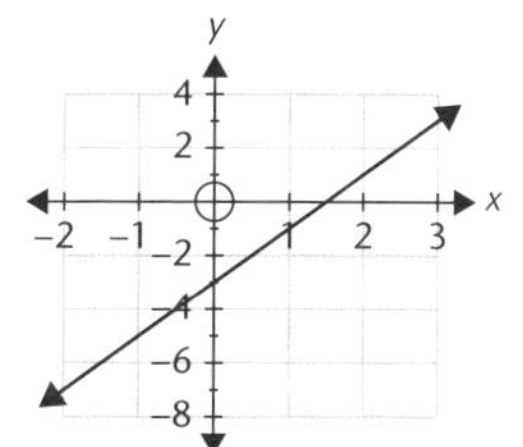

Example C

Some of the ordered pairs for the relation $x^2 + y^2 = 25$ are (3, 4), (–5, 0) (3, –4) (4, –3).

If we graph the whole set of ordered pairs for this relation we get the circle graph shown at right.

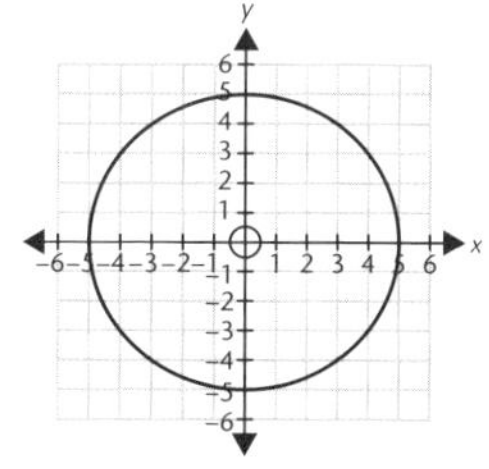

Functions

If a relation is a set of ordered pairs (x, y) then a function is a special type of relation where, for every value of x, there is only one value of y.

The notation $f(x)$ is often used for functions. This enables us to find $f(a)$ which is the value of the function (the y-value) when $x = a$.

For example: the relation (2, 3), (2, 4), (3, 3), (3, 4) is not a function because there are two y-values for each of the x-values, 2 and 3.

The relation $y = 2x - 3$ (Example B above) is a function. This function could be written as $f(x) = 2x - 3$ and a graph of $y = f(x)$ would be the straight line graphed in Example B above.

The relation $x^2 + y^2 = 25$ (Example C above) is not a function because there is more than one y-value for each x-value; eg the points (3, 4) and (3, –4) are on the circle.

The vertical line test for graphs of functions

To determine whether a graphed relation is a function we can use the **vertical line test**:

- If a **vertical line** (a line drawn parallel to the y-axis) is drawn through any part of the graph and it cuts the graph more than once then the relation is **not** a function.

Below are some relations illustrating how the vertical line test is applied.

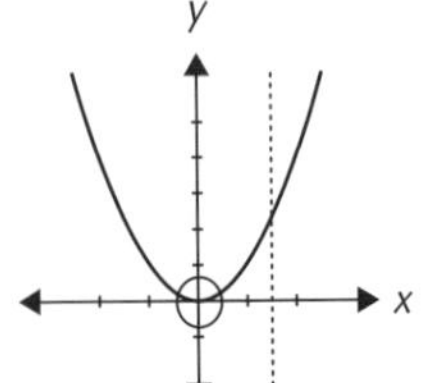

The vertical line cuts the graph once only so this parabola is a function.

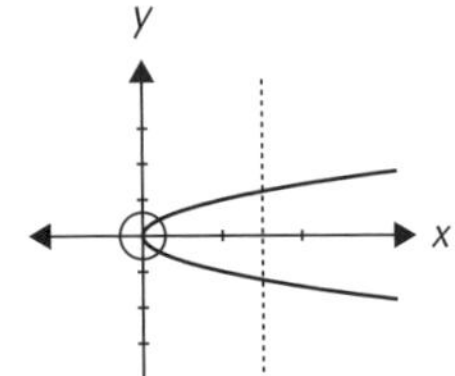

The vertical line cuts the graph more than once so this parabola is not a function.

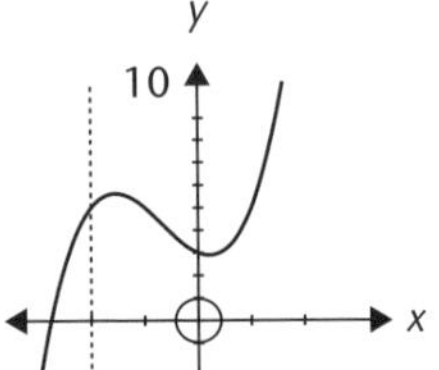

The vertical line cuts the graph once only so this cubic is a function.

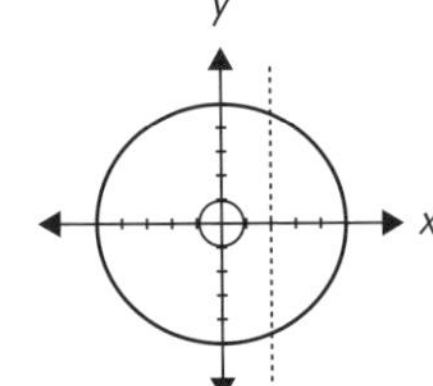

The vertical line cuts the graph more than once so this circle is not a function.

Domain and range of a function

- The numbers which can be entered into a function are called the **domain** of the function.
- The numbers which result from the function rule being applied are called the **range** of the function.

Example D

Find the domain and range and draw the graph of the function: $y = x^2$, $-2 \le x \le 2$.

Solution

The domain is the set $-2 \le x \le 2$ (the x-values for which the function f is defined).

The ordered pair corresponding to any x-value is (x, y) where $y = x^2$.

The graph is drawn by setting up a **table** of sensibly chosen x-values (which are values from the domain) and calculating the corresponding y-values (which are values in the range).

x	–2	–1	0	1	2
y	4	1	0	1	4

The points in the table are **plotted** with the domain on the horizontal axis. Where sensible, the points are joined by a smooth curve. The range is the set of all values on the vertical axis through which a horizontal line can be drawn which will touch or cut the graph. The range shown by the graph is $0 \le y \le 4$.

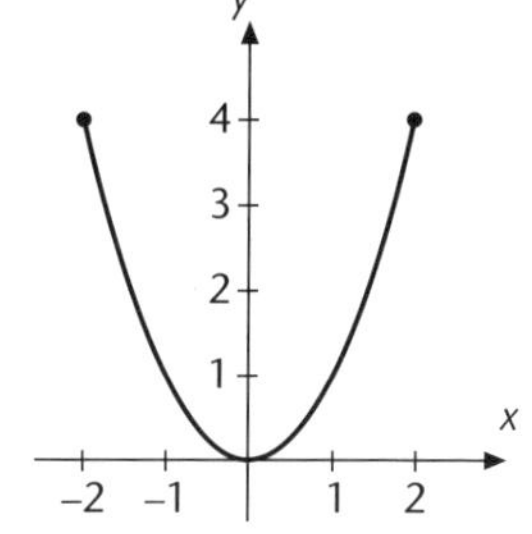

Note: The range $0 \le y \le 4$ can be written using **interval notation** as [0, 4].

Example E

Q. The graph of the function $f(x) = \frac{1}{2}x - 3$ is shown at right.

1. State the domain of the function expressing this in set and interval notation.
2. State the range of the function expressing this in set and interval notation.

A. 1. The domain for this linear function is the set of real numbers as any real number can be halved and have 3 subtracted from it. This is implied by the arrows on either end of the graph, although these are not always included. The domain can be written as $x \in R$ (set notation) or as $(-\infty, \infty)$ in interval notation.

2. The range for this linear function is also the set of real numbers. The range can be written as $y \in R$ (set notation) or as $(-\infty, \infty)$ in interval notation.

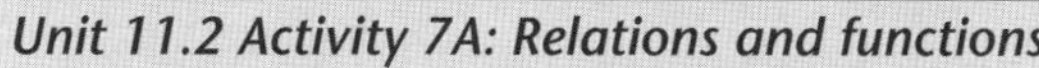

Unit 11.2 Activity 7A: Relations and functions

1. Which of the following relations are functions?

a.

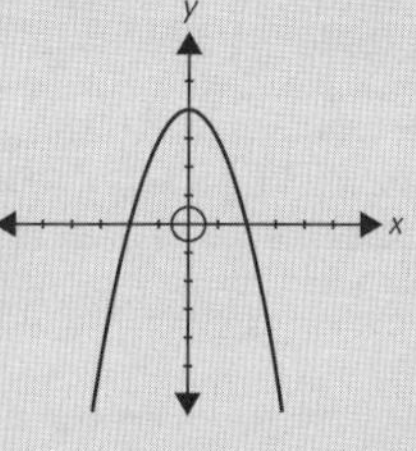

b.

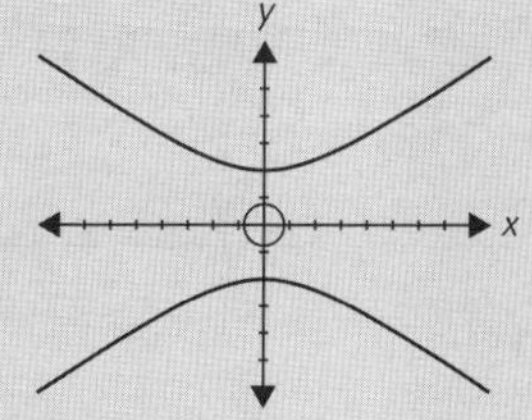

c.

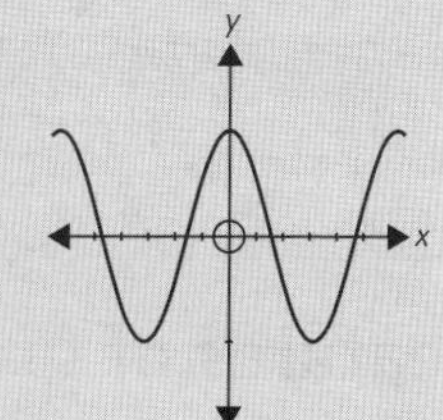

d.

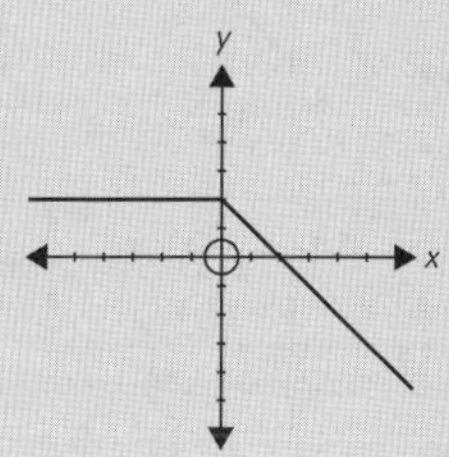

e.

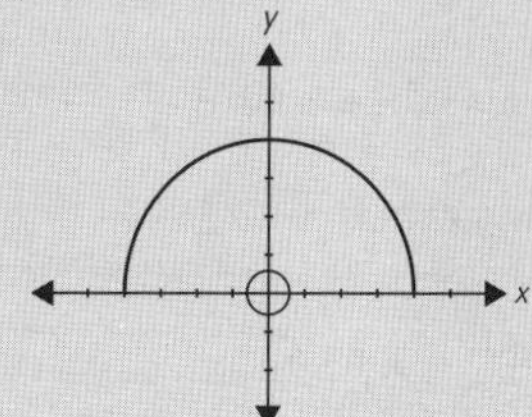

f.

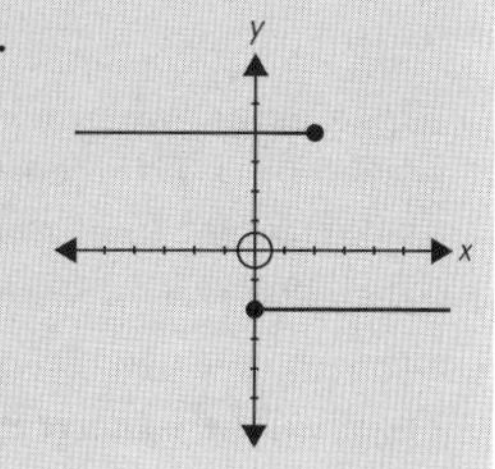

g. 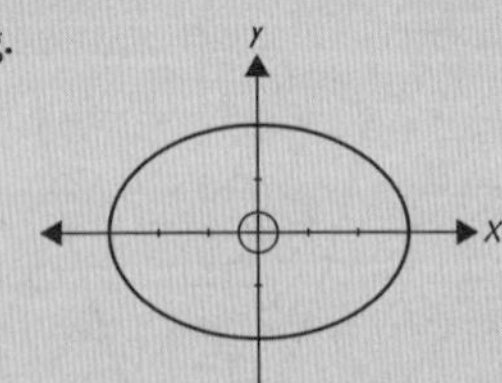

h. 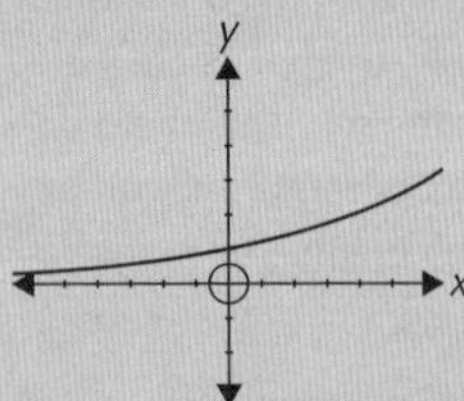

i. 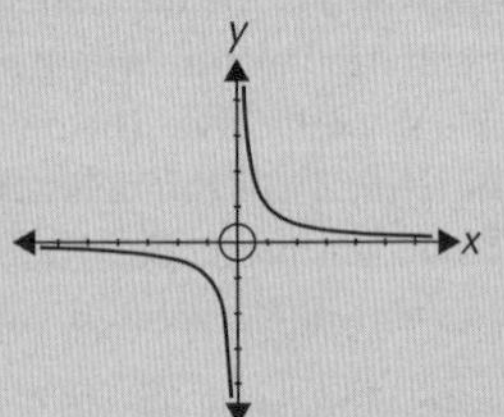

2. Which *one* of the following relations is not a function?
   a. (2, 3), (3, 4), (5, 6), (6, 7)
   b. (2, 3), (4, 4), (3, 3), (5, 4)
   c. (1, 3), (3, 3), (2, 5), (3, 4)
   d. (–1, –3), (1, –1), (–3, –5), (–5, –3)
3. For each of the following functions state the:
   i. domain
   ii. range.
   a. (2, 3), (3, 4), (4, 5), (5, 6)
   b. 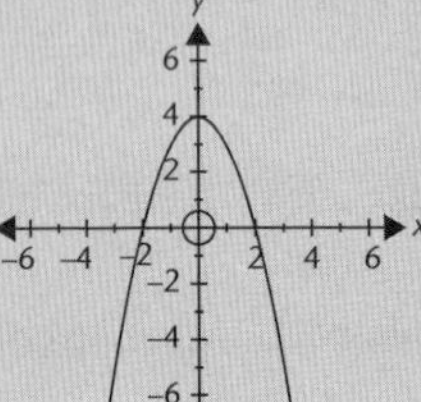

   c. 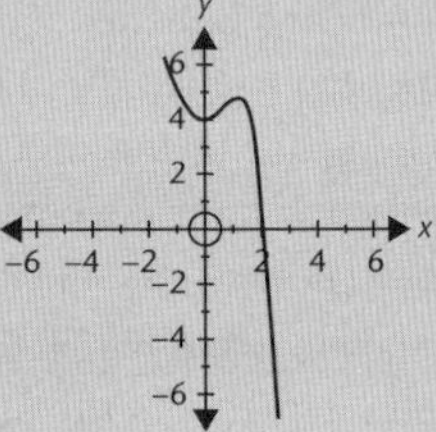

   d. 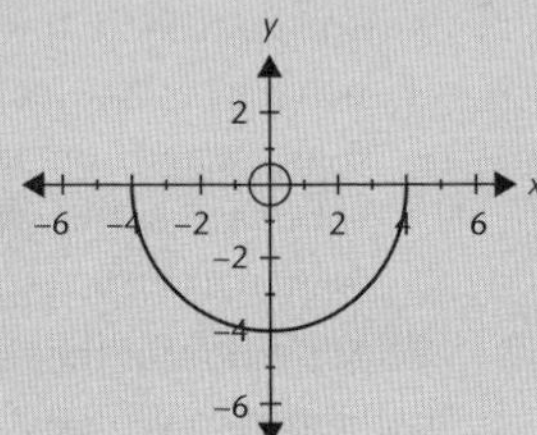

   e. 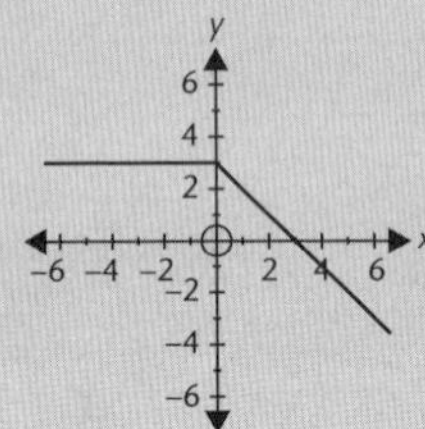

   f. 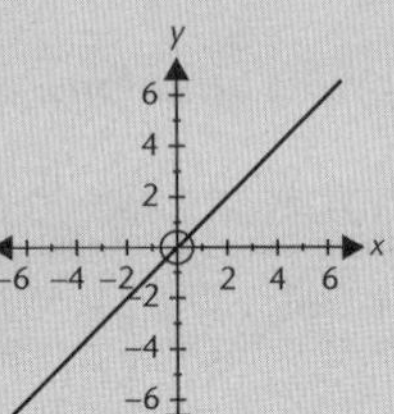

   g. 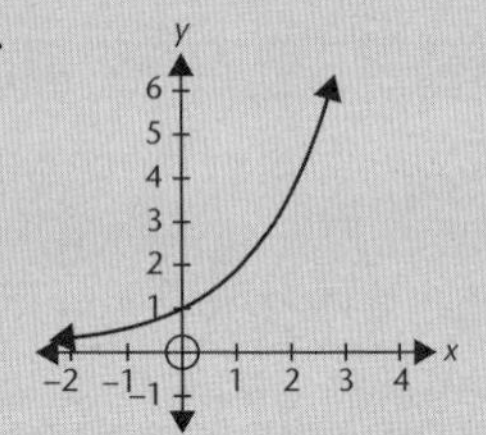

Features of graphs

The following terms are used when describing important features of graphs.

- The point(s) where a graph cuts the **x-axis** is called the ***x*-intercept(s).** x-intercepts can be found by solving y = 0 or $f(x) = 0$.
- The point where a graph cuts the **y-axis** is called the ***y*-intercept**. There will be only one y-intercept on the graph of a function. y-intercepts can be found by solving $x = 0$ or $f(0)$.

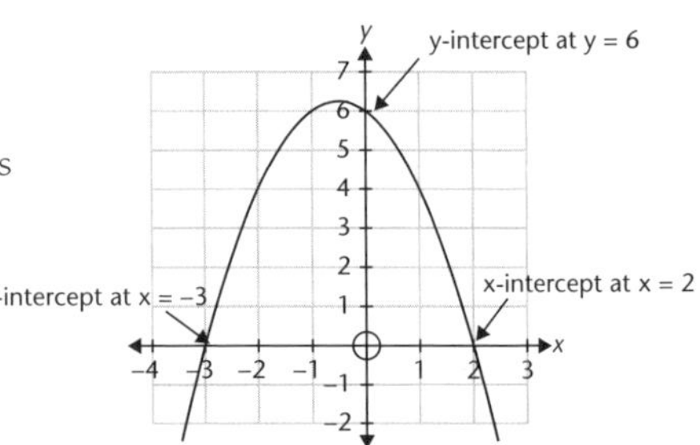

- A function is **increasing** if its values in its range increase as the domain values increase.

An increasing function for x >−4

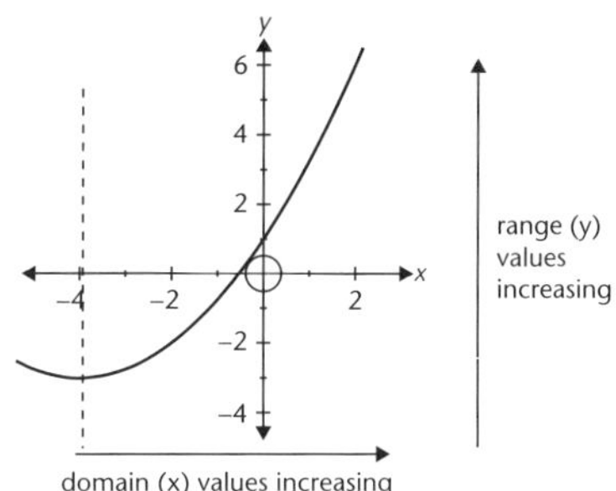

- A function is **decreasing** if its values in its range (y) decrease as the domain (x) values increase.

A decreasing function for x >−2

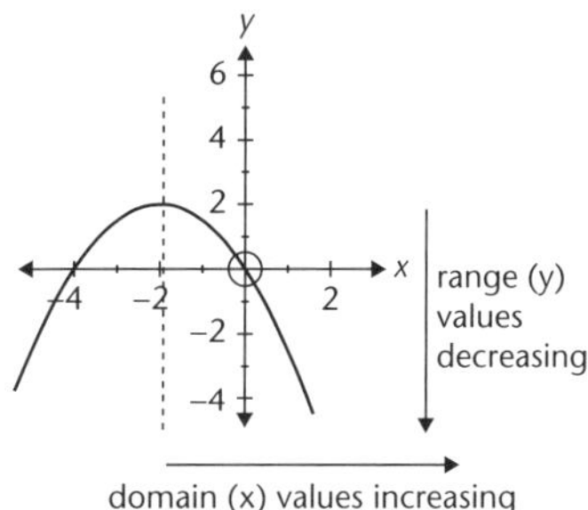

- A function has a **maximum turning point** on its graph when it changes from being an increasing function to being a decreasing function.

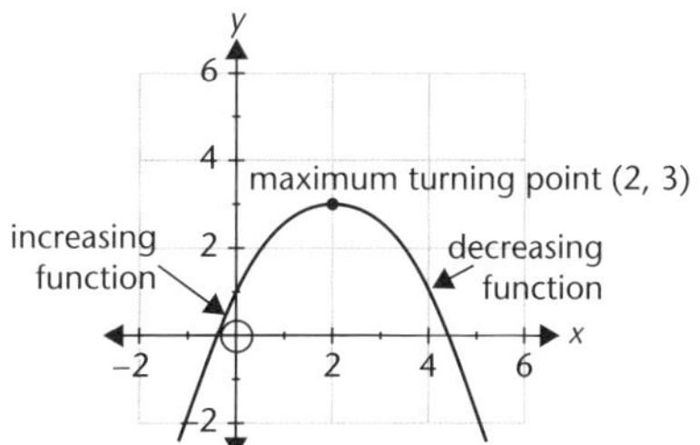

- A function has a **minimum turning point** on its graph when it changes from being a decreasing function to being an increasing function.

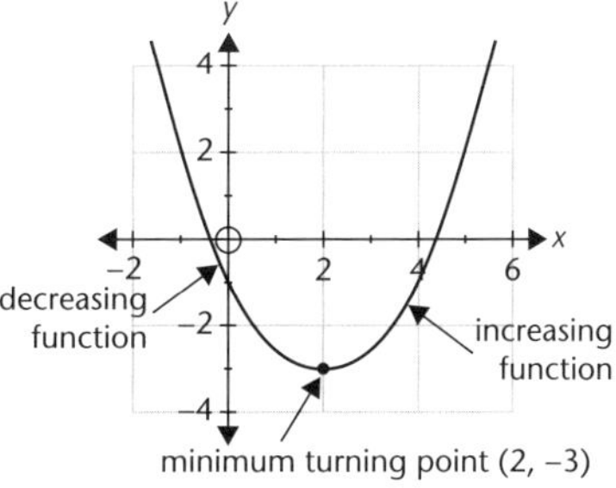

- A function is neither increasing nor decreasing at the turning point.
- Finding the **nature of a turning point** means finding whether the turning point is a maximum or minimum point.

Example E

Q. The function $y = f(x) = -x^2 - 2x + 8$ is graphed at right:

State

1. the coordinates of the x-intercepts.
2. the coordinates of the y-intercept.
3. the coordinates of the turning point.
4. the nature of the turning point.
5. $f(0)$ and what this represents.
6. the values of x for which $f(x)$ is
   - **a.** increasing
   - **b.** decreasing

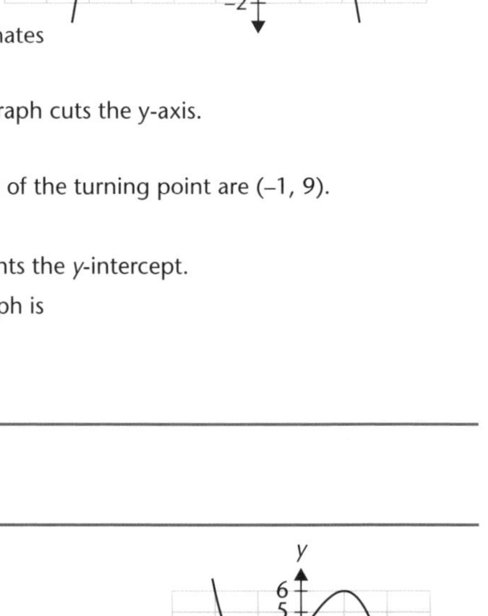

A. **1.** The x-intercepts are the points where

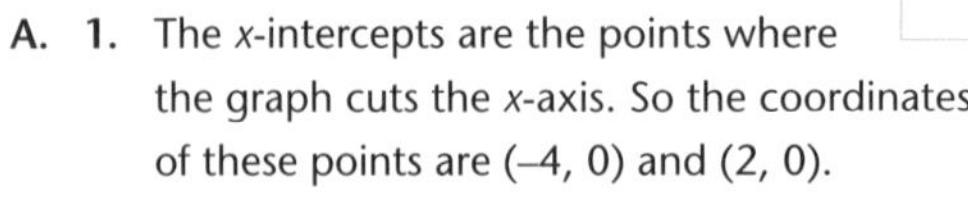

the graph cuts the x-axis. So the coordinates of these points are (–4, 0) and (2, 0).

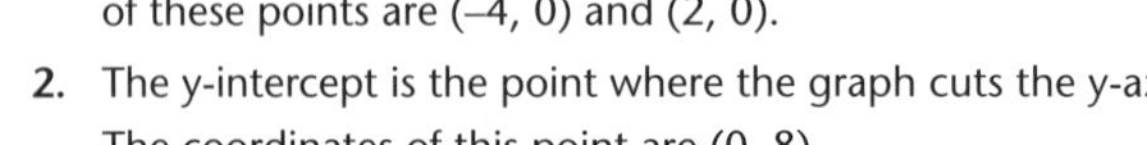

2. The y-intercept is the point where the graph cuts the y-axis. The coordinates of this point are (0, 8).

3. Reading from the graph the coordinates of the turning point are (–1, 9).

4. The turning point is a maximum point.

5. $f(0) = -(0)^2 - 2 \times 0 + 8 = 8$. This represents the y-intercept.

6. The turning point is at $x = -1$ so the graph is

- **a.** increasing for $x < -1$
- **b.** decreasing for $x > -1$

Example F

Q. The cubic function graphed at right has turning points at (-1, 2) and (1, 6).

Find the values of x for which $f(x)$ is

- **a.** increasing
- **b.** decreasing.

A. **a.** The function is increasing for $-1 < x < 1$ (y values increasing as x values increasing)

b. The function is decreasing for $x < -1$ and for $x > 1$. (y values decreasing as x values increasing)

Note: For the cubic graph in example F the minimum turning point at (–1, 2) is not the minimum value of the function and so is called a **local minimum point**. Similarly, the maximum turning point at (1, 6) is called a **local maximum point**.

Unit 11.2 Activity 7B: Functions and graphs

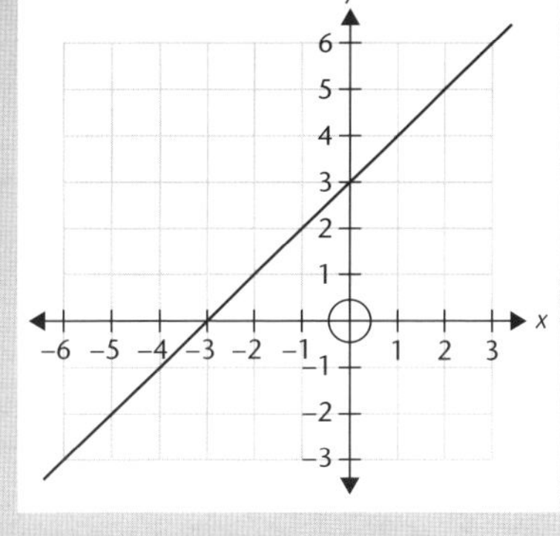

1. For the linear function graphed at right, find:

a. the coordinates of the x-intercept.

b. the coordinates of the y-intercept.

c. the values of x for which the function is increasing.

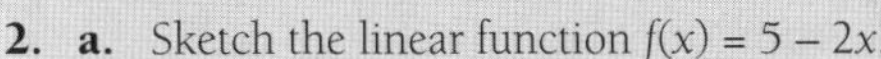

2. a. Sketch the linear function $f(x) = 5 - 2x$.

b. For the function $f(x) = 5 - 2x$ find:

i. the coordinates of the x-intercept.

ii. the coordinates of the y-intercept.

iii. the values of x for which the function is increasing.

iv. $f(1)$

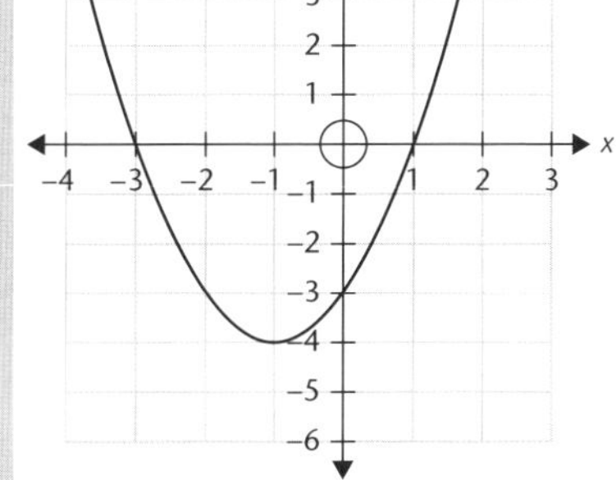

3. The quadratic function $f(x) = x^2 + 2x - 3$ is graphed at right.

For the function $f(x) = x^2 + 2x - 3$ find:

a. the coordinates of the x-intercepts.

b. the coordinates of the y-intercept.

c. the coordinates of the turning point.

d. the nature of the turning point.

e. the values of x for which the function is increasing.

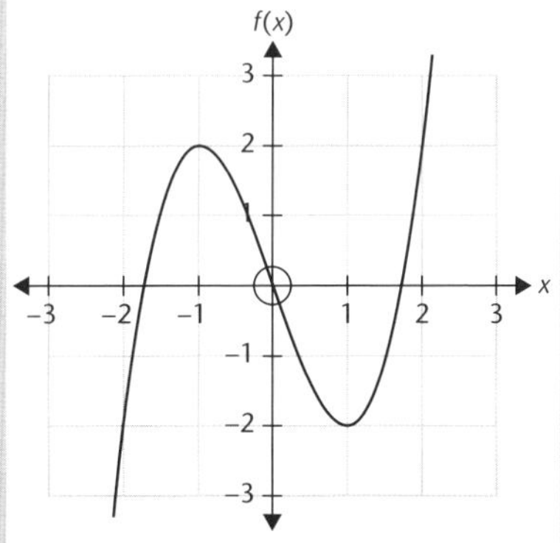

4. The graph of the cubic function $f(x) = x^3 - 3x$ is given at right.

The function has turning points at $x = 1$ and $x = -1$.

For this function find:

a. $f(1)$.

b. $f(-1)$.

c. $f(0)$.

d. the coordinates of the maximum turning point.

e. the coordinates of the minimum turning point.

f. the values of x for which $f(x)$ is

i. increasing

ii. decreasing.

5. The cubic function graphed at right has turning points at (–2, 2) and (1, 4).
Find:

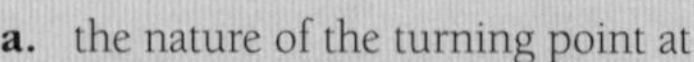

a. the nature of the turning point at:
   i. (-2, 2)
   ii. (1, 4).

b. the values of x for which $f(x)$ is increasing.

c. the values of x for which $f(x)$ is decreasing.

d. the domain of the function.

e. the range of this function.

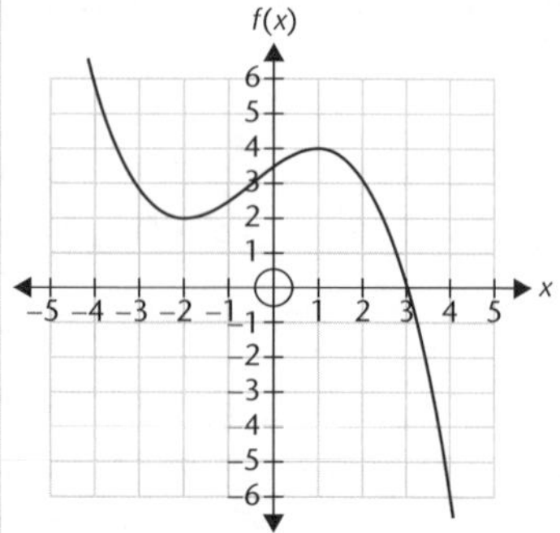

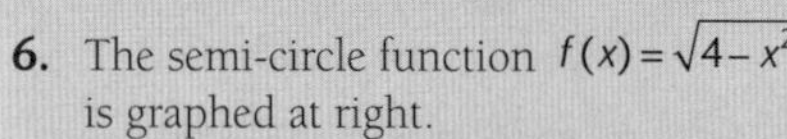

6. The semi-circle function $f(x) = \sqrt{4 - x^2}$ is graphed at right.

a. State the x-intercepts.

b. State the y-intercept.

c. Give the coordinates of the turning point and state the nature of the turning point.

d. Give the domain of the function in interval notation.

e. Give the range of the function in interval notation.

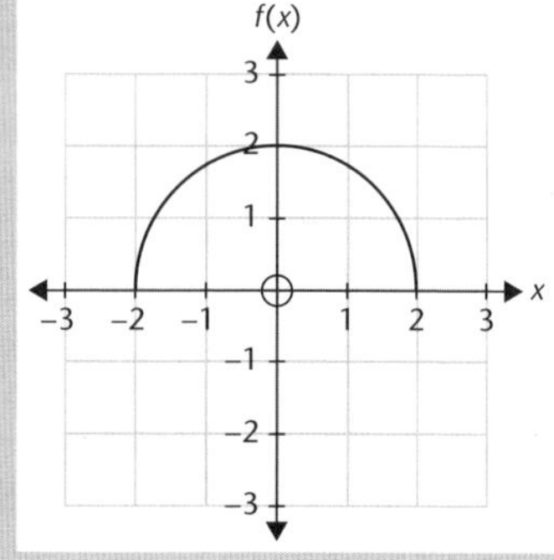

Unit 11.2 Graphs and Functions
Topic 8: Coordinate geometry

This Topic introduces some analytical geometry (also known as 'coordinate geometry') as set out under the heading 'Graphs and Functions' on p. 15 of the Syllabus, in particular:

- The distance between two points.
- The midpoint between two points.

The distance between two points

To find the **distance between two points**, a **right-angled triangle** is used in which the line joining the two points forms the **hypotenuse**. The distance required is found using **Pythagoras' theorem**.

Example A

The distance between $A(-1, 2)$ and $B(2, 7)$ is the length of the hypotenuse on the **triangle** shown.

From the diagram:

- The *horizontal* displacement of B from A is 3.
- The *vertical* displacement of B from A is 5.

By Pythagoras' theorem:

$$AB^2 = 3^2 + 5^2$$
$$= 9 + 25$$
$$= 34$$
$$\therefore\ AB = \sqrt{34}$$
$$= 5.83 \text{ (2 dp)}$$

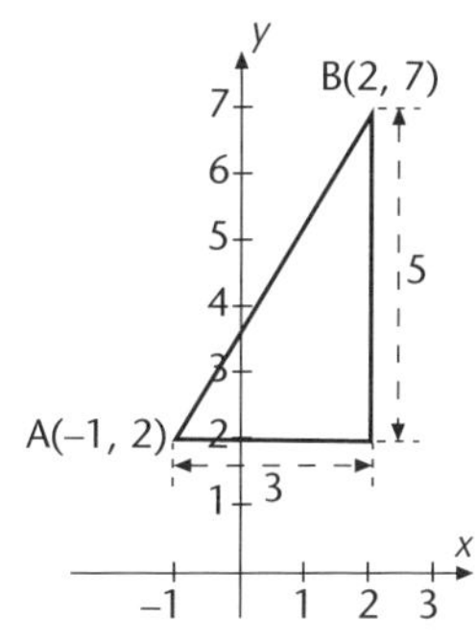

The method shown above can be generalised:

The distance, d, between the points with **coordinates** (x_1, y_1) and (x_2, y_2) is:

$$d = \sqrt{(x_1 - x_2)^2 + (y_1 - y_2)^2}$$

Note: In the formula for distance between two points, it does not matter which point is (x_1, y_1) and which is (x_2, y_2).

Example B

The distance between (2, 3) and (–1, 4) is found by letting (x_1, y_1) be (2, 3), and (x_2, y_2) be (–1, 4) and substituting.

$$d = \sqrt{(x_1 - x_2)^2 + (y_1 - y_2)^2}$$
$$= \sqrt{(2 - -1)^2 + (3 - 4)^2} \qquad \text{[substituting]}$$
$$= \sqrt{3^2 + (-1)^2}$$
$$= \sqrt{10}$$
$$= 3.16 \text{ (2 dp)}$$

A variety of problems can be solved using the formula for distance between a pair of points.

Example C

Show that (1, 2), (2, 6) and (6, 5) are the vertices of an **isosceles** triangle.

Solution

In an isosceles triangle, two of the sides are of equal length.

Length of line joining (1, 2) and (2, 6) is $\sqrt{(2-1)^2 + (6-2)^2} = \sqrt{17}$

Length of line joining (1, 2) and (6, 5) is $\sqrt{(6-1)^2 + (5-2)^2} = \sqrt{34}$

Length of line joining (2, 6) and (6, 5) is $\sqrt{(6-2)^2 + (5-6)^2} = \sqrt{17}$

Two of the lines are of equal length and thus the triangle is isosceles.

Unit 11.2 Activity 8A: Distance between points

1. Find the distance between each of the following pairs of points:

a. (–1, 2) and (3, 6) **b.** (8, 5) and (2, –3) **c.** (–3, 5) and (2, 17)

d. (–1, 4) and (–3, –2) **e.** (2, –5) and (–2, 0)

2. The diagram alongside shows the various points of interest in a town relative to its centre, which is at (0, 0). Find the distance between the following points (all distances are in kilometres; give answers to the nearest tenth of a kilometre):

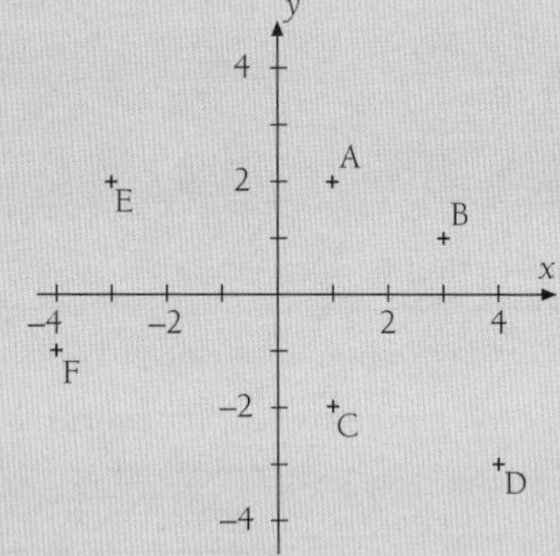

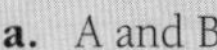

a. A and B

b. E and B

c. B and C

d. E and D

e. A and F

3. A tramper is 14 km west and 15 km north of base at the end of the first day's tramping. At the end of the second day, the tramper is 8 km east and 25 km north of base. How far did the tramper travel during the second day?

4. After a day's sailing since leaving Manus Island the ferry *Neville* is 40 km west and 65 km north of the island. At the same time, the ferry *Cosmo* is 32 km east and 75 km north of the island. What is the distance between them?

5. **a.** Draw the triangle A(5, 1), B(0, 2) and C(4, 6).

b. Show that this triangle is isosceles but not equilateral.

6. Show that the triangle with vertices (1, 1), (7, 3) and (3, 7) is isosceles (has two sides of equal length).

7. Which of the points (0, 1), (–1, 2) or (–2, 5) forms an isosceles triangle with (6, 4) and (1, 9)?

8. **a.** Find the distance between two points on a map which are 1 cm east, 2 cm north of a church and 2 cm west, 6 cm north of the same church.

b. Find the centre of a circle which has a diameter running from (3, 5) to (–1, 7).

9. Find x so that the points $(x, x + 1)$, $(x + 2, x + 3)$ and $(x + 3, 2x + 4)$ form a right-angled triangle.

10. The points $(a, 10)$ and $(–2, 6)$ are 5 units apart. Find the value(s) of a.

Midpoint of a line segment

The **midpoint** of a **line segment** is the point halfway between the end-points of the line segment.

Example D

The diagram shows that the midpoint of the line segment joining (1, 3) and (5, 7) is (3, 5).

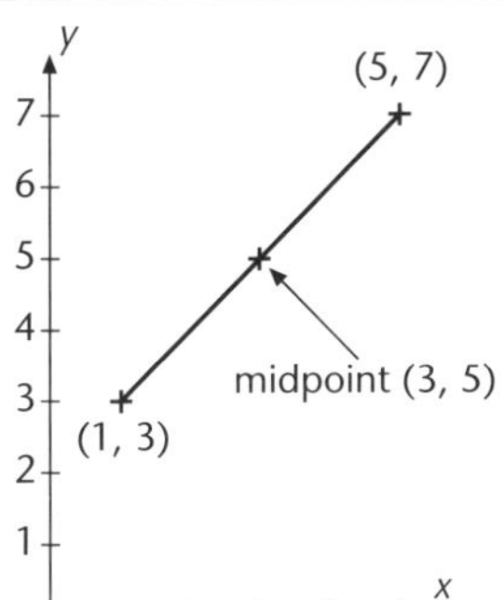

Note: The x-coordinate is $\frac{1 + 5}{2} = 3$.

The y-coordinate is $\frac{3 + 7}{2} = 5$

The result in Example D can be generalised:

> The coordinates of the midpoint of the line segment joining (x_1, y_1) and (x_2, y_2) are: $\left(\frac{x_1 + x_2}{2}, \frac{y_1 + y_2}{2}\right)$

Example E

Using the formula, the coordinates of the midpoint of the line segment joining (–1, 5) and (2, 7) are $\left(\frac{-1 + 2}{2}, \frac{5 + 7}{2}\right) = (\frac{1}{2}, 6)$

Unit 11.2 Activity 8B: Midpoints of line segments

1. Find the coordinates of the midpoint of the line segment joining:

a. (5, 8), (2, 4) **b.** (–3, –4), (–1, 8) **c.** (5, –7), (3, 1)

d. (2, 1), (1, –1) **e.** (–1, 5), (2, 3) **f.** $(3a + 2, a)$, $(a, a – 2)$

2. The diameter of a circle is PQ, where P = (1, 4) and Q = (7, –2). Find the coordinates of the centre of the circle.

3. The midpoint of the line segment joining A(–1, 3) and B is the point (3, –2). Find the coordinates of B.
4. A villager is 3 km east and 5 km north of her village. Another villager is 1 km west and 1 km south of the same village. A third villager is midway between these two villagers. Give her location relative to the village.
5. A ship is 80 km west and 40 km north of a small island which is 20 km west and 20 km north of a port. A canoe is midway between the island and the ship. Give the position of the canoe
   a. relative to the island.
   b. relative to the port.
6. Find the distance between the midpoint of the line segment joining (–1, –2) and (1, 4) and the midpoint of the line segment joining (3, 14) and (7, 12).
7. A circle has centre (2, –7). PQ is a diameter of the circle, where P is the point (a, –5) and Q is the point (8, b). Find the values of a and b.

Unit 11.2 Graphs and Functions

Topic 9: The gradient of a straight line

This Topic continues the analytical geometry as set out under the heading 'Graphs and Functions' on p. 15 of the Syllabus, in particular:

- Finding the equation of a line.

The gradient of a straight line

For any pair of points on a non-vertical straight line, the ratio $\frac{\text{vertical change}}{\text{horizontal change}}$ has a constant value and is called the **gradient** (or **slope**) of the line. Gradient has the symbol $\boldsymbol{m}$.

$$m = \frac{\text{vertical change}}{\text{horizontal change}}$$

Note: If a line is vertical its gradient is undefined.

Example A

The points A(–2, 2) and B(2, 4) lie on the straight line shown.

From the diagram, going *from* A *to* B, the *vertical* change is 2 and the *horizontal* change is 4. The gradient of AB, the straight line going through A and B, is:

$$\frac{\text{vertical change}}{\text{horizontal change}} = \frac{2}{4} = \frac{1}{2}$$

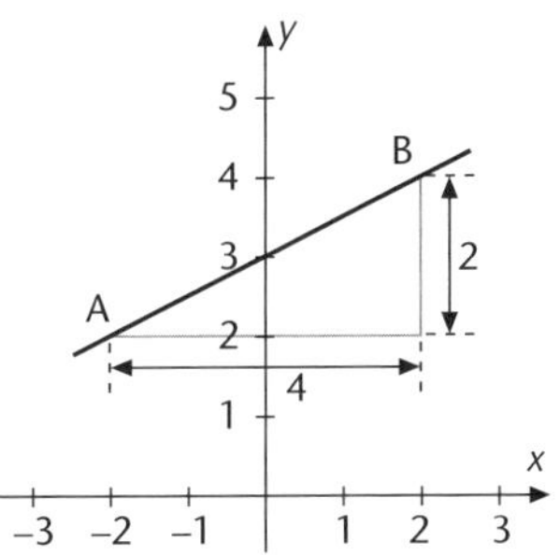

By convention:

- For a vertical change, *upward is positive* and *downward is negative*.
- For a horizontal change, *to the right is positive* and *to the left is negative*.

Note: The gradient in Example A can be calculated by going *from* B *to* A (instead of A to B).

In this case:

- the vertical change is –2 [going down]
- and the horizontal change is –4 [going to the left]

The gradient of the line is now:

$$\frac{\text{vertical change}}{\text{horizontal change}} = \frac{-2}{-4} = \frac{1}{2}\text{, the same value as before.}$$

The procedures in Example A can be used to show that the gradient (m) of the line passing through the points (x_1, y_1) and (x_2, y_2) is:

$$m = \frac{y_2 - y_1}{x_2 - x_1}$$

Note: The gradient of a straight line is equal to the change in the y-value as x increases by 1.

Example B

Find the gradient of the line joining (2, –3) and (–3, 2).

Solution

Let (x_1, y_1) be (2, –3) and (x_2, y_2) be (–3, 2).

$$m = \frac{y_2 - y_1}{x_2 - x_1}$$

$$= \frac{2 - (-3)}{(-3) - 2} \qquad \text{[substituting]}$$

$$= \frac{5}{-5} = -1$$

Note: i. In the formula for gradient, it does not matter which point is (x_1, y_1) and which is (x_2, y_2).

ii. If the point (a, b) is on a line with gradient $\frac{k}{1}$ then so is the point $(a + 1, b + k)$

Example C

Find k so that the line joining $(3, k)$ and (5, 6) has gradient 4.

Solution

a. Algebraic solution:

Let (x_1, y_1) be $(3, k)$ and (x_2, y_2) be (5, 6). The gradient of the line joining $(3, k)$ and (5, 6) is $\frac{6 - k}{5 - 3}$ [using $m = \frac{y_2 - y_1}{x_2 - x_1}$]

Setting this equal to 4 gives:

$$\therefore \frac{6 - k}{5 - 3} = 4 \qquad \text{[substituting]}$$

$$\therefore \frac{6 - k}{2} = 4$$

$$\therefore 6 - k = 8 \qquad \text{[multiplying by 2]}$$

$$\therefore k = -2 \qquad \text{[solving for } k\text{]}$$

b. Graphical solution:

i. Plot (5, 6)

ii. Use the gradient $4 = \frac{4}{1}$ to get another point on the line $(5 + 1, 6 + 4) = (6, 10)$ [see previous notes]

iii. Draw the line through (5, 6), (6, 10)

iv. Read k, the y-value corresponding to $x = 3$, off the line: $k = -2$

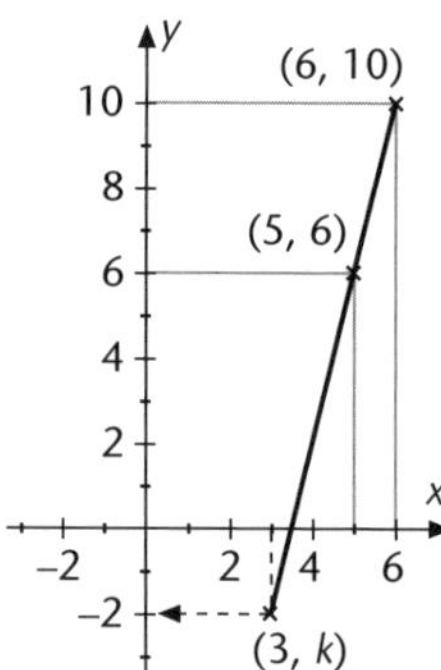

Collinearity

Three or more points A, B, C,... are said to be **collinear** if they all lie on the same straight line.

Three points A, B, C can be shown to be collinear if it can be shown the gradient of AB is the same as the gradient of BC or AC.

Example D

Show that the points A(–3, 5), B(0, 6) and C(6, 8) are collinear.

Solution

Gradient AB $= \dfrac{6-5}{0--3} = \dfrac{1}{3}$ $\qquad [m = \dfrac{y_2 - y_1}{x_2 - x_1}]$

Gradient BC $= \dfrac{8-6}{6-0} = \dfrac{2}{6} = \dfrac{1}{3}$ $\qquad [m = \dfrac{y_2 - y_1}{x_2 - x_1}]$

Since gradient AB = gradient BC ⇒ AB // BC. However, since the point B is common to both lines, AB is the same line as BC, ie the three points A, B, C are collinear.

Unit 11.2 Activity 9A: The gradient of a straight line

1. For each of the following pairs of points find the gradient of the line which goes through them.

a. (–1, 2), (2, 3) **b.** (2, 5), (6, 8) **c.** (–2, 1), (0, 3)

d. (–5, 2), (–6, 4) **e.** (3, 2), (3, 4) **f.** (4, 2), (3, 2)

2. Show that the line going through (1, 3) and (3, 7) has the same gradient as the line going through (2, 8) and (4, 12).

3. Show that the line going through (0, 2) and (2, 8) is parallel to the line going through (1, –1) and (3, 5). [**Note:** if lines are **parallel** they have the same gradient.]

4. Prove that the line going through (2, 5) and (3, 6) has a different gradient to the line going through (–2, 3) and (1, 12).

5. Find out if the line going through (–1, 2) and (2, 8) is parallel to the line going through (–2, 2) and (3, 8).

6. Find out if the line going through (3, 2) and (4, 0) is parallel to the line going through (1, 3) and (–1, 7).

7. Find k so that the line joining (2, k) and (3, 5) has gradient 2.

8. Find k so that the line joining (3, k) and (5, 8) has gradient 2.

9. Find k so that the line joining (–1, 5) and (3, k) has gradient –2.

10. Find k so that the line joining (–2, 6) and (3, k) has gradient 1.

11. Find x so that the line joining (x, 3) and (5, 8) has gradient $\dfrac{5}{4}$.

12. Find x so that the line joining (5, 8) and (x , 12) has gradient $\dfrac{1}{2}$.

13. Find x so that the line joining (2, 5) and (x, 10) has gradient $\frac{-1}{3}$.

14. (a, b) lies on a straight line having gradient $\frac{2}{3}$. If a is increased by 6, what does b increase by?

15. (a, b) lies on a straight line having gradient –2. If a is increased by 2, what does b change by?

16. Show that the three points A(–5, –9), B(1, 3), C(–1, –1) are **collinear** (lie on the same straight line). Plotting points is NOT sufficient.

17. Prove the points A(1, 5), B(–1, –1) and C(4, 14) are collinear.

Drawing a line given its gradient and a point

The gradient of a line can be used to 'step out' from a given point to another point on the line.

Example E

The straight line going through the point (2, 3) with gradient $\frac{1}{2}$, is drawn as follows:

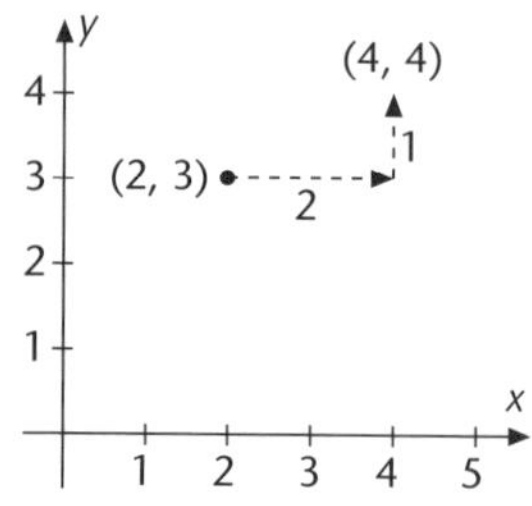

Plot the point (2, 3). Since the gradient is $\frac{1}{2}$, step out from (2, 3) to another point on the line, by making a horizontal change of 2 and a vertical change of 1.

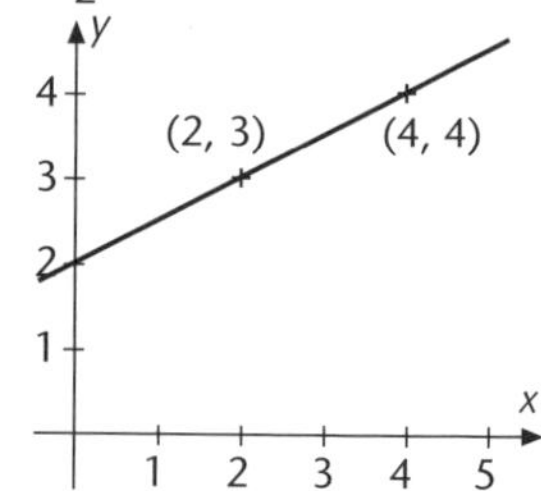

This gives the point

(2 + 2, 3 + 1) = (4, 4).

The straight line is now drawn through (2, 3) and (4, 4).

Unit 11.2 Activity 9B: Drawing straight lines

1. Draw the straight line going through (0, 1) with gradient $\frac{2}{5}$.
2. Draw the straight line going through (–2, 4) with gradient $\frac{3}{4}$.
3. Draw the straight line going through (3, 2) with gradient $\frac{1}{4}$.
4. Draw the straight line going through (4, 3) with gradient $-\frac{2}{3}$.
5. Draw the straight line going through (2, 1) with gradient $-\frac{1}{2}$.
6. Draw the straight line going through (–2, –2) with gradient 0.
7. Draw the straight line going through (–3, 0) with gradient undefined.

Unit 11.2 Graphs and Functions

Topic 10: Equation of a straight line

Analytical geometry can be used to represent geometrical shapes in a numerical way and to extract information from the representation of those geometrical shapes. In this Topic we use analytical geometry to:

- Find the equation of a line.

Introduction

A straight line is an infinite set of points obeying a rule (its equation).

The **equation of a straight line** can be found using various methods.

Example A

A straight line goes through (0, 3) with gradient $\frac{3}{4}$. If (x, y) is a point on the line, a right-angled triangle can be drawn as shown in the diagram.

From the diagram, the *horizontal* change from (0, 3) to (x, y) is x.

The *vertical* change from (0, 3) to (x, y) is $y - 3$.

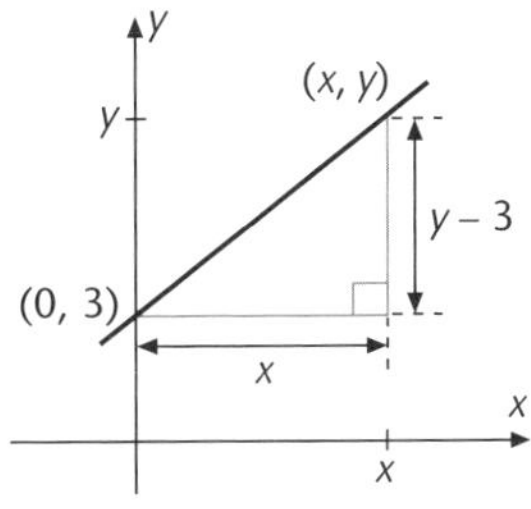

$\therefore$ gradient $= \frac{y-3}{x}$

$\therefore \frac{y-3}{x} = \frac{3}{4}$ [as gradient $= \frac{3}{4}$]

$\therefore y - 3 = \frac{3}{4}x$ [multiplying by x]

$\therefore y = \frac{3}{4}x + 3$ [adding 3]

So the equation of the line is $y = \frac{3}{4}x + 3$.

Generally the equation of any non-vertical straight line can be written in the form:

> $y = mx + c$ where m is the gradient and c is the y-intercept.

The graph of a straight line can be drawn directly from the equation of the line.

Note: It is worth recalling that the **x-intercept** of a graph is the point where the graph cuts the x-axis (at this point, $y = 0$). Similarly, the **y-intercept** of a graph is the point where the graph cuts the y-axis (where $x = 0$).

Example B

Draw the line $y = -\frac{2}{3}x - 2$.

Solution

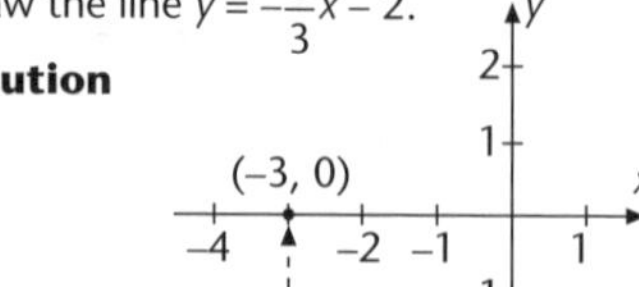

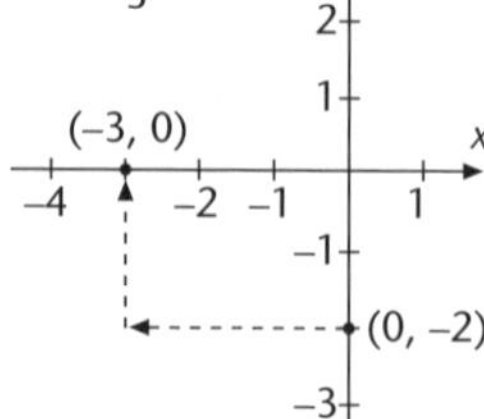

The y-intercept (0, –2) is plotted. The gradient is $-\frac{2}{3}$, so a horizontal change of –3 and a vertical change of 2 from the y-intercept arrives at the point (–3, 0).

A straight line is drawn through the two points (0, –2) and (–3, 0). This is the required line.

Note: A horizontal change of 3 and a vertical change of –2 could have been used, arriving at the point (3, –4).

The equation of a **vertical line** is of the type $x = k$, where k is the x-intercept.
The equation of a **horizontal line** is of the type $y = c$, where c is the y-intercept.

Example C

Find the value of k so that $(k, 3)$ lies on the line $y = \frac{1}{2}x + 1$.

Solution

1. **Graphical solution:**

   Draw the graph of $y = \frac{1}{2}x + 1$ and locate 3 on the y-axis.

   From 3 on the y-axis, go across to the line $y = \frac{1}{2}x + 1$. Now go down to the x-axis and read the value which is $x = 4$.

   Hence $k = 4$.

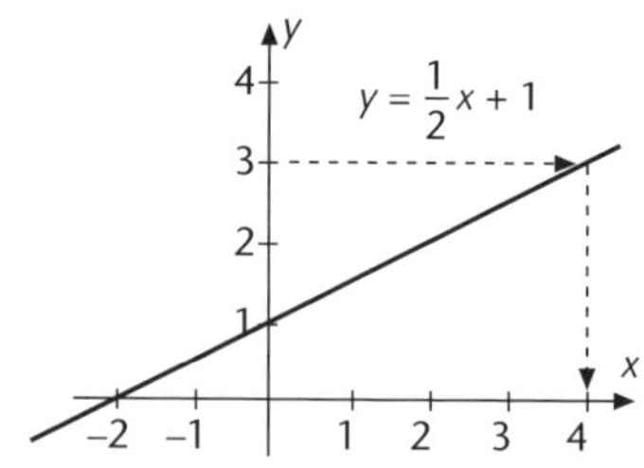

2. **Algebraic solution:**

   Substitute $x = k$ and $y = 3$ (obtained from the coordinates of the point) directly into the equation. This gives:

   $3 = \frac{1}{2}k + 1$

   $2 = \frac{1}{2}k$ [subtracting 1]

   $k = 4$ [multiplying by 2 and rearranging]

Unit 11.2 Activity 10A: Straight line equations

1. Draw graphs of each of the following straight lines:

a. $y = 2x + 1$ **b.** $y = 3x - 4$ **c.** $y = \frac{x}{4} + 1$

d. $y = \frac{2}{3}x + 2$ **e.** $y = -x + 5$ **f.** $y = \frac{-3}{4}x + 3$

g. $y = 3 - \frac{2}{5}x$ **h.** $y = 2$ **i.** $x = -3$

j. $y + 5 = 0$ **k.** $x + 1 = 0$ **l.** $y = -1 - x$

2. For each line in question 1, find the gradient.

3. **a.** $(2, k)$ lies on the graph of $y = 2x + 1$. Find k.
b. $(-3, k)$ lies on the graph of $y = 3x$. Find k.
c. $(2, k)$ lies on the graph of $y = 3$. Find k.
d. $(k, 5)$ lies on the graph of $y = 3x + 2$. Find k.
e. $(k, -6)$ lies on the graph of $y = 2x + 2$. Find k.
f. $(k, -8)$ lies on the graph of $y = 3x + 1$. Find k.
g. $(1\frac{1}{3}, k)$ lies on the graph of $y = \frac{x+5}{2}$. Find k.
h. $(d, 3)$ lies on the graph of $y = 3x + 1$. Find d.
i. $(r, -2)$ lies on the graph of $y = \frac{2}{3}x - 1$. Find r.
j. $(p, -1\frac{1}{2})$ lies on the graph of $y = \frac{1}{3}x + 1$. Find p.

4. Find the x- and y-intercepts of the following straight lines:

a. $y = 2x - 4$ **b.** $y = 3x + 6$ **c.** $y = \frac{2}{3}x - 4$

d. $y = -x + 2$ **e.** $y = 5 - \frac{2}{3}x$ **f.** $y = 2x$

The equation of a straight line when the gradient and a point on the line are known

Example D

Find the equation of the line going through $(-2, 2)$ with gradient 2.

Solution

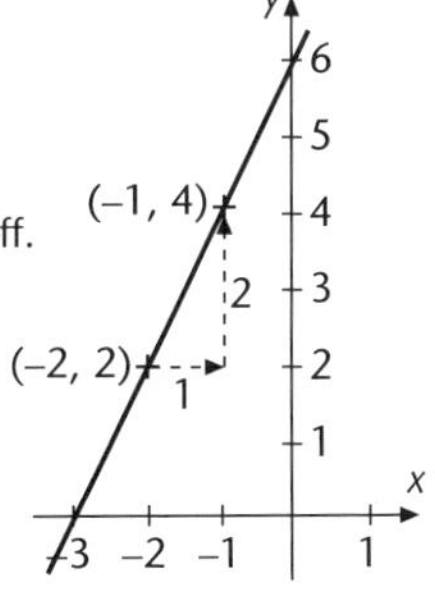

1. **Graphical solution:**

The point $(-2, 2)$ is plotted and the gradient used to find another point on the line. The line is drawn and the y-intercept, 6, is read off. The equation of the line is:

$y = 2x + 6$. [substituting in $y = mx + c$]

2. **Algebraic solution:**

Since the gradient is 2, the equation is $y = 2x + c$.
The point $(-2, 2)$ lies on the line and so will satisfy the equation:

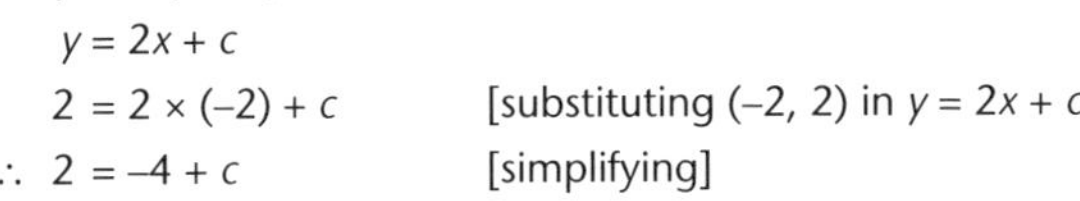

$y = 2x + c$

$2 = 2 \times (-2) + c$ [substituting $(-2, 2)$ in $y = 2x + c$]

$\therefore\ 2 = -4 + c$ [simplifying]

$\therefore\ 6 = c$ [adding 4]

$\therefore$ the equation is $y = 2x + 6$

Using the algebraic method in example D, it can be shown that a straight line with gradient m going through a point (x_1, y_1) has the equation:

$$y - y_1 = m(x - x_1)$$

If the equation of a straight line has to be found quickly or involves awkward numbers then the formula above is very useful.

Example E

The equation of the straight line going through (3, 5) with gradient –2 is:

$y - 5 = -2(x - 3)$	[substituting into $y - y_1 = m(x - x_1)$]
$\therefore\ y - 5 = -2x + 6$	[expanding]
$\therefore\ y = -2x + 11$	[adding 5]

Unit 11.2 Activity 10B: Line equations using gradient and point

Find the equation of the following straight lines:

1. passes through (1, 2) with gradient 1.
2. passes through (2, 7) with gradient 2.
3. passes through (1, 1) with gradient 3.
4. passes through (–1, –1) with gradient 4.
5. passes through (2, –3) with gradient $\frac{1}{2}$.
6. passes through (3, 4) with gradient $\frac{2}{3}$.
7. passes through (–1, 5) with gradient –2.
8. passes through (4, –1) with gradient $\frac{-3}{4}$.
9. passes through (4, –4) with gradient 0.
10. Prove that the equation of a straight line with gradient m, passing through the point (x_1, y_1) has the equation $y - y_1 = m(x - x_1)$.

The equation of a straight line going through two points

The equation is found by using the two points to find the gradient of the line.

Example F

Find the equation of the line going through (3, 3) and (6, 5).

Solution

1. **Graphical solution:**

   The points (3, 3) and (6, 5) are plotted.

   From the horizontal and vertical changes, the gradient of the line is $m = \frac{2}{3}$ and the y-intercept is $c = 1$. By comparison with the line $y = mx + c$, the equation of the line is $y = \frac{2}{3}x + 1$.

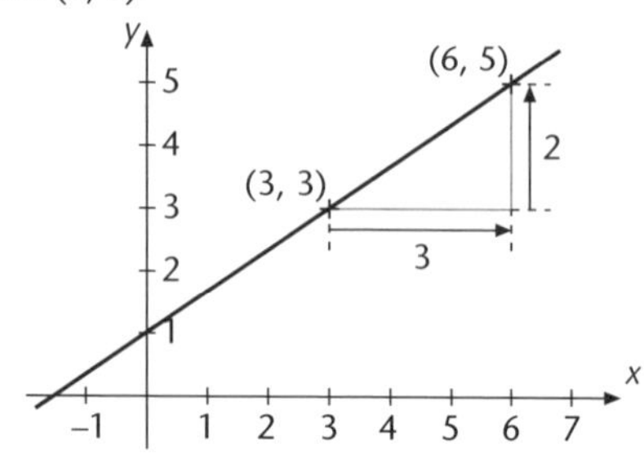

2. **Algebraic solution:**

The gradient of the line is:

$$m = \frac{5-3}{6-3} \qquad \left[\text{substituting into } \frac{y_2 - y_1}{x_2 - x_1}\right]$$

$$= \frac{2}{3}$$

The equation of the line is:

$$y - 5 = \frac{2}{3}(x - 6) \qquad [\text{substituting into } y - y_1 = m(x - x_1)]$$

$$\therefore\ y - 5 = \frac{2}{3}x - 4 \qquad [\text{expanding}]$$

$$\therefore\ y = \frac{2}{3}x + 1 \qquad [\text{adding } 5]$$

Note: The point (3, 3) could have been used for (x_1, y_1), giving the same result.

Using the algebraic method of the above example it is possible to show that a straight line going through the points (x_1, y_1) and (x_2, y_2) has the equation:

$$\boxed{\frac{y - y_1}{x - x_1} = \frac{y_2 - y_1}{x_2 - x_1}}$$

Medians

If ABC is a triangle then the **median** through A is the line joining A to the midpoint M of BC (the median is the line AM shown on the diagram alongside). The medians through B and C are similarly defined.

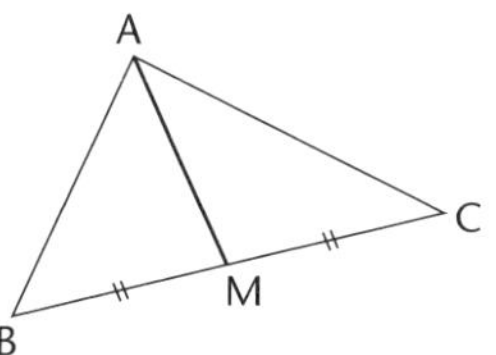

Example G

Consider the triangle ABC whose vertices are A(0, 3), B(4, 8) and C(6, 6). Find the equation of the median through A.

Solution

First it is necessary to find the coordinates of the midpoint of BC.

The midpoint M has coordinates $\left(\frac{4+6}{2}, \frac{8+6}{2}\right) = (5, 7)$ $\qquad \left[\left(\frac{x_1 + x_2}{2}, \frac{y_1 + y_2}{2}\right)\right]$

$\therefore$ the gradient of the line joining A(0, 3) and M(5, 7) is

$$\frac{7-3}{5-0} = \frac{4}{5} \qquad \left[m = \frac{y_2 - y_1}{x_2 - x_1}\right]$$

Hence the equation of the median through A is

$$y - 3 = \frac{4}{5}(x - 0) \qquad [y - y_1 = m(x - x_1)]$$

$$y = \frac{4}{5}x + 3$$

Unit 11.2 Activity 10C: Line equations using two points

Find the equation of the line going through:

1. (1, 0) and (2, 1) **2.** (2, 7) and (3, 9) **3.** (–2, 1) and (2, 4)

4. (–2, 6) and (3, 1) **5.** (–6, –8) and (3, –2) **6.** (1, 4) and (5, 4)

7. (–2, 1) and (4, –2) **8.** (4, –1) and (12, –7)

9. Find the equation of the straight line representing a road on a map, which passes through (2, 5) and (–2, –7).

10. Find the equation of the median of the triangle whose vertices are (0, 0), (0, 4) and (4, 0) and which passes through the point (4, 0).

11. Find the equation of the median of the triangle A(1, –4), B(2, 3) and C(6, 1), which passes through A.

12. A triangle has vertices P(–4, –2), Q(8, –2) and R(–6, 6). Show that the median which passes through R also passes through the origin (0, 0).

Other forms of the straight line equation

The equation of a straight line can be written in different ways, each of which represents the same infinite set of points on a graph. One important form of the straight line equation is called the **general form**.

> The general form of the equation of a straight line is $ax + by + c = 0$, where $a > 0$ and a, b, c are integers.

Example H

Rearrange the equation $y = \frac{3}{4}x + 2$ into general form.

Solution

$$y = \frac{3}{4}x + 2$$

$$\therefore\ 4y = 3x + 8 \qquad \text{[multiplying by 4]}$$

$$\therefore\ 3x - 4y + 8 = 0 \qquad \text{[subtracting } 4y \text{ and swapping sides]}$$

Any equation which can be arranged into the form $ax + by + c = 0$ or $y = mx + c$ is a straight line.

Example I

$3x - 2y + 5 = 6x - 4$ is the equation of a straight line because it can be rearranged into the form $y = mx + c$:

$$3x - 2y + 5 = 6x - 4$$

$$\therefore\ -2y = 3x - 9 \qquad \text{[subtracting } 3x \text{ and 5]}$$

$$\therefore\ y = \frac{-3}{2}x + 4\frac{1}{2}, \text{ which is of the form } y = mx + c.$$

Thus $3x - 2y + 5 = 6x - 4$ is the equation of a straight line with gradient $= \frac{-3}{2}$ and y-intercept $= 4\frac{1}{2}$.

Note: $3x - 2y + 5 = 6x - 4$ can also be rearranged into general form $3x + 2y - 9 = 0$.

Unit 11.2 Activity 10D: Other forms of the straight line equation

1. Which of the following are equations of straight lines?

a. $y = \frac{2}{x}$ **b.** $y = 2x^2 - 3$ **c.** $y = 3x + 1$

d. $2x + 3y = 6$ **e.** $y = |x| + 1$ **f.** $y = x^3 - 3$

g. $y = x^2 - 5x$ **h.** $y = 2y + 3x - 2$ **i.** $y = 2^x$

2. Each of the following is the equation of a straight line. In each case write it in the form: $y = mx + c$.

a. $3y - 2x = 5$ **b.** $2y - 3x = -7$ **c.** $4y - 3x + 5 = 0$

d. $3y - 2x = y + x - 4$ **e.** $3y - \frac{2}{3} = 4x + \frac{5}{2}$ **f.** $\frac{x}{2} + \frac{y}{3} = 1$

3. a. Show that any equation of the form $ax + by + c = dx + ey + f$, where a, b, c, d, e, and f are real numbers, is that of a straight line.

b. Which of the following are straight line equations?

i. $2x^2 + 3y = 2$ **ii.** $2y - x + 7 = 4x$ **iii.** $y = 2^x + 3$

iv. $y = x^2 + 3$ **v.** $\frac{4}{x} + y = 2$ **vi.** $3x + 4y = 1 - x - y$

Linear modelling (problems in context)

Example J

A tank contains 3 litres of water. More water is run into the tank at a constant **rate** of 0.5 litres per hour.

1. Fill in the table to show the amount of water in the tank at each time.

Time (t) in hours	0	1	2	3	4	5
Volume (V) in L	3					

2. Plot V against t and find the equation relating V to t.

3. If the tank can hold 6.3 L, how long will it take to fill the tank?

Solution

1.

t	0	1	2	3	4	5
V	3	3.5	4	4.5	5	5.5

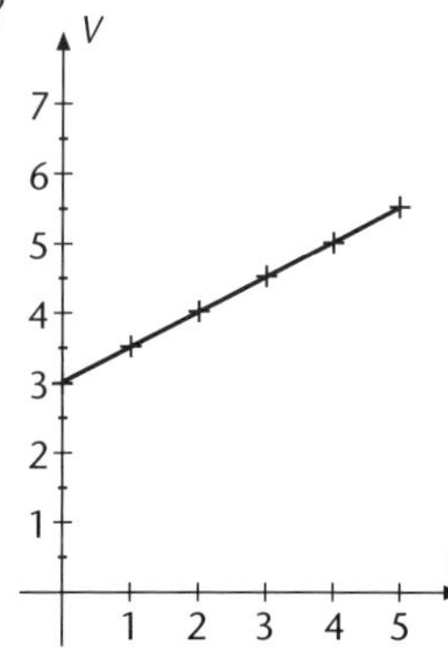

2. The graph is a straight line with gradient $\frac{1}{2}$ and V-intercept 3, hence $V = \frac{1}{2}t + 3$.

3. When the tank is full $V = 6.3$.

$\therefore\ 6.3 = \frac{1}{2}t + 3$ [substituting into $V = \frac{1}{2}t + 3$]

$\frac{1}{2}t = 3.3$ [subtracting 3]

$\therefore\ t = 6.6$ hours, so the tank fills in 6.6 hours.

In the previous example, the gradient of the straight line is the rate at which the tank fills. In many similar situations, the gradient is the **rate of change**.

Unit 11.2 Activity 10E: Linear modelling

1. An elastic band initially of length 20 cm is stretched with a steadily increasing force. Each increase in force of 1 newton causes an increase in length of 0.125 cm.

a. Copy and fill in this table:

Forces applied in newtons (F)	0	1	2	3	4
Length of band in cm (L)					

b. Draw a graph of length against force and find the equation relating the length of the band to the stretching force.

c. The band will snap when it reaches a length of 27.6 cm. Find the force that will cause the band to snap.

2. A tank contains 10 litres of water. A small hole is drilled in it and water escapes at a constant rate of 0.12 litres per minute.

a. Draw a graph of volume of water, v, against time, t.

b. Find the equation relating volume to time.

c. At what time is the volume 5.2 L?

3. A pupil reads 0.8 pages of a book per minute. At the start of a reading lesson she begins reading at the top of page 13.

a. Draw a graph of how many pages she has read, P, against time, t.

b. Find the equation relating how many pages she has read with the time she has been reading during the lesson.

c. What is the maximum number of pages the book can have if she finishes it in a 1-hour reading lesson?

4. A company car whose initial purchase price is $20 000 is depreciated in value at the rate of 25 c a kilometre.

a. Find the equation relating value of the car, v, to the distance, d, it has travelled.

b. How far will the car have to travel before it has a value of $15 000?

5. A tank is filled at a constant rate. 10 minutes after filling is started, the tank contains 4.8 litres of water. After 35 minutes the tank contains 7.3 litres of water.

a. Find the rate at which the tank is being filled.

b. Find the initial volume of fluid in the tank.

c. Find how long it takes to fill, if the tank has a maximum capacity of 60 L.

6. Find the equation of the volume, y, of water in a tank if after 2 minutes there are 11 litres in it and the tank is filling at the rate of 4 litres a minute. (Let x represent the number of minutes since filling commenced.)

7. A car is 30 km from a town 10 minutes after the start of its journey. 25 minutes later it is 65.42 km from the town.
   a. Find the equation relating distance, in kilometres, d, from the town to time travelled in hours, t.
   b. How close was the car to the town when it began its journey?

8. a. A student with careless habits was given the task of recording weights of some specimens on different days. It is well known that the weight increases at a constant rate with time. The student forgets his scales on two of the days and guesses the weights. Here are the results which the student hands in:

weight (g)	0.55	0.68	0.95	1.05	1.35	1.55	1.75
day	1	2	3	4	5	6	7

   Which were the days he forgot his scales?

   b. The same student from the previous problem correctly weighed another specimen on 5 different days. He recorded the weights but not the days he weighed them on. He remembers that he did the last one yesterday and the previous weighing 4 days before that. The specimen increases its weight at a steady rate. The weights he has recorded are 0.77, 1.25, 1.49, 1.85, 2.33. Yesterday was December 30. What were the dates of his previous readings?

Unit 11.2 Activity 10F: Equation of a straight line—multiple choice

1. The gradient and the y-intercept of the straight line with equation $y = 3 - 2x$ are, respectively:

   A. 3, 2 **B.** 3, –2 **C.** –2, 3 **D.** 2, 3

2. The equation of the straight line graphed at right is:

   A. $3x + 5y = -15$ **B.** $5x - 3y = 15$
   C. $3x - 5y = 15$ **D.** $5x + 3y = -15$

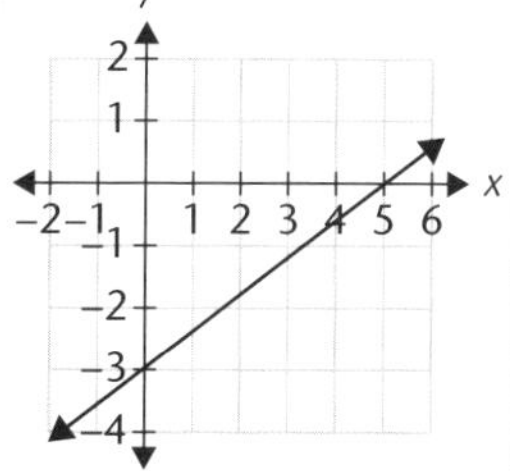

3. The gradient of the straight line going through the points (2, 3) and (5, 7) is:

   A. $\frac{4}{3}$ **B.** $\frac{3}{4}$
   C. $-\frac{4}{3}$ **D.** –1

4. The equation of the straight line with a gradient of 3 and going through the point (–1, 4) is:

   A. $y = 3x + 7$ **B.** $y = 3x + 1$ **C.** $y = 3x - 13$ **D.** $3x - y = 1$

5. The equation of the line passing through the point (2, 3) and parallel to the x-axis is:

   A. $x + y = 5$ **B.** $y = 3$ **C.** $x = 2$ **D.** $y - x = 1$

6. A candle burns at a rate of 6 cm per hour. If the candle was 25 cm high when it was first lit, which one of the following equations would represent the height of the candle (H cm) after t hours?

A. $H = 6t - 25$ **B.** $t = 25 - 6H$ **C.** $H = 6(25 - t)$ **D.** $H = 25 - 6t$

7. Which one of the following lines is parallel to the y-axis?

A. $y = x$ **B.** $y = -x$ **C.** $y = 2$ **D.** $x = -3$

8. The equation of the line graphed is:

A. $y = \frac{3}{4}x + 1$ **B.** $y = \frac{2}{3}x + 1$

C. $y = \frac{3}{2}x + 1$ **D.** $y = \frac{3}{4}x - 1$

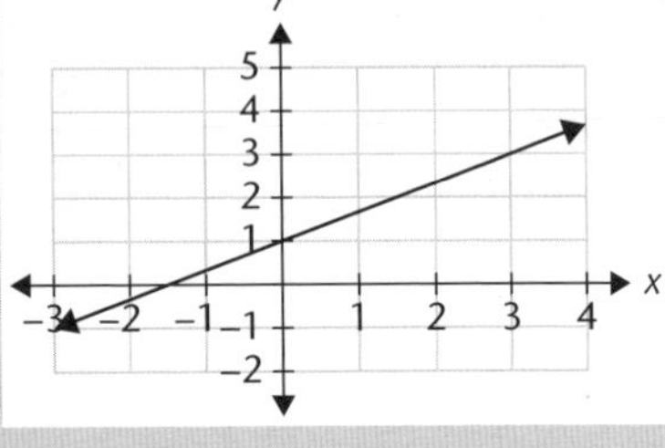

9. The x-intercept of the line $2x = 3y - 6$ is:

A. -3 **B.** -2

C. 2 **D.** $\frac{3}{2}$

10. Which one of the following points is *not* on the straight line $x + 2y = -3$?

A. $(5, -4)$ **B.** $(-1, -2)$ **C.** $(0, -1.5)$ **D.** $(3, -3)$

Unit 11.2 Graphs and Functions
Topic 11: Parallel and perpendicular lines

In this topic we use analytical geometry methods to:
- Find the equation of parallel and perpendicular lines.
- Find the equation of a perpendicular bisector.
- Find the equation of an altitude.

Parallel lines

Two lines are **parallel** if they have the *same gradient*.

> If $y = m_1x + c_1$ is parallel to $y = m_2x + c_2$,
>
> then $m_1 = m_2$.

Example A

Find the equation of the line parallel to $y = 2x + 3$ which goes through (2, 5).

Solution

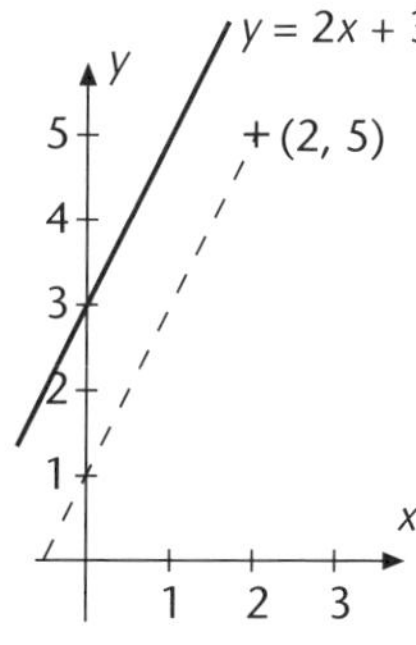

1. Graphical solution:
   The line required, shown dotted, has gradient 2 because it is parallel to the line $y = 2x + 3$.
   From the graph, the y-intercept of the line is 1.
   The equation of the line is found by substituting $m = 2$ and $c = 1$ in $y = mx + c$, to give $y = 2x + 1$.
2. Algebraic solution:
   The equation of the straight line is found by substituting $m = 2$ and $(x_1, y_1) = (2, 5)$
   into $y - y_1 = m(x - x_1)$ giving:
   $y - 5 = 2(x - 2)$
   $\therefore\ y - 5 = 2x - 4$ [expanding]
   $\therefore\ y = 2x + 1$ [adding 5]

Note: The **graphical method** is simpler but is limited to problems involving ***integral*** gradients and intercepts.

Perpendicular lines

Perpendicular lines meet at **right angles** (90°). The following example *compares* the gradients of two perpendicular lines and shows that the product of the gradients of the two perpendicular lines is –1.

Example B

Draw the line $y = 2x + 1$ and the line perpendicular with the same y-intercept. Find the gradient of the perpendicular line.

Solution

The perpendicular line (dotted) goes through (0, 1) and (2, 0). Going from (0, 1) to (2, 0), the gradient of the perpendicular line is:

$$\frac{y_2 - y_1}{x_2 - x_1} = \frac{0 - 1}{2 - 0} = \frac{-1}{2}$$

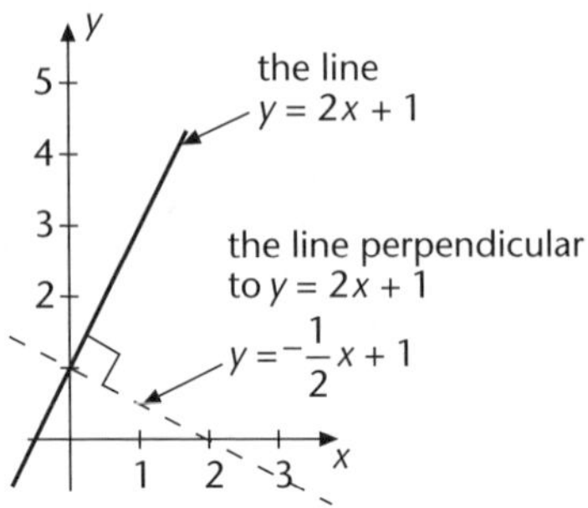

Note: (gradient of $y = 2x + 1$) × (gradient of perpendicular line) is $2 \times \frac{-1}{2} = -1$

The above result is true for any two perpendicular lines (except horizontal and vertical lines) and can be written:

> If m_1 and m_2 are the gradients of two perpendicular lines,
> then $m_1 \times m_2 = -1$

In other words, $m_2 = \frac{-1}{m_1}$, or $m_1 = \frac{-1}{m_2}$ (the gradient of one perpendicular line is equal to the opposite reciprocal of the gradient of the other perpendicular line).

Note: In the case of horizontal lines (gradient = 0) and vertical lines (gradient undefined) this rule cannot be applied since $m_1 \times m_2$ has no value.

Example C

Find the equation of the line perpendicular to $y = \frac{2}{3}x + 1$, going through the point (2, 1).

Solution

The gradient of $y = \frac{2}{3}x + 1$ is $\frac{2}{3}$ [comparing with $y = mx + c$]

$\therefore$ the gradient of the new line is $-\frac{3}{2}$. [since $\frac{2}{3} \times -\frac{3}{2} = -1$]

$\therefore\ y - 1 = -\frac{3}{2}(x - 2)$ [substituting into $y - y_1 = m(x - x_1)$]

$\therefore\ 2y - 2 = -3\ (x - 2)$ [multiply by 2]

$\therefore\ 2y - 2 = -3x + 6$ [expanding]

$\therefore\ 2y = -3x + 8$ [adding 2]

$3x + 2y - 8 = 0$ [adding $3x$, subtracting 8]

Note: The equation could be written in the form $y = -\frac{3}{2}x + 4$. [dividing $2y = -3x + 8$ by 2]

Unit 11.2 Activity 11A: Parallel and perpendicular lines

1. Find the gradients of lines perpendicular to lines with the following gradients:

a. 3 **b.** 10 **c.** −2 **d.** −5 **e.** $\frac{1}{2}$ **f.** $\frac{3}{4}$

g. $\frac{-2}{3}$ **h.** $\frac{-1}{6}$ **i.** 1.5 **j.** −3.2 **k.** 0 **l.** 0.001

2. Find the equation of the line:

a. parallel to $y = 2x - 1$ going through (3, 4).

b. parallel to $y = \frac{x}{2} + 5$ going through (–1, 4).

c. parallel to $y = 6 - 3x$ going through (2, 1).

d. parallel to $y = 4 - \frac{2}{3}$ going through (3, 4).

e. parallel to $2x - 3y + 11 = 0$ going through (2, –1).

f. parallel to $3x + y = 2y - 4x + 7$ going through (–2, –7).

3. Find the equation of the line:

a. perpendicular to $y = 3x + 2$ going through (6, 4).

b. perpendicular to $y = 2x - 7$ going through (4, 5).

c. perpendicular to $y = \frac{-1}{2}x$ going through (1, –3).

d. perpendicular to $y = \frac{2}{3}x + 4$ going through (4, –3).

e. perpendicular to $3y - 2x + 11 = 0$ going through (2, –2).

f. perpendicular to $4x + 3y = 15$ going through (1, –3).

g. perpendicular to $2x - 3y = 4x + 2y - 11$ going through (2, 10).

h. perpendicular to $4x - 2y = 5y + 7$ going through (–1, 0).

4. Find k if the lines $kx - 2y + 5 = 0$ and $3x + 4y + 1 = 0$ are:

a. perpendicular **b.** parallel

Coordinate geometry and the straight line

Many problems involve a number of aspects of coordinate geometry, along with the equation of a line.

Example D

Find the equation of the line parallel to $3x - 4y = 5$ and passing through the midpoint of the line segment AB, where A = (–1, 8) and B = (–3, –4).

Solution

The midpoint of AB is

$$\left(\frac{-1 + (-3)}{2}, \frac{8 - 4}{2}\right) = (-2, 2)$$

The line $3x - 4y = 5$ can be rearranged into $y = mx + c$ form

$$-4y = -3x + 5 \qquad \text{[subtracting } 3x\text{]}$$

$$y = \frac{3}{4}x - \frac{5}{4} \qquad \text{[dividing by } -4\text{]}$$

which has gradient $\frac{3}{4}$ (so any line parallel to it will also have gradient $\frac{3}{4}$).

The equation of the line with gradient $\frac{3}{4}$ and passing through the point (–2, 2) is:

$$y - 2 = \frac{3}{4}(x - (-2)) \qquad [y - y_1 = m(x - x_1)]$$

$$y - 2 = \frac{3}{4}x + \frac{3}{2} \qquad \text{[expanding]}$$

$$y = \frac{3}{4}x + \frac{7}{2} \qquad \text{[adding 2]}$$

The **perpendicular bisector** of a line segment is a line which cuts the line segment in half and is at right angles to it.

Example E

Find the equation of the perpendicular bisector of the line segment joining (–3, 5) and (1, 3).

Solution

The midpoint is

$$\left(\frac{-3+1}{2}, \frac{5+3}{2}\right) = (-1, 4)$$

The line passing through (–3, 5) and (1, 3) has gradient

$$\frac{3-5}{1-(-3)} = \frac{-2}{4} = \frac{-1}{2}$$

∴ the perpendicular bisector has gradient 2 $\qquad [\frac{-1}{2} \times 2 = -1]$

∴ the equation of the perpendicular bisector is

$$y - 4 = 2(x - (-1)) \qquad [y - y_1 = m(x - x_1)]$$

$$y - 4 = 2x + 2 \qquad \text{[expanding]}$$

$$y = 2x + 6 \qquad \text{[adding 4]}$$

Unit 11.2 Activity 11B: Coordinate geometry and straight lines

1. Find the equation of the line parallel to $y = 2x + 3$ going through the midpoint of the line segment joining (2, 3) and (4, 5).
2. Find the equation of the line parallel to $y = x + 11$ going through the midpoint of the line segment joining (–2, –1) and (6, 9).
3. Find the equation of the line parallel to $y = \frac{(x+3)}{2}$ going through the midpoint of the line segment joining $(\frac{-1}{2}, 3)$ and $(2\frac{1}{2}, -1)$.
4. Find the equation of the line perpendicular to the line segment joining (–4, 5) and (–2, 1) and going through its midpoint.
5. Find the equation of the line perpendicular to the line segment joining (2, 0) and (–4, 6) and going through its midpoint.
6. Find the equation of the perpendicular bisector of the line segment joining (2, 4) and (2, 6).

7. Find the equation of the perpendicular bisector of the line segment joining the origin and (4, 0).

8. Find the equation of the perpendicular bisector of the line segment joining (0, 6) and (6, 0).

9. Find the equation of the perpendicular bisector of the line segment joining (5, 2) and (–3, 4).

10. A triangle has vertices (0, 0), (0, b) and (a, 0). Find the condition in terms of a and b that the median of the triangle passing through (0, 0) is also the perpendicular bisector of the line segment joining (a, 0) and (0, b).
[Remember: a median of a triangle is a straight line which passes through a vertex of a triangle and the midpoint of the opposite side.]

Altitudes

If ABC is a triangle then the **altitude** through A is the line perpendicular to BC which goes through A (the line AP shown in the diagram alongside is the altitude through A).

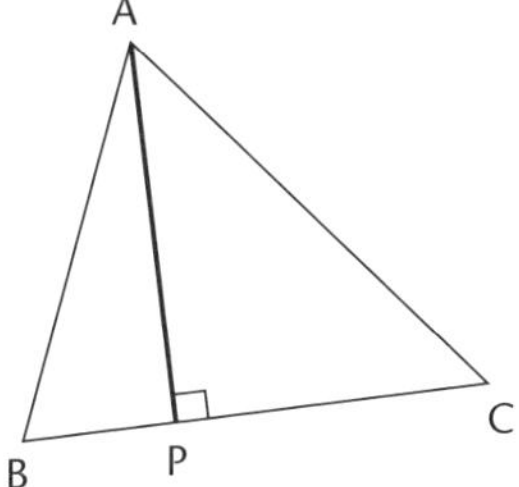

Example F

1. Find the equation of the altitude which passes through A(0, 4) in the triangle ABC, where B is (0, 0) and C is (–9, 3).

2. Find the distance of A from BC.

Solution

1. gradient of BC is

$$\frac{3-0}{-9-0} = \frac{-1}{3} \qquad [m = \frac{y_2 - y_1}{x_2 - x_1}]$$

Since the altitude through A is perpendicular to BC it must have gradient 3. $\qquad [m_2 = \frac{1}{m_1}]$

Hence the equation of the altitude is:

$$y - 4 = 3(x - 0) \qquad \text{[using } y - y_1 = m(x - x_1)\text{]}$$

$$\therefore\ y = 3x + 4 \qquad \text{[rearranging]}$$

2. The equation of BC is:

$$y - 0 = \frac{-1}{3}(x - 0) \qquad \text{[using } y - y_1 = m(x - x_1)\text{]}$$

$$\therefore\ y = \frac{-1}{3}x$$

Finding the point of intersection of the lines,

$y = \frac{-1}{3}x$ and $y = 3x + 4$ gives the point (–1.2, 0.4) [solving simultaneously]

The distance of A(0, 4) from this point of intersection is:

$$\sqrt{(0 - -1.2)^2 + (4 - 0.4)^2} = 3.795 \text{ (3 dp)}$$

Distance of a point from a line

The perpendicular distance from a point P(A, B), to the line $ax + by + c = 0$ is the shortest possible distance from this point to the line. It can be shown that this distance is $\left|\frac{aA + bB + c}{\sqrt{a^2 + b^2}}\right|$, where $\left|\frac{aA + bB + c}{\sqrt{a^2 + b^2}}\right|$ is the **modulus** or **absolute value** of $\frac{aA + bB + c}{\sqrt{a^2 + b^2}}$.

Note: The absolute value signs around $\frac{aA + bB + c}{\sqrt{a^2 + b^2}}$ mean that, no matter whether $\frac{aA + bB + c}{\sqrt{a^2 + b^2}}$ is positive or negative, the positive value of $\frac{aA + bB + c}{\sqrt{a^2 + b^2}}$ is selected. For example, $|2| = 2$ and $|-2| = 2$.

Example G

Find the distance of the point (0, 4) from the line $y = \frac{-1}{3}x$.

Solution

First the line $y = \frac{-1}{3}x$ has to be put in the form $ax + by + c = 0$.

$\frac{1}{3}x + y = 0$ [rearranging]

$x + 3y = 0$ [multiplying both sides of the equation by 3]

Thus $a = 1, b = 3, c = 0$.

Now substitution is made into the distance formula giving:

$\text{distance} = \left|\frac{1 \times 0 + 3 \times 4 + 0}{\sqrt{1^2 + 3^2}}\right| = 3.795$ [substituting into $\left|\frac{aA + bB + c}{\sqrt{a^2 + b^2}}\right|$]

Note: This is the same as the value obtained in Example F.

It can be seen that this method is much simpler than the method outlined in Example F, which involved finding first the coordinates of the point where the altitude met the opposite side of the triangle. The formula for the distance between two points was then used.

At time of writing this formula is not included in the tables of formulae distributed at NCEA examinations, so it will be necessary to memorise this formula if it is to be used in the examination.

Unit 11.2 Activity 11C: Altitude and distance of a point from a line

1. A triangle ABC is formed by A(2, –2), B(1, 5) and C(–5, –1).

- **a.** Prove that the triangle is isosceles.
- **b.** Find the equation of the altitude which goes through A.

2. Consider the triangle A(9, 0), B(1, –1), C(3, 8). Find the equation of the altitude passing through B.

3. A triangle PQR has vertices P(3, 0), Q(15, 5), R(15, 0).

- **a.** Show that triangle PQR is not a right-angled isosceles triangle.
- **b.** Find the equation of the altitude going through R.

4. A triangle ABC has vertices A(1, 4), B(–2, –3) and C(6, 1). Find the equation of the altitude from A to the line BC.

5. Find the perpendicular distance from (3, 4) to the line $3x + 4y + 10 = 0$.

6. Find the perpendicular distance of (24, –7) from $24x - 7y + 50 = 0$.

7. Prove the perpendicular distance of (–3, 2) from $y = \frac{-3}{2}x$ is $\frac{5}{\sqrt{13}}$.

8. Determine whether the shortest distance from (1, 2) to $y = x + 7$ is $3\sqrt{2}$ or $\frac{3}{\sqrt{2}}$.

Unit 11.2 Activity 11D: Parallel and perpendicular lines—multiple choice

1. A straight line perpendicular to another straight line with gradient $\frac{4}{3}$ will have a gradient of:

A. $\frac{3}{4}$ **B.** $-\frac{3}{4}$ **C.** $\frac{4}{3}$ **D.** $-\frac{4}{3}$

2. Straight lines parallel to the straight line with equation $3x + 5y = 15$ will have a gradient of:

A. 5 **B.** **3** **C.** $-\frac{3}{5}$ **D.** $\frac{5}{3}$

3. Straight lines perpendicular to the straight line with equation $2x - 3y = 7$will have a gradient of:

A. $\frac{3}{2}$ **B.** $-\frac{2}{3}$ **C.** $\frac{2}{3}$ **D.** $-\frac{3}{2}$

4. Which one of the following straight lines is parallel to the straight line with equation $y = 5 - 2x$?

A. $4x + 2y = 5$ **B.** $6x - 3y = 8$ **C.** $2x - y = 3$ **D.** $5x - y = 10$

5. Which one of the following lines is parallel to the line with equation $2y = 3x - 4$?

A. $y = 3x + 4$ **B.** $2y = 2x - 1$ **C.** $2y = 3x + 1$ **D.** $3y = 3x + 4$

6. Which one of the following straight lines is perpendicular to the straight line $2x + y = 5$?

A. $x - y = -5$ **B.** $x - 2y = 5$ **C.** $-2x -y = 7$ **D.** $x + 2y = 3$

7. The equation of the straight line that is perpendicular to the line $y = 3 - \frac{1}{2}x$ and going through the point (–1, 1) is:

A. $y = 2x + 1$ **B.** $y = 2x + 3$ **C.** $y = 3 - 2x$ **D.** $y = -2x - 1$

8. Which one of the following is the equation of a straight line that is perpendicular to the straight line with equation $y = \frac{9}{2}$?

A. $y = -\frac{2}{9}$ **B.** $x = -\frac{2}{9}$ **C.** $\frac{2}{9}x + y = 3$ **D.** $y = \frac{-2x}{9}$

9. The equation of the perpendicular bisector of the points (1, 2) and (3, 4) is:

A. $y = 5 - x$ **B.** $y = x + 1$ **C.** $y = -x + 1$ **D.** $y = x - 1$

10. The gradient of the perpendicular bisector of the points (3,-1) and (-2, k) is $-\frac{3}{2}$. The value of k is:

A. $-\frac{7}{3}$ **B.** $-\frac{13}{3}$ **C.** $-\frac{1}{3}$ **D.** $\frac{7}{3}$

Unit 11.2 Graphs and Functions

Topic 12: The intersection of two straight lines

In this Topic we use analytical geometry (also known as coordinate geometry) methods to:

- Find coordinates of the point of intersection of two lines.

The point where two straight lines meet can be found *graphically* or *algebraically*. An algebraic solution involves solving two **simultaneous equations**.

Example A

Find the point of **intersection** of the lines $y = x + 2$ and $y = 3x$.

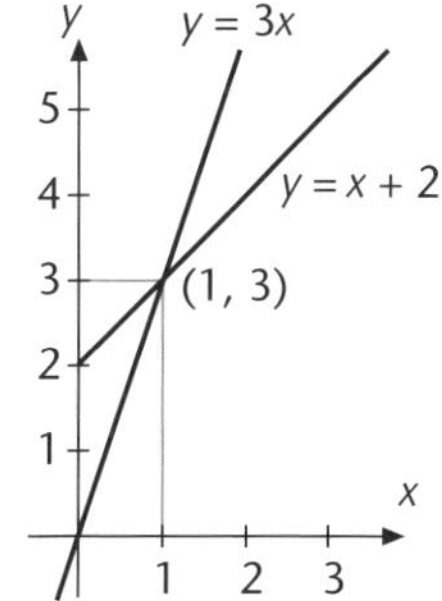

Solution

1. **Graphical solution:**
   The lines $y = 3x$ and $y = x + 2$ are drawn and the point of intersection (1, 3) is read off.
2. **Algebraic solution:**
   The equations of the two lines are solved simultaneously by elimination.

$$y = 3x \quad \text{.... (1)}$$
$$y = x + 2 \quad \text{.... (2)}$$
$$0 = 2x - 2 \quad \text{[subtracting (2) from (1)]}$$
$$2x = 2 \quad \text{[rearranging]}$$
$$\therefore\ x = 1$$
$$\therefore\ y = 3 \quad \text{[substituting } x = 1 \text{ into the equation of either line]}$$

$\therefore$ The point of intersection of $y = 3x$ and $y = x + 2$ is (1, 3).

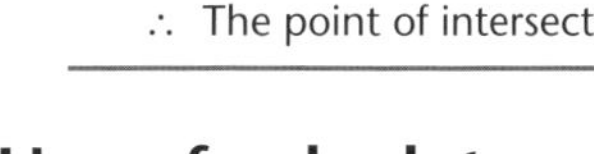

Use of calculators

The solution of simultaneous equations is enormously simplified by the use of graphical calculators. The graphs of the straight lines are drawn and the point of intersection read off.

On calculators with equation solving mode, there are even more direct ways of solution of simultaneous equations.

Concurrence

Two or more lines are **concurrent** if they all pass through the same point.

Example B

Show that $y = 3x$, $y = 2x + 1$ and $y = x + 2$ are concurrent.

Solution

From example A, $y = 3x$ and $y = x + 2$ pass through (1, 3). Substituting (1, 3) into $y = 2x + 1$ gives $3 = 2 \times 1 + 1$, which is true. Hence (1, 3) lies on $y = 2x + 1$ also. Thus all three lines pass through (1, 3).

Unit 11.2 Activity 12A: Intersection of straight lines

1. Find the point of intersection of the following pairs of straight lines:

a. $y = 2x,\ y = 3x - 1$

b. $y = x + 3,\ y = 2x + 3$

c. $y = 3,\ y = x + 1$

d. $x = 2,\ y = x - 1$

e. $y = \frac{1}{2}x + 1,\ y = 2x + 4$

f. $y = x,\ 2x + 3y = 5$

g. $2x - 3y = 1,\ y = x - 1$

h. $3x + 2y = 11,\ 2x - y = 5$

i. $2x + 5y = 11,\ 3y = 4x - 9$

j. $3x + 2y - 7 = 0,\ 5x - 6y - 12 = 0$

2. Show that the following sets of straight lines are concurrent and find their point of concurrence:

a. $x = 1,\ y = 4,\ y = x + 3$

b. $y = 2x + 3,\ y = 3x + 2,\ y = 5x$

c. $y = \frac{x}{2} + 1,\ y = x,\ 3y = x + 4$

3. Find k if the following sets of lines are concurrent:
$y = x - 1,\ y = 2x + 1,\ y = kx + 2$

4. Prove that the following sets of lines are not concurrent:

a. $4y - 3x - 6 = 0,\ 4y + 3x + 6 = 0,\ x = 2$

b. $y = x + 1,\ y = 2x + 3,\ y = 3$

c. $y = x,\ y = x + 3,\ y = 2x - 1$

5. Prove $2x - 3y = 5$ and $4x = 6y - 7$ are lines which never intersect.

6. On a map, a line joining points A(2, 2) and B(5, 11) meets the line joining the points C(0, 16) and D(8, 0). Where do these lines meet?

7. Find the length of the perimeter of the triangle formed by the following straight lines:
$y = x,\ y = 2x + 3,\ y = 3x + 1$.

8. a. Prove that the triangle formed by the lines $y = x + 3,\ y = 3 - x$ and $y = 2x + 6$ is right-angled.

b. Find the area of this triangle.

9. Investigate the claim that the lines $x - 2y + 3 = 0,\ 3x - 4y + 7 = 0$ and $t(x - 2y + 3) - u(3x - 4y + 7) = 0$ are always concurrent, where t and u are constants.

Unit 11.2 Graphs and Functions

Topic 13: Inequalities

This Topic relates to the last bullet point listed under 'Basic algebra' (Syllabus p.4) and also to the 4th bullet point under the heading 'Graphs and functions' (Syllabus p. 15):

- Solving linear inequalities and representing solutions on a plane.

This Topic is repeated from Unit 1 Topic 15.

Inequalities

Solving linear inequations

A linear **equation** has the form $mx + c = 0$ where m and c are constants.

A linear **inequation** has a similar form but the equals sign (=) is replaced by one of the inequality signs:

- $>$, greater than.
- $\geq$, greater than or equal to.
- $<$, less than.
- $\leq$, less than or equal to.

The process for solving linear inequations is the same as that for solving linear equations except for the following rule:

The inequality sign is reversed when multiplying or dividing an inequation by a negative number.

An inequality can be graphed on a number line or written in interval notation.

For example : $x \geq 3.5$.

On a number line:

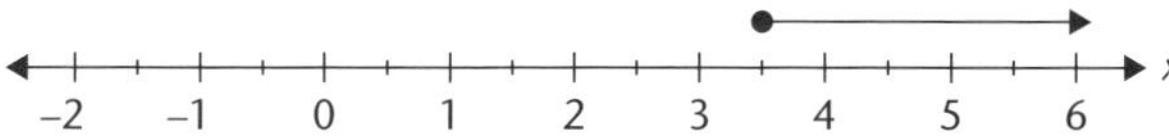

● means the value is included and ○ means the value is not included

This inequality can be expressed in **interval notation** as $[3.5, \infty)$ where the square bracket, [, means the value is included and the round bracket,), means the value is not included.

Note: Round brackets are always used with ∞ and $-\infty$.

Example A

Q. For each of the following inequations:

1. $3x + 2 \leq 14$

2. $4x - 3 > 6x - 7$

i. Solve for x.

ii. Show the solutions on a number line.

iii. Write the answer in interval notation.

A. 1. i. $3x + 2 \leq 14$: subtract 2 from both sides

$3x \leq 12$: divide both sides by 3

$x \leq 4$

ii. $x \leq 4$

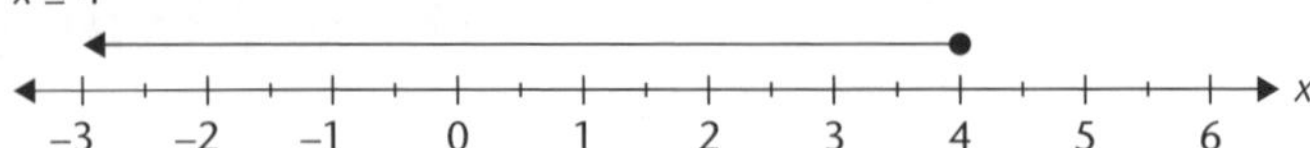

iii. $(-\infty, 4]$

2. i. $4x - 3 > 6x - 7$: subtract $6x$ from both sides

$-2x - 3 > -7$: add 3 to both sides

$-2x > -4$: divide both sides by -2; we need to reverse the inequality sign

$x < 2$

ii. $x < 2$

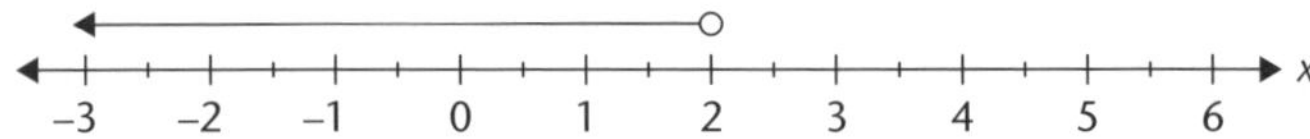

iii. $(-\infty, 2)$

Unit 11.2 Activity 13A: Inequations

1. For each of the following inequations:

   i. Solve for x.

   ii. Show the solutions on a number line.

   iii. Write the answer in interval notation.

   a. $5x \leq 10$ **b.** $-7 > 6x + 5$ **c.** $9 \geq 5 - x$

   d. $\frac{x}{7} \geq \frac{3}{14}$ **e.** $\frac{x}{3} + 4 > 1.5$ **f.** $15 - 2x < 8$

2. Solve for x, giving your answer in interval notation.

   a. $3x + 5 > 0$ **b.** $-3x \geq 9$ **c.** $\frac{x}{6} \leq \frac{7}{2}$ **d.** $5 - 6x \leq -7$

   e. $\frac{2x+1}{2} \geq 5$ **f.** $\frac{x-3}{-2} < 7$ **g.** $\frac{4-x}{5} \geq -2$ **h.** $\frac{3-2x}{2} \leq -1$

3. Solve for a.

   a. $3a - 2 > a - 5$ **b.** $2a - 3 > 5a - 7$ **c.** $5 - 2a \geq a + 4$

   d. $7 - 3a \leq 5 - a$ **e.** $3(a-1) > a + 2$ **f.** $3(a + 2) > 4 - a$

   g. $4(a + 1) < 3(2 + a)$ **h.** $3 + 2(a - 5) \leq 5 - 3(a + 1)$

4. Solve for b.

   a. $5 + \frac{b+3}{2} > 1$ **b.** $3 - \frac{3-b}{4} \leq -1$ **c.** $\frac{b}{2} - \frac{b}{6} < 4$

   d. $\frac{b+2}{3} + \frac{b-3}{4} \leq 1$ **e.** $\frac{1-b}{2} + \frac{b+2}{3} \leq 1$ **f.** $\frac{b+1}{3} - \frac{b}{6} \geq \frac{2b-3}{2}$

Regions

On a set of Cartesian axes (the Cartesian plane) a **linear equation** defines a set of points that are in a straight line. A **linear inequation** defines a **region** of the Cartesian plane.

Example B

Q. 1. For each of the following, graph the region defined by the inequation

a. $x \geq 2$ **b.** $x + y > 3$

c. $2x + 3y \geq 6$

2. Find the equation of the boundary line and hence state the inequality that defines the following shaded region.

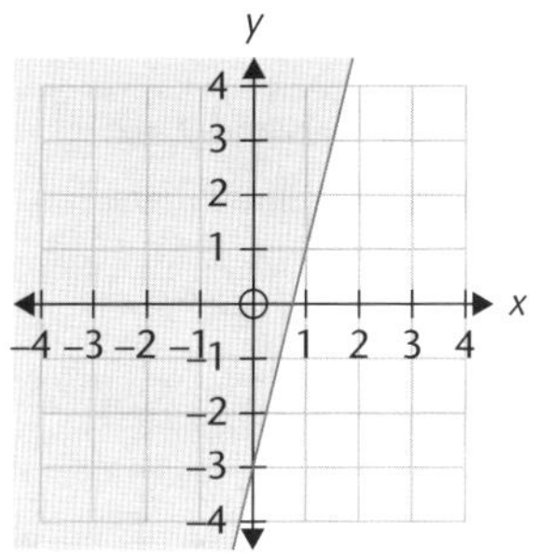

A. 1. a. We need to graph all the points that have an x-coordinate that is **greater than or equal to 2**.

The points that have an x-coordinate **equal to 2** will be the points on the straight line $x = 2$ and this line forms a boundary for the region.

The points that have an x-coordinate greater than 2 will be all the points to the right of the line $x = 2$, shaded opposite.

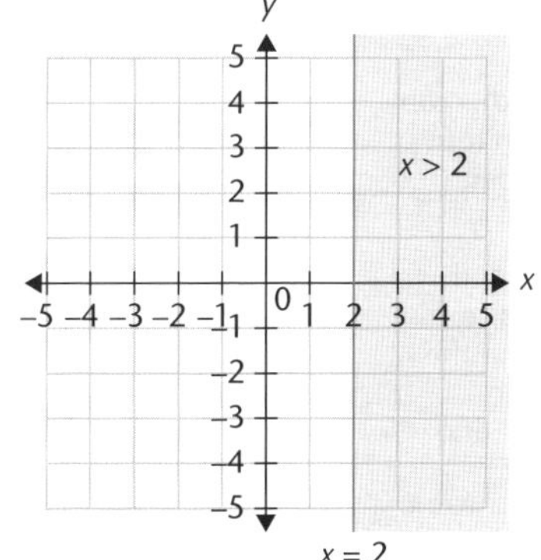

The region defined by $x \geq 2$ includes all points on the line $x = 2$ and the shaded region $x > 2$.

b. Two points on the line $x + y = 3$ are (0, 3) and (3, 0).

The straight line $x + y = 3$ can be drawn through these two points.

For the region $x + y > 3$ the points on the line $x + y = 3$ are **not** included so we draw this line as a dotted line.

Testing that the origin is in the region: substituting (0, 0) into $x + y > 3$, gives $0 + 0 > 3$ which is **not** true. The region is the side of the line not containing the origin.

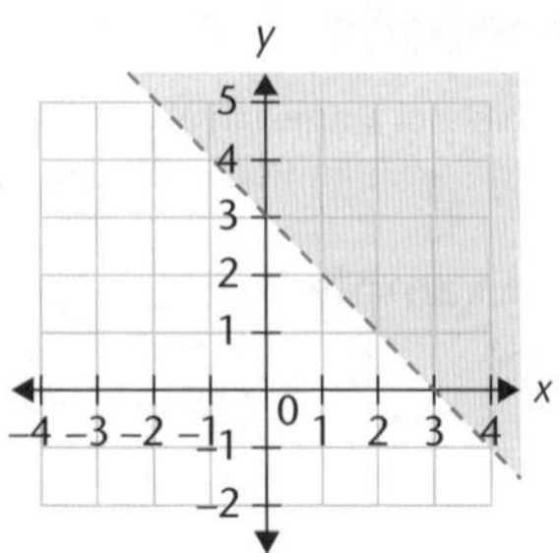

c. To graph the inequality $2x + 3y \leq 6$ we first graph the equality part as the boundary; the straight line $2x + 3y = 6$.

The line $2x + 3y = 6$ has intercepts with the axes $x = 3$ and $y = 2$.

One side of the line $2x + 3y = 6$ will contain all the points where $2x + 3y < 6$ and the other $2x + 3y > 6$.

To find which side of this line is the 'less than' side we choose a point that is not on the line and test the inequation with the x and y values.

For example: Choosing (1, 4) as the point.

Testing: substituting in $2x + 3y$ gives $2 \times 1 + 3 \times 4 = 14$. Is this less than 6? No.

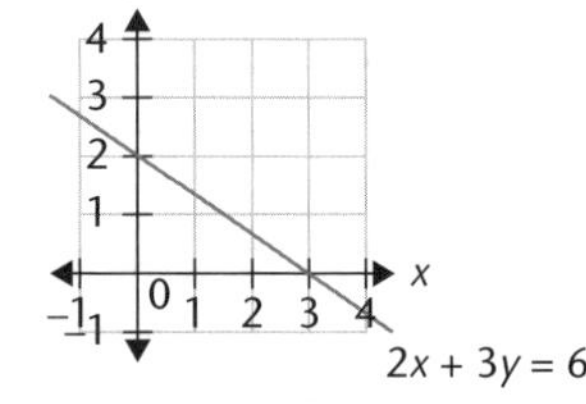

Note. So the point (1, 4) is not on the side of the line that includes all the points $2x + 3y < 6$.

The side required will be the other side of the line.

The region $2x + 3y \leq 6$ is the shaded region and the line.

Note: If the boundary line of an inequation does not go through the origin then it is usually convenient to choose the point (0, 0) as the testing point.

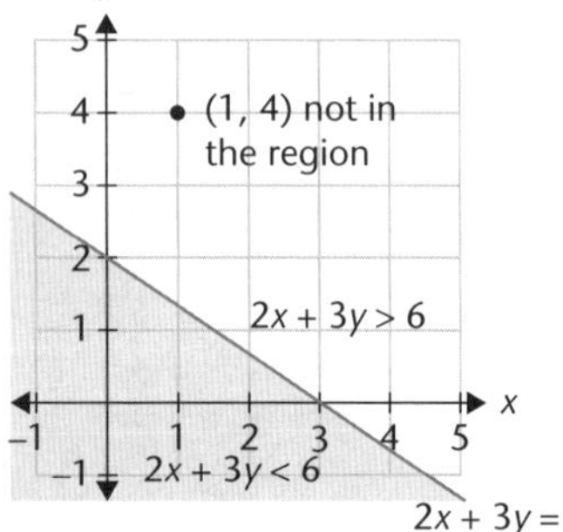

2. Looking at the boundary line of the shaded region we can see that it goes through the points (0, –3) and (1, 1). This means that the line has a gradient of $\frac{1-(-3)}{1-0} = \frac{4}{1} = 4$ and the y-intercept is –3 so the equation of the line is $y = 4x-3$.

The origin (0, 0) is in the region so testing $y \leq 4x-3$ gives $0 \leq 4 \times 0-3$ or $0 \leq -3$ which is **not true**. So the region must be $y \geq 4x-3$.

Unit 11.2 Activity 13B: Regions

1. Sketch the region defined by each of the following inequalities:

   a. $x \geq 0$ b. $y < -2$

   c. $x + y \leq 5$ d. $2x + y \geq 6$

   e. $4x - 3y \leq 24$ f. $y \geq x$

   g. $y \leq 5x - 3$

2. Find the equation of the boundary line and hence state the inequality that defines each of the following shaded regions:

a.

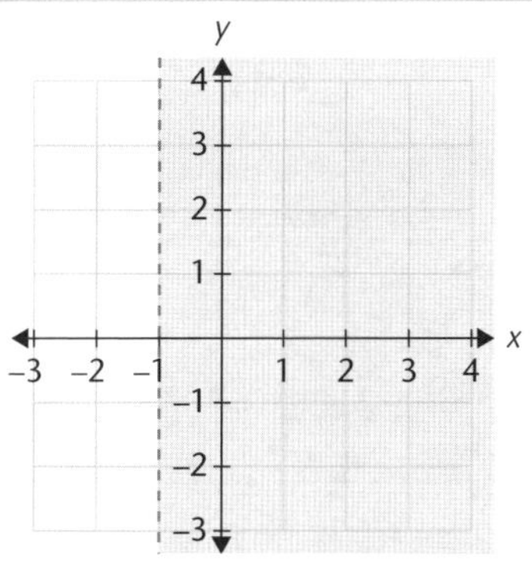

b.

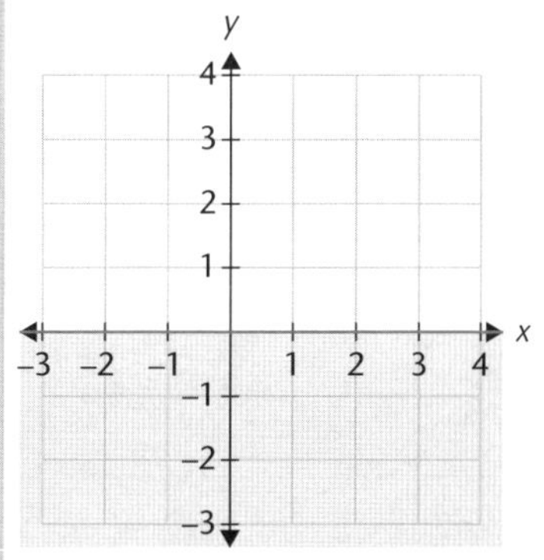

c.

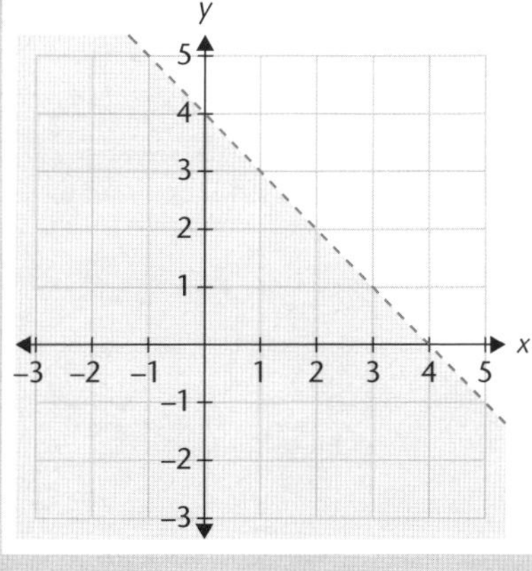

d.

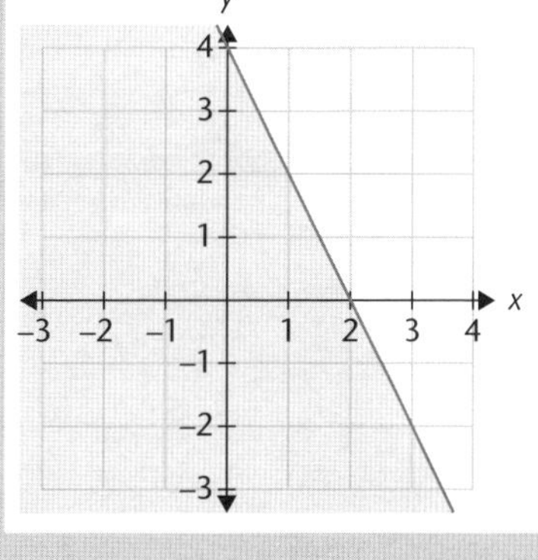

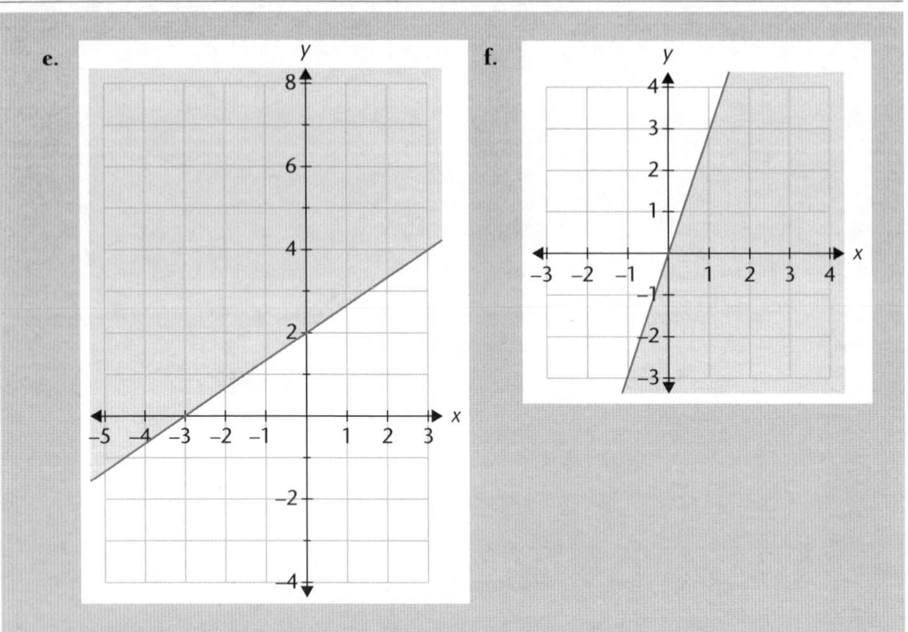
e.
y
8
6
4
2
x
−5
−4
−3
−2
−1
1
2
3
−2
−4
f.
y
4
3
2
1
x
−3
−2
−1
1
2
3
4
−1
−2
−3

Unit 11.2 Graphs and Functions

Topic 14: Solution of quadratic equations

This Topic relates to the content relating to quadratic equations, as set out under the heading 'Graphs and functions' (Syllabus p. 15), with a review of polynomials which were introduced in Topic 4, p. 161. In particular, it covers:

- Solving quadratic equations that can be factorised.
- Solving quadratic equations requiring the use of the quadratic formula.
- Exploring the nature of the roots of a quadratic.
- Solving linear/non-linear simultaneous equations.
- Choosing algebraic techniques and strategies to solve problems.

Review

A **polynomial** is an algebraic expression in which each term is a multiple of a power of the same variable. In a polynomial these powers are *whole numbers*. The **degree** of a polynomial is the *highest* power of the variable present in any of its terms. Function notation such as $P(x)$ can be used to denote a polynomial whose variable is x.

Example A

1. $P(x) = 2x^4 - 3x^3 + 2x^2 + 5x - 7$ is a polynomial function with five terms: $2x^4$, $-3x^3$, $2x^2$, $5x$ and -7. The polynomial has degree 4 because the highest power of x is 4, which occurs in the term $2x^4$.

2. $P(x) = (x + 5)^2$ is a polyn]omial of degree 2, because it expands to $x^2 + 10x + 25$.

In particular, this topic deals with **quadratics**, which are polynomials of degree 2.

Unit 11.2 Activity 14A: Polynomials

1. Which of the following are polynomials?

a. $2x^2 + 3x - 1$ **b.** $4x^3 - 5x + 2$ **c.** $x^2 - 3$

d. $\dfrac{1}{x^2+1}$ **e.** $x^2 + x^{-1}$ **f.** $\frac{1}{2}x^2 + 3x - 2\frac{1}{3}$

g. $\sqrt{x}$ **h.** 7 **i.** $(x + 3)^2$

j. $4(x + 3)^2$

2. What is the degree of each of the following polynomials?

a. $2x^3 - x + 3$ **b.** $x^5 - 2x^4 + x^3 + 2x^2 - 4x + 3$ **c.** $\frac{1}{3}x^6 - 4x + 2$

d. $12x^4 - 3x^2 + x$ **e.** $3x + 4$ **f.** $(x + 3)(x + 5)$

g. $x(x + 3)^2$ **h.** $(2x^2 - 1)^2$ **i.** $4 - x^2$

j. 5

3. Which of the following polynomials are quadratics? Which are cubics?

a. $2x^2 + 3x + 5$ **b.** $8x^3 - 4x + 5$ **c.** $2 - 3x^2$

d. $2x^4 + x^2 + 1$ **e.** $(x + 1)^2$ **f.** $(x + 3)(2x + 1)$

g. $x^2(x - 1)$ **h.** $x^6 - 5x^2 + 7$

Calculations with polynomials

The following algebraic skills are used when sketching graphs of factorised polynomials.

Different values for the variable can be substituted into a polynomial.

Example B

If $h(x) = (x - 3)(x + 4)(2x - 1)$, then

1. $h(-2)$ is the value of the polynomial when $x = -2$. It can be found by substituting $x = -2$ into the formula for $h(x)$.

$h(-2) = (-2 - 3)(-2 + 4)(2 \times -2 - 1)$ [substituting]

$\therefore\ h(-2) = -5 \times 2 \times -5$ [simplifying]

$= 50$

2. $h(\frac{1}{2})$ is the value of the polynomial when $x = \frac{1}{2}$. It can be found by substituting $x = \frac{1}{2}$ into the formula for $h(x)$.

$h(\frac{1}{2}) = (\frac{1}{2} - 3)(\frac{1}{2} + 4)(2 \times \frac{1}{2} - 1)$ [substituting]

$= -2\frac{1}{2} \times 4\frac{1}{2} \times 0$ [simplifying]

$= 0$

Factorised polynomial equations (of degree 2 or more) are solved by setting each factor equal to zero.

Example C

$g(x) = x(x + 4)(x - 2)$. Solve the equation $g(x) = 0$.

Solution

$g(x) = 0$

$\therefore\ x(x + 4)(x - 2) = 0$ [this is already in factored form]

$\therefore$ either $x = 0$, $x + 4 = 0$ or $x - 2 = 0$ [setting each factor equal to 0]

$\therefore\ x = 0$, $x = -4$ or $x = 2$ [solving $x + 4 = 0$ and $x - 2 = 0$]

Unit 11.2 Activity 14B: Substitution and solving equations

1. $g(x) = (x + 2)(x - 1)$

a. Find: **i.** $g(2)$ **ii.** $g(4)$ **iii.** $g(-1)$ **iv.** $g(-2)$ **v.** $g(1\frac{1}{2})$

b. Solve $g(x) = 0$

c. Express $g(x)$ in expanded form.

2. $P(x) = (2x - 1)(x - 4)(x + 1)$

Find: **a.** $P(1)$ **b.** $P(-2)$ **c.** $P(\frac{1}{2})$ **d.** $P(1\frac{1}{3})$ **e.** $P(-2.2)$ **f.** Solve $P(x) = 0$

3. Solve these equations:

a. $(x + 1)(x - 3) = 0$ **b.** $(x - 3)(x + 4)(x - 5) = 0$ **c.** $x(x - 1) = 0$

d. $3(x + 2)(x - 1) = 0$ **e.** $(x + 4)^2 = 0$ **f.** $(x - 3)(x + 4)^2 = 0$

g. $\frac{(x-1)(x+2)}{4}=0$ **h.** $\frac{1}{2}(x+1)(x-1)(x-4)=0$ **i.** $x(x+1)^2=0$

j. $0.3x^2(x-1)=0$ **k.** $2x^2(x-7)(3-x)(5x+4)=0$

4. $h(x)=(x-1)(x+4)$, $g(x)=(x+3)(x+4)$, $J(x)=(3x+2)(x-1)(x-5)$

a. Find: **i.** $h(2)$ **ii.** $h(-3)$ **iii.** $h(\frac{1}{2})$ **iv.** $h(2\frac{1}{3})$ **v.** $h(-1\frac{1}{9})$

b. Find: **i.** $g(0)$ **ii.** $g(-\frac{1}{2})$ **iii.** $g(4)$

c. Find: **i.** $J(3)$ **ii.** $J(-2)$ **iii.** $J(\frac{-1}{3})$

d. Solve: **i.** $h(x)=0$ **ii.** $g(x)=0$ **iii.** $J(x)=0$

5. $f(x)=(x-1)(x-2)(x-4)$

$g(x)=(x-1)(x-2)(x+1)$

$h(x)=(x-2)(x+1)(x+3)$

a. Find:

i. $f(1)$ **ii.** $f(2\frac{1}{2})$ **iii.** $g(3)$ **iv.** $h(2.3)$ **v.** $h(-3.2)$

b. Solve these equations:

i. $f(x)=0$ **ii.** $g(x)=0$ **iii.** $h(x)=0$

Quadratic functions

A **quadratic function** has the form $y = ax^2 + bx + c$ where a, b and c are **constant** real numbers. A quadratic function is a **polynomial** of **degree** 2.

The graphs of quadratic functions are always **parabolas**. This means the graphs will be similar to those shown.

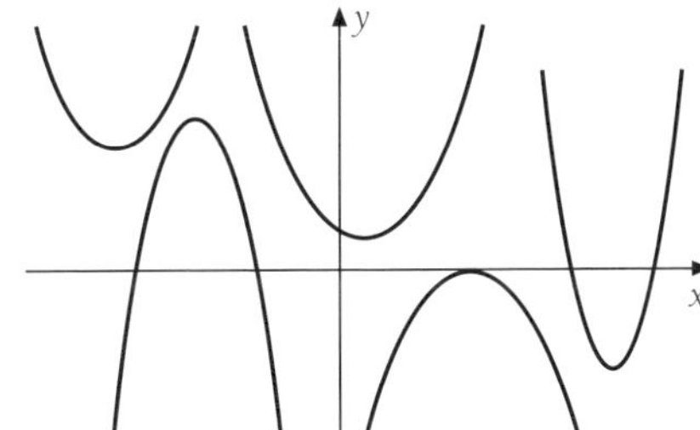

Example D

The functions $y = 2x^2 - 3x + 4$, $y = 5 - \frac{x^2}{2}$ and $y = \frac{2}{3}(x-1)^2$ are all quadratic because they are of the type $y = ax^2 + bx + c$ (or they can be expanded and/or **rearranged** into functions of the type $y = ax^2 + bx + c$).

A **quadratic equation** is an equation which is (or can be) written in the form $ax^2 + bx + c = 0$.

Example E

The equations $3x^2 - 5x + 1 = 0$, $3(x+1)^2 - 2 = 0$ and $4x^2 - 5x = 2x^2 - 11x + 4$ are quadratic equations.

Solution of quadratic equations by factorisation

Quadratic equations can sometimes be solved by factorising and where factorisation is possible this method should be used.

Example F

Solve the equation $x^2 + 4x + 3 = 0$

Solution

$$x^2 + 4x + 3 = 0$$

$\therefore\ (x + 3)(x + 1) = 0$ [factorising]

$\therefore\ x + 3 = 0$ or $x + 1 = 0$

$\therefore\ x = -3$ or $x = -1$ [solving $x + 3 = 0$ and $x + 1 = 0$]

Ensure the quadratic expression equals zero before factorising.

Example G

Solve the equation $x^2 + 5x = 4x + 2$.

Solution

$$x^2 + 5x = 4x + 2$$

$\therefore\ x^2 + x = 2$ [subtracting $4x$]

$\therefore\ x^2 + x - 2 = 0$ [subtracting 2]

$\therefore\ (x + 2)(x - 1) = 0$ [factorising]

$\therefore\ x = -2$ or 1 [solving $x + 2 = 0$ and $x - 1 = 0$]

Example H

Solve $3x^2 + x - 4 = 0$

Solution

$$3x^2 + x - 4 = 0$$

$\therefore\ (3x + 4)(x - 1) = 0$ [factorising]

$\therefore\ x = -1\frac{1}{3}$ or $x = 1$ [solving $3x + 4 = 0$ and $x - 1 = 0$]

Unit 11.2 Activity 14C: Solution of quadratic equations by factorisation

Solve the following quadratic equations.

1. $A^2 - 3A - 4 = 0$

2. $A^2 + 3A - 10 = 0$

3. $B^2 - 2B - 8 = 0$

4. $C^2 + 13C + 36 = 0$

5. $A^2 - 3A - 18 = 0$

6. $B^2 - 16 = 0$

7. $x^2 = x + 12$

8. $U^2 + 6U = 8 - U$

9. $2Z^2 + 11Z = Z^2 + 9Z$

10. $5R^2 - 3R + 4 = 4R^2 + R + 16$

11. $(x + 2)^2 = 9$

12. $3x^2 + 2x - 1 = 0$

13. $4x^2 + 8x + 3 = 0$

14. $\dfrac{3}{x+1} = x - 1$

15. $(x + 1)^2 - 2(x + 1) - 3 = 0$

Solution of quadratic equations by formula

Sometimes a quadratic equation has a solution, yet cannot be factorised readily. In these cases the **quadratic formula** is used.

The solution to the general quadratic equation, $ax^2 + bx + c = 0$ is:

$$x = \frac{-b \pm \sqrt{b^2 - 4ac}}{2a}$$

Note: The solutions of an equation are often called the **roots** of the equation.

Example I

Find the solutions to the equation $2x^2 + 8x - 3 = 0$, correct to 2 decimal places.

Solution

Comparing $2x^2 + 8x - 3 = 0$ with $ax^2 + bx + c = 0$ gives $a = 2$, $b = 8$, $c = -3$.

$$x = \frac{-8 \pm \sqrt{8^2 - 4 \times 2 \times (-3)}}{4} \quad \text{[substituting into } x = \frac{-b \pm \sqrt{b^2 - 4ac}}{2a}\text{]}$$

$$\therefore\ x = \frac{-8 \pm \sqrt{88}}{4}$$

$$\therefore\ x = \frac{-8 + \sqrt{88}}{4} \text{ or } \frac{-8 - \sqrt{88}}{4}$$

$$\therefore\ x = \frac{-8 + 9.381}{4} \text{ or } \frac{-8 - 9.381}{4} \quad \text{[working to 3 dp]}$$

$$\therefore\ x = 0.35 \text{ or } -4.35 \text{ (2 dp)} \quad \text{[rounding to 2 dp]}$$

A quadratic equation may need to be rearranged into the form $ax^2 + bx + c = 0$ first, as the following example shows.

Example J

Solve $(x + 1)^2 = x + 5$ to 2 decimal places.

Solution

$$(x + 1)^2 = x + 5$$

$$\therefore\ x^2 + 2x + 1 = x + 5 \quad \text{[expanding } (x + 1)^2\text{]}$$

$$\therefore\ x^2 + x - 4 = 0 \quad \text{[collecting terms on one side]}$$

$$\therefore\ x = \frac{-1 \pm \sqrt{1^2 - 4 \times 1 \times (-4)}}{2} \quad \text{[substituting } a = 1, b = 1, c = -4 \text{ in formula]}$$

$$\therefore\ x = \frac{-1 \pm \sqrt{17}}{2}$$

$$\therefore\ x = 1.56, -2.56 \text{ (2 dp)}$$

Unit 11.2 Activity 14D: Solution of quadratic equations by formula

1. Solve each of these equations giving answers correct to two decimal places:

a. $x^2 + 4x + 1 = 0$ **b.** $x^2 + 8x + 3 = 0$ **c.** $2x^2 - 7x + 1 = 0$

d. $3x^2 + 6x - 7 = 0$ **e.** $2R^2 + R - 8 = 0$ **f.** $x^2 + 1.1x - 0.3 = 0$

g. $2.3x^2 - 3.5x - 1.2 = 0$ **h.** $3A^2 + 5A = 2A^2 - 3A + 2$

i. $8U^2 + 3U + 5 = 2U^2 + 11U + 7$ **j.** $\frac{1}{3}x^2 + 1\frac{1}{2}x + \frac{1}{12} = 0$

2. Solve each of these equations to two decimal places:

a. $(x + 3)^2 = 8$ **b.** $(2x - 3)^2 = 7$ **c.** $3(x - 1)^2 + 1 = 6$

d. $x + 1 = \frac{3}{x}$ **e.** $(2x + 1)^2 = x^2 - x + 5$ **f.** $\frac{x+1}{3} + \frac{1}{x} = 2$

g. $\frac{2x+1}{x+3} = \frac{x+4}{x+1}$ **h.** $3(x + 1)^2 - 6(x + 1) + 2 = 0$ **i.** $\frac{x+1}{x^2+x+1} = 0.2$

j. $\frac{x^2+x+1}{x^2-x+1} = 3$

Some calculators have an equation-solving mode which makes it possible to solve quadratic equations – discuss with your teacher.

Word problems

The solution to many word problems can be found using a quadratic equation.

Example K

The length of a small room is 3 m more than its width. The room has an area of 72 m^2. Find the width of the room.

Solution

Let the width of the room be w.

$\therefore$ the length is $w + 3$	[length is 3 m more than width]
$\therefore w(w + 3) = 72$	[width × length = area]
$\therefore w^2 + 3w = 72$	[expanding]
$\therefore w^2 + 3w - 72 = 0$	[rearranging]
$\therefore w = 7.12$ or -10.12	[solving the quadratic equation using $a = 1$, $b = 3$, $c = -72$]
$\therefore$ the width is 7.12 metres	[reject –10.12 metres because it is impossible to have a negative length]

Unit 11.2 Activity 14E: Word problems

For each of these problems write an equation then solve.

1. A room is 2 m longer than it is wide. It has an area of 100 m^2. Find the width and the length.

2. A room is twice as long as it is wide. It has an area of 200 m^2. Find its width.

3. A triangle has a base which is 2 m longer than its height. It has an area of 10 m^2. Find the height of the triangle.

4. A triangle has a base which is 4 m longer than its height. Its area is 15 m^2.

Find the length of the base.

5. John is 29 years older than Peter. The product of their ages is 1 272. Find their ages.

6. The product of two consecutive natural numbers is 3 782. Find the numbers.

7. A village woman has a garden with an area of 100 m^2. The garden is in the shape of a square. She increases all the sides of this garden by a certain amount and now has an area of 150 m^2. How much did she increase each side by?

8. The sum of the area and perimeter of a square gives a value of 15. Find the perimeter of the square.

9. A box is constructed so that its length is one metre more than its height which in turn is one metre more than its width. It has a total surface area of 16 m^2. Find the lengths of all sides.

10. Find any number which when added to its reciprocal gives 5.

11. The sum of the first n natural numbers is $\frac{1}{2}n(n + 1)$.

a. Find how many natural numbers you have to add to get a sum of 496.

b. How many do you have to add to exceed 1 000?

12. The height of a right-angled triangle is one metre more than its base. The hypotenuse of the triangle is 4 m. Find the lengths of the base and height.

13. A girl is asked by her friend how much she gets paid per hour. She answers in the following way: "The number of hours I work each week is one more than double my hourly pay. I get paid K153 per week. Work it out yourself."

How much does she get paid per hour?

14. A square room is extended so that it is 2.5 m wider and 3.7 m longer. If its new area is 54.4 m^2, find its original length and width.

Solving simultaneous linear and non-linear equations

To solve a linear equation and a quadratic equation simultaneously, **substitute** an expression for one of the variables of the linear equation into the quadratic equation.

Example L

Solve $y = x + 1$ and $y^2 - 3x = 13$ simultaneously.

Solution

Substitute for y (the subject of the linear equation) in $y^2 - 3x = 13$.

$\therefore\ (x + 1)^2 - 3x = 13$ [substituting $(x + 1)$ for y in the quadratic]

$\therefore\ x^2 + 2x + 1 - 3x = 13$ [removing brackets]

$\therefore\ x^2 - x - 12 = 0$ [simplifying]

$\therefore\ (x - 4)(x + 3) = 0$ [factorising]

$\therefore$ either $(x - 4) = 0$ or $(x + 3) = 0$

$\therefore\ x = 4$ or $x = -3$

When $x = -3$, $y = -3 + 1 = -2$ [substituting into $y = x + 1$]

When $x = 4$, $y = 4 + 1 = 5$ [substituting into $y = x + 1$]

If necessary, one of the variables will need to be made the subject of the linear equation.

Example M

Give the coordinates of the points where the line $2x - 3y = 6$ cuts the circle $x^2 + y^2 = 9$.

Solution

The equations $2x - 3y = 6$ and $x^2 + y^2 = 9$ need to be solved simultaneously.

First make one variable (x here) the subject of the linear equation.

$$2x - 3y = 6$$

$$\therefore\ 2x = 3y + 6 \qquad \text{[adding } 3y\text{]}$$

$$\therefore\ x = \frac{3y+6}{2} \qquad \text{[dividing by 2]}$$

Now substitute this expression for x in the second equation.

$$\therefore\ \left(\frac{3y+6}{2}\right)^2 + y^2 = 9 \qquad \text{[substituting for } x \text{ in (2)]}$$

$$\therefore\ \frac{9y^2 + 36y + 36}{4} + y^2 = 9 \qquad \text{[expanding]}$$

$$\therefore\ 9y^2 + 36y + 36 + 4y^2 = 36 \qquad \text{[multiplying by 4]}$$

$$\therefore\ 13y^2 + 36y = 0 \qquad \text{[simplifying]}$$

$$\therefore\ y(13y + 36) = 0 \qquad \text{[factorising]}$$

$$\therefore\ y = 0 \text{ or } 13y + 36 = 0$$

$$\therefore\ y = 0 \text{ or } y = \frac{-36}{13}$$

The line cuts the circle at (3, 0) and $\left(\frac{-15}{13}, \frac{-36}{13}\right)$.

When $y = 0$, $x = 3$ and when $y = \frac{-36}{13}$, $x = \frac{-15}{13}$. [substituting into $x = \frac{3y+6}{2}$]

Unit 11.2 Activity 14F: Simultaneous linear and non-linear equations

Solve each of the following pairs of simultaneous equations:

1. $y = x - 1$, $x^2 + y^2 = 1$

2. $y = 2x + 1$, $x^2 + y^2 = 1$

3. $y = x + 2$, $y^2 + 2x^2 = 4$

4. $y = 2x - 3$, $x^2 + y^2 = 9$

5. $2y - 3x = 6$, $x^2 + y^2 = 9$

6. $y = x + 1$, $x^2 + y^2 = 5$

7. $y = x - 3$, $x^2 + y^2 = 29$

8. $x - 5y = 12$, $y^2 = 2x$

9. $4x + 3y = 25$, $xy = 12$

10. $2x + y = 7$, $xy = 6$

11. $x^2 + y^2 = 13$, $y = x + 1$

12. $x^2 + y^2 = 5$, $y + x = 3$

13. $xy = 8$, $x - y = 2$

14. $y^2 = 3x$, $3x - 2y = 8$

15. $y = \frac{1}{x+1}$, $2y + 3x = 4$

16. Give the coordinates of the points where the line $y = x - 1$ cuts the circle $x^2 + y^2 = 25$.

17. Find the x-coordinates of the points where the line $3x - 2y = 6$ cuts the parabola $y = (x - 1)(x - 3)$.

18. Give the coordinates of the points where the line $x + 5y = 2$ cuts the curve $y = 2x^2 - 3x - 26$.

The nature of the roots of a quadratic equation

The roots of the equation $ax^2 + bx + c = 0$ are the **x-intercepts** of the parabola $y = ax^2 + bx + c$ (the points where the graph cuts the x axis). The sketch shows several quadratic functions and their roots and shows that quadratic equations can have *two*, *one* or *no* real roots.

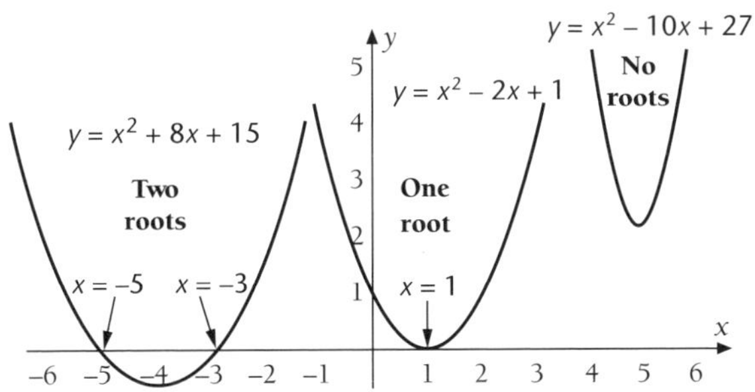

The formula for finding the roots of $ax^2 + bx + c = 0$ is $x = \dfrac{-b \pm \sqrt{b^2 - 4ac}}{2a}$.

However, if $b^2 - 4ac$ (the expression under the square root sign) is negative, no real roots can be found. Consequently, the sign of $b^2 - 4ac$ is very important in working out whether a quadratic has a solution.

The **discriminant** of the equation $ax^2 + bx + c = 0$ is: $b^2 - 4ac$

The discriminant can be used to check the number of roots that a particular quadratic equation has. The equation $ax^2 + bx + c = 0$ will have:

Two distinct real roots if $b^2 - 4ac > 0$
One real root if $b^2 - 4ac = 0$
No real roots if $b^2 - 4ac < 0$

The reason for this result is that:

i. If $b^2 - 4ac > 0$ then there are two roots, namely $\dfrac{-b + \sqrt{b^2 - 4ac}}{2a}$ and $\dfrac{-b - \sqrt{b^2 - 4ac}}{2a}$.
(Note that if $b^2 - 4ac$ is a perfect square, the roots are rational.)

ii. If $b^2 - 4ac = 0$, the root will be $x = \dfrac{-b}{2a}$ by substituting into the formula.

iii. If $b^2 - 4ac < 0$ then the expression under the $\sqrt{\ }$ sign is negative and no real roots exist.

The **nature** of the roots refers to how many roots an equation has and whether or not the roots are real.

Example N

For each equation find the discriminant $b^2 - 4ac$ and the nature of the roots:

1. $9x^2 + 3x + 0.25 = 0$ **2.** $x^2 + x + 1 = 0$ **3.** $x^2 + 4x + 1 = 0$

Solution

1. $b^2 - 4ac = 3^2 - 4 \times 9 \times 0.25$
$= 0$, one real root

2. $b^2 - 4ac = 1^2 - 4 \times 1 \times 1$
$= -3$, no real roots

3. $b^2 - 4ac = 4^2 - 4 \times 1 \times 1$
$= 12$, two real roots

Example O

Find k so that $2x^2 - 3x + k = 0$ has exactly one root.

Solution

If the equation has exactly one root then the discriminant is zero:

$$b^2 - 4ac = 0$$

$\therefore\ (-3)^2 - 4 \times 2 \times k = 0$ [substituting $a = 2$, $b = -3$, $c = k$]

$\therefore\ 9 - 8k = 0$ [simplifying]

$\therefore\ 9 = 8k$

$\therefore\ k = 1.125$

Unit 11.2 Activity 14G: The nature of the roots of a quadratic equation

1. For each of the following quadratic equations:

i. find the discriminant, and **ii.** state the nature of the roots.

a. $2x^2 + 3x + 1 = 0$ **b.** $4x^2 - 5x + 2 = 0$

c. $U^2 + U + 4 = 0$ **d.** $a^2 + 8a + 16 = 0$

e. $2x^2 - 2x + 3 = 0$ **f.** $3x^2 + x = x^2 + 2x - 3$

g. $2x^2 + 4x = 5 - x - x^2$ **h.** $(2x + 1)^2 = 2x - 3$

i. $x + \frac{1}{x} = 2$ **j.** $2x - \frac{3}{x} = 4$

2. a. Find k so that $kx^2 - 2x + 4 = 0$ has one root.

b. Find P so that $3x^2 - 8x - P = 0$ has one root.

c. Find q so that $4x^2 + 3qx + 2 = 0$ has one root.

d. Find all values of k so that $kx^2 - 5x + 6 = 0$ has two roots.

e. Find all values of p so that $x^2 - 3x + p = 0$ has no roots.

3. Find a value for k so that the quadratic equation $x^2 + (2k + 3)x + (k^2 + 2) = 0$ has rational roots.

4. A company employs a consultant to analyse its profit. The consultant finds that the company's profit is determined by the formula,

$P(t) = 10\,000(t^2 - 6.2t + k)$, where P is in dollars and t is in years, and $P(t) \geq 0$. Find the smallest possible value of k (so that $P(t) \geq 0$).

Unit 11.2 Graphs and Functions

Topic 15: Graphing quadratic functions

'Sketching and writing quadratic, cubic and absolute value functions' is the third bullet point under the heading 'Graphs and functions' (Syllabus p. 15) and this Topic deals with quadratic functions. In particular, it covers:

- Drawing quadratics that can be factorised or put in the form $y = \pm(x - a)^2 + b$.
- Drawing polynomials in factorised form.
- Identifying and interpreting features of graphs, including intercepts, symmetry, maxima and minima.

Parabolas

Parabolas are the graphs of **quadratic functions**, which are functions of the type $y = ax^2 + bx + c$ where a, b and c are constant numbers.

Graphs of quadratics $y = ax^2 + bx + c$ where $a = \pm1$

The graphs of functions of the type $y = ax^2 + bx + c$, where $a = \pm1$, can be obtained by filling in a table of values (as in Chapter 9) and joining the resulting points by a smooth curve.

Example A

The graphs of $y = x^2$ (graph I), $y = (x + 3)^2 + 2$ (graph II) and $y = 4 - (x - 2)^2$ (graph III) are shown. Each of the equations of these three graphs can be expanded and simplified to the form $y = ax^2 + bx + c$ where $a = \pm1$ (as shown below).

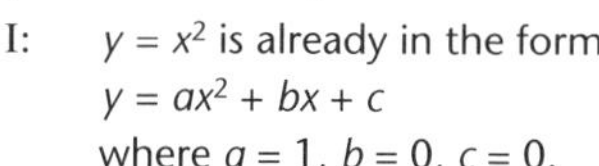

I: $y = x^2$ is already in the form $y = ax^2 + bx + c$ where $a = 1$, $b = 0$, $c = 0$.

II: $y = (x + 3)^2 + 2$

$= (x + 3)(x + 3) + 2$

$= x^2 + 6x + 9 + 2$ [expanding]

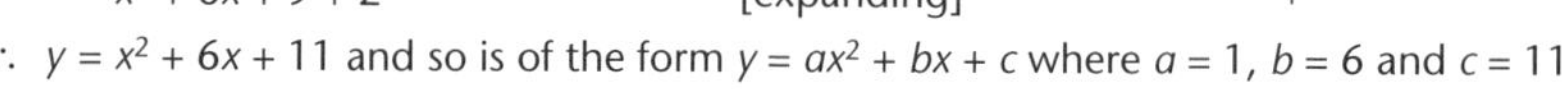

$\therefore$ $y = x^2 + 6x + 11$ and so is of the form $y = ax^2 + bx + c$ where $a = 1$, $b = 6$ and $c = 11$

III: $y = 4 - (x - 2)^2$

$= 4 - (x - 2)(x - 2)$

$= 4 - (x^2 - 4x + 4)$ [expanding]

$= 4 - x^2 + 4x - 4$

$\therefore$ $y = -x^2 + 4x$ and so is of the form $y = ax^2 + bx + c$ where $a = -1$, $b = 4$ and $c = 0$

Note: If a (the coefficient of x^2) is negative, the graph is inverted.

Graphs of $y = \pm(x - a)^2 + b$

Example A illustrates that any graph of the type $y = \pm(x - a)^2 + b$ is **congruent** (identical in size and shape) to the graph of $y = x^2$, and has a **vertex** (turning point) at (a, b).

For example, $y = (x + 3)^2 + 2$ has a vertex at $(-3, 2)$ which is a **local minimum**, and $y = 4 - (x - 2)^2$ (which can be written as $y = -(x - 2)^2 + 4$) has a vertex at $(2, 4)$, which is a **local maximum**.

> As a general rule, $y = (x - a)^2 + b$ has a local minimum at (a, b) and $y = -(x - a)^2 + b$ has a local maximum at (a, b).

- Graphs of functions of the type $y = (x - a)^2 + b$ are translations of the graph of $y = x^2$, by a units horizontally and b units vertically.
- This translation can be expressed in vector form as translation by the vector $\begin{pmatrix} a \\ b \end{pmatrix}$.
- Similarly, graphs of functions of the type $y = -(x - c)^2 + d$ are translations of the graph of $y = -x^2$, by c units horizontally and d units vertically, ie by the vector $\begin{pmatrix} c \\ d \end{pmatrix}$.

By identifying the translations of $y = \pm x^2$, the equations of parabolic graphs can be written down.

Example B

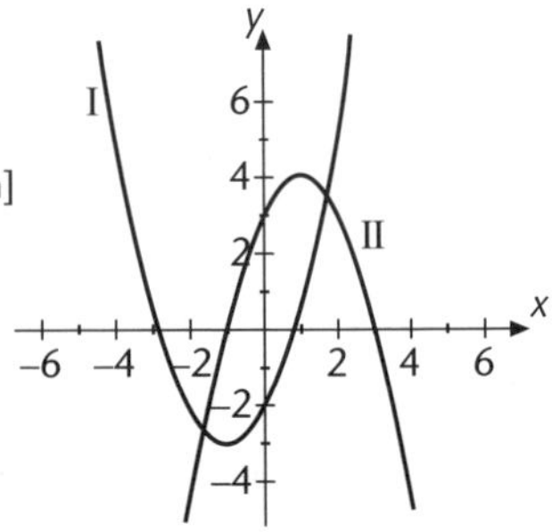

The graph of function I is congruent to $y = x^2$, so its equation is of the form:

$$y = (x - a)^2 + b$$

where $a = -1$ [the horizontal translation]

and $b = -3$ [the vertical translation]

So the equation of graph I is:

$$y = (x + 1)^2 - 3$$

The graph of function II is congruent to $y = -x^2$, so its equation is of the form:

$$y = -(x - a)^2 + b$$

where $a = 1$ [the horizontal translation]

and $b = 4$ [the vertical translation]

So the equation of graph II is:

$$y = -(x - 1)^2 + 4$$

Unit 11.2 Activity 15A: Graphs of $y = \pm(x - a)^2 + b$

1. Express each of the functions in the form $y = ax^2 + bx + c$, and write down the values of a, b and c.

a. $y = (x + 2)^2 + 4$ **b.** $y = (x - 3)^2 - 1$ **c.** $y = 8 - x^2$

d. $y = 5 - (x - 2)^2$ **e.** $y = (x - 4)^2 - 11$

2. Copy and complete the table of values for each of the given functions, then use your table to draw the graph of the function.

a. $y = x^2 + 2$

x	–3	–2	–1	0	1	2	3
y			3	2	3		

b. $y = -x^2 - 1$

x	–3	–2	–1	0	1	2	3
y			–2	–1	–2		

c. $y = (x - 1)^2$

x	−3	−2	−1	0	1	2	3
y			4	1	0	1	

d. $y = -(x + 2)^2$

x	−5	−4	−3	−2	−1	0	1
y			−1	0	−1		

3. Sketch the graph of each of the following functions:

a. $y = (x - 1)^2 + 3$ **b.** $y = (x - 3)^2 - 2$

c. $y = -(x + 2)^2 + 4$ **d.** $y = 5 - (x - 3)^2$

4. For each of the graphs in question 3, identify the coordinates of the vertex and state whether it is a local maximum or minimum.

5. For each of the graphs in question 3, identify any symmetries.

6. Write down the equations of the functions whose graphs **I**, **II**, **III**, **IV**, **V** are shown. Each graph is congruent to that of $y = x^2$.

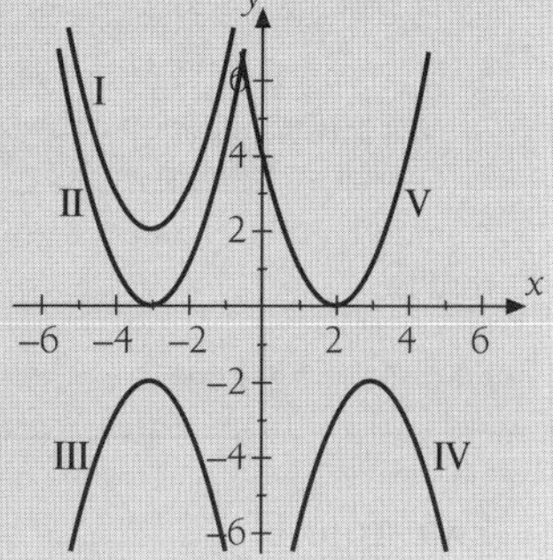

7. Write down the equations of these graphs which are all **congruent** to $y = x^2$ (ie can be written in the form $y = \pm(x - a)^2 + b$).

a.

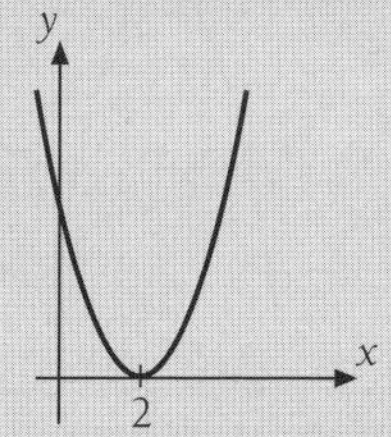

b.

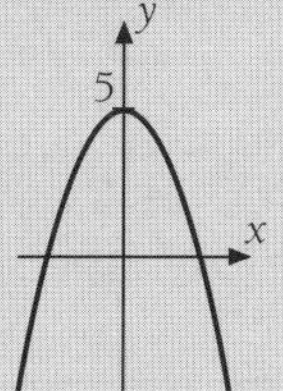

c.

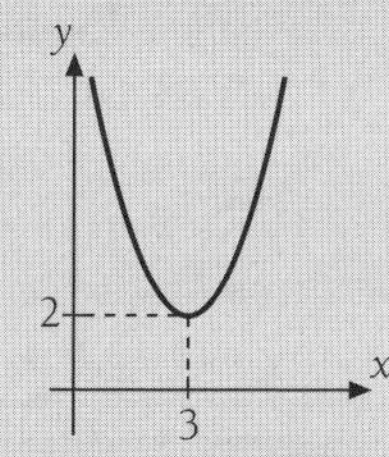

d.

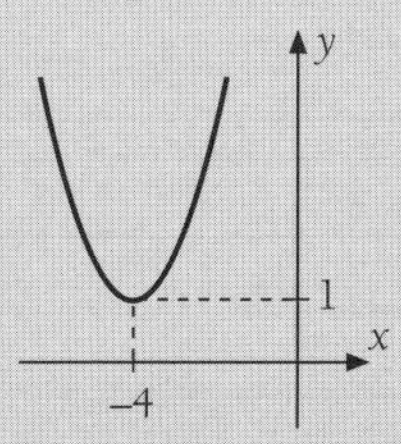

e.

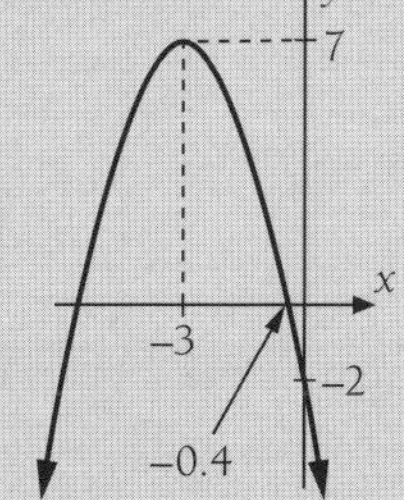

f.

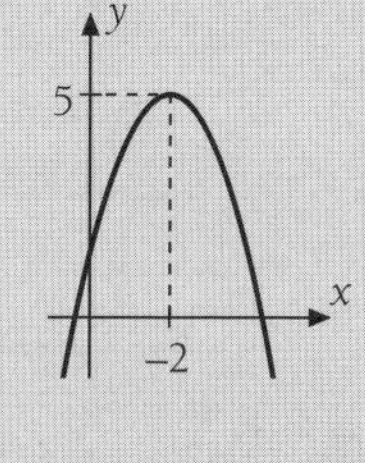

Graphs of quadratics in factorised form

Quadratic functions which are factorised are easily sketched, by plotting intercepts and finding the vertex using the axis of symmetry. Functions whose graphs are congruent to $y = x^2$ have factorised form $y = (x - a)(x - b)$ where a and b are real numbers.

Example C

The graph of $y = (x - 1)(x - 3)$ is shown (the parabola) along with the axis of symmetry (the dotted vertical line $x = 2$).

Note the following important points:

- The **y-intercept** is 3, which is found by setting $x = 0$ in the equation. This gives $y = (0 - 1)(0 - 3) = 3$.
- The **x-intercepts** are –1 and 3 which are found by setting $y = 0$ in the equation.
- The **axis of symmetry** is midway between the x-intercepts. Its equation is $x = \frac{1 + 3}{2}$, which is $x = 2$.
- The coordinates of the **vertex** are (2, –1). The vertex is found by substituting $x = 2$ (the equation of the axis of symmetry) into the equation. This gives $y = (2 - 1)(2 - 3) = -1$.

A smooth curve is drawn through the intercepts and vertex, so that a symmetrical graph results.

If a graph is **congruent** to $y = -x^2$, its factorised equation will be of the form $y = -(x - a)(x - b)$ or $y = (x - a)(b - x)$.

Example D

Write down the equation in factorised form of the parabola shown.

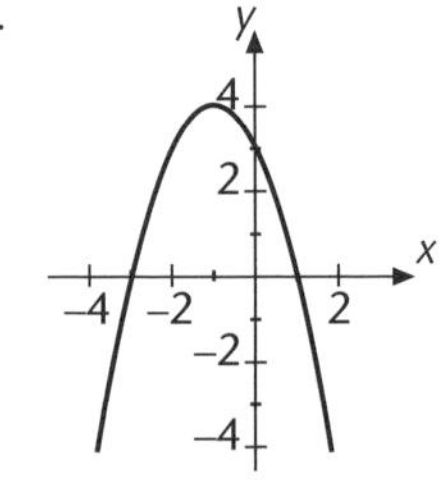

Solution

The graph is congruent to $y = -x^2$ and has x-intercepts –3 and 1.

The equation of the parabola is therefore:

$y = -(x + 3)(x - 1)$

Alternatively, since $-(x - 1) = 1 - x$, the equation can be written:

$y = (x + 3)(1 - x)$.

Unit 11.2 Activity 15B: Graphs of quadratics in factorised form

1. **a.** Copy and complete in this table of values for the function $y = (x + 3)(x - 1)$.

x	–4	–3	–2	–1	0	1	2
y							

   b. Use your table to draw the graph of $y = (x + 3)(x - 1)$.

   c. What are the coordinates of the x-intercepts of the graph you have drawn?

   d. What are the coordinate of y-intercept of the graph you have drawn?

e. What is the equation of the axis of symmetry of this graph?

f. What are the coordinates of the vertex of this graph?

g. Write the equation of this parabola in the form $y = ax^2 + bx + c$.

2. a. Copy and complete the table and draw the graph of $y = (x + 2)(2 - x)$.

x	−3	−2	−1	0	1	2	3
y	−5	0					

b. What is the equation of the axis of symmetry of this graph?

c. What are the coordinates of the vertex of this graph?

d. What are the coordinates of the x-intercepts of this graph?

e. Write the equation of this parabola in the form $y = ax^2 + bx + c$.

3. Write the equation in factored form for each of the parabolas (**I**, **II**, **III**) shown. Each is congruent to $y = x^2$.

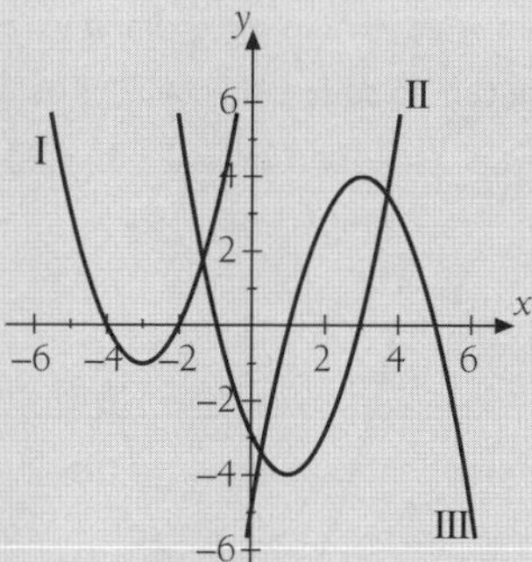

Graphs of quadratics which can be factorised

Quadratic functions which can be factorised are easily sketched, using the techniques discussed in the previous section.

Example E

$y = x^2 - 6x + 5$ can be factorised to give $y = (x - 1)(x - 5)$.

The y-intercept is found by making $x = 0$:

$$\therefore\ y = 5$$

The x-intercepts are found by making $y = 0$:

$$\therefore\ (x - 1)(x - 5) = 0$$

$$\therefore\ x = 1 \text{ or } x = 5$$

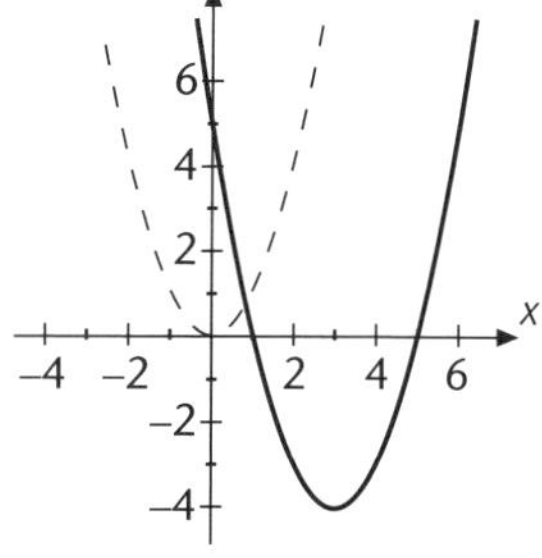

The axis of symmetry lies midway between the x-intercepts and hence is the vertical line through $x = 3$.

The turning point will lie on the axis of symmetry and will have vertical coordinate:

$$y = (3 - 1)(3 - 5) = -4 \qquad \text{[substituting } x = 3 \text{ into } y = (x - 1)(x - 5)\text{]}$$

$\therefore$ the turning point is $(3, -4)$

Note: $y = x^2 - 6x + 5$ can be written as $y = (x - 3)^2 - 4$ which is in the form $y = a(x - b)^2 + c$ and thus could be sketched using the method described in Example A.

Unit 11.2 Activity 15C: Graphs of quadratics which can be factorised

Draw a neat graph for each of the following functions, showing clearly:

i. y-intercept. **ii.** x-intercepts. **iii.** turning points.

1. $y = x^2 - 2x - 3$ **2.** $y = x^2 - 4x + 3$ **3.** $y = -(x^2 - 2x - 8)$

4. $y = x^2 - 6x + 8$ **5.** $y = x^2 - 3x - 4$ **6.** $y = -x^2 - 4x + 5$

Graphs of $y = ax^2 + bx + c$ when $a \neq 1$

When b, c are both zero, the graphs are of the type $y = ax^2$.

Example F

Sketches of $y = x^2$, $y = 2x^2$, $y = \frac{1}{3}x^2$ and $y = -\frac{1}{2}x^2$ are drawn by plotting a few points and joining them. (Take care to square the x values before multiplying by the coefficient.

For example, if $x = 1$ then $2x^2 = 2 \times 1^2 = 2$.)

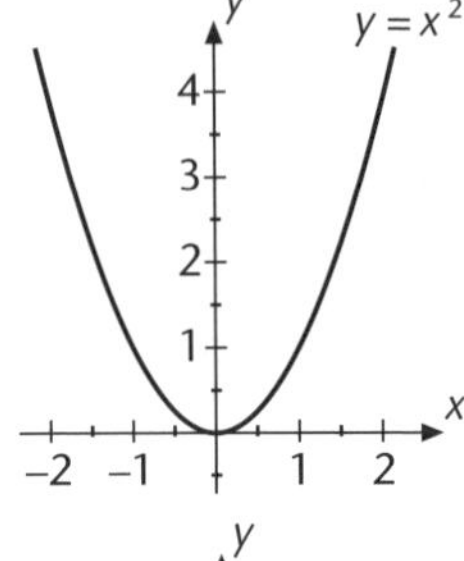

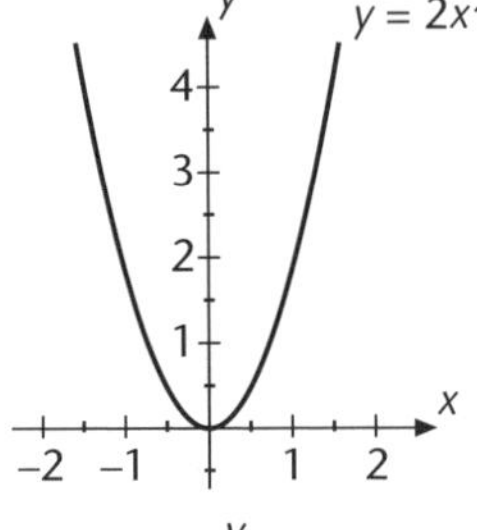

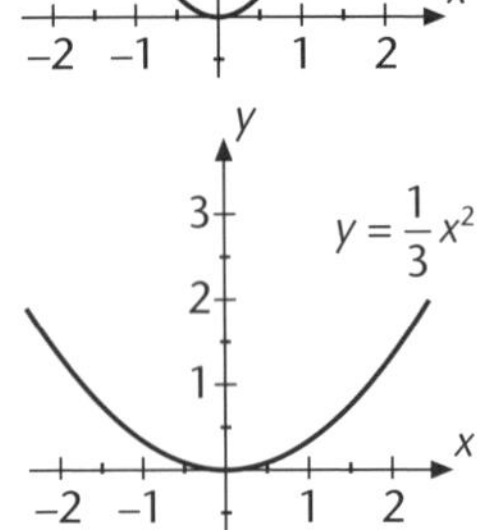

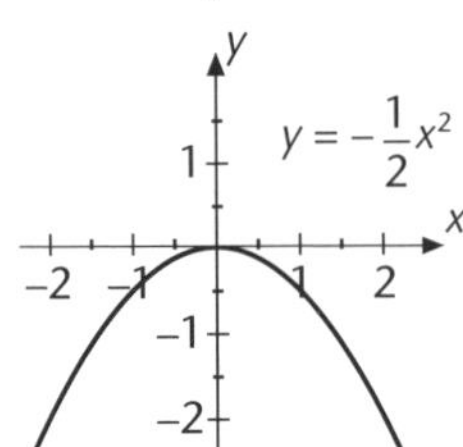

Note: The graph of $y = 2x^2$ is the graph of $y = x^2$ *stretched vertically*.

The graph of $y = \frac{1}{3}x^2$ is the graph of $y = x^2$ *compressed vertically*.

The graph of $y = -\frac{1}{2}x^2$ is the graph of $y = x^2$ *compressed vertically* and *inverted*.

Unit 11.2 Activity 15D: Graphs of $y = ax^2$

Sketch the graphs of the following quadratic functions.

1. $y = 3x^2$ **2.** $y = \frac{1}{2}x^2$ **3.** $y = -x^2$ **4.** $y = \frac{2}{3}x^2$ **5.** $y = -4x^2$

6. Write down the equations of the parabolas shown in the graphs below.

a.

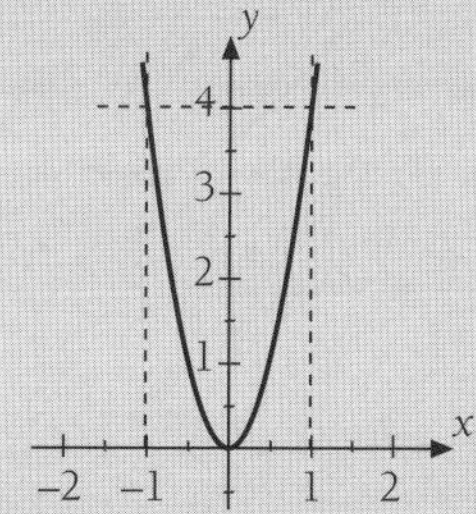

b.

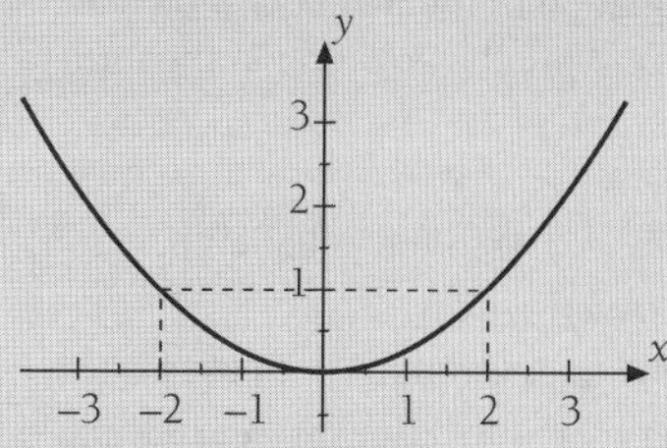

c.

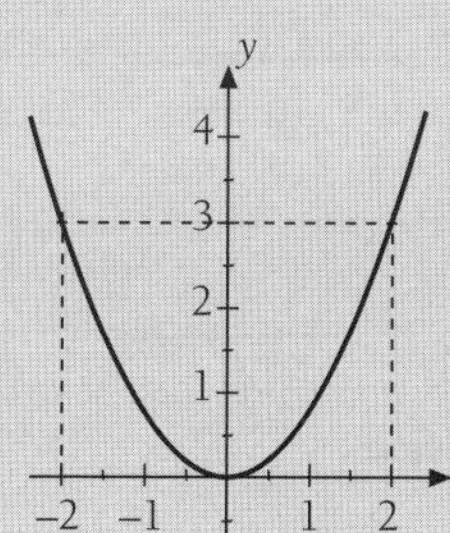

d.

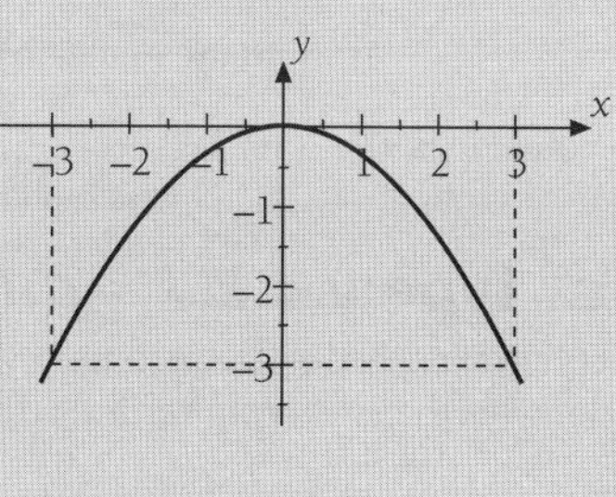

Graphs of type $y = a(x - b)^2 + c$

All quadratic functions can be expressed in the form $y = a(x - b)^2 + c$.

Quadratic functions of the type $y = a(x - b)^2 + c$ are sketched by finding the **intercepts**, **turning points** and **maximum** and **minimum values** as shown in the following example.

Example G

For the graph of $y = 2(x - 1)^2 + 3$,
the minimum value of y occurs when $x = 1$
(since $(x - 1)^2 > 0$ except when $x = 1$).

Substituting $x = 1$ in $y = 2(x - 1)^2 + 3$ gives a minimum value for y of 3.

Therefore (1, 3) is the turning point.

The y-intercept occurs when $x = 0$

$\therefore\ y = 2(0 - 1)^2 + 3$ [substituting $x = 0$]

$y = 5$

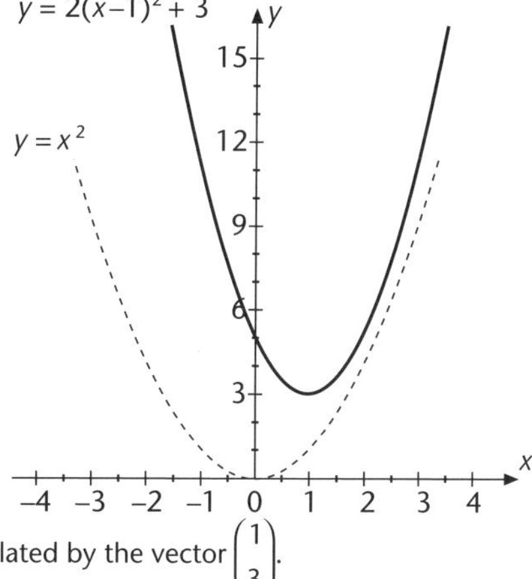

Note: The graph of $y = 2(x - 1)^2 + 3$ is the graph of $y = x^2$ which has been:

- Stretched vertically by a factor of 2.
- Shifted to the right by 1.
- Shifted vertically upward by 3, ie translated by the vector $\begin{pmatrix}1\\3\end{pmatrix}$.

The equation of a parabola can also be written down by applying the above techniques.

Example H

Find the equation of the parabola shown in the diagram. Write the equation in the form $y = a(x - b)^2 + c$.

Solution

The vertex is (–2, 4) so the equation is:

$y = a(x + 2)^2 + 4$.

The parabola goes through (0, 5) thus:

$\therefore\ 5 = a(0 + 2)^2 + 4$ [substituting $x = 0$, $y = 5$ into $y = a(x + 2)^2 + 4$]

$\therefore\ 5 = 4a + 4$

$\therefore\ a = \frac{1}{4}$ [solving the equation]

Hence the equation is $y = \frac{1}{4}(x + 2)^2 + 4$.

Graphs of type $y = k(x - a)(x - b)$

Quadratic functions expressed in factored form $y = k(x - a)(x - b)$ can be sketched as discussed previously, with the factor, k, causing vertical stretch or **compression**.

Example I

Sketch the graph of $y = 2(x - 3)(x + 1)$

Solution

- y-intercept is $y = 2(0 - 3)(0 + 1) = -6$ [substituting $x = 0$]
- x-intercepts are 3, –1 [solving $2(x - 3)(x + 1) = 0$]
- Axis of symmetry is $x = \frac{3 - 1}{2} = 1$ [average of x-intercepts]
- Vertex is $x = 1$, $y = 2(1 - 3)(1 + 1) = -8$ which gives the point (1, –8). The graph is as shown.

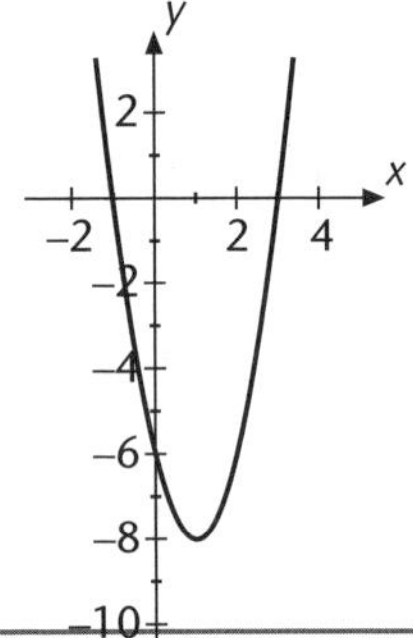

Note: The graph of $y = 2(x - 3)(x + 1)$ is the graph of $y = (x - 3)(x + 1)$ stretched vertically.

When intercepts are known, the graph of a quadratic can be written using factored form.

Example J

Write down the equation of the parabola shown.

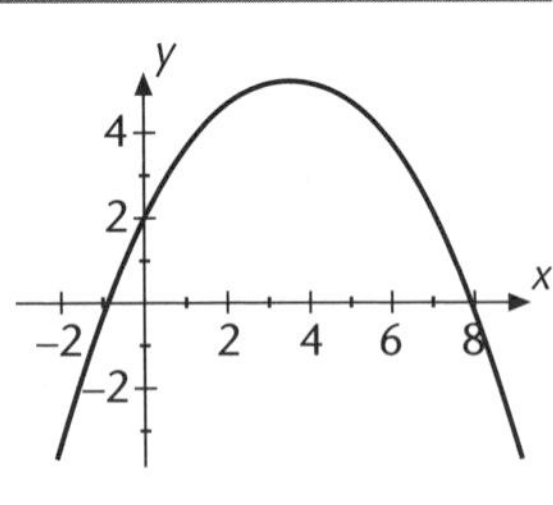

Solution

x-intercepts are –1 and 8 so the equation is of the form:

$y = k(x + 1)(x - 8)$

y-intercept is 2, so

$2 = k(0 + 1)(0 - 8)$ [substituting $x = 0$, $y = 2$]

$2 = -8k$ [simplifying]

$k = -\frac{1}{4}$

$\therefore$ equation is $y = -\frac{1}{4}(x + 1)(x - 8)$

Note: 1. k is negative since graph is inverted.

2. The graph of $y = -\frac{1}{4}(x + 1)(x - 8)$ is the graph of $y = -(x + 1)(x - 8)$ compressed vertically.

Use of graphical calculators

Students using such calculators will find the task of sketching graphs enormously simplified. Once drawn, the use of the TRACE or G-Solv function will enable easy location of such features as maximum and minimum points, intercepts, etc.

Unit 11.2 Activity 15E: Graphs of $y = ax^2 + bx + c$

1. By plotting the vertex and y-intercept of the parabolas, sketch their graphs.

a. $y = (x - 1)^2 + 2$ **b.** $y = -(x + 2)^2 + 2$ **c.** $y = 3(x + 2)^2 - 4$

d. $y = -\frac{1}{2}(x + 2)^2 + 2$ **e.** $y = -2 - 2(x + 1)^2$

2. Write the equations of the following parabolas in the form $y = a(x - b)^2 + c$.

a.

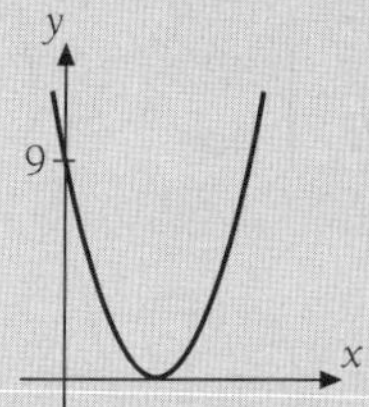

b.

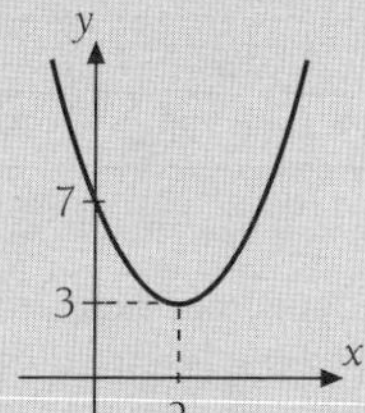

c.

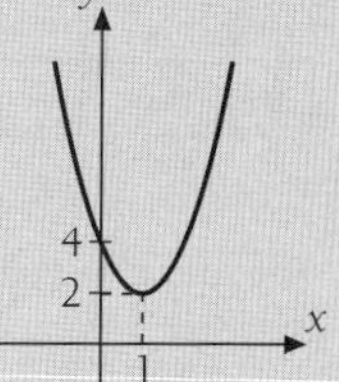

d.

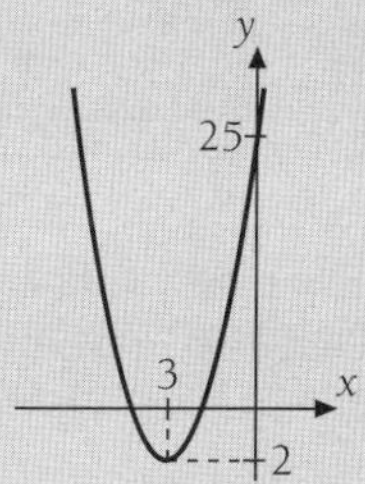

e.

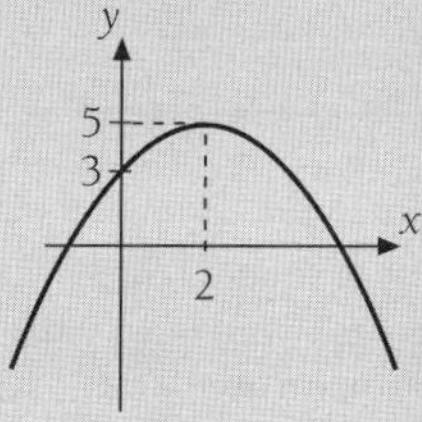

3. Express each of the following functions in the form $y = ax^2 + bx + c$ and make a neat sketch of its graph.

a. $y = 2(x - 1)(x - 5)$

b. $y = 0.5(x - 4)(x + 2)$

c. $y = 3(x^2 - 2x - 3)$

4. a. i. Find k if the parabola $y = k(x - 2)(x - 4)$ has a y-intercept of $(0, 16)$.

ii. Find the coordinates of the turning point.

b. Find B if the parabola $y = B(3 - x)(x + 5)$ has a turning point at $(-1, 4)$.

c. Find L if $y = 25 - L(x - 2)^2$ has x-intercepts of $-\frac{1}{2}$ and $4\frac{1}{2}$.

5. Find equations of each of the following parabolas:

a. x-intercepts at $(-3, 0)$, $(1, 0)$ and y-intercept at $(0, -3)$.

b. x-intercepts at 1 and 6, and y-intercept at 4.

c. Turning point at $(1, 5)$ and y-intercept at $(0, 3)$.

d. Turning point at $(-1, 2)$ and y-intercept at $(0, 5)$.

Quadratic modelling (problems in context)

Quadratic functions arise in many practical contexts.

Example K

Draw the graph for the area of a circle when graphed against radius.

Solution

This table gives some radii and the corresponding area.

[Area = πR^2 when R is radius.]

radius (R)	0	1	2	3	4	5
area (πR^2)	0	3.1	12.6	28.3	50.3	78.5

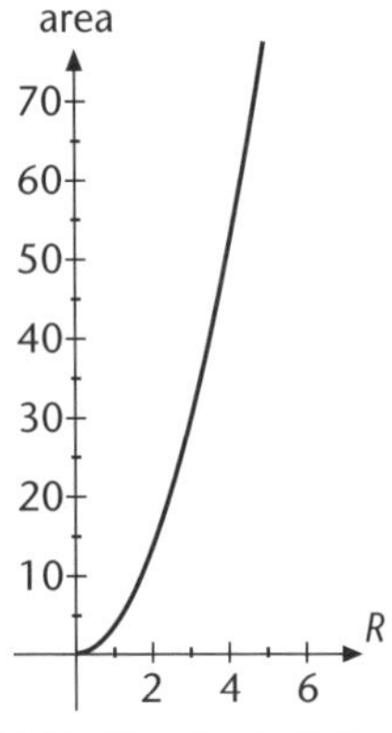

The graph is 'half a parabola' of the type from Example F. There is only half a parabola since a negative radius is not possible.

Example L

A gardener has to mow a number of square lawns. The time it takes to mow a lawn includes a fixed time, independent of the size of the lawn where the lawnmower is taken off the truck, etc. This time is 5 minutes. Each square metre takes 1.5 seconds to mow. The gardener does not mow within one metre of the fence as there are gardens there.

1. Find the time it takes to mow a lawn whose fences on each side measure:

   a. 20 m b. 40 m c. 60 m

2. Find the equation which represents the time it takes to mow lawns of this type.
3. Draw a sketch of a graph of time to mow against length of fence.

Solution

1. a. This diagram illustrates the situation:

   Area to be mowed is $(20 - 2)^2$.

   $\therefore$ time is $\dfrac{1.5\,(20-2)^2}{60} + 5$ minutes [$1.5\,(20 - 2)^2$ gives time in seconds, dividing by 60 converts into minutes]

   $= 13.1$ minutes

   b. and c. Similarly the time taken to mow a lawn with fences of length 40 m is 41.1 minutes and to mow a lawn with fences of length 60 m is 89.1 minutes.

2. Using similar reasoning to that in **a.** for a garden whose side fence is of length x gives the diagram below:

   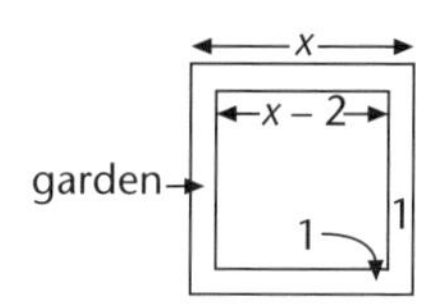

   The time spent using the mower is:

   $1.5(x - 2)^2$ seconds

   $= \dfrac{1.5\,(x-2)^2}{60}$ minutes

   Total time is $T = \dfrac{(x-2)^2}{40} + 5$ minutes [simplifying and adding constant of 5 minutes to include time taken for fixed tasks]

3. The graph is half a parabola as it is not possible to have a lawn whose fence length would be less than 2. [Why not?]

   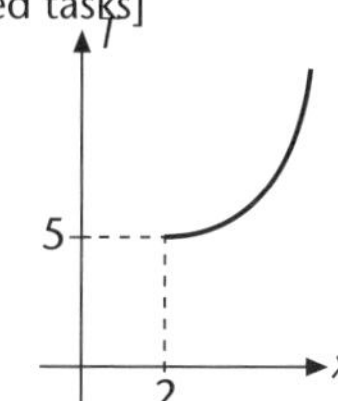

Unit 11.2 Activity 15F: Quadratic modelling

1. Draw the graph of the area of a semi-circle against radius.

2. Draw the graph of the area of a circle against diameter.

3. The cost of painting a wall is 30t per square metre. Draw the graph of cost in kina of painting squares of various side lengths against side length.

4. A square shed of side length 5 m is in the centre of a square field. Draw a graph of the area of grass in the field against length of side of the field.

5. A square shed of side length 8 m is surrounded by a moveable square fence. The shed is in the middle of the area bounded by the fence. Draw a graph of the total area bounded by the fence against the distance of the fence from the shed. [See below.]

   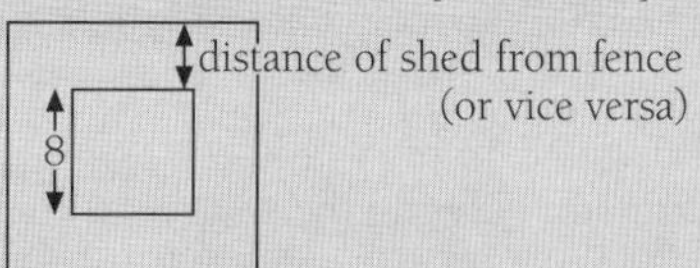

6. The minimum distance of a villager from a straight road is 5 km which she achieves after 2 hours travel. Initially her distance from the road is 7 km. The path she traces out as she travels is parabolic. Find a formula which gives her distance from the road in terms of time.

   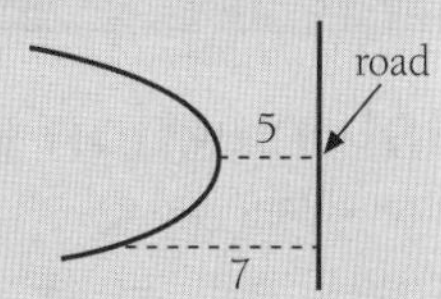

7. A wire AD is 8 metres in length. A length BD is removed from it to give AB. 3 metres are removed from BD to give BC [see diagram]. AB and BC are now joined at right angles to form two sides of a rectangle.

   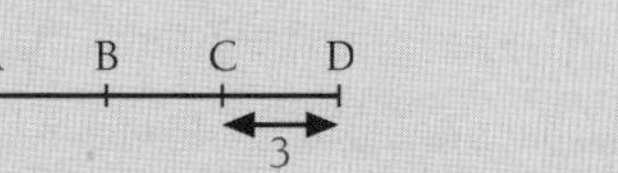

a. Copy and complete this table:

BD(x)	AB	BC	area of rectangle
3	5	0	0
$3\frac{1}{2}$	$4\frac{1}{2}$	$\frac{1}{2}$	2.25
4	4	1	4
$4\frac{1}{2}$			
5			
6			
7			
8			

b. Draw a graph of area against x.

c. Write down a formula which relates the area of the rectangle to the length x of BD.

d. What is the maximum area of the rectangle?

8. The formula for the area of a rectangle made from 10 m of wire is $A = x(5 - x)\,\text{m}^2$.

a. Sketch a graph of A against x.

b. What is the greatest possible value of A?

c. How would the graph change if an area of $x\,\text{m}^2$ was removed from the rectangle corresponding to x?

9. Joseph is digging a cylindrical swimming pool. The cost of digging such a pool is K25 per **cubic metre**. The pool must have radius of at least 2 m and a depth of exactly 2 m. Joseph has a maximum of K6 000 to spend on the digging of the pool.

a. The formula for the volume of a cylinder is $\pi r^2 h$,where r is the radius of the base and h is the height of the cylinder. Find a formula for the cost of the pool in terms of the radius.

b. Draw a graph of cost against radius.

c. Find the maximum radius of the pool.

d. What is the minimum cost of digging such a pool, to the nearest kina?

Intersection of a quadratic and a straight line

A quadratic and a straight line can meet in 0, 1 or 2 points as shown below:

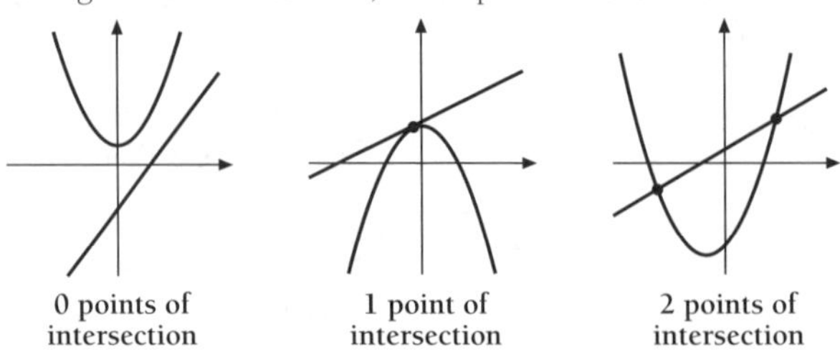

The point(s) of **intersection** can be found graphically or algebraically.

Example M

Find the coordinates of the point(s) of intersection of the straight line $y = 3x - 6$ with the parabola $y = x^2 - 6x + 8$.

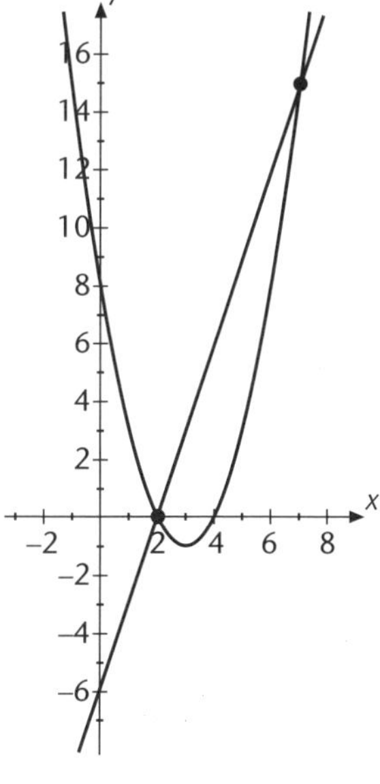

Solution

1. **Graphically**

   $y = x^2 - 6x + 8$ can be expressed in factored form as
   $y = (x - 4)(x - 2)$ which has x-intercepts 4, 2
   and y-intercept 8
   $y = 3x - 6$ is a straight line with gradient 3
   and y-intercept -6
   The line and the curve meet at (2, 0) and (7, 15).

2. **Algebraically**

   Equating y values to get the point of intersection gives
   $x^2 - 6x + 8 = 3x - 6$
   $x^2 - 9x + 14 = 0$ [rearranging]
   $(x - 7)(x - 2) = 0$ [factorising]
   $x = 2$ or $x = 7$
   When $x = 2$, $y = 3 \times 2 - 6$ [substituting in $y = 3x - 6$]
   $= 0$
   When $x = 7$, $y = 3 \times 7 - 6$ [substituting in $y = 3x - 6$]
   $= 15$
   ∴ Points of intersection are (2, 0) and (7, 15).

Unit 11.2 Activity 15G: Intersection of parabola and line

Note: In order to gain 'high achievement' an equation has to be formed and solved.

1. Find the coordinates of the points of intersection, if any, of the parabola $y = x^2$ with the following straight lines:

a. $y = 4$ **b.** $x = 2$ **c.** $y = -9$ **d.** $y = x$
e. $y = 2x$ **f.** $y = 3x$ **g.** $y = 2x + 3$ **h.** $y = 2x - 1$
i. $y = 4 - 3x$ **j.** $y = x + 1$

2. Find the coordinates of the point(s) of intersection of:
$y = x - 1$ with $y = x^2 - 1$.

3. Find the coordinates of the point(s) of intersection of:
$y = 3x - 9$ and $y = x^2 - 11x + 24$.

4. Find the coordinates of the point(s) of intersection of:
$y = x^2 + x + 1$ with $y = 5x - 3$.

5. Find the coordinates of the point(s) of intersection of:
$y = x^2 + x + 2$ and $x + y = 5$.

6. Prove that the line $y = 2x + 4$ is a tangent to the curve $y = 3x^2 - 4x + 7$.

Unit 11.2 Activity 15H: Graphing quadratic functions—multiple choice

1. Compared to the graph of $y = x^2$ the graph of $y = 2(x - 3)^2 + 5$ is:

A. the same shape but translated 3 units to the right and 5 units up.
B. stretched vertically and translated 3 units to the right and 5 units up.
C. stretched vertically and translated 3 units to the left and 5 units down.
D. compressed vertically and translated 3 units to the right and 5 units up.

2. The graph of $y = x^2 + 3x - 10$ has x and y intercepts, respectively:

A. $x = -5, x = 2, y = -10$ **B.** $x = -5, x = -2, y = -10$
C. $x = 5, x = -2, y = -10$ **D.** $x = 5, x = 2, y = 10$

3. For the graph of $y = 3(x + 2)^2 - 4$ the coordinates of the turning point and y-intercept, respectively, are:

A. $(2, -4)$ and 8 **B.** $(-2, -4)$ and 0
C. $(-2, -4)$ and 8 **D.** $(2, 4)$ and 0

4. The graph at right has the equation $y = a(x - 2)^2 + 3$. The value of a is:

A. 2 **B.** 1
C. $\frac{1}{2}$ **D.** 4

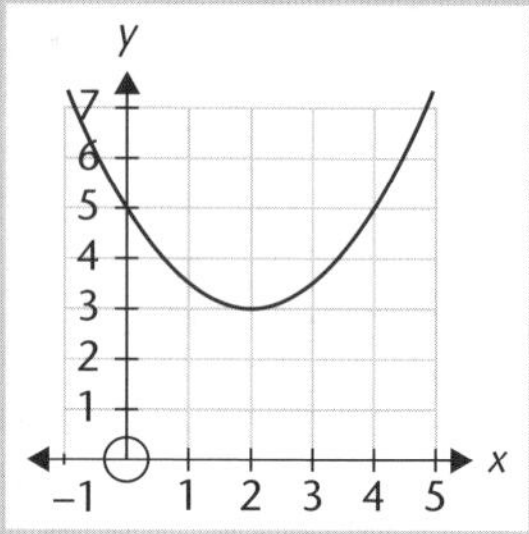

5. The equation of the quadratic graphed below right is:

A. $y = -(x + 2)^2 + 2$ **B.** $y = -x^2 + 2$
C. $y = -2x^2 - 2$ **D.** $y = -2x^2 + 2$

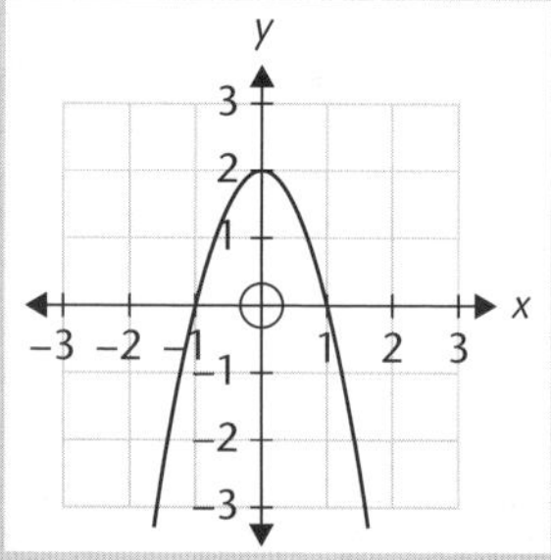

6. The maximum value of the quadratic $y = -(x + 1)^2 + 4$ is:

A. 1 **B.** 2
C. 3 **D.** 4

7. For the quadratic with equation $y = 2(x - 3)^2 - 5$ the axis of symmetry and the y-intercept, respectively, are:

A. $x = 3$ and 4 **B.** $x = 3$ and 13
C. $x = -3$ and 5 **D.** $x = -3$ and 13

8. For the quadratic with equation $y = 2(x + 5)(x - 3)$ the axis of symmetry and the minimum value, respectively, are:

A. $x = -1$ and -32 **B.** $x = -1$ and -16
C. $x = 1$ and -24 **D.** $x = 1$ and -12

Unit 11.2 Graphs and Functions

Topic 16: Cubic graphs

'Sketching and writing quadratic, cubic and absolute value functions' is the third bullet point under the heading 'Graphs and functions' (Syllabus p. 15). This Topic deals with cubic graphs. It covers:

- Drawing factorised polynomials.
- Identifying and interpreting features of graphs, including intercepts, symmetry, and behaviour of graphs at large values of x and y.

Cubic graphs

Cubic graphs are graphs of **cubic functions** which are polynomials of **degree** 3.

Cubic graphs can be written in the form:

$$y = ax^3 + bx^2 + cx + d \quad \text{where } a, b, c \text{ and } d \text{ are numbers}$$

The graph of the simplest cubic, $y = x^3$, is drawn by plotting points.

Example A

The graph of $y = x^3$ can be drawn by filling in a table then plotting the points:

x	–10	–2	–1	0	1	2	10
x^3	–1 000	–8	–1	0	1	8	1 000

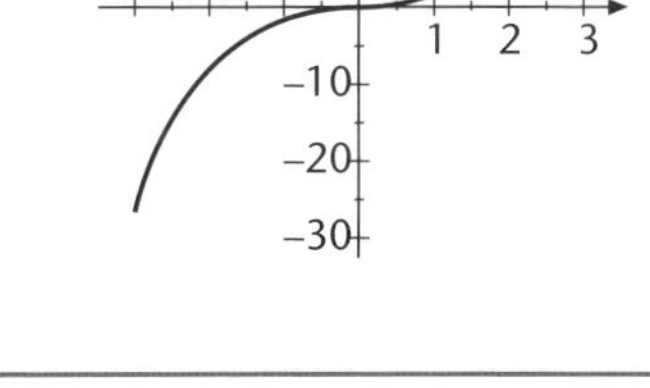

Notice that the graph has **point symmetry** about the point (0, 0).

The x-intercept is $x = 0$ and the y-intercept is $y = 0$.

The graph of $y = x^3$ can be transformed in a similar way to the transformations of $y = x^2$ in the previous chapter.

As $x \to \infty$ (x takes on large positive values), $y \to \infty$ also.

As $x \to -\infty$ (x takes on large negative values), $y \to -\infty$ also.

Cubic graphs in factored form are sketched by finding the x- and y-intercepts and sketching the curve through them.

Example B

The graph of $y = (x - 1)(x - 3)(x - 4)$ can be sketched by finding the intercepts:

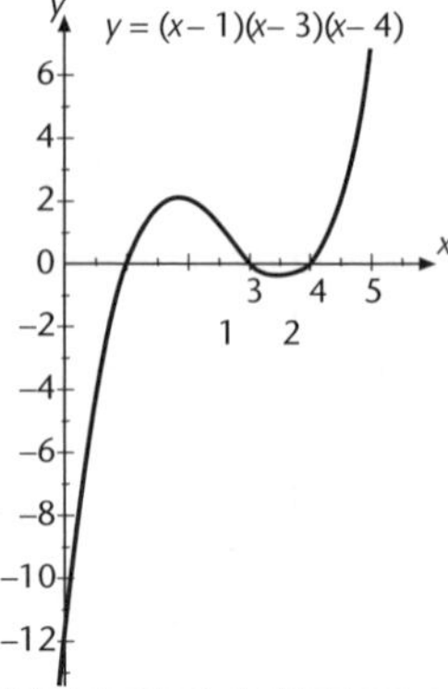

Substituting $x = 0$ in $y = (x - 1)(x - 3)(x - 4)$, gives:

$$y = (0 - 1)(0 - 3)(0 - 4)$$
$$= (-1) \times (-3) \times (-4)$$
$$= -12$$

The x-intercepts are found by substituting $y = 0$, giving:

$$(x - 1)(x - 3)(x - 4) = 0$$
$$\therefore\ x = 1, 3 \text{ or } 4$$

As $x \to \infty$, $y \to \infty$ and as $x \to -\infty$, $y \to -\infty$.

Example C

For the graph of $y = (x + 1)^2(2 - x)$,

the y-intercept is 2. [substituting $x = 0$]

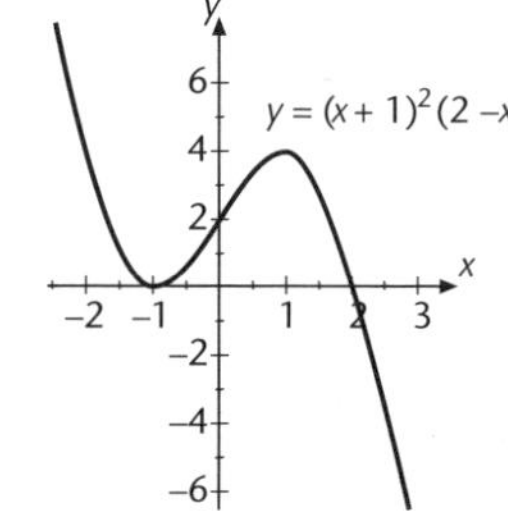

The x-intercepts are found by setting $y = 0$, giving $(x + 1)^2(2 - x) = 0$, so the x-intercepts are -1 and 2.

As $x \to \infty$, $y \to \infty$ and as $x \to -\infty$, $y \to -\infty$.

Note: 1. The graph *touches* the x-axis at -1. In **cubics** like $y = (x + 1)^2(2 - x)$ where one of the factors of the polynomial is a perfect square, the graph will *touch* the x-axis rather than cut it.

2. In the second bracket the coefficient of x is negative which means that when $y = (x + 1)^2(2 - x)$ is expanded, the coefficient of x^3 is negative. Thus for large positive x, y is negative.

By considering intercepts, the equation of a cubic can be written down.

Example D

What is the equation of the graph shown?

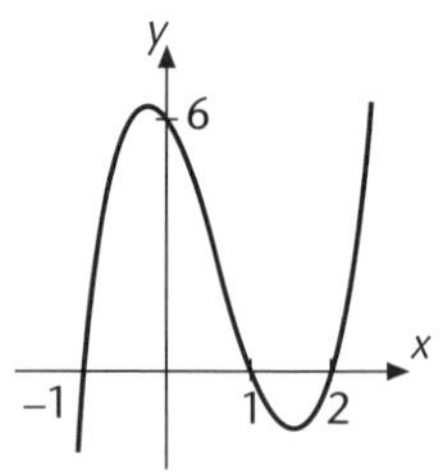

Solution

The graph has x-intercepts at -1, 1 and 2 so the equation of the graph is $y = k(x + 1)(x - 1)(x - 2)$, where k is a constant.

The graph has a y-intercept of 6.

Substituting $x = 0$, $y = 6$ in $y = k(x + 1)(x - 1)(x - 2)$, gives:

$$6 = k(0 + 1)(0 - 1)(0 - 2) \quad \text{[substituting]}$$
$$\therefore\ 6 = 2k$$
$$\therefore\ k = 3$$

Hence the equation of the graph is $y = 3(x + 1)(x - 1)(x - 2)$.

Unit 11.2 Activity 16A: Cubic graphs

1. Sketch the graphs of the following cubic functions:
   - **a.** $y = (x - 1)(x - 2)(x + 4)$
   - **b.** $y = (x + 3)(x + 2)(x - 1)$
   - **c.** $y = (2 - x)(2 + x)(x - 1)$
   - **d.** $y = x(x - 1)(x + 3)$
   - **e.** $y = x(x - 1)^2$
   - **f.** $y = (x - 2)^2(x - 3)$
   - **g.** $y = (x - 1)^2(2 - x)$
   - **h.** $y = -(2 - x)(x + 1)(x - 3)$

2. Write down the equations of the cubic functions whose graphs are shown below:

a.

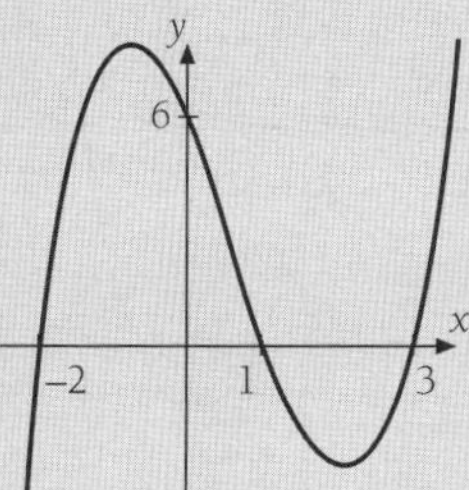

b.

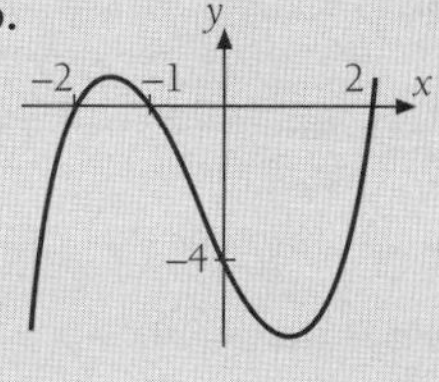

c.

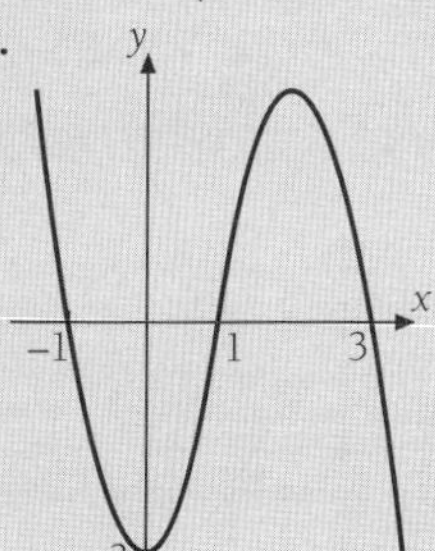

d.

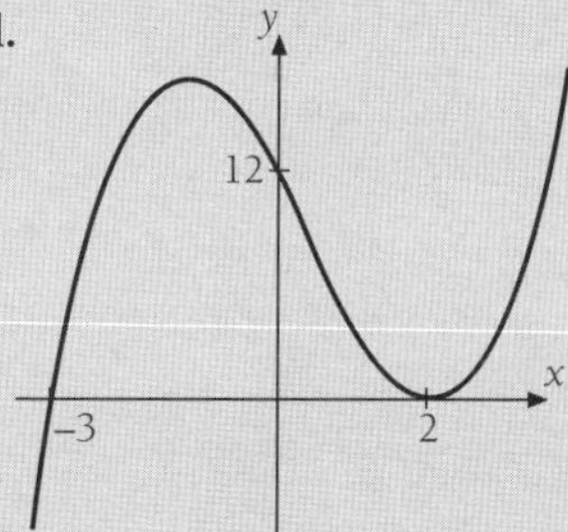

3. Draw the graph of the volume of a box against its length. Its height is two metres less than the length and its width is one metre less than the length.

4. Draw a graph of the volume of a square-based box against side length of the base. Its height exceeds the base length by one metre.

5. Records were kept through the first few months of this year of the kina value above K1.50 of one kilogram of a commodity. The results of this survey are shown in the graph.

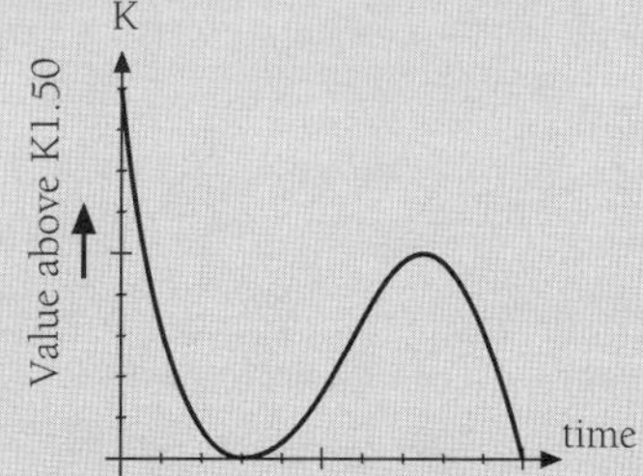

   - **a.** If the horizontal marks represent the passage of one month then what were the dates when the value of one kg was K1.50?
   - **b.** What was the value at the beginning of the year? (You may assume that each vertical mark represents one kina.)
   - **c.** When was the value of one kilogram in excess of K5.50?
   - **d.** What is an equation which would describe such a graph?

Unit 11.2 Activity 16B: Cubic graphs—multiple choice

1. For the graph of the cubic with equation $y = (x - 2)(x + 2)(x + 3)$ the x-intercepts and the y-intercept, respectively, are:

A. $x = 2, -2, -3$ and $y = 12$ **B.** $x = 2, -2, 3$ and $y = -12$

C. $x = 2, -2, -3$ and $y = -12$ **D.** $x = 2, -2, 3$ and $y = 12$

2. The equation for the cubic graphed at right is:

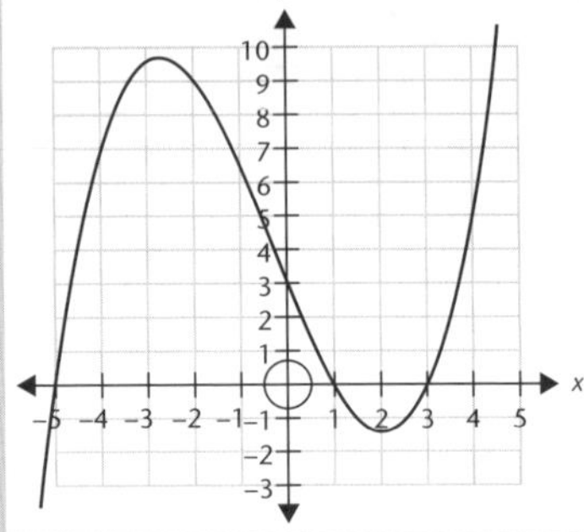

A. $y = (x - 3)(x - 1)(x + 5)$

B. $y = -\frac{1}{5}(x - 3)(x - 1)(x + 5)$

C. $y = (x + 3)(x + 1)(x - 5)$

D. $y = \frac{1}{5}(x - 3)(x - 1)(x + 5)$

3. The number of times that the cubic with the equation $y = x^3 - 3$ cuts the x-axis is:

A. 0   **B.** 1

C. 2   **D.** 3

4. Which one of the following cubics touches the x-axis at $x = 2$?

A. $y = (x - 1)(x + 2)^2$ **B.** $y = -(x - 1)(x - 2)^2$

C. $y = (x - 2)(x + 1)^2$ **D.** $y = -(x + 2)(x - 1)^2$

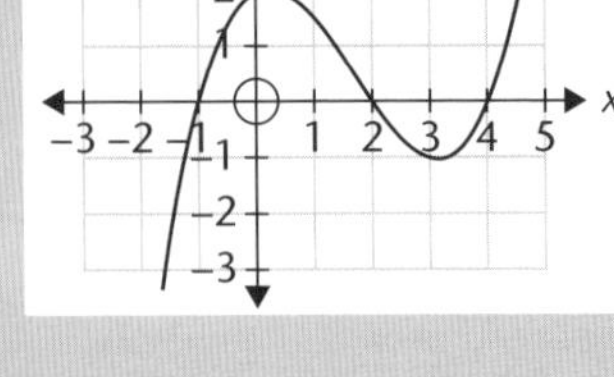

5. The graph of $y = k(x - 4)(x + 1)(x - 2)$ is graphed at right. The value of k is:

A. $\frac{1}{4}$ **B.** $\frac{1}{2}$

C. 1   **D.** 2

6. Which one of the following points is not on the graph of the cubic $y = \frac{1}{3}(x^2 - 1)(x - 9)$?

A. (0, 3)   **B.** (2, −7)   **C.** (3, −16)   **D.** (4, 25)

7. The number of times that the cubic with the equation $y = -2(x + 1)^2(x - 3)$ cuts the x-axis is:

A. 0   **B.** 1

C. 2   **D.** 3

8. The equation for the cubic graphed at right is:

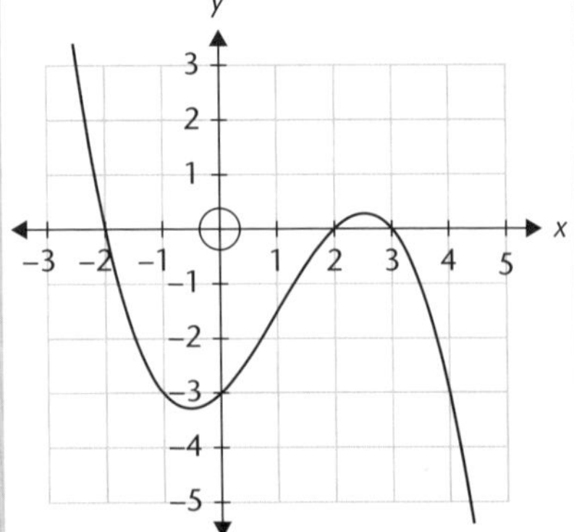

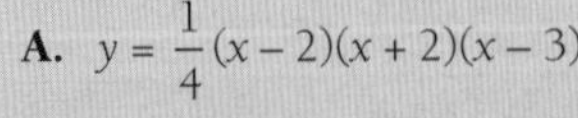

A. $y = \frac{1}{4}(x - 2)(x + 2)(x - 3)$

B. $y = -\frac{1}{4}(x + 3)(x - 2)(x + 2)$

C. $y = \frac{1}{4}(2 - x)(2 + x)(x - 3)$

D. $y = -\frac{1}{3}(x - 3)(x - 2)(x + 2)$

Unit 11.2 Graphs and Functions

Topic 17: Absolute value functions

'Sketching and writing quadratic, cubic and absolute value functions' is the third bullet point under the heading 'Graphs and functions' (Syllabus p. 15). This Topic deals with absolute functions. It covers:

- The absolute value of a number.
- Graphs of absolute value functions.

The absolute value of a number

The absolute value of a real number can be defined as the distance that the real number is from the origin (zero). All absolute values are positive (or zero).

For example, on the number line at right: the absolute value of –3 is 3 because –3 is 3 units from the origin.

This is written as $|-3| = 3$

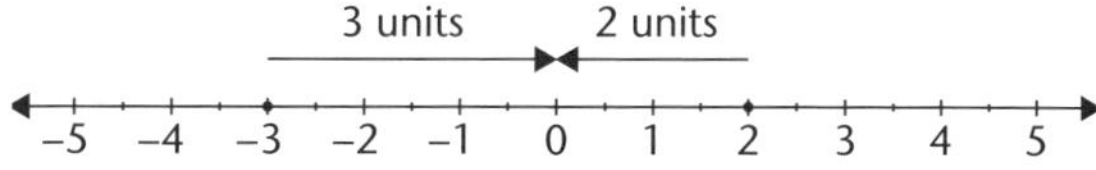

The absolute value of 2 is 2 because 2 is two units from the origin.

This is written as $|2| = 2$

Note: The absolute value of a number is sometimes written as *abs*(*x*). This notation is often used on calculators.

In general:

If $x \geq 0$ then $|x| = x$
If $x < 0$ then $|x| = -x$

Example A

Q. If $x = -3$ and $y = 5$ find each of the following:

1. $|x| + |y|$
2. $|x + y|$
3. $|x| \times |y|$
4. $|x \times y|$

A.

1. $|x| + |y| = |-3| + |5| = 3 + 5 = 8$
2. $|x + y| = |-3 + 5| = |2| = 2$
3. $|x| \times |y| = |-3| \times |5| = 3 \times 5 = 15$
4. $|x \times y| = |-3 \times 5| = |-15| = 15$

Unit 11.2 Activity 17A: Absolute value of a number

1. Evaluate:

a. $|5|$ **b.** $|-3.4|$ **c.** $\left|-1\frac{5}{6}\right|$ **d.** $|-0.0056|$

e. $|2.5|$ **f.** $|-\sqrt{2}\,|$ **g.** $|0|$ **h.** $|-1024|$

2. a. Copy and complete the following table:

a	b	$\|a\| + \|b\|$	$\|a + b\|$
2	7		
–2	7		
2	–7		
–2	–7		

b. Based on your results in part a., is it *always true* that $|a| + |b| = |a + b|$ if a and b are real numbers?

3. a. Copy and complete the following table:

a	b	$\|a\| - \|b\|$	$\|a - b\|$
8	3		
–8	3		
8	–3		
–8	–3		

b. Based on your results in part a., is it *always true* that $|a| - |b| = |a - b|$ if a and b are real numbers?

4. a. Copy and complete the following table:

a	b	$\|a\| \times \|b\|$	$\|a \times b\|$
3	7		
–3	7		
3	–7		
–3	–7		

b. Based on your results in part a., is it *always true* that $|a| \times |b| = |a \times b|$ if a and b are real numbers?

5. a. Copy and complete the following table:

a	b	$\|a\| \div \|b\|$	$\|a \div b\|$
12	4		
–12	4		
12	–4		
–12	–4		

b. Based on your results in part a., is it *always true* that $|a| \div |b| = |a \div b|$ if a and b are real numbers?

6. Evaluate each of the following:

a. $|-5| \times 3 + |4| \times |-5|$

b. $(3 + |-2|)(|5|-|-3|)$

c. $(5 + |-1|)^2$

d. $(-|-3| \times |-2|)^2$

e. $26 + |-25| \div |-5| \times (-|-3|)$

f. $-\sqrt{|-121|}$

g. $|15| \div |-3| \times |0.2| - |-1|$

h. $-|16| + |-16| \div |16| \times |-16|$

Graphs of absolute value functions

The graph of $y = |x|$

A table of values for $y = |x|$ is given below:

x	−3	−2	−1	0	1	2	3
y	3	2	1	0	1	2	3

When these points are plotted we get the graph at right:

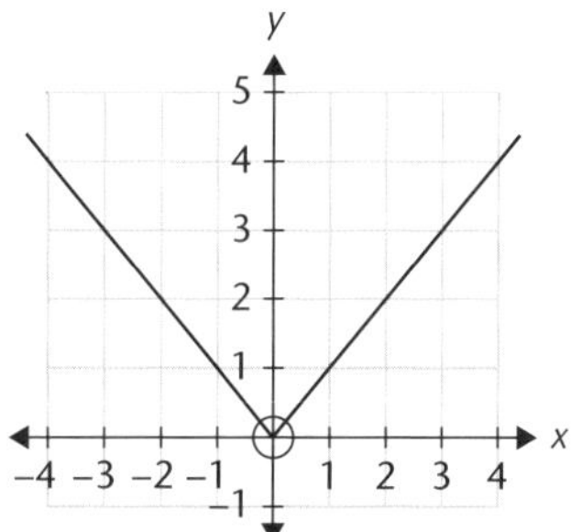

Considering this graph we can say:

If $x \geq 0$ then the graph is $y = x$.

If $x < 0$ then the graph is $y = -x$.

The 'vertex' of this graph is at (0, 0).

Graphs of the form $y = a|x|$ or $y = |ax|$; $a \in R$

If $x \geq 0$ then the graph is $y = ax$.

If $x < 0$ then the graph is $y = -ax$.

a changes the gradient of the graph.

If a is negative then the graph will have a 'negative' V shape.

Some graphs of this form are given at right:

A $y = |x|$

B $y = -|x|$

C $y = 2|x|$

D $y = 5|x|$

E $y = -\frac{1}{2}|x|$

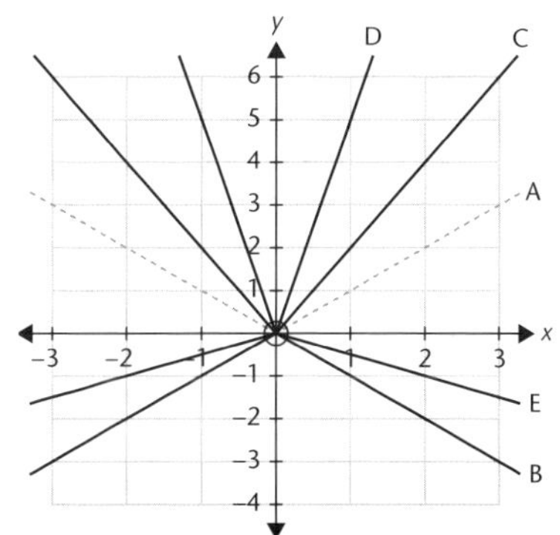

Note: all of these graphs have their vertex at (0, 0).

Graphs of the form $y = |x| + k$; $k \in R$

A table of values for $y = |x| + 2$ is given below:

x	−3	−2	−1	0	1	2	3
y	5	4	3	2	3	4	5

When these points are plotted we get the graph at right:

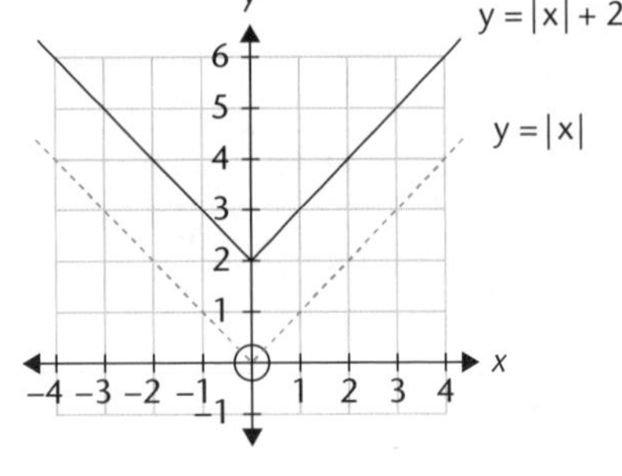

Considering this graph we can say:

If $x \geq 0$ then the graph is $y = x + 2$.

If $x < 0$ then the graph is $y = -x + 2$.

In this example the graph of $y = |x|$ is translated 2 units up (parallel to the y-axis). The 'vertex' of this type of graph is at $(0, k)$.

Some graphs of this type are given at right:

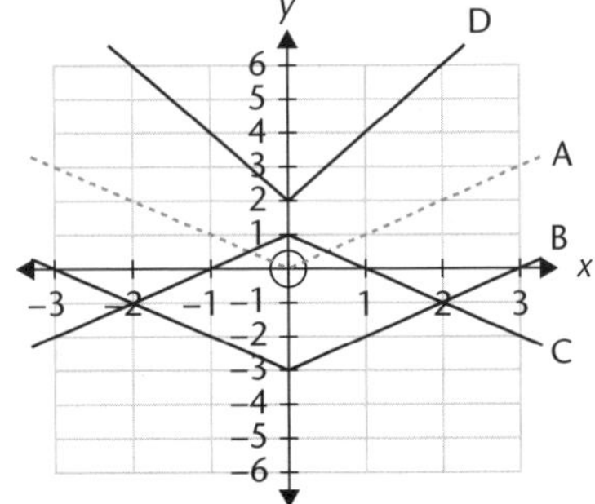

A $y = |x|$

B $y = |x| - 3$

C $y = -|x| + 1$

D $y = 2|x| + 2$

Graphs of the form $y = |x - h|$

A table of values for $y = |x - 2|$ is given below:

x	−3	−2	−1	0	1	2	3	4	5
y	5	4	3	2	1	0	1	2	3

When these points are plotted we get the graph at right:

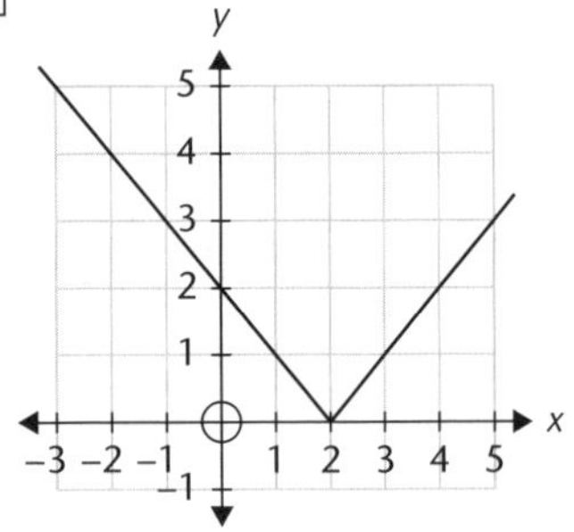

Considering the graph we can say:

If $x - 2 \geq 0$, ie $x \geq 2$, then the graph is $y = x - 2$.

If $x - 2 < 0$, ie $x < 2$, then the graph is $y = -(x - 2) = 2 - x$.

The vertex of this graph is at $(2, 0)$. In this example the graph of $y = |x|$ is translated h units to the right.

The vertex of this type of graph is at $(h, 0)$.

Graphs of the form $y = a|x - h| + k$

A table of values for $y = |x + 3| - 1$ is given below:

x	−6	−5	−4	−3	−2	−1	0	1	2
y	2	1	0	-1	0	1	2	3	4

When these points are plotted we get the following graph:

Considering the graph we can say:

If $x + 3 \geq 0$, ie. $x \geq -3$ then the graph is $y = (x + 3) - 1 = x + 2$.

If $x + 3 < 0$, ie. $x < -3$ then the graph is $y = -(x + 3) - 1 = -4 - x$.

Note : the vertex for this graph is at $(-3, -1)$.

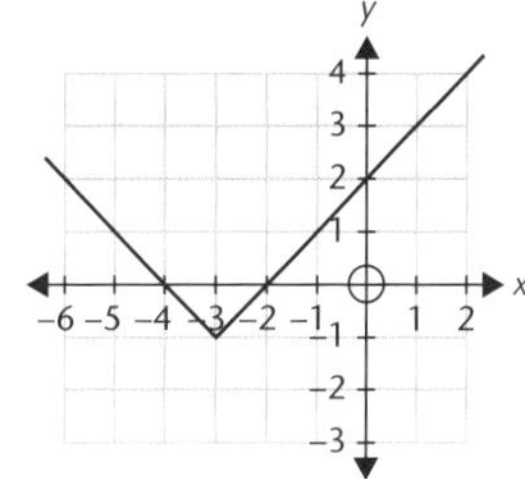

In general:

> The graph of $y = a|x - h| + k$ has a vertex at (h, k).
>
> If $x - h \geq 0$ ie $x \geq h$ then the graph is $y = a(x - h) + k = ax + (k - ah)$.
>
> If $x - h < 0$ ie $x < h$ then the graph is $y = a \times -(x - h) + k = -ax + (k + ah)$.

Example B

Q. Sketch the graph of $y = 2|x - 2| + 3$.

A. The vertex of the graph is at $(2, 3)$.

If $x - 2 \geq 0$ ie. $x \geq 2$ then the graph is $y = 2(x - 2) + 3 = 2x - 1$.

If $x - 2 < 0$ ie. $x < 2$ then the graph is $y = 2 \times -(x - 2) + 3 = -2x + 7$.

Using this information to sketch the graph:

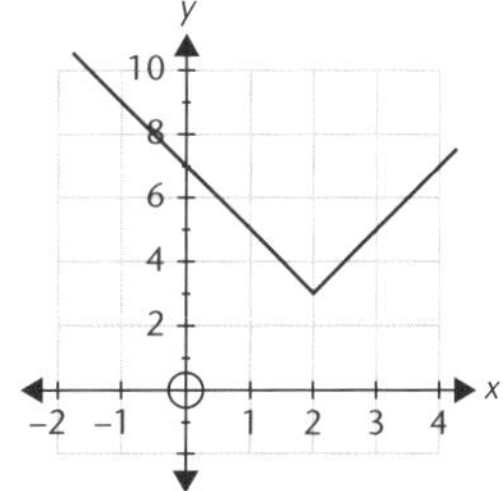

Note: The graphs of $y = 2x - 1$ and $y = -2x + 7$ can be sketched in full then the unwanted parts erased.

Example C

Q.

1. Sketch the graph of the function $y = 5 - |x - 3|$.
2. State the coordinates of the x and y-intercepts of this graph.

A.

1. The equation of the function can be written as $y = -|x - 3| + 5$. The vertex of this graph will be at $(3, 5)$ and it will be a 'negative' graph.

   If $x - 3 \geq 0$, ie. $x \geq 3$ then the graph will be $y = -(x - 3) + 5 = 8 - x$.

   If $x - 3 < 0$, ie. $x < 3$ then the graph will be $y = -[-(x - 3)] + 5 = x + 2$.

   The graph is given at right:

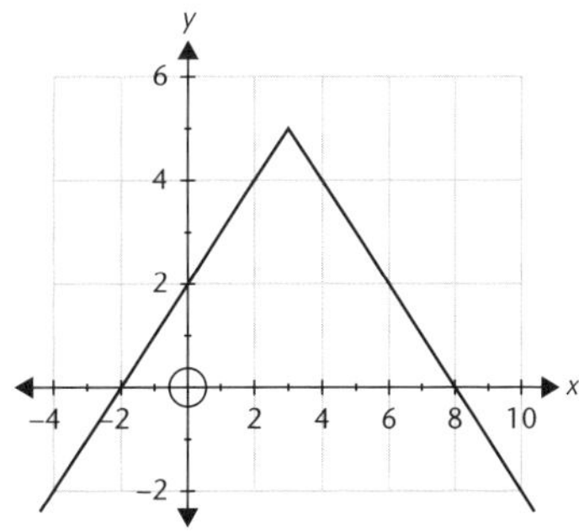

2. The x-intercepts are the points $(-2, 0)$ and $(8, 0)$. The y-intercept is the point $(0, 2)$.

Unit 11.2 Activity 17B: Graphs of absolute value functions

1. On the same set of axes sketch the graphs of:

a. $y = |x|$

b. $y = 2|x|$

c. $y = -|2x|$

2. On the same set of axes sketch the graphs of:

a. $y = |x| - 1$

b. $y = 2|x| - 1$

c. $y = 3 - |x|$

3. On the same set of axes sketch the graphs of:

a. $y = |x + 1|$

b. $y = 2|x - 1|$

c. $y = -|x + 2|$

4. On the same set of axes sketch the graphs of:

a. $y = |x - 1| + 1$

b. $y = |x + 2| - 3$

c. $y = |x + 1| + 4$

5. **a.** Sketch the graph of $y = 2|x - 1| + 1$.

b. Give the coordinates of the y-intercept.

6. **a.** Sketch the graph of the function $y = -|x + 2| + 3$.

b. Give the coordinates of the y-intercept.

c. Give the coordinates of the x-intercepts.

d. State the equation of the function for $x \geq -2$.

e. State the equation of the function for $x < -2$.

7. **a.** Sketch the graph of the function $y = 2|x - 3| - 5$.

b. Give the coordinates of the y-intercept.

c. Give the coordinates of the x-intercepts.

d. State the equation of the function for $x \geq 3$.

e. State the equation of the function for $x < 3$.

8. **a.** Sketch the graph of the function $y = 4 - \frac{1}{2}|x + 2|$.

b. Give the coordinates of the y-intercept.

c. Give the coordinates of the x-intercepts.

d. State the equation of the function for $x \geq -2$.

e. State the equation of the function for $x < -2$.

Unit 11.3 Managing Data
Topic 1: Basic data display

Unit 11.3 focuses on statistics, probability, permutations and combinations. Statistics are used in many situations in the world around us. Topic 1 looks at surveys and the decisions that can be made from them:

- Justifying inferences made about the population, including the use of statistical graphs.

Note that in statistics, the word 'population' means all the items that are the subject of the survey (see Topic 6, p. 299).

Introduction

Once **data** from a survey have been collected they are often *displayed* so that **deductions** can be made about the data. The following example shows **raw data** and is in the typical form of a **tally chart** combined with a **frequency table**.

Example A

In a survey on the attitudes of 25 maths pupils, a student obtained the following data which **ranked** maths by preference:

Ranking	Tallies	Frequency
1	|||	3
2	~~||||~~ |	6
3	~~||||~~ ||	7
4	~~||||~~	5
5	|||	3
6	|	1

Tallies are vertical lines used to record how often values occur (a tally is put beside a value each time it occurs). When four lines have been put down the fifth is usually drawn as a horizontal or diagonal line through the previous four. This gives blocks of five tallies which are easy to count.

The **frequency** of any possible value is the number of times it occurs in a survey. It can be found by counting the number of tallies for that value.

Discrete data have gaps between possible values. This means that there are no other values possible between the values obtained in the data.

Example B

In Example A, the data obtained are discrete because the only possible values are 1, 2, 3, 4, 5 and 6. There are no other values possible between 1 and 2 (such as 1.2), or between 2 and 3, etc.

Display of discrete, ungrouped data

Bar graphs are the most basic and useful visual display of discrete data. All bars on a bar graph should be of equal width, with the same size gaps between them. The heights of the bars are proportional to the frequency.

Bar graphs are drawn using the following general rules:

- Frequency is usually on the vertical axis.
- The bars are separated when dealing with discrete data.
- The axes must be labelled and, where appropriate, a title should be given to the bar graph.

Example C

On a bar graph, the data in Example A appear as shown:

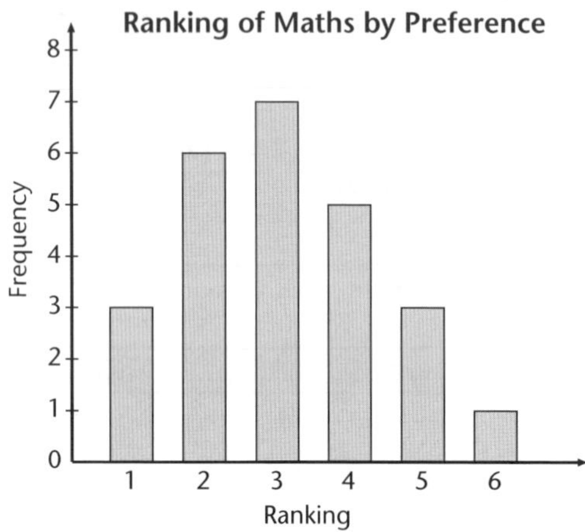

Note: There is a bar for each possible value that the **ungrouped** data can take.

Display of discrete grouped data

Grouped data are displayed in groups of values (intervals) to show a meaningful **trend**.

Example D

A student collected the following data about marks (%) in a mathematics test. Rather than recording each mark separately, the student grouped the marks in intervals of size 10. A bar graph representing the data is also shown.

Marks	Tallies	Frequency
30 – 39	|	1
40 – 49	|||	3
50 – 59	𝍸	5
60 – 69	𝍸 𝍸	10
70 – 79	|||	3
80 – 89	||	2
90 – 99	|	1

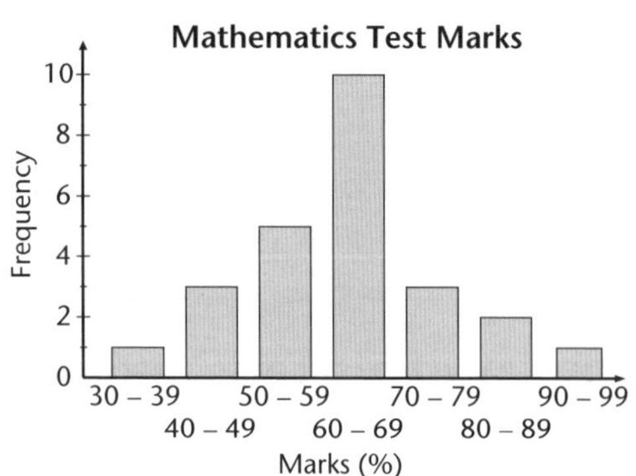

Note: The major trends are the concentration of most marks between 50 and 80 and the approximate symmetry of the distribution of marks.

Relative frequency

The **relative frequency** of a value is the **proportion** of total occurrences of that value. This is calculated by dividing the frequency of the value by the total frequency.

Example E

Determine the relative frequencies of the data in Example D and display these in a table.

Solution

Marks	Relative frequency
30–39	0.04
40–49	0.12
50–59	0.20
60–69	0.40
70–79	0.12
80–89	0.08
90–99	0.04

[frequency of 30–39 is 1, total frequency is 25

$\therefore$ relative frequency of 30–39 is $\frac{1}{25} = 0.04$, and so on]

Note: The total of the relative frequencies is always 1.

Continuous data

Continuous data have no gaps between possible values. This means that for any two possible data values, another possible value between them can always be found. Such data are displayed on bar graphs called **histograms**. The bars in histograms *touch* because the data are continuous. The *area* of the bars is proportional to the frequency.

Example F

A student collected data about the heights of pupils in her class. The results are displayed below, both in a table and on a histogram.

Height (cm)	Tallies	Frequency
150 < 155	𝍷	1
155 < 160	𝍷𝍷𝍷	3
160 < 165	𝍷𝍷𝍷𝍷	4
165 < 170	𝍸 𝍸 𝍷	11
170 < 175	𝍷𝍷𝍷𝍷	4
175 < 180	𝍷𝍷	2
180 < 185	𝍷𝍷	2

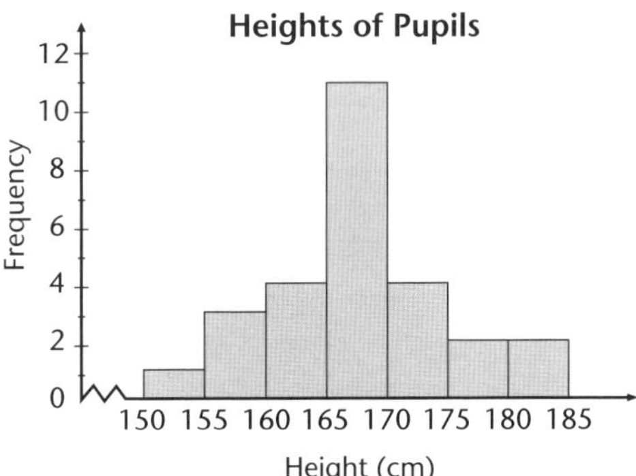

Note: 150 < 155 means
$150 \le \text{height} < 155$ etc.

Note: The **compression** of the bottom axis is shown using a **concertina** (looks like a w or m). This is a frequently used technique in statistics. It is acceptable only when done to the horizontal axis of bar graphs and histograms, but should not be done to the vertical axis (showing the frequency).

Non-equal intervals

Some data have groups or intervals which are not equal in size (width). When these data are displayed, *it is important to keep the area of the bars proportional to the frequency*.

Example G

A survey was taken of the times between successive vehicles passing under a bridge. The following chart shows the distribution of these times, and a histogram displays the results.

Time (s)	Tallies	Frequency
0 < 1	卌 卌 卌 卌 卌 III	28
1 < 2	卌 卌 II	12
2 < 4	卌 卌 II	12
4 < 8	卌 卌 II	12
≥ 8	III	3

Notes:

1. $0 < 1$ means $0 \le \text{time} < 1$ etc.
2. The height of the bar above the interval 2 to 4 is *half* that of the bar above 1 to 2 (even though their frequencies are both equal to 12) because the interval $2 < 4$ is *twice* as long as the interval $1 < 2$.
   Likewise, the height of the bar above the interval 4 to 8 is one quarter that of the interval 1 to 2 because the interval is four times as long.
3. A maximum value (12 seconds) is *chosen* for time because one is not given.

Pie graphs

Pie graphs are widely used in many fields of study and in the media to show *proportions*.

Example H

The relative proportions of different types of crops planted on a nation's arable land are shown:

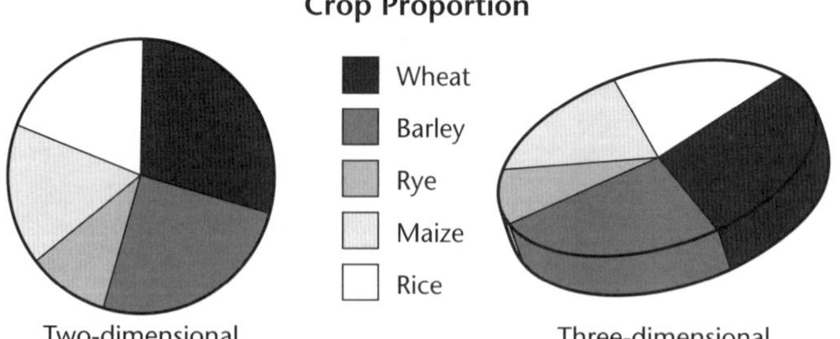

Two-dimensional

Three-dimensional

The pie graph shows that about $\frac{1}{3}$ of the nation's arable land is planted in wheat, about $\frac{1}{4}$ in barley, and the remainder is planted in other crops.

Often the actual **percentages** are included on the pie graph.

Example I

Draw a pie graph showing the proportions of magazine types in a bookstore. They have the following numbers of each brand on sale:

Cosmopolitan	45
More	35
Listener	25
TV Guide	15
Others	30
Total	150

Solution

The **sector** angles are calculated using proportion. The relative frequency of *Cosmopolitan* is $\frac{45}{150}$.

Therefore, the sector angle for *Cosmopolitan* is $\frac{45}{150} \times 360° = 108°$ [360° in a full circle]

The **percentage** for *Cosmopolitan* is $\frac{45}{150} \times 100\% = 30\%$

Similar calculations lead to the following table of values for sector angles and percentages:

Magazine	Frequency	Percentage	Sector Angle
Cosmopolitan	45	30.0	108°
More	35	23.3	84°
Listener	25	16.7	60°
TV Guide	15	10.0	36°
Others	30	20.0	72°

Using these sector angles the pie graph is completed:

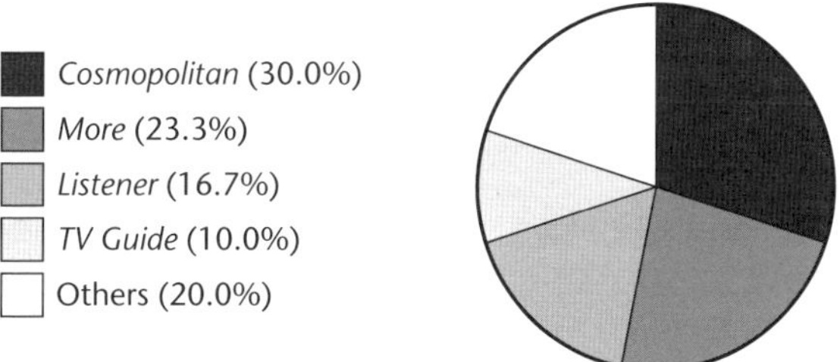

Stem and leaf diagrams

Stem and leaf diagrams are similar to tally charts, but provide more information. Each data value is separated in two parts called a **stem** (the left part) and **leaf** (the right part, which is always a single digit).

Example J

Draw a stem and leaf diagram of the following set of numbers.

{13, 25, 12, 34, 47, 53, 17, 12, 10, 48, 36, 37, 27, 29, 14, 9, 2, 24, 42, 42, 39, 33, 45, 50, 38}

Stems are to be tens and leaves are to be ones.

Solution

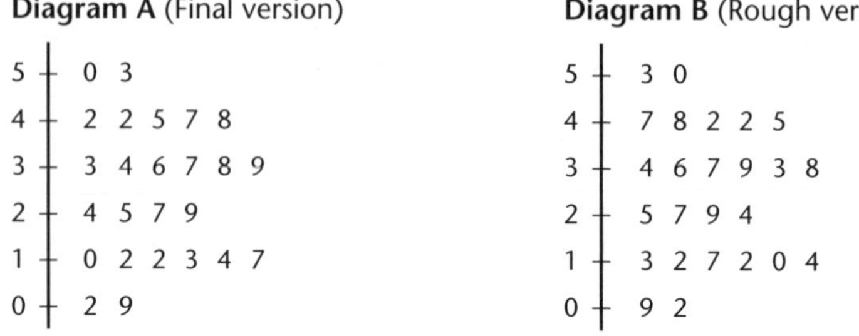

Diagram A is largely self-explanatory. Numbers from 10 to 19 are placed beside the 1, in order of size and so on. For each number, the digit in the tens place is called the *stem*, the digits in the ones place are called the *leaves*.

1 | 0 2 2 3 4 7 represents the numbers 10, 12, 12, 13, 14, 17 from the given set.

2 | 4 5 7 9 represents the numbers 24, 25, 27, 29, etc.

Note: One technique that works well is to plot an unordered stem and leaf diagram first (see diagram B above) then put it in order.

Where necessary, a key is used to explain the data values.

Example K

Two stem and leaf charts with keys are shown:

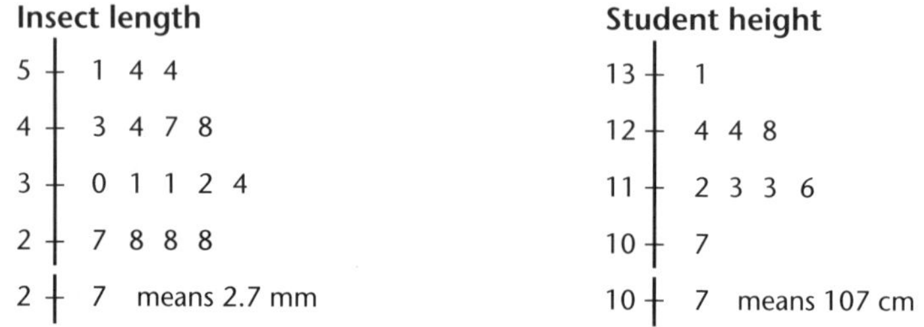

Scatter diagrams

Scatter diagrams give a quick and quite reliable method for detecting relationships when dealing with paired data.

In cases where there is a linear relationship it is possible to find what is called the **line of best fit** using a technique called *regression*. This topic is outside the syllabus, but a reasonable approximation to this line may be achieved by eye. A line is drawn through the data which has as about as many points on one side of it as it does on the other. Common sense must prevail when drawing such a line.

Example L

The following table gives widths and lengths of leaves from a sample taken from a species of tree.

Length (cm)	12.5	13.2	12.1	14.3	15.6	14.7	12.8	13.6	13.8	14.0
Width (cm)	1.6	2.1	1.4	2.6	3.4	2.9	1.8	2.5	2.6	3.0

The scatter diagram for this data appears as shown:

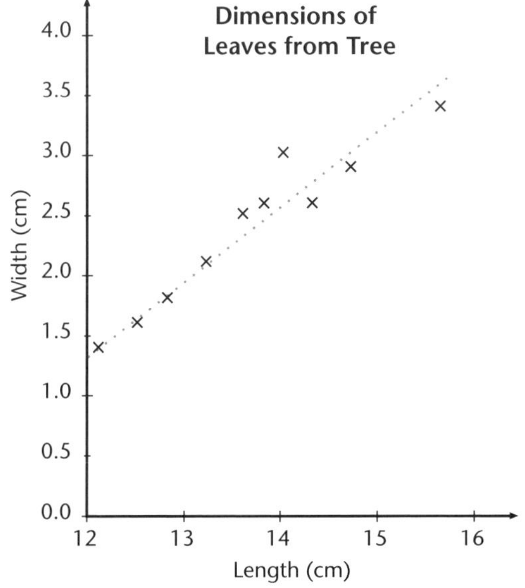

1. Inspecting this graph reveals an obvious trend for width to increase as length increases.
2. The relationship also appears to be linear as all plotted points lie close to the straight line dotted in.

Correlation

When there is a clear linear relationship with positive gradient (as in Example L), the two variables are said to be **positively correlated**. As the horizontal variable increases, so does the vertical variable. This does not necessarily mean that one variable is influencing the other.

Sometimes scatter diagrams appear as in the graph alongside. In this case there is a linear trend with negative gradient. The variables are said to be **negatively correlated** (as the horizontal variable increases, the vertical variable decreases).

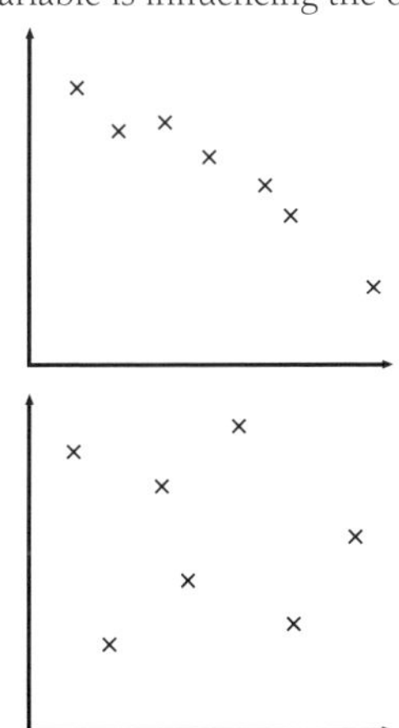

Often scatter diagrams have the appearance shown in the graph alongside.

In this graph there is no obvious pattern to the distribution of points and the variables are said to be **uncorrelated**.

On occasion the scatter diagram will reveal a trend, but one which is clearly not linear. One example is shown alongside.

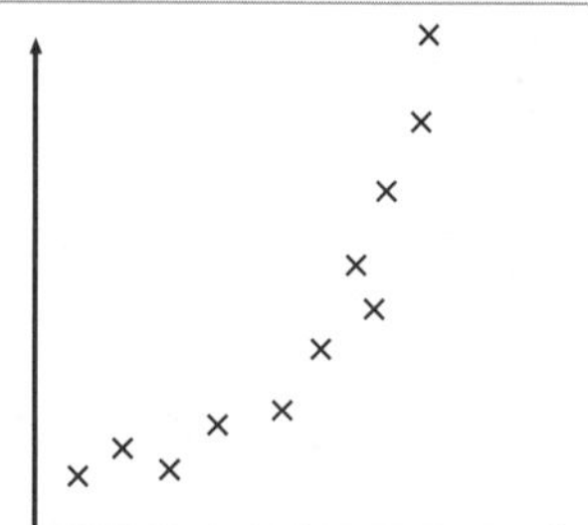

Using computers

Spreadsheets can make short work of data display. With only a little practice, it is possible to produce a great variety of data displays very quickly and neatly. The series of displays in Example M was produced with an Excel spreadsheet in the space of ten minutes.

Example M

Demonstration of some of the different types of data display available with spreadsheets.

Mark	0	1	2	3	4	5	6	7	8	9	10
Frequency	5	7	6	9	12	13	8	5	4	3	2

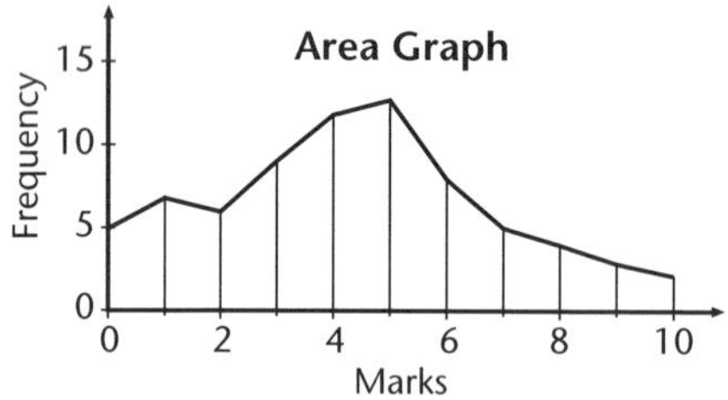

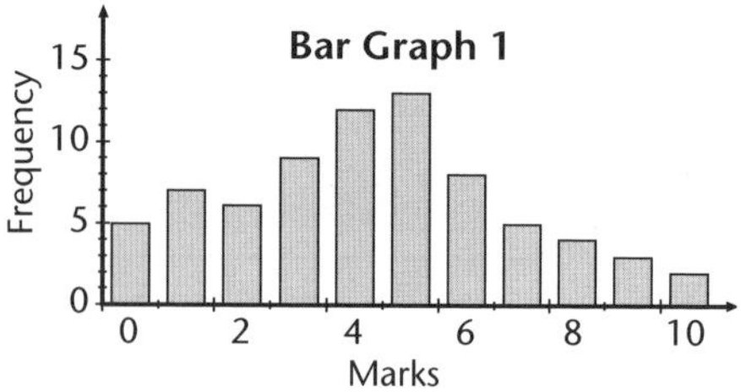

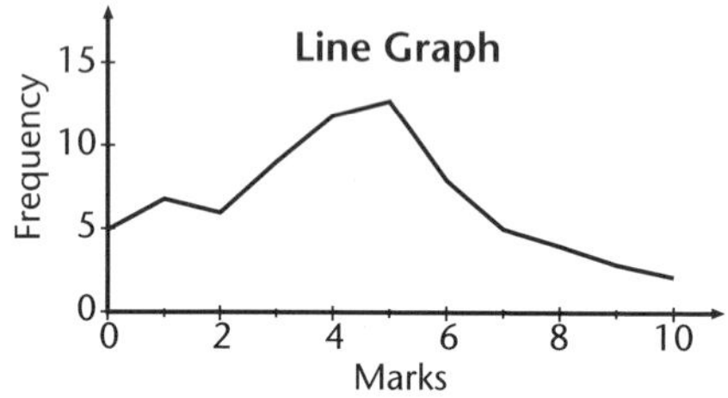

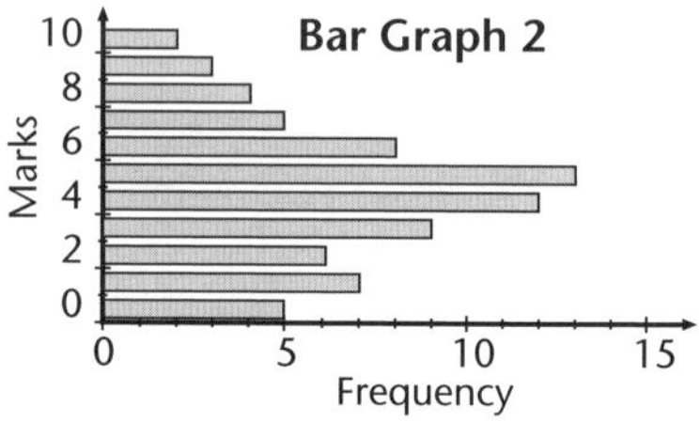

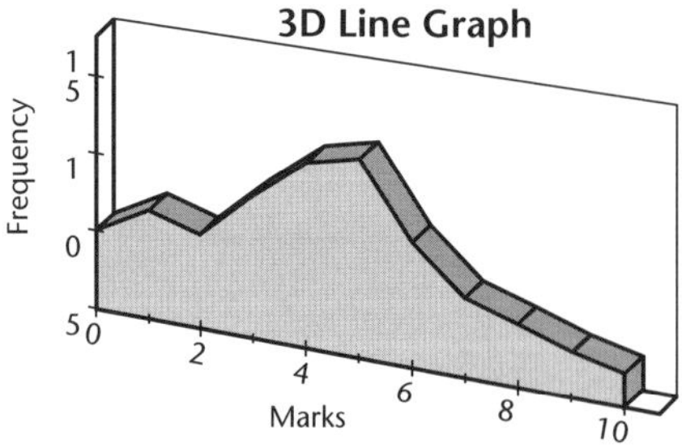

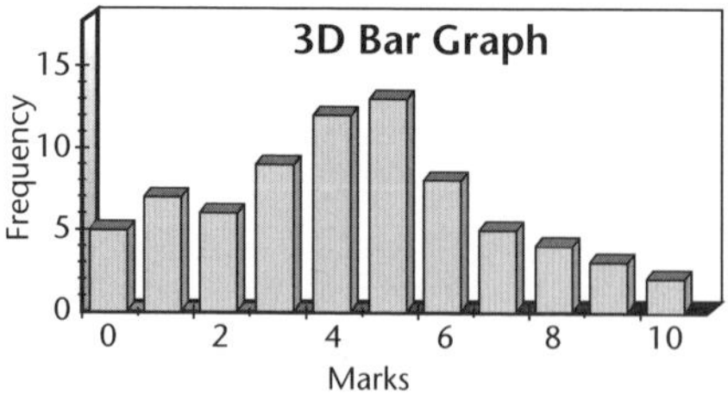

Area Graph

Bar Graph 1

Line Graph

Bar Graph 2

3D Line Graph

3D Bar Graph

Pie Graph

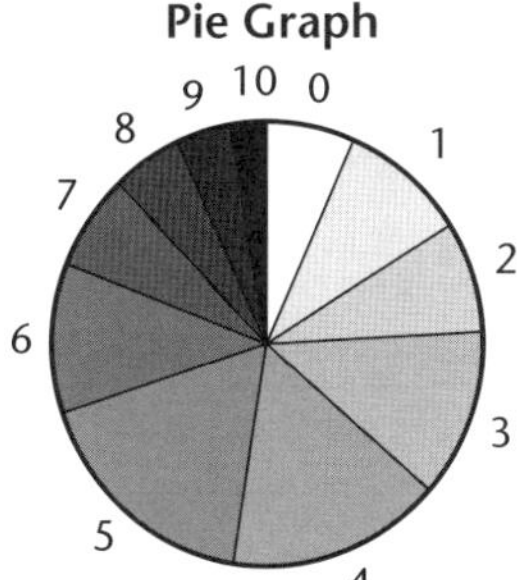

3D Pie Graph

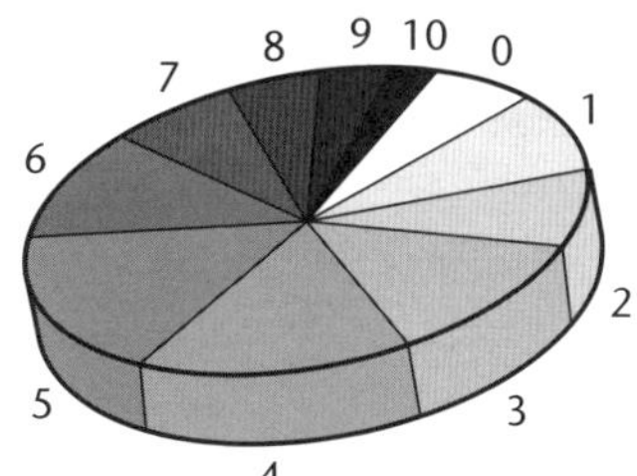

Unit 11.3 Activity 1A: Basic data display

1. Gymnasts in a competition scored the following marks out of 10:
6, 5, 4, 5, 7, 3, 6, 5, 5, 7, 6, 5, 4, 7, 5, 3, 5, 8, 5, 3, 6, 7.
Draw a bar graph to show these results.

2. a. Set up a tally chart as below for the set of numbers and complete it.

Number	Tallies	Frequency
20	I	1
21	III	3
⋮		
46	I	1

Numbers

27	21	40	37	30	36	24	31	43
36	34	26	40	33	31	43	37	27
34	29	26	46	42	38	40	36	35
34	32	43	23	24	30	40	33	25
21	25	34	39	28	36	40	34	28
21	29	35	31	45	25	43	20	41

b. Use the tally chart you have set up to draw a bar graph.

c. Set up another tally chart with the data grouped in fives.

i. Draw a bar graph for this tally chart.

ii. Comment on what trends or patterns are revealed by each graph.

iii. Which is the more suitable graph and why?

3. 'Statistics is a branch of mathematics which in some countries is considered to be so important that it is studied as a subject in its own right. The new sixth form syllabus recognises this by devoting far more time to this topic.'

   Analyse this paragraph by setting up a tally chart as below in order to determine the relative frequencies of the vowels in the English language.

Vowel	Tallies	Frequency	Relative Frequency
a			
e			
i			
o			
u			

   Using your tally chart, draw a suitable bar graph.

4. Which of the following variables are *discrete* and which are *continuous*?

   a. Year of birth
   b. Weight
   c. Sex
   d. Exact age
   e. Number in family
   f. Shoe size
   g. Surname
   h. Amount of time spent watching TV
   i. Age to nearest minute
   j. Number in families of children in Grade 11 classes

5. Measurements are made of the heights of 100 children. The results are given in the table below:

Height (cm)	Frequency
$160 \le h < 161$	2
$161 < 162$	3
$162 < 163$	12
$163 < 164$	28
$164 < 165$	26
$165 < 166$	12
$166 < 167$	9
$167 < 168$	5
$168 < 169$	3

   Draw a histogram showing the distribution of heights.

6. A fisherman weighed a sample of his catch and obtained the following data:

Weight (kg)	Frequency
$0 < 5$	35
$5 < 10$	22
$10 < 15$	16
$15 < 20$	12
$20 < 30$	8
$30 < 50$	4
$50 < 100$	18

Note: $0 < 5$ means $0 \le W < 5$, etc.

Draw a histogram which accurately shows the distribution of fish in his catch.

7. A boy in Grade 12 took a survey of the heights of his classmates. His data are shown in the following chart:

Height (cm)	Frequency
$130 < 140$	4
$140 < 150$	8
$150 < 155$	7
$155 < 160$	5
$160 < 170$	7
$170 < 180$	4
$180 < 190$	2
$190 \le 200$	1

Note: $130 < 140$ means $130 \le \text{height} < 140$, etc.

The histogram he drew to show this set of data is drawn below.

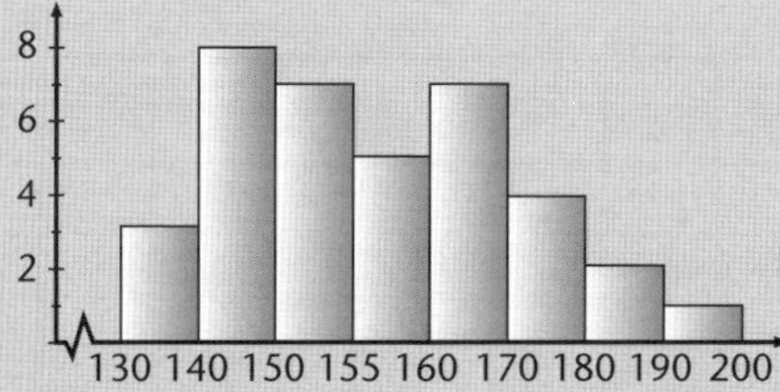

a. Write down all the things that are wrong with the histogram.

b. Redraw the histogram so that it more accurately represents the distribution of heights in the class.

8. The pie chart below shows the proportions of a large company's major expenses:

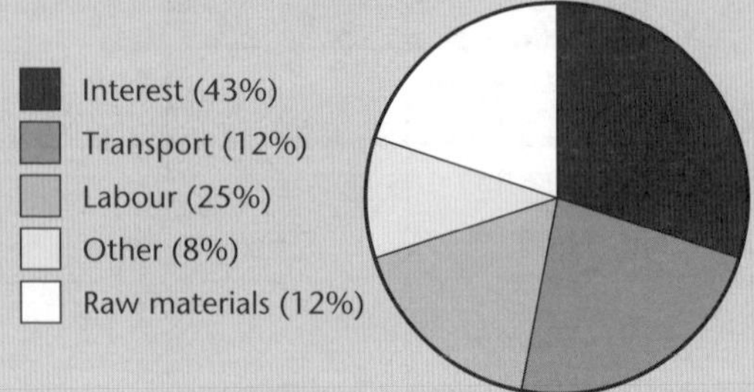

a. If total expenses were K63 479 000, find the interest and labour costs.

b. How large should the sector angle be for transport?

c. Sketch a bar graph showing this same distribution.

9. A computer sales manager asked her assistant to draw a large pie graph showing the distribution of the different makes of computer they stocked.

These are given in the table below:

Make	HP	IBM	Toshiba	Apple	Others
Number	65	42	23	12	8

Draw a pie graph of the type the assistant should have drawn.

10. Draw a stem and leaf diagram with stems being units and leaves being tenths for this set of data:

3.1, 3.3, 2.2, 1.4, 0.2, 3.4, 3.4, 4.5, 4.2, 3.7, 5.6, 4.4, 4.2, 3.5, 3.8, 3.6, 2.6, 2.5, 2.5

11. Find out if there is any correlation between the number of pages in children's books and their price by drawing a scatter diagram for the following sample:

Number of pages	32	32	24	32	16	32	24	16	24	32	24
Price (K)	9.50	10.00	7.00	6.00	6.50	4.50	5.50	17.00	15.00	5.00	5.00

Number of pages	160	32	32	32	32	32	32	32
Price (K)	17.00	3.50	2.00	3.00	3.50	2.00	3.50	2.50

12. The following table gives the lifts in the 'snatch' and the 'clean and jerk' for competitors in the 76 kg class at a world weightlifting championship. Find out if performance in the two lifts is correlated.

Snatch	110	110	120	115	117	125	122	130	130	135	145
C & J	142	145	140	155	157	160	165	160	157	160	165

Unit 11.3 Managing Data

Topic 2: Measures of central tendency and spread

Remember that in statistics, the term 'population' means all the items that are the subject of a survey (see Topic 6, p. 299). In Topic 2 we look at selecting a sample and using it to make inferences about the population:

- Calculating appropriate sample statistics, selecting from mean, median, quartiles, standard deviation and proportion.
- Making an inference about the population.
- Justifying inferences made about the population, including the use of statistical graphs and the interpretation of sample statistics.

Introduction

Two important concepts in the **analysis** of statistical data are central tendency and spread or dispersion.

Central tendency measures the 'central value' of a **sample** or **population**.

Spread or **dispersion** measures how the data are spread out from the centre in a sample or population.

The central value gives no information about how the values are spread around that centre, as the following example shows.

Example A

Bar graphs have been drawn for each of the following sets of values. For each set, *x* represents the possible values and *f* represents the frequency with which each possible value occurs.

1.

x	1	2	3	4	5
f	2	2	2	2	2

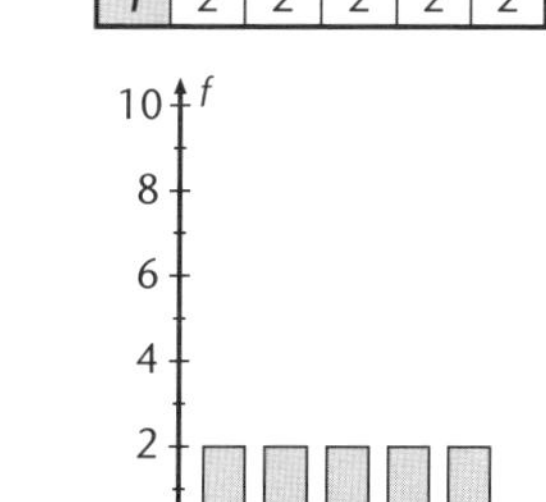

Central value is 3. Values are *evenly* spread around the centre.

2.

x	1	2	3	4	5
f	5	0	0	0	5

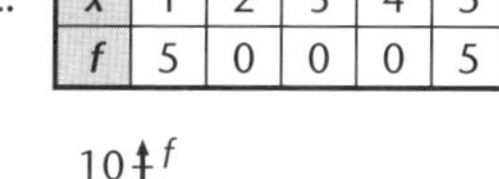

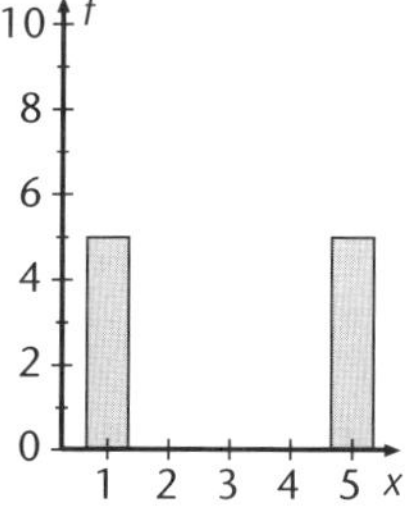

Central value is 3. Values are the *most* spread because more values are further from the centre than **1.** or **2.**.

3.

x	1	2	3	4	5
f	0	0	10	0	0

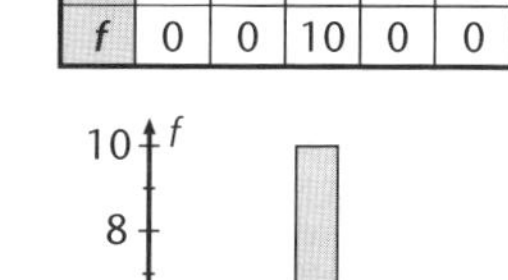

Central value is 3. Values are the *least* spread because they are all at the centre, **3.**

Measures calculated from the data in a sample are called **sample statistics**. These are often used to estimate measures in the population from which they are sampled (these measures are called **population parameters**), a process called **inference**.

Measures of central tendency

Measures of central tendency, such as the **mean** and the **median**, give information about 'middle' values of a **distribution**.

The mean

The **mean** is the sum of all the values in a data set divided by the total number of values.

The mean of a *sample* is given the symbol $\bar{x}$ (read 'x bar') and is often called the **average**. In ungrouped data, where $x_1, x_2, \ldots, x_n$ are the different values in the sample and n is the number in the sample, the **sample mean** is given by the formula:

$$\boxed{\bar{x} = \frac{\sum x}{n}} \qquad \textbf{Note: } \frac{\sum x}{n} \text{ means } \frac{\sum_{i=1}^{n} x_i}{n} = \frac{x_1 + x_2 + \ldots + x_n}{n}$$

Example B

For the sample 1, 2, 7, 12, 13 the mean is $\bar{x} = \frac{1+2+7+12+13}{5} = 7$.

In a sample of grouped data, each value x_i has a frequency f_i and the following formula for the sample mean is more appropriate:

$$\boxed{\bar{x} = \frac{\sum fx}{\sum f}} \qquad \textbf{Note: } \frac{\sum fx}{\sum f} \text{ means } \frac{\sum_{i=1}^{n} f_i x_i}{\sum_{i=1}^{n} f_i} = \frac{f_1 x_1 + f_2 x_2 + \ldots + f_n x_n}{f_1 + f_2 + \ldots + f_n}$$

The mean of a whole *population* has the symbol μ (pronounced 'mew') and is found using the same method as above, when all values in a population are known. Sample means are often used to estimate **population means**.

Example C

The following results are a **random sample** of 75 marks from a Grade 11 History test, where x stands for the mark scored out of 10 and f for the frequency.

x	0	1	2	3	4	5	6	7	8	9	10
f	1	1	2	5	12	18	16	10	5	4	1

The mean mark for the sample is:

$$\bar{x} = \frac{\sum_{i=1}^{n} f_i x_i}{\sum_{i=1}^{n} f_i} \qquad \text{[where } f_i \text{ is the frequency of the } i\text{th } x \text{ value, } x_i\text{]}$$

$$= \frac{1\times0+1\times1+2\times2+5\times3+12\times4+18\times5+16\times6+10\times7+5\times8+4\times9+1\times10}{1+1+2+5+12+18+16+10+5+4+1}$$

$$= \frac{410}{75}$$

$$= 5.47$$

The sample mean $\bar{x} = 5.47$ is an estimate of the **population mean** μ.

The mean is the most commonly used measure of central tendency in applications of statistics.

Unit 11.3 Activity 2A: The mean

1. Find the means of the following sets of numbers:

a. –5, 12, 13, 5, 6, 11, 12, 27 **b.** 49, 23, 47, 68, 113, 226, 14

c.

number	3	4	5	6
frequency	13	22	47	25

2. A student needs to have a mean mark of 58% to get into a course. The student is sitting six subjects and in the first five subjects has got marks of 47, 51, 53, 49, 68.

What mark will need to be obtained in the sixth subject to get into the course?

3. Write down a set of 5 different numbers whose mean is 23.

Medians and modes

The **median** is the middle value of a population or sample (whose values are listed in numerical order). If there are two middle values the median is found by taking the mean of these two values.

Example D

Find the median of the numbers: **a.** 3, 1, 2 **b.** 2, 3, 4, 1

Solution

a. Placing 3, 1, 2 in numerical order gives 1, 2, 3.

The median is 2 because it is the middle number.

b. In numerical order, the numbers 2, 3, 4 and 1 are 1, 2, 3, 4.

Because there is an even number of values there are two middle values, namely 2 and 3. The median is found by taking the mean of these two values. Thus the median is $\frac{2+3}{2} = 2.5$.

It can be tedious to list a large sample or population in order of size. The following example shows an alternative method.

Example E

Find the median of the values in Example C.

Solution

There are 75 values in Example C. The middle number of these 75 values in order is the 38th value. From the frequency table the 38th value must be 5.

(There are 1 + 1 + 2 + 5 + 12 = 21 values 4 or less and 39 values 5 or less.)

Hence the median is 5.

The mean is generally easier to find than the median. However, if unusually large or small values occur, the mean can be a false indicator of the centre of the sample. In such cases, the median is preferred because it is not affected by these extreme values.

The **mode** is the most frequently occurring value in a data set. In Example C, the mode is 5 because it occurs 18 times.

The mode can be a very inaccurate measure of central tendency and hence caution should be applied in using it. It can be used in circumstances where one value is very much more common than all others, such as in Example C.

Quartiles

The **quartiles** are two more measures associated with the central tendency of samples and populations.

The **lower quartile** is the value which 25% of the sample or population is less than or equal to. This value is the *median* of the lower half of the values.

The **upper quartile** is the value which 75% of the sample or population is less than or equal to. This value is the *median* of the upper half of the values.

Example F

Find the lower and upper quartiles of the following data 1, 3, 5, 4, 9, 7, 1, 4, 4.

Solution

The values are listed in order: 1, 1, 3, 4, 4, 4, 5, 7, 9.

The quickest way of finding the upper quartile is to find the median of the upper half of the values (excluding the median). In this case, the median of 4 is excluded and the upper half of the values is 4, 5, 7, 9. The median of these values is 6, so the upper quartile of the data is 6. Similarly, the lower half of the values is 1, 1, 3, 4 whose median is 2, so the lower quartile of the data is 2.

Note: If you are using a computer, be aware that most **spreadsheets** (including the most commonly used, EXCEL) do not use this method to calculate quartiles. As a result, there are usually small differences between quartiles calculated using the method described above and those calculated using a computer.

Example G

Calculate the lower quartile of the set {1, 2, 3, 4, 5, 6, 7, 8} using the method above and using a spreadsheet.

Solution

In this case the number of data values is even (8), so the median lies between the two middle values 4 and 5 (and is therefore naturally excluded when finding upper and lower quartiles). The lower half of the data is {1, 2, 3, 4} whose median is 2.5, so the lower quartile of the set is 2.5.

Using both StarOffice and Microsoft EXCEL, a lower quartile of 2.75 is obtained.

Outliers

Outliers are values which are well away from the centre and also from most of the data. Outliers affect the mean, but not the median.

Example H

Find the mean and median income of one thousand people, one of whom earns $1 000 000, while the other 999 people earn nothing.

Solution

The mean income is $\frac{1\,000\,000 + 999 \times 0}{1\,000} = \$1\,000$

The median income is $0.

Unit 11.3 Activity 2B: Medians, modes and quartiles

1. Find the median, mode, upper and lower quartiles of each of the following sets of numbers.

a. 23, 32, 13, 16, 73, 42

b. 4.3, 3.12, 4, 2.7, 4.3

c.

number	0	1	2	3	4	5
frequency	11	23	42	55	27	8

d. 64, 4, 34, 13, 78, 13, 50, 62, 73, 42, 14, 45, 69, 50, 85, 63, 56, 88, 0, 77, 91, 32, 31, 85, 34, 45, 12, 47, 41, 70

2. Write down sets of 13 different numbers which have:

a. a median of 46.

b. a median of 53.

c. an upper quartile of 79.

d. a lower quartile of 87.

Measures of spread

The **range** is found by subtracting the smallest value in a sample or population from the largest value. The range is the simplest of the **measures of spread** but provides no information about how spread out most values are. The following example shows this.

Example I

In Example A, the range of **1.** is 5 – 1 = 4, the range of **2.** is 5 – 1 = 4 and the range of **3.** is 3 – 3 = 0

Note: **1.** and **2.** have the same range, yet as Example A shows, different spreads.

The **interquartile range** is the difference between the upper and lower quartiles. The interquartile range for the data in Example F is:

$$\text{upper quartile} - \text{lower quartile} = 6 - 2 = 4$$

The **variance** of a sample (symbol s^2) is the mean of the squares of the distance of the data values from the centre (the mean). If x refers to the values in the sample, f refers to the respective frequencies of those values and x is the mean of the sample, then the formula for the variance is:

$$\text{variance} = \frac{\sum f(x-\overline{x})^2}{\sum f - 1} = \frac{\sum^{n} f_i(x_i - \overline{x})^2}{\sum_{i=1}^{n} f_i - 1}$$

Note: For ungrouped data, the variance can be written: variance $= \frac{\sum (x-\overline{x})^2}{n-1}$.

With practice, the variance can be easily found from the formula. In the following example the variance calculation is done directly by first finding the mean then using the variance formula. In a second method, a table is used to calculate the variance.

Example J

The following table is a sample taken from a much larger population.

x	1	2	3	4	5
f	4	3	2	3	4

Find the variance of the sample values.

Direct Solution

$$\text{Mean} = \frac{4\times1+3\times2+2\times3+3\times4+4\times5}{4+3+2+3+4} \qquad \left[\text{substituting in } \bar{x} = \frac{\sum fx}{\sum f}\right]$$

$$= \frac{48}{16}$$

$$= 3$$

$$\text{Variance} = \frac{\sum f(x-\bar{x})^2}{\sum f}$$

$$= \frac{4(1-3)^2 + 3(2-3)^2 + 2(3-3)^2 + 3(4-3)^2 + 4(5-3)^2}{(4+3+2+3+4)-1}$$

$$= \frac{4(-2)^2 + 3(-1)^2 + 2(0)^2 + 3(1)^2 + 4(2)^2}{16-1}$$

$$= \frac{16+3+0+3+16}{15}$$

$$= 2.5\dot{3}$$

Table Solution

x	f	$f.x$	$(x-\bar{x})$	$(x-\bar{x})^2$	$f.(x-\bar{x})^2$
1	4	4	−2	4	16
2	3	6	−1	1	3
3	2	6	0	0	0
4	3	12	1	1	3
5	4	20	2	4	16
Totals	16	48			38

$$\bar{x} = \frac{\sum fx}{\sum f} = \frac{48}{16} = 3 \qquad s^2 = \frac{\sum f(x-\bar{x})^2}{\sum f - 1} = \frac{38}{15} = 2.5\dot{3}$$

Note: The calculation of $\bar{x}$ must be done before the table can be completed.

The **standard deviation of a sample** (symbol s) is the square root of the variance.

Example K

In Example J, the standard deviation is:

$s = \sqrt{2.533}$

$s = 1.59$ (2 dp)

When dealing with an entire population, the symbol used for the **population standard deviation** is the Greek letter σ, pronounced 'sigma'. This is calculated in the same way as the standard deviation of a sample (by taking the square root of the variance formula given above). A **sample standard deviation**, s, is often used to estimate a population standard deviation, σ.

Use of scientific calculators

The standard deviation is the most important measure of spread. Calculation of the standard deviation follows directly from calculation of the variance and is easily calculated with a calculator which has a statistical mode.

Example L

Use a calculator to find the mean and standard deviation of the data in Example J.

Solution

Using the latest model *Casio fx–82*, press:

[mode] [2]	[this puts the calculator in a statistical mode]
[shift] [mode] [1] [=]	[this sequence of key strokes clears the memory]
[1] [shift] [;] [4] [M+]	[this puts four 1's into the memory]
[2] [shift] [;] [3] [M+]	[this puts three 2's into the memory]

... continue this process until all data are entered.

To calculate the mean, press:

[shift] [2] which brings up a menu showing

$\bar{x}$	$x\sigma_n$	$x\sigma_{n-1}$
1	2	3

Press [1] to obtain the mean and press [3] for the standard deviation.

Note: By pressing [2], another standard deviation is obtained whose value is close to that obtained by pressing [3].

In most situations it does not matter which button is used. If the number of values in a sample is larger than about 20 the difference is usually insignificant.

Estimating the mean and standard deviation

Often data are given in such a way that it is impossible to work out the mean and standard deviation exactly. However, a good **estimate** of the mean and standard deviation can be found. The following example illustrates a typical method.

Example M

The table shows the length L of some steel bars.

Length (m)	Frequency
$0 \le L < 1$	23
$1 \le L < 3$	42
$3 \le L < 7$	63
$7 \le L < 12$	22
$12 \le L < 20$	18
$20 \le L < 30$	9
$30 \le L < 50$	5
$50 \le L < 100$	2

Although the exact values of all the data are not known, an approximation can be found by letting all values in any interval take on the value of the *midpoint* of that interval. Thus all values in the interval $0 \le$ length ≤ 1 are assumed to be 0.5; all values in the interval $1 \le L \le 3$ are assumed to be 2, etc. Putting these values into a calculator or using a table as in Example M gives:

$\bar{x} = 8.06$ (2 dp) and $s = 10.75$ (2 dp).

Unit 11.3 Activity 2C: Estimating means and standard deviations

1. Find estimates of the mean and standard deviation for the following:

a.

x	0–2	2–5	5–10	10–15	15–30	30–50
f	2	16	25	32	12	3

b.

x	$-2 < -1$	$-1 < 1$	$1 < 5$	$5 < 7$	$7 < 9$	$9 < 12$
f	6	3	2	1	2	9

Note 0–2 or $0 < 2$ means $0 \le x < 2$ etc.

2. Find the approximate mean and standard deviation of the data shown on these histograms:

a.

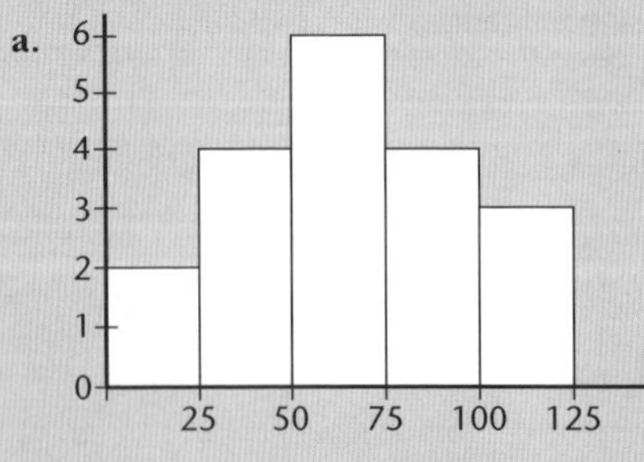

b.

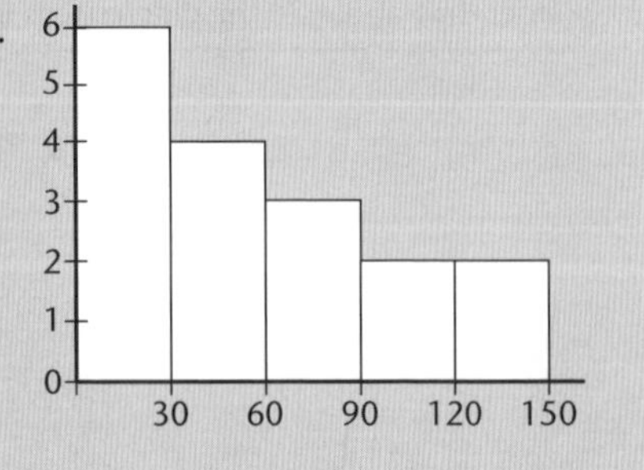

3. The pair of graphs below shows the distribution of the weights of two crops of the same vegetable.

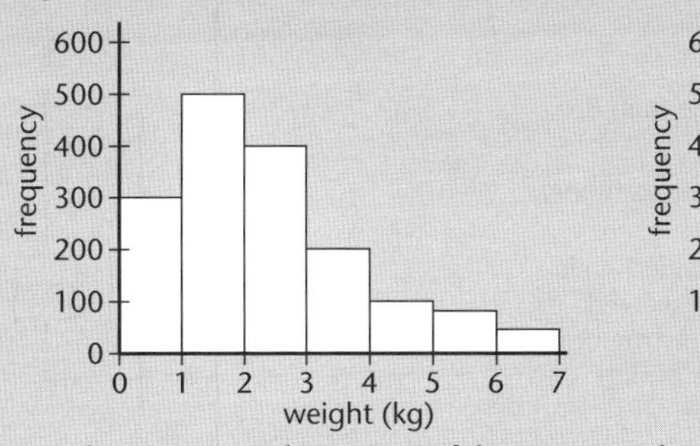

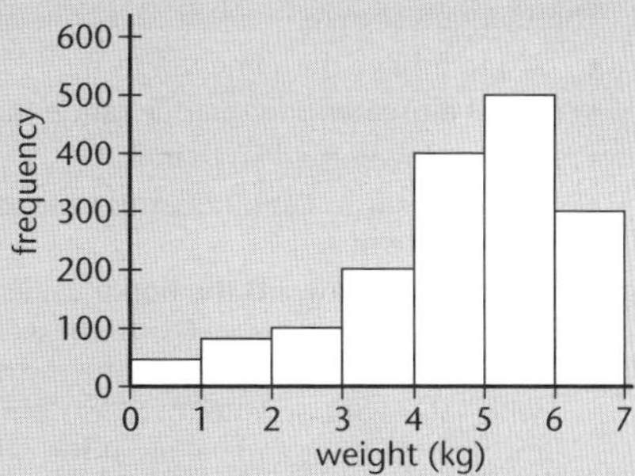

a. Obtain numerical estimates of the mean weight of each crop and also the standard deviations of the weights of each crop.

b. What would the values of the mean and standard deviation of the weights of each crop be if:

i. the weight of every vegetable had been increased by 1 kg?

ii. the weight of every vegetable had been doubled?

iii. there had been twice as many vegetables of every weight in each of the original crops?

Box plots

Box plots are a one-dimensional form of data display useful for getting an overview of the distribution of data. They are also called **box-and-whisker plots**.

Example N

Draw a box plot of data whose maximum value is 13.6 kg, minimum value is 5.3 kg, upper quartile is 10.2 kg, lower quartile is 8.7 kg and median is 9.1 kg.

Solution

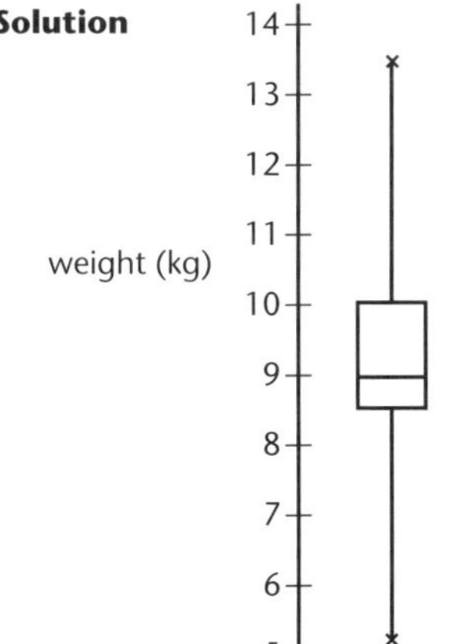

- The rectangle has its upper side at the level of the **upper quartile**, the lower side at the level of the **lower quartile**. A further parallel line is drawn through the rectangle at the **median**. This part of the diagram is the 'box' (the width of the box has no significance).
- The 'whiskers' are the two lines which emanate from the box to the maximum and minimum values.
- This box plot shows data which is slightly 'skewed' towards the upper end. This means that data above the median is more spread out than the data below it.

Box plots of different sets of data when placed beside each other provide excellent means of comparing the two sets of data (these plots should be drawn using boxes of equal width).

Example O

Box plots are drawn for two data sets, A and B.

This diagram shows:

- A and B have the same range.
- A has a greater interquartile range than B.
- B has a lower median than A.
- B has maximum and minimum values higher than those of A.
- B is 'skewed' toward the upper values while A is more symmetrical. The term 'skewed' refers to a lack of symmetry in a distribution, with values at one 'end' of a distribution more spread out and values at the other 'end' grouped more closely together.

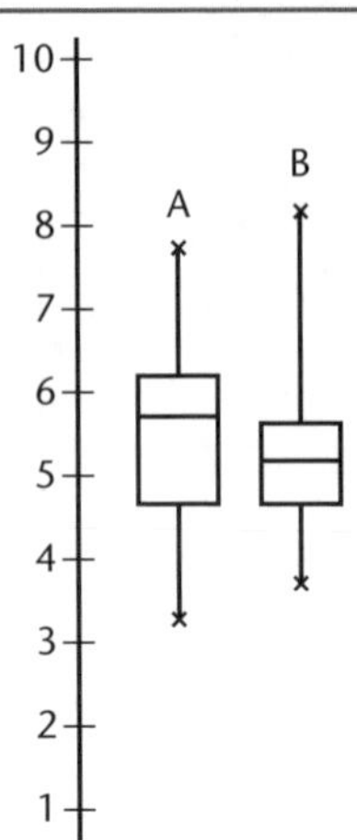

Unit 11.3 Activity 2D: Box plots

1. Draw a box plot of these data:
   13, 41, 15, 35, 38, 24, 29, 35, 31, 38, 31, 49, 55

2. **a.** Draw box plots to compare these marks obtained by 2 different groups of pupils sitting the same exam:

   Group A – 90, 12, 28, 5, 5, 74, 93, 1, 53, 18, 4, 89, 49, 29, 68, 87, 47, 92, 24, 84, 6, 2, 2, 16, 0, 61, 25

   Group B – 3, 50, 76, 86, 84, 0, 73, 28, 15, 40, 10, 18, 5, 86, 53, 74, 9, 87, 75, 67, 87, 40, 81, 3, 99, 61, 88

   b. Comment on any differences shown in the plots.

Unit 11.3 Managing Data

Topic 3: Cumulative frequency graphs

In Topic 3 we continue the focus on statistics and in particular the analysis and interpretation of data, in particular, the 3rd bullet point under the heading 'Statistics' (Syllabus p. 16):

- Represent data using cumulative frequency graphs.

Cumulative frequency graphs and percentiles

Cumulative frequency graphs are constructed for grouped data and are used to give estimates of the median, the upper and lower quartiles and percentiles.

For the following data collected about the weight of pupils in a class:

Weight (kg)	Frequency
$45 \leq x < 50$	2
$50 \leq x < 55$	6
$55 \leq x < 60$	9
$69 \leq x < 65$	8
$65 \leq x < 70$	4
$70 \leq x < 75$	1

A separate table has to be constructed to show the cumulative frequency:

Weight (kg)	Cumulative frequency	
< 45	**0**	There are no data values less than 45 kg
< 50	**2**	There are 2 data values less than 50 kg
< 55	2 + 6 = **8**	There are 8 data values less than 55 kg
< 60	2 + 6 + 9 = **17**	There are 17 data values less than 60 kg
< 65	2 + 6 + 9 + 8 = **25**	There are 25 data values less than 75 kg
< 70	2 + 6 + 9 + 8 +4 = **29**	There are 29 data values less than 70 kg
< 75	**30**	There are 30 data values (all) that are less than 75 kg

Note: The number of data values in the last row of the second column should be the sum of all the frequencies.

Below is the cumulative frequency curve for the frequency distribution above. The variable (weight values) is placed on the horizontal axis and the cumulative frequencies on the vertical axis. The points (45, 0), (50, 2), (55, 8) ... (75, 30) are plotted and straight lines are drawn to join the points. The graph should always have the 'stretched- s' shape.

Graph 1

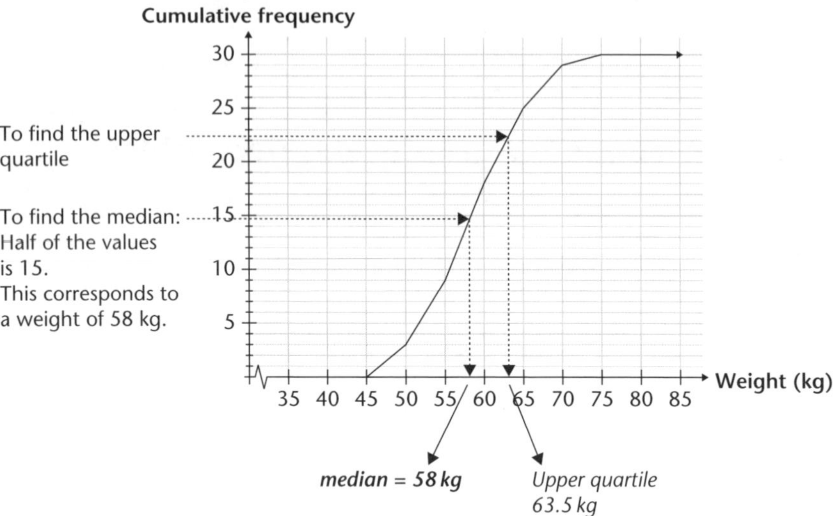

A percentage cumulative frequency table and curve can also be drawn:

Table

Weight (kg)	Cumulative frequency	Percentage cumulative frequency (%)
< 45	0	$\frac{0}{30} \times 100 = 0$
< 50	2	$\frac{2}{30} \times 100 = 6.7$
< 55	8	$\frac{8}{30} \times 100 = 26.7$
< 60	17	56.7
< 65	25	83.3
< 70	29	96.7
< 75	30	100

Graph 2

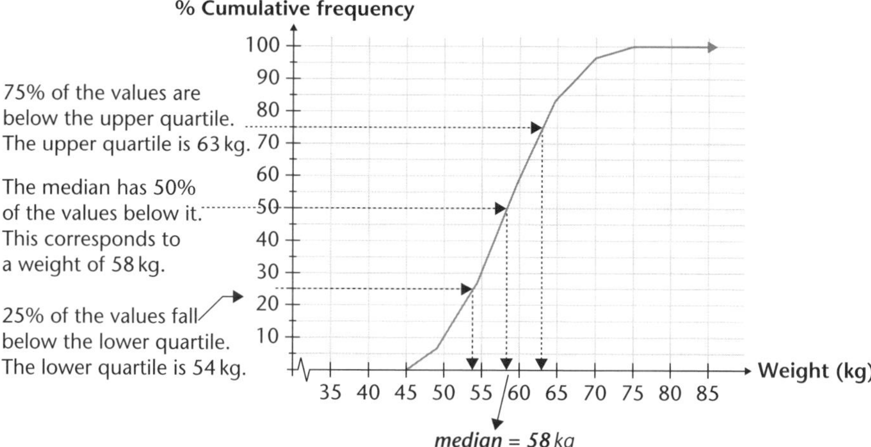

Reading statistics from a cumulative frequency curve

We can read estimates of statistics from the cumulative frequency curve that we would not be able to calculate if we do not have the actual data values.

The **median** is the middle value in a set of ordered data. For our example, using Graph 1, 15 values would have a value less than the median so going across to the graph at 15 and then down to the *weight* axis gives the median as 58 kg.

The **upper and lower quartiles** can also be read from the graphs. 75% of the data values are less than the upper quartile, so moving horizontally at 22.5(Graph 1) or 75% (Graph 2) across to the graph then down to the *weight* axis gives an upper quartile of about 63.5 kg. Similarly, the lower quartile can be read as approximately 54 kg.

This data set could be described as: 'The weight of students in this class was centred around 58 kg (median) with the lightest 25% having a weight of 54 kg or less (lower quartile) and the heaviest 25% having a weight of 63.5 kg or more (upper quartile)'.

Percentiles

Percentiles are frequently used to describe sets of data.

For the data graphed above we could make statements such as: '90% of the students in the class had a weight less than 67 kg. This figure is called the **90th percentile**'; or 'the lightest 10% of the class had a weight of 50kg or less.' Here we are quoting the **10th percentile**.

- The median is also called the 50th percentile.
- The lower quartile is also called the 25th percentile.
- The upper quartile is also called the 75th percentile.

Unit 11.3 Activity 3A: Cumulative frequency graphs

1. A percentage cumulative frequency curve, shown below, has been constructed for the heights of a basketball squad.

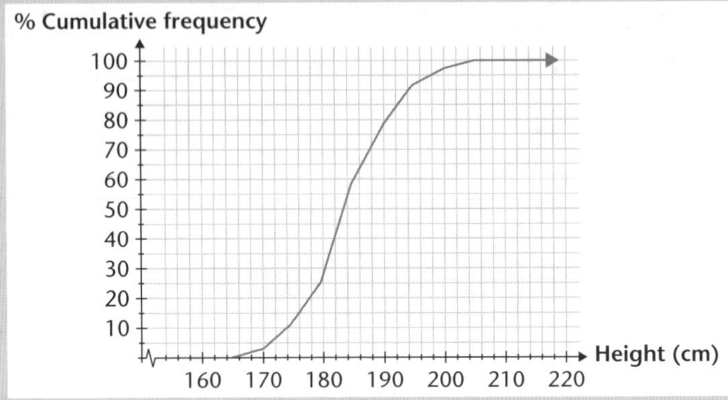

a. From the graph find estimates of:

i. The median height.

ii. The upper quartile.

iii. The lower quartile.

iv. The 80th percentile.

v. The 10th percentile.

b. Copy and complete the following statements:

i. Half of the heights of the basketballers are greater than …

ii. 25% of the basketballers have a height that is less than …

iii. … of the basketballers have a height that is less than 190 cm.

iv. …% of the basketballers have a height that is greater than 186 cm.

2. a. Copy and complete the cumulative frequency table and graph for the data given below:

Maximum daily temperature (°C)	Frequency
15 –	1
20 –	5
25 –	14
30 –	5
35 –	3
40 –	2

Maximum daily temperature (°C)	Cumulative frequency
< 15	0
< 20	
< 25	
< 30	
< 35	
< 40	
< 45	30

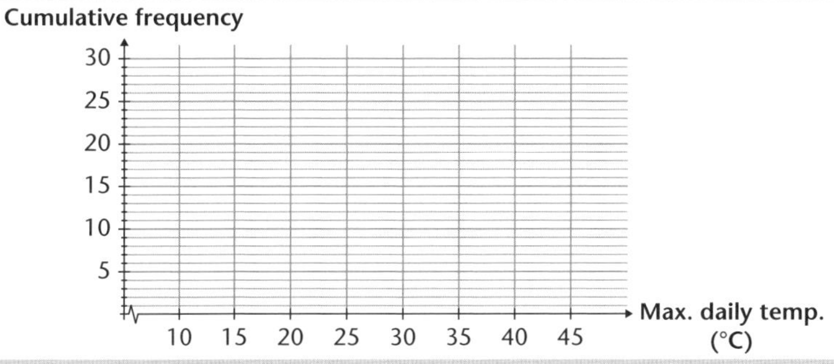

b. From your graph estimate:

i. The median maximum daily temperature.

ii. The 80th percentile.

iii. The 10th percentile.

iv. The interquartile range.

c. Write a sentence that interprets each of the statistics from **b** parts **i** to **iv**.

3. The speed of vehicles travelling along a section of highway where the speed limit is 100km/hr has been recorded and given in the table below:

Speed (km/hr)	Number of cars (Frequency)
50 –	8
60 –	18
70 –	84
80 –	116
90 –	212
100 –	150
110 –	10
120 – < 130	2

a. How many vehicles were included in this survey?

b. What percentage of the vehicles were travelling at speeds equal to or greater than 100 km/hr?

c. Construct a percentage cumulative frequency table for the data.

d. Construct a percentage cumulative frequency graph from your table.

e. From your graph find estimates of the given statistic, and write a sentence to interpret:

i. The median. **ii.** The lower quartile.

iii. The upper quartile. **iv.** The 90th percentile.

v. The 10th percentile. **vi.** The interquartile range.

4. The following frequency table gives the time it takes to swim 100 m for the forty members of a swimming squad.

Time to complete 100m swim (secs)	Number of swimmers
50 –	1
55 –	7
60 –	17
65 –	11
70 –	2
75 – < 80	2

a. Construct a percentage cumulative frequency table for the data.

b. Construct a percentage cumulative frequency graph from your table.

c. From your graph find estimates of, and write a sentence to interpret:

i. The median. **ii.** The lower quartile.

iii. The upper quartile. **iv.** The 90th percentile.

v. The 10th percentile. **vi.** The interquartile range.

Unit 11.3 Managing Data

Topic 4: Time series graphs

The Syllabus mentions the ways in which statistics can enable us to make informed decisions about many issues in business, science, agriculture and other aspects of the world around us. The intelligent use of data enables us to communicate, justify, predict and critically analyse findings, and draw conclusions and make recommendations. Time series graphs are especially useful and in this Topic we look at their application to currency, sales, crop production and travel.

Time series graphs

When numerical data is collected over **consecutive time intervals** a line graph joining the data points is the appropriate type of graph. This is called a **time series graph.**

The consecutive time intervals can be seconds, minutes, hours, days, weeks, months, years, etc.

The following time series graph shows the exchange rate for one Papua New Guinea kina against the Australian **dollar** for the years 1988 to 2008.

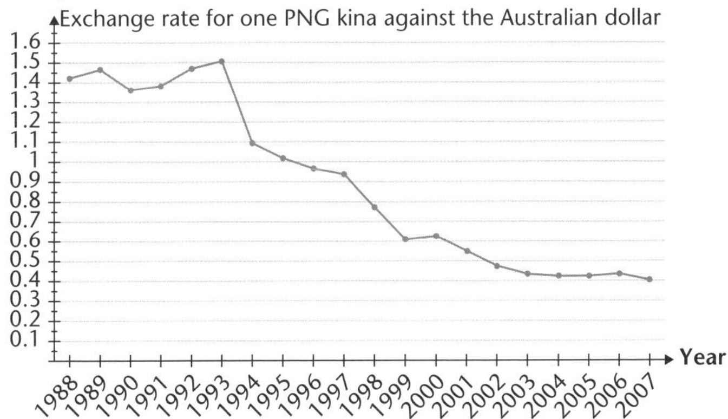

A continuous line joins the points on a time series graph so that any upward or downward **pattern**, called a **trend**, can be seen.

The graph above shows a slight upward trend for the years 1988 to 1993 then a steady downward trend from 1993 to 2003. After 2003 the exchange rate has remained about the same at approximately 0.42.

If there is a **repeated pattern** apparent on a graph then the graphed data is said to be **seasonal**. Seasonal data is often seen on graphs for sales.

The graph at right shows a seasonal pattern that is repeated every four quarters. There also appears to be a downward trend to the data.

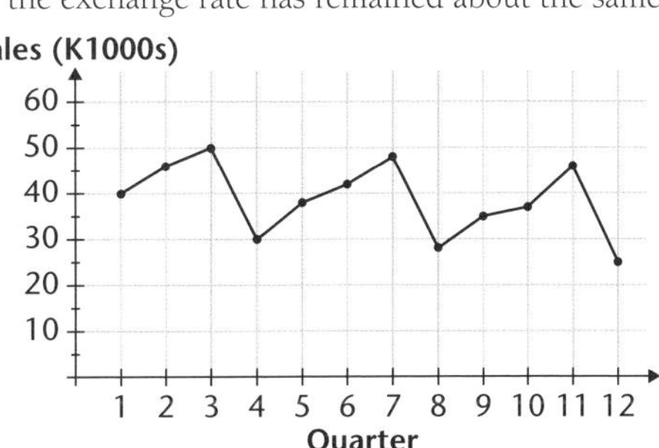

Time series graphs are used to **predict** into the future. A **trend line** can be applied to the data so that predictions can be made. Using the trend line to predict within the given data set is called **interpolation**.

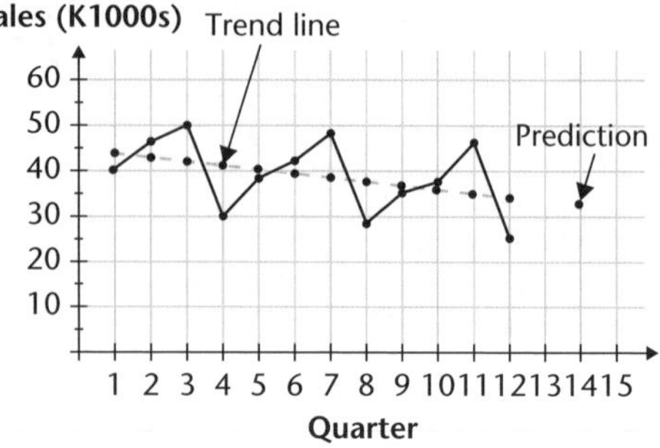

For example, if the trend line is used to predict the sales for quarter 4 then the prediction would be K41 000.00. This would be a prediction that ignores the seasonal fluctuations; called a de-seasonalised prediction.

Using the trend line to predict sales outside the given data set is called **extrapolation**. For example, using the trend line to predict the sales (de-seasonalised) for quarter 14 gives the value K33 000.

Some time series graphs show no trend or seasonal pattern; these graphs are described as showing **random fluctuations**. On graphs that show random fluctuations any particular highs or lows should be mentioned.

Example L

Kendra recorded the number of hours that she worked on each of 12 days.

The graph shows only a random pattern (no trend or seasonal fluctuations). The most hours that she worked in a day was 12 hours and on two days she worked the minimum of 4 hours.

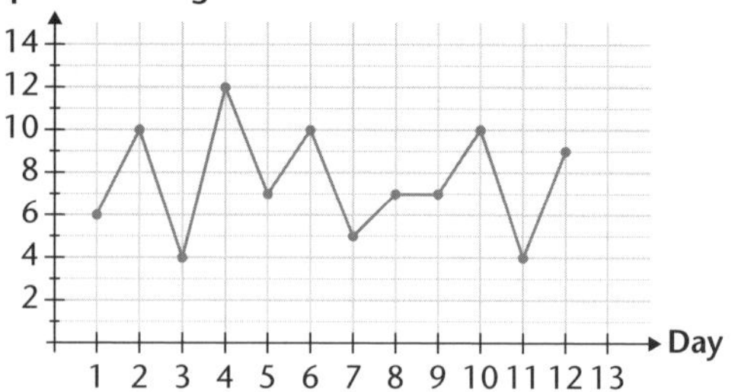

Unit 11.3 Activity 4A: Time series graphs

1. a. Construct a time series graph for the data given at right:

Estimated volume (tonnes) of cured vanilla bean exports in PNG for the years 1997–2006.

Year	Volume (tonnes)
1997	1
1998	4
1999	8
2000	24
2001	50
2002	160
2003	200
2004	215
2005	68
2006	26

b. Comment on the graph.

2. The time series graph below shows the estimated rice production in PNG for the years 1980–2000.

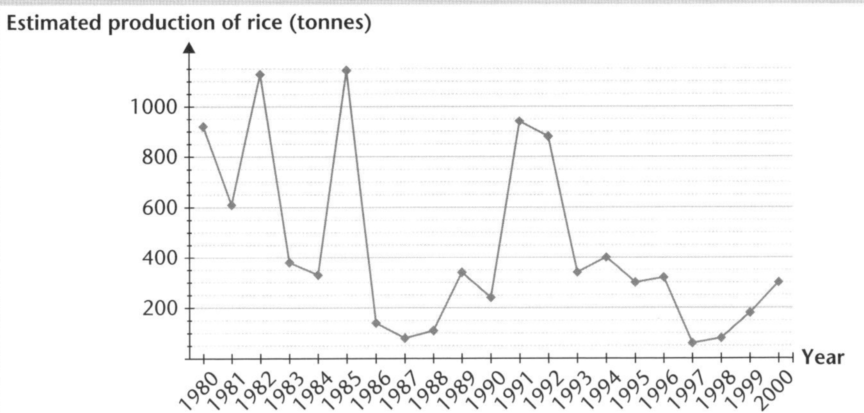

a. Does the graph show: **i.** A random pattern?

ii. A seasonal pattern?

iii. A trend?

b. Determine the year of: **i.** Highest production.

ii. Lowest production.

c. Describe the trend over the 21 years.

d. Use the trend over the last four years to predict the rice production in the year 2001.

3. The graph below shows the volume (tonnes) of fruit and vegetable imports from Australia and New Zealand for the years 1983–2003.

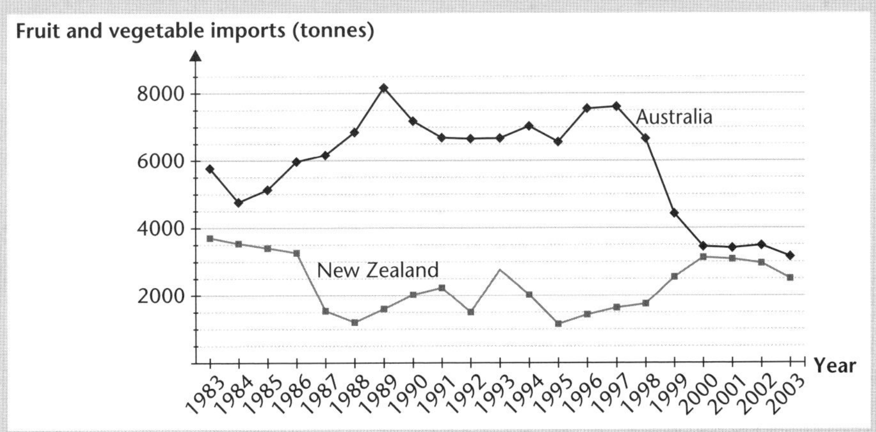

a. Describe the pattern of the data for: i. Australia.
ii. New Zealand.

b. Compare the pattern for each of the countries over the years 2000 to 2003.

c. Use the data to predict the fruit and vegetable imports from Australia and New Zealand in 2004.

4. The following graph shows small holder coffee production and sales of food in the Kainantu area, Eastern Highlands Province, 1982–1984.

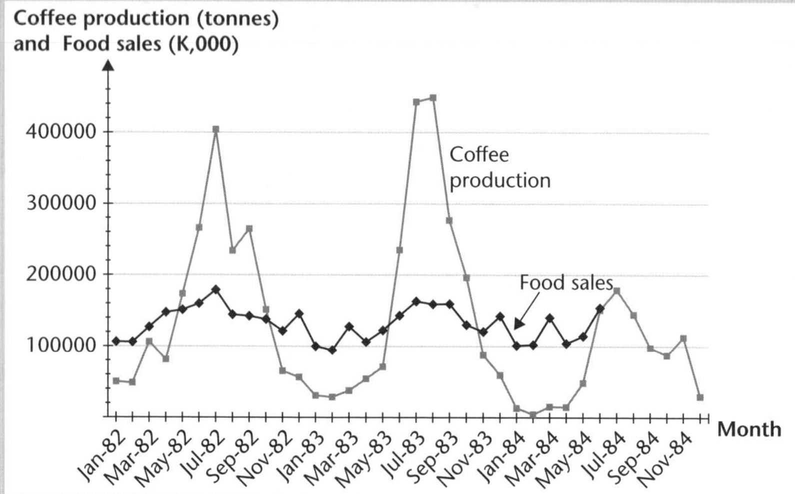

a. Describe the pattern in the coffee production graph.

b. Describe the pattern in the food sales graph.

c. What do you notice when comparing the coffee production pattern and the food sales pattern?

5. The following graph shows the average price per year, in toea, of sweet potato in Port Moresby and Madang.

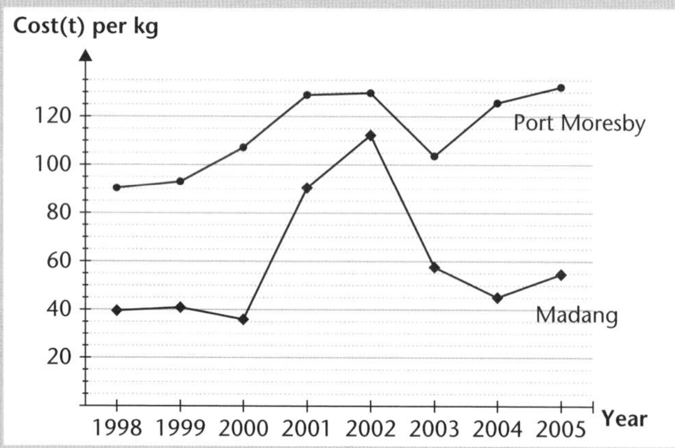

a. Describe the pattern in the Port Moresby data.

b. Describe the pattern in the data from Madang.

c. Compare the data for Port Moresby and Madang.

6. The graph below shows the distance travelled by a PMV owner in the first four weeks of operation. Day 1 is a Sunday.

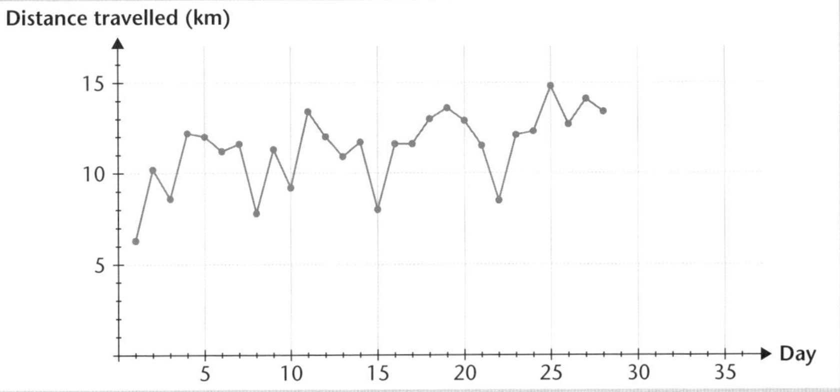

a. Describe the pattern in this time-series graph.

b. A line that can be used to predict the trend in this data goes through the points (5, 10) and (15, 11.5). Locate this line on the graph using a ruler.

c. Use this line to predict the distance the PMV will travel on:

i. Sunday of the 5th week.

ii. Wednesday of the 5th week.

iii. Saturday of the 5th week.

d. Which of the predictions from part c is likely to be closer to the actual value? Why?

7. On the following diagram the quarterly unemployment rates, as a percentage, are graphed for the years 2004 to 2007 inclusive:

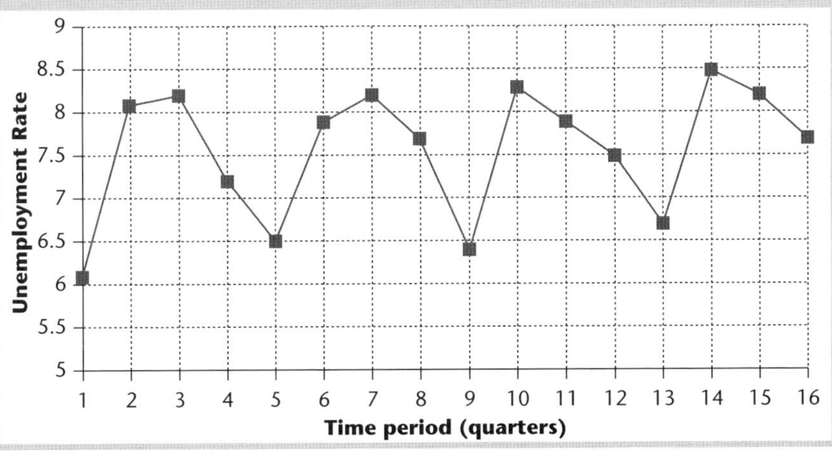

Quarter 1 is the months January to March of 2004, quarter 2 is the months April to June 2004 … to quarter 16 which is the months October to December 2007.

a. Can you see a pattern in the graph?

b. What is this pattern called?

c. How often does the pattern repeat itself?

d. In which months of the year is the unemployment rate at its lowest?

e. In which months of the year is the unemployment rate at its highest?

f. Is there a trend apparent in the data?

g. Use the graph to predict the unemployment rate in time period 17.

Unit 11.3 Managing Data

Topic 5: Analysis of data—interpreting statistics

Continuing the emphasis on the use of statistics to aid our decision-making in many aspects of the world around us, Topic 5 looks at the use of statistical methods and information, and covers:

- Interpreting statistical information and answering straightforward questions.
- Commenting on the validity of a statement with respect to features of statistical information.

Introduction

In everyday life statistical information is presented in many different forms, such as tables, graphs and **summary statistics** (averages, spread, etc). This chapter deals with techniques for interpreting such information and commenting on its features.

Interpreting graphs and statistics

Reading tables

Often data is presented in a table, and, although a suitable graph can be drawn from the table, the data can be analysed just as easily from the figures in the table.

Example A

A. The life expectancy at birth for the years 1995 and 2011 is given in the table below for 15 countries.

Life expectancy at birth (years)

Country	1995	2011
PNG	61	65
Australia	78	82
Indonesia	66	71
Solomon Islands	70	74
Fiji	66	71
Malaysia	69	74
India	60	67
Thailand	70	74
China	69	75
Russia	65	66
Turkey	68	73
South Africa	64	49
Samoa	68	72
Uganda	42	53
Cuba	75	78

a. What does 'life expectancy at birth' mean?

b. Which of the countries had the lowest life expectancy at birth in:

   i. 1995?

   ii. 2011?

c. For which country did the life expectancy at birth decrease from 1995 to 2011?

d. Which of these countries had the greatest increase in life expectancy at birth from 1995 to 2011?

e. Which of these countries had an increase in life expectancy at birth of more than 5 years in the period from 1995 to 2011?

f. What is the percentage change in life expectancy at birth for China?

g. List some factors that you think might affect the life expectancy at birth.

A. a. Life expectancy at birth means the number of years you are expected to live before you die. A child born in 2011 in PNG is expected to live for 65 years.

b. i. Uganda had the lowest life expectancy at birth in 1995.

ii. South Africa had the lowest life expectancy at birth in 2011.

c. The life expectancy at birth decreased in South Africa from 1995 to 2011.

d. Uganda had the greatest increase (11 years).

e. India, China and Uganda.

f. Percentage change for China = $\frac{75-69}{69} \times \frac{100}{1} = 8.7\%$

g. Nutrition, access to medical treatment, education, immunisation.

The distribution of a set of data

If a set of data is graphed we can comment on the overall shape, called the distribution of the data. We use the terms symmetric, positively skewed and negatively skewed.

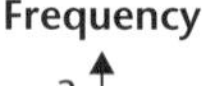

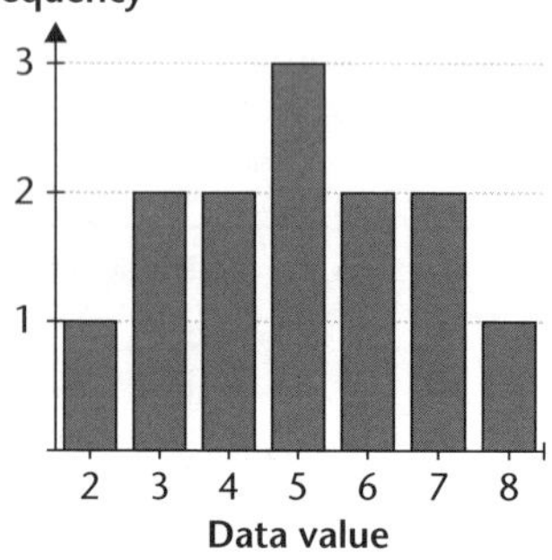

A **symmetric** distribution of a set of data appears equally spread on either side of a central value.

For this **symmetric data set** the mean, median and mode are all the same value (5). The graph is **symmetric** about the value 5.

A **positively skewed** set of data would appear on a graph to be stretched to the right and the mean (19) would be greater than the median and mode (both in the 10–15 **class interval**). This set of data has an extreme value in the 60–65 interval.

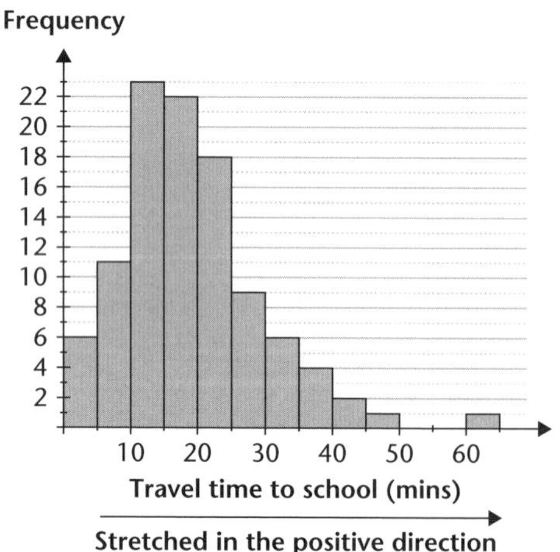

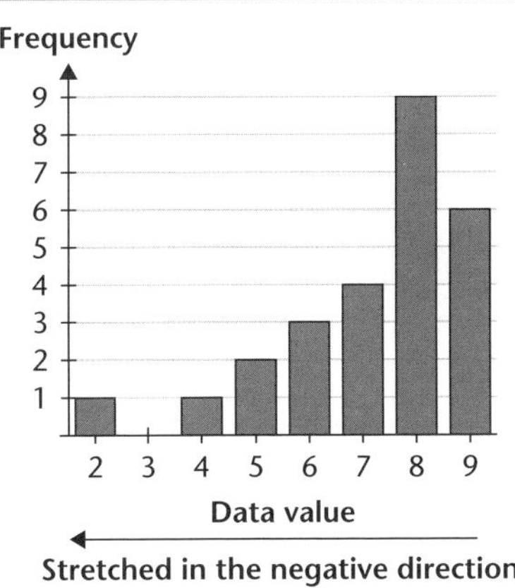

A **negatively skewed** set of data appears stretched to the left.

For this **negatively skewed data set** the mean is less than the median and mode because the mean is dragged towards the extreme small values (4 and 2).

In cases where there are extreme data values and the distribution of the data is skewed the most suitable measure of centre to use is the **median** or the **mode**.

The distribution of a set of data can also be seen from a **stem and leaf plot.**

Looking sideways at the stem and leaf plot below, we can see a frequency graph:

Stem	Leaf
1	6 7 8 8
2	2 3 3 4 4 6 7 8 9
3	1 2 5 8
4	0 4 6
5	1

The distribution of this set of data can be described as **positively skewed**; it is stretched towards the larger values.

Boxplots can also display the fact that a distribution is symmetric or skewed.

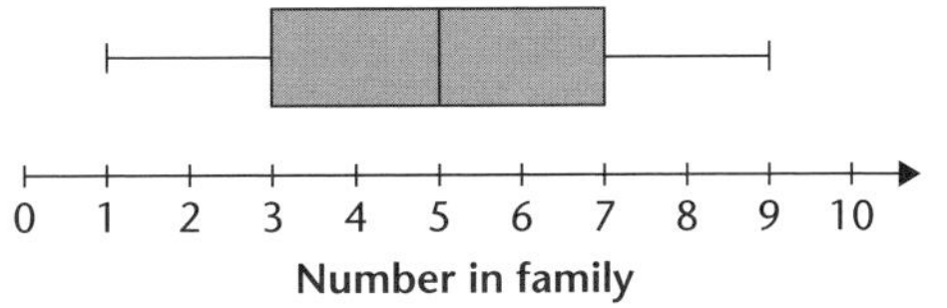

A symmetrically distributed set of data will have a symmetric boxplot; the median will be in the centre of the 'box' and the whiskers will be the same length.

The boxplot for a **positively skewed set of data** has a longer right whisker and the median is not in the centre of the 'box'; the whole boxplot appears stretched to the right:

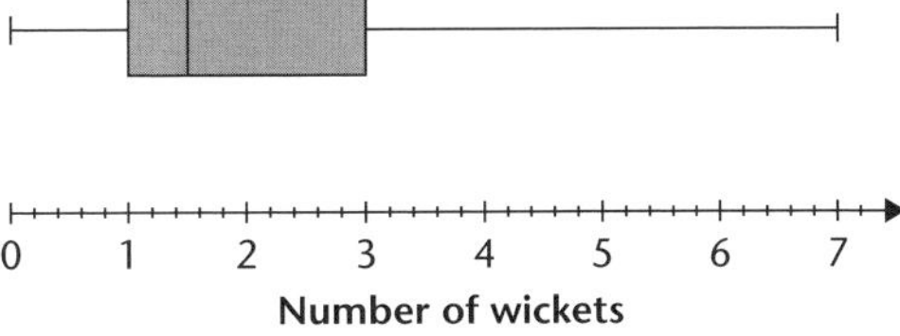

The boxplot for a **negatively skewed set of data** has a longer left whisker and generally appears stretched to the left.

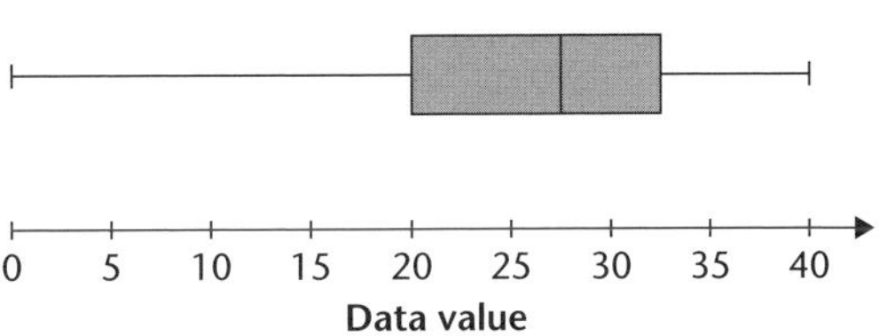

When comparing two sets of data any statements made should be supported by the relevant statistic.

Example B

Q. The mathematics results for two classes are displayed on the boxplots below. Comment and compare the two sets of data.

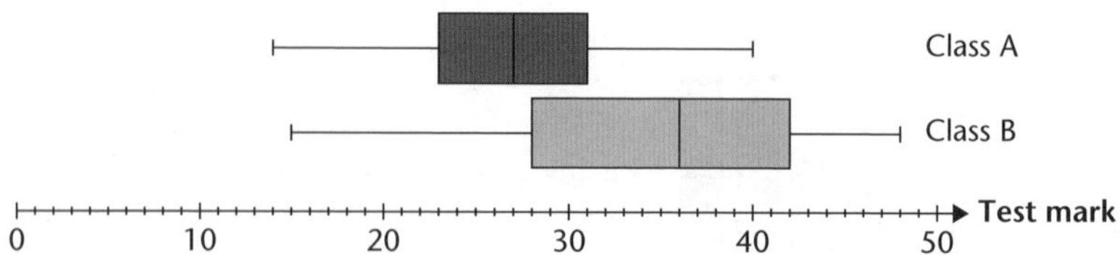

A. The distribution of test marks in class A appears to be symmetric about a median of 27. The middle 50% of test results are spread over 8 marks (IQR); the lowest mark is 14 and the highest is 40, giving a range of 26.

The distribution of test marks in class B appears to be negatively skewed with a greater spread of marks below the median mark. Half of the students in class B had a test mark of 36 or more (median) and the middle 50% of marks were spread over 14 marks (IQR). The lowest mark in class B was 15 and the highest was 48, giving a range of 33.

In general the marks were lower in class A than class B (comparing the medians: 27 compared to 36) but the students in class A had less varied results (comparing IQR's: 8 compared to 14. Also comparing ranges; 26 compared to 33).

More than 75% of the students in class B had a mark greater than the median of class A.

More than 25% of class B had a mark greater than the highest mark in class A.

Unit 11.3 Activity 5A: Interpreting graphs and statistics

1. The data given below gives the under-5 mortality rate for the years 1990 and 2009, for thirteen countries.

Under-5 mortality rate (deaths per 1000 live births)

Country	1990	2009
PNG	91	68
Australia	9	5
Indonesia	86	39
Solomon Islands	38	36
Fiji	22	18
Malaysia	18	6
India	118	66
Thailand	32	14
China	46	19
Russia	27	12
Turkey	84	20
South Africa	62	62
Samoa	50	25

a. What does the figure 68, for PNG in 2009, mean?

b. Which of these countries had the greatest decrease in their under-5 mortality rate from 1990 to 2009?

c. Which country had no change in their under-5 mortality rate from 1990 to 2009?

d. Which countries had an under-5 mortality rate greater than 50 in:

i. 1990?

ii. 2009?

e. What is the percentage change in under-5 mortality for PNG from 1990 to 2009?

2. In a village the number of chickens per household was counted to give the following data:

4 4 4 5 5 5 5 5 6 6 6 7 7 7 8 8 9 10 11 13

a. Construct a suitable graph for the data.

b. Calculate the median and the upper and lower quartiles.

c. Construct a boxplot for the data.

d. Use your graph and statistics to comment on the distribution of the data.

3. The following graph gives the PNG production of export tree crops in the years 1975 and 2005.

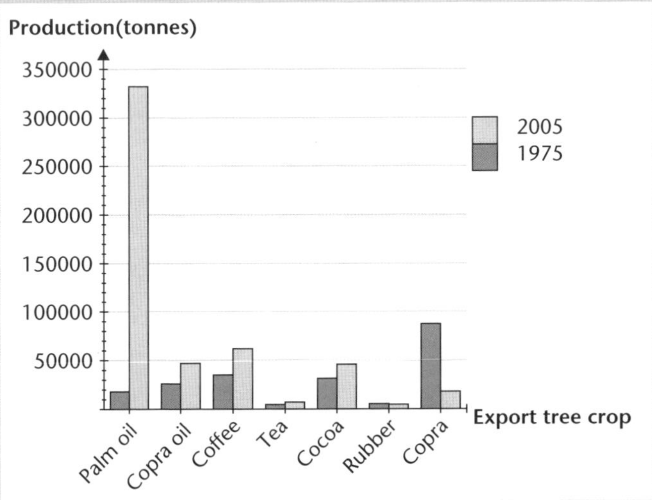

Comment on the change, from 1975 to 2005, in production for:

a. Palm oil.

b. Coffee.

c. Copra.

4. The number of avocados per tree was counted and the data displayed on the graph below:

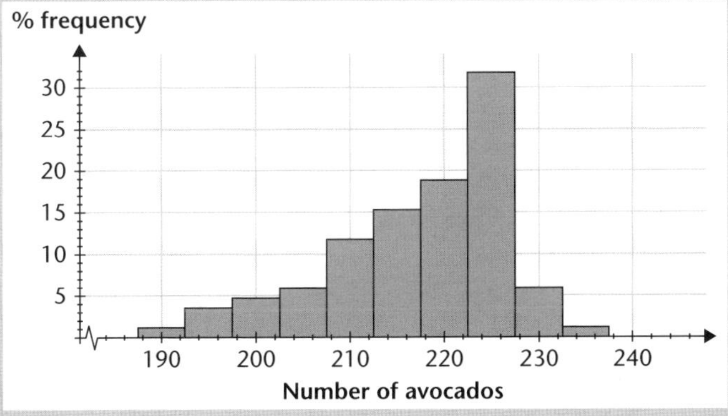

A boxplot showing the median, upper and lower quartiles of the data is given below:

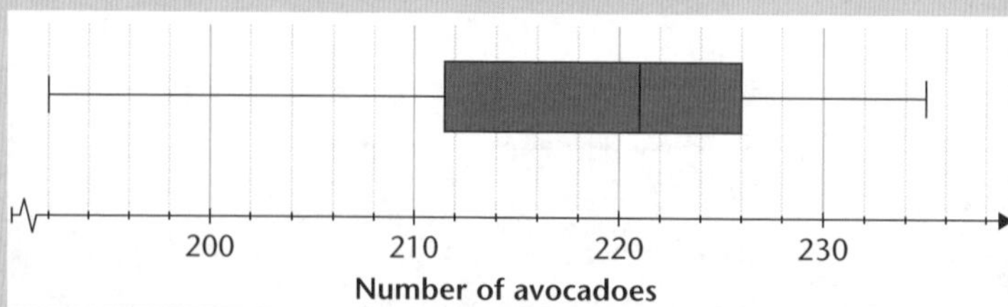

Comment on the distribution of the data, quoting statistics from the boxplot.

5. A team has recorded the number of points scored in each game of rugby over the last three years. The relevant statistics are displayed on the boxplots below:

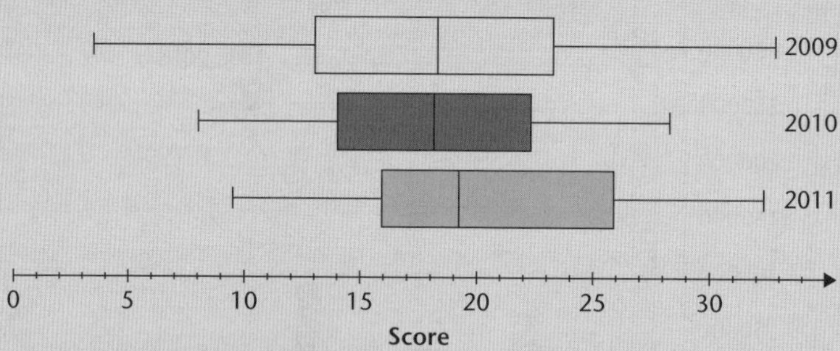

Comment on the team's scoring over the last three years.

6. The table below gives the male and female literacy rate (%) for 14 countries for the period 2004–2008:

Literacy (%)	Male	Female
PNG	65	69
Indonesia	97	96
India	88	74
China	99	99
Malaysia	98	99
Russia	100	100
Turkey	99	94
South Africa	96	98
Samoa	99	100
Cuba	100	100
Uganda	89	86
Afghanistan	49	18
Egypt	88	82
Philippines	94	96

a. What does a male literacy rate of 65% for PNG mean?

b. In which of these countries is the male literacy rate less than the female literacy rate?

c. In which countries is there a difference in male/female literacy of more than 5%?

d. Which country has the greatest difference in male/female literacy?

e. Construct boxplots for male literacy and female literacy on the same axis.

f. Comment and compare the boxplots from part e, supporting your statements with the relevant statistic.

Unit 11.3 Managing Data

Topic 6: Conducting a statistical investigation

Topic 6 covers the area of data collection, analysis and interpretation with the emphasis on how a statistical investigation may be planned, organised and carried out.

Populations and samples

A **survey** is a common method of collecting data for investigations that involve things that affect us in everyday life. Governments, businesses and individuals collect data from surveys so that they can make informed decisions and plan for the future.

Before data is collected a decision needs to be made about whether data for the investigation will be collected from the **whole population** or a **sample** of the population.

In some **statistical investigations** it is possible to collect data for the whole population. The National Statistical Office of Papua New Guinea conducts a survey of the whole population of PNG, called a **census**, every 10 years. This census collects data relating to population distribution and housing, and provides valuable information for future planning. Because the whole population is surveyed it should provide accurate information, but it is a very time-consuming and costly exercise, and given the isolation of many rural areas and the ruggedness of the PNG landscape, the census data is not always as reliable as it should be.

Population does not always mean 'all the people in PNG'. When we use the term in statistics, 'population' generally means all the people or things that the statistical investigation (or survey) and its **conclusions** would apply to. For example, if you want to investigate a theory related only to your school then the 'population' is all the students who attend your school. But if you want to investigate something that applies to all school children in PNG then the 'population' would be all the school children in PNG (including the students in your school).

In most statistical investigations surveying the whole population is either too time-consuming or too costly. In these cases a **sample** is chosen to represent the population.

If data is to be collected from a sample then a **representative sample** must be selected; in other words, a sample that represents the population must be chosen so that reliable conclusions about the population can be made. Conclusions from a sample, however, will not be as accurate as conclusions made from the whole population.

When planning a survey, the type of data that will be collected and how the conclusions are going to be presented also need to be considered. There are two types of data that we commonly deal with and there are appropriate ways in which each type of data should be organised (tables etc.), displayed (graphs) and interpreted (conclusions).

Choosing a sample

In an investigation where a sample is chosen to represent the population, the sample must be chosen so that the results will not show **bias** towards a particular **outcome**. If the sample is biased towards a particular outcome then any conclusions that are made from the investigation will not apply to the whole population and much time and money will have been wasted.

If the purpose of a survey is to get an accurate indication of how the population of PNG is going to vote at the next election, then surveying people from only one village would not provide information that represents all of PNG: there may be a local candidate from this village and all the residents plan to vote for him or her regardless of the political party that they represent; or the people in the village may concentrate on one particular local issue (eg improvements to their school) and so they may all vote for the party that is likely to provide this.

There are many ways of choosing an **unbiased** sample. One way is a **simple random sample**, which is a sample chosen so that every person or object in the population has an equal chance of being chosen. If the **sampling** is done at random then it is possible to draw valid conclusions about the population from the survey.

Random means that there is no pattern or plan (no bias) in choosing one element of the population over another. There are various simple means of randomly selecting a sample. The first one we will look at involves the use of **random numbers**.

Random number tables

Random number tables appear as a list of the digits from 0 to 9. The digits are generated by a computer and are grouped in fives so that the table is easier to read. You can start anywhere in the table and move across or down:

39634	62349	74088	65564	16379	19713	39153	69459	17986	24537
14595	35050	40469	27478	44526	67331	93365	54526	22356	93208
30734	71571	83722	79712	25775	65178	07763	82928	31131	30196
64628	89126	91254	24090	25752	03091	39411	73146	06089	15630
42831	95113	43511	42082	15140	34733	68076	18292	69486	80468
80583	70361	41047	26792	78466	03395	17635	09697	82447	31405
00209	90404	99457	72570	42194	49043	24330	14939	09865	45906
05409	20830	01911	60767	55248	79253	12317	84120	77772	50103
95836	22530	91785	80210	34361	52228	33869	94332	83868	61672
65358	70469	87149	89509	72176	18103	55169	79954	72002	20582
72249	04037	36192	40221	14918	53437	60571	40995	55006	10694
41692	40581	93050	48734	34652	41577	04631	49184	39295	81776
61885	50796	96822	82002	07973	52925	75467	86013	98072	91942
48917	48129	48624	48248	91465	54898	61220	18721	67387	66575

If we want to randomly choose 10 different numbers from 1 to 50 (inclusive) then we would need to look at two digits at a time.

Starting in the top left corner and working down we would have the numbers 39, 14, 30, (64 is ignored because it is larger than 50), 42, (80 ignored), (00 ignored), 5 which appears as 05, (95, 65, 72 ignored), 41, (61 ignored), 48, returning to the top and choosing the next two digits, (63, 59, 73, 62, 83, 58 ignored), 20, 40, (83 ignored) and 35.

The ten random numbers between 1 and 50 inclusive are 39, 14, 30, 42, 5, 41, 48, 20, 40, 35.

Some rules for using random number tables:

- Choose a starting point and a direction and continue with this for the entire sample; don't jump around the table!
- Ignore numbers that are larger than your population size.
- Ignore numbers that are repeated.

Example A

Q. A teacher wants to choose a simple random sample of three students from a class of 32 students. She has a class list of the students in alphabetical order. How can she get this simple random sample of three students using a random number table?

A. The population size in this case is 32 and the sample size is 3. She could assign a number from 1 to 32 to the students on the list.

The teacher will need to choose, from the random number table, three numbers in the range 1 to 32 inclusive. On the table she will look at the digits two at a time, ignoring any numbers that are outside the range 1 to 32 and any repeated numbers.

Starting in the second row of the table

14595 35050 40469 27478 44526 67331 93365 54526 22356 93208

and moving across, the numbers are:

14, (59), (53), (50), (50), (40), (46), (92), (74), (78), (44), (52), (66), (73), **31**, (93), (36), (55), (45), **26**

(Numbers in brackets are ignored because they are outside the range 1–32.)

Going down the alphabetical list of students, she would choose the 14th, 26th and 31st student.

Other random number generators

There are some other objects that will generate sets of random numbers:

- A carefully constructed spinner. A spinner could be made specifically for a situation. For example, if you needed to select samples of 3 objects from a population of 8 then a spinner would be appropriate for this situation. However, the physical limitations of making a spinner would mean that 10 is probably the maximum size that will give reliably **random samples**.
- Selecting numbered objects from a bag. Care needs to be taken that the sample is random; the objects should all be the same shape, not visible to the selector and returned to the bag after each selection and mixed well. There are physical limits to the use of this method also.
- A set of numbered cards. You will need a card for every object in the population so this method is limited to a population of about 100. Playing cards can sometimes be used, taking care to allocate the cards to members of the population.
- A coin. This can be used if there are only two possible outcomes or the size of the population is 2.
- A **die**. A die can be used if the number of possible outcomes (or the size of the population) is a factor of 6.
- Lists of random numbers can also be obtained from a calculator or computer that has a random number generator.

Using a calculator or computer to produce random numbers

1. **A scientific calculator:** most random number generators on scientific calculators produce a three decimal place random number between 0 (included) and 1.

   These numbers can be used in a similar manner to a random number table; ignore the decimal point and use as many digits as needed.

2. **A graphics calculator:** will generate a list of random numbers within a given range, one at a time or in groups.

   Using a TI83 calculator key □| to get to the probability menu (PRB) choose 5:randInt(⊆; type in the range of numbers, 1 to 25 inclusive, and a random number in the range will be generated each time you press ⊆.

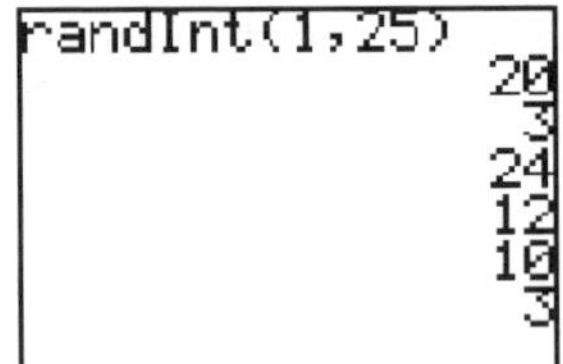

   Alternatively, if you want a group of numbers in a range then you type 1,25 for the range and then ,5) for the number in the group; 5 in this case. Groups of five random numbers in the range 1 to 25 inclusive will be generated each time you press ⊆.

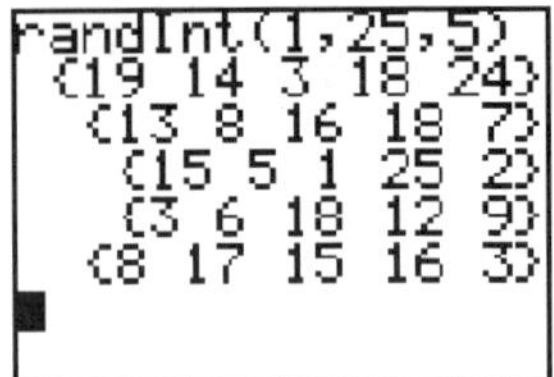

3. **A computer:** will generate a table of random numbers within a given range.

   The following table was generated using a spreadsheet on the computer typing = RANDBETWEEN(1,25) in the first cell and copying this into the other cells.

10	5	1	5	25	1	18	6	14	2
7	21	9	7	3	10	24	17	21	25
14	11	22	7	13	19	16	2	1	25
2	15	7	21	15	1	7	15	8	13
9	22	2	2	11	13	9	20	2	16
22	10	1	6	17	4	22	5	6	6
7	20	8	15	6	25	25	8	16	10

Sample size

If a sample is to be used to represent a population then it is important to consider the size of the sample. A sample that is too small will not provide information from which reliable conclusions about the population can be made. For example, measuring the height of a group of three fifteen-year-olds would not give a very reliable estimate of 'the height of fifteen-year-olds'.

A sample that is too large could be a waste of time and money and would not provide any more information about the population than could be provided by a smaller sample.

Usually samples are used to estimate statistics (mean, median, mode, maximum, minimum, range, etc) of a population. If the sample size is too small then any statistics that are based on this sample may not give a good estimate of the population statistics.

Designing a questionnaire for a statistical investigation

A questionnaire is a set of questions designed to obtain information from people. Questionnaires can be conducted in person, by mail or over the phone. Questionnaires conducted in person are more effective, especially where the questioner is recording the answers. There is often a large non-response rate for mail or phone questionnaires.

A questionnaire relies on the person being surveyed giving truthful or accurate answers to the questions and this is influenced by:

- The directness of the questions. (The person being surveyed should know immediately what form the answer should take and not need to ask questions like 'What do you mean by ...?')
- The convenience of the survey. (A few direct questions in a logical order will get a more positive and accurate response than a long questionnaire where the relevance of some questions is not obvious.)
- The relevance of the questions. (The purpose of the survey should be stated to the person being surveyed so they can judge the relevance of what is being asked.)
- The questions not being leading questions. (Avoid questions with value judgments built in, such as 'Do you agree that criminals should be put in jail?')
- The questions not being of a private nature where giving truthful answers will embarrass the person being questioned or the questioner. (Some people might not want to tell you their age, their weight, their HIV/AIDS status, their income, etc.)

If you have designed a questionnaire to collect information it is a good idea to test it on a few friends first to see if it provides the required information.

Collecting numerical data for a statistical investigation

Numerical data can be collected if it involves some characteristic of the local community that can be counted or measured.

Numerical data is either discrete or continuous data and each of these types has an appropriate method for organisation (tables) and display (graphs). Statistics such as the mean, median, mode, range, interquartile range and standard deviation can be found for numerical data, and boxplots can be used to display these statistics.

Steps in a statistical investigation

For your investigation you will need to:

1. State the purpose of your investigation.
2. Decide who/what will be surveyed. Will you be using a sample of a larger population or are you surveying the whole population?
3. Decide how you are going to choose a random sample if you are using a sample of the population.
4. Decide how the information will be obtained. Will it be by questionnaire, observation, measuring or counting?
5. Collect the data.
6. Identify the type of data that you will be collecting, eg categorical, discrete numerical or continuous numerical.
7. Use the work from this unit to find the appropriate tables, graphs and statistics for the analysis of the data that you have collected.
8. Write a report on your investigation mentioning all of the items 1 to 7 above. Also mention any problems that you encountered, how you dealt with the problems and how these may have affected the data collected and the final conclusions.

Analysing your data

After you have collected your data you should:

1. **Organise your data in a suitable table.**

   For categorical data you would use a frequency table. For numerical data you could use a stem and leaf diagram or a frequency table.

2. **Display the data using a suitable graph.**

   For categorical data this could be a bar graph or a pie chart. For discrete numerical data this could be a **column graph**. For continuous numerical data this could be a histogram.

3. **Comment on the features of the graph.**

   For all graphs the most frequently occurring category or number is mentioned, and possibly also the least frequently occurring category or number. If there are any outstanding data values (too big, too small) then these should be mentioned.

   For column graphs and histograms the shape of the distribution of the data should be mentioned (negatively skewed, symmetric, positively skewed) and any outliers observed.

4. **Find relevant statistics.**

   For categorical data this may be finding percentages; eg 77% of the students walked to school. For numerical data this would be finding the mean and median, range, interquartile range and standard deviation and then displaying the relevant statistics on a boxplot.

5. **Interpret the statistics.**

   For categorical data, state what the percentages mean; eg if you found that 77% of the students walked to school then you could say 'Most of the students walked to school'.

For numerical data you will need to use common-usage language to interpret the mean and median; eg the average (the mean) age of marriage is 20 for women but half (the median) of the women were married by the age of 19.

Similarly for the measures of spread; eg the age of marriage for women was spread over 30 years (the range) with the middle half of the women being married over a spread of 7 years (the interquartile range) from 17 years (the lower quartile) to 24 years (the upper quartile).

6. Write a conclusion briefly summarising what you have found.

Unit 11.3 Activity 6A: Statistical investigations

1. A plantation has its mango trees planted as shown on the map below. An investigation into the production (output) of the trees and the relevance of each tree's position on its production is made using a random sample of trees.

Map of mango plantation

(with tree numbers)

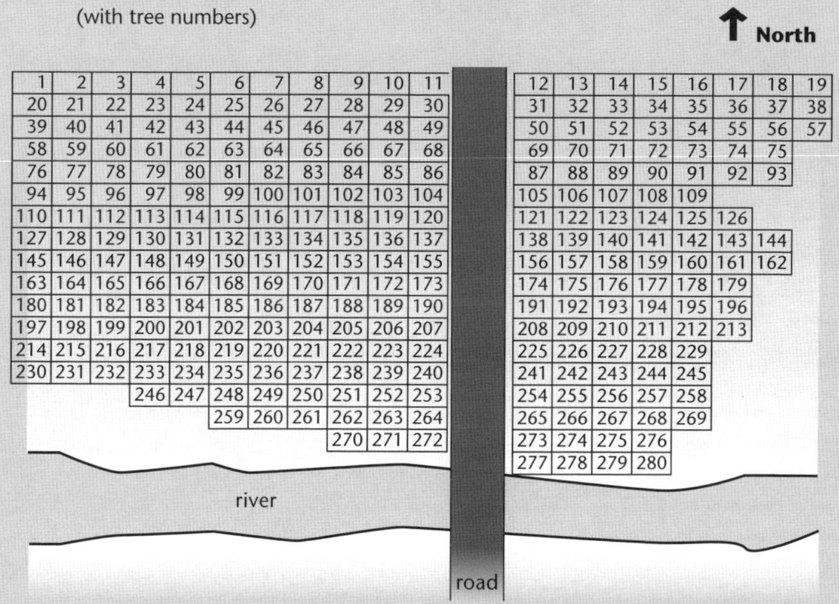

North

West of road:

1	2	3	4	5	6	7	8	9	10	11
20	21	22	23	24	25	26	27	28	29	30
39	40	41	42	43	44	45	46	47	48	49
58	59	60	61	62	63	64	65	66	67	68
76	77	78	79	80	81	82	83	84	85	86
94	95	96	97	98	99	100	101	102	103	104
110	111	112	113	114	115	116	117	118	119	120
127	128	129	130	131	132	133	134	135	136	137
145	146	147	148	149	150	151	152	153	154	155
163	164	165	166	167	168	169	170	171	172	173
180	181	182	183	184	185	186	187	188	189	190
197	198	199	200	201	202	203	204	205	206	207
214	215	216	217	218	219	220	221	222	223	224
230	231	232	233	234	235	236	237	238	239	240
			246	247	248	249	250	251	252	253
					259	260	261	262	263	264
								270	271	272

East of road:

12	13	14	15	16	17	18	19
31	32	33	34	35	36	37	38
50	51	52	53	54	55	56	57
69	70	71	72	73	74	75	
87	88	89	90	91	92	93	
105	106	107	108	109			
121	122	123	124	125	126		
138	139	140	141	142	143	144	
156	157	158	159	160	161	162	
174	175	176	177	178	179		
191	192	193	194	195	196		
208	209	210	211	212	213		
225	226	227	228	229			
241	242	243	244	245			
254	255	256	257	258			
265	266	267	268	269			
273	274	275	276				
277	278	279	280				

river

road

a. How many mango trees are in this plantation?

b. Use the random number table on page 300 to select a random sample of 30 mango trees. How are you going to do this? (Write down your 30 numbers from the table that will enable you to randomly choose 30 trees from this plantation.)

We do not have the output data for all the trees in the plantation. Below is a table giving the number of mangoes produced by a sample of 35 trees from the plantation. The number of each tree is also given so its position in the plantation is known. Use this sample to answer the following questions.

Tree	Output	Tree	Output	Tree	Output
6	49	218	183	142	131
25	78	233	187	158	125
40	82	248	208	176	113
65	85	259	217	194	165
80	97	272	235	210	134
98	104	12	44	229	169
111	111	37	91	241	128
133	125	51	75	256	168
149	149	74	103	268	214
172	109	88	89	276	253
188	140	103	146	280	260
197	141	123	112		

c. Produce a suitable table for this data.

d. Construct a suitable graph for this data and comment on the shape of the distribution of the data.

e. Find two measures of centre for this sample and interpret these statistics.

f. Find some measures of spread for this data and interpret these statistics.

g. Construct a boxplot for the data and comment on any features of the boxplot.

h. An agricultural adviser claims that trees planted on the right of the road will have a greater output than those planted on the left. Examine the data in the table and construct parallel boxplots to enable you to comment on this statement.

i. A local villager says that the closer to the river a tree is, the greater its output. Examine the data in the table and construct parallel boxplots to enable you to comment on this statement.

Unit 11.3 Managing Data
Topic 7: Permutations and combinations

The focus of Unit 11.3 is on statistics, probability, permutations and combinations. Topic 7 relates to the content set out in the two bullet points beneath the heading 'Permutation and combination' (Syllabus p. 16):

- Discussing powers versus factorial.
- Applying permutation and combination to real-life problems.

The multiplication principle

If Sae has four different T-shirts, three different pairs of shorts and two different pairs of shoes, how many different outfits has Sae got to wear if he has to wear a T-shirt *and* a pair of shorts *and* a pair of shoes?

We can draw a diagram to illustrate this:

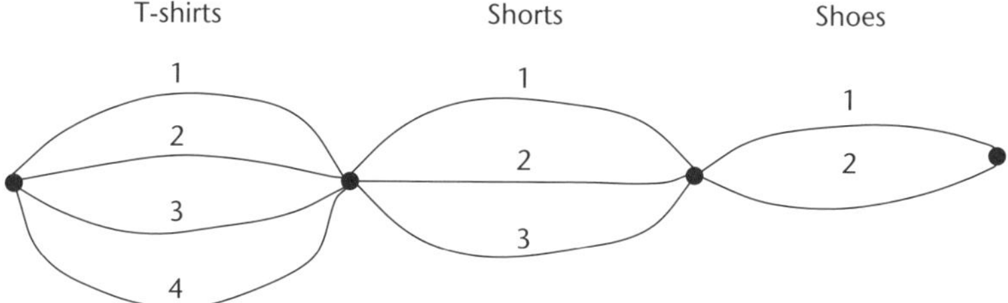

One choice could be T-shirt 3 and shorts 1 and shoes 2. Going through all the possible outfits reveals that there are $4 \times 3 \times 2 = 24$ different possibilities.

The **multiplication principle** is associated with the word '**and**'.

The addition principle

Consider the problem: How many different numbers can be formed from the digits 2, 4 and 8 where each number is used only once in a number?

We can have one-digit numbers: 2, 4 or 8 (3 altogether).

Or we can have two-digit numbers: 24, 42, 28, 82, 48 or 84 (6 altogether).

Or we can have three-digit numbers: 248, 284, 428, 482, 814 or 842 (6 altogether).

Each of the operations 'one-digit numbers', 'two-digit numbers' and 'three-digit numbers' are **mutually exclusive** operations; they do not occur at the same time. So the total of numbers formed from the digits is $3 + 6 + 6 = 15$

The **addition principle** is associated with the word '**or**'.

Permutations (arrangements)

When investigating the number of ways objects can be arranged, each possible arrangement is called a **permutation** (or an **arrangement**). The words 'permutation' and 'arrangement' mean the same thing. With permutations the order of the objects is important.

Consider the possible number of arrangements, in a row, of the three letters A, B and C. We can list these arrangements:

ABC, ACB, BAC, BCA, CAB, CBA

There are six possible arrangements.

Rather than **listing** all the possible arrangements, we can look at the problem as having three positions to fill:

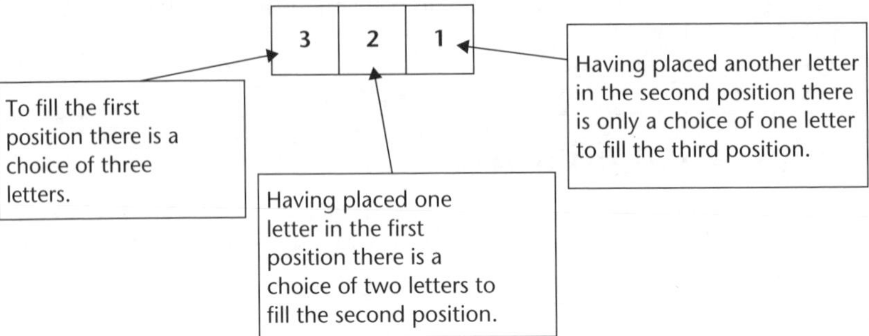

So the number of possible arrangements of three different letters is $3 \times 2 \times 1 = 6$; as we found when we listed all the possible permutations.

Try listing the number of arrangements of four different objects, A, B, C and D, in a row. You will find that there are 24 different arrangements. In this case we have four positions to fill:

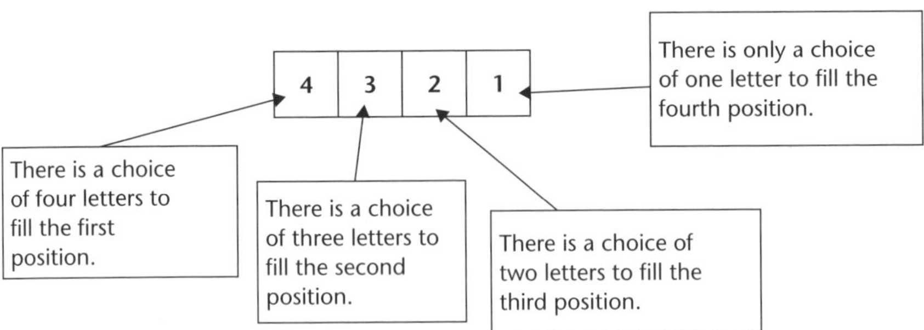

So the number of possible arrangements of four different letters is $4 \times 3 \times 2 \times 1 = 24$, as found when all the possible permutations are listed.

Similarly, if we had, say, seven different objects to be arranged in a row, the number of possible permutations would be $7 \times 6 \times 5 \times 4 \times 3 \times 2 \times 1 = 5040$.

Factorial notation

Factorial notation is a convenient way of writing products like $4 \times 3 \times 2 \times 1$

$4 \times 3 \times 2 \times 1$ is written as 4! ; said as 'four factorial'.

Similarly $7! = 7 \times 6 \times 5 \times 4 \times 3 \times 2 \times 1$.

In general:

$$n! = n \times (n-1) \times (n-2) \times (n-3) \ldots\ldots\ldots\ldots \times 3 \times 2 \times 1$$

You will find that most calculators have a factorial button.

The number of permutations of *n* objects taking *r* at a time

Consider the case where you have four different objects A, B, C and D. How many ways can these four objects be arranged two at a time?

We can list the possibilities:

AB, BA, AC, CA, AD, DA, BC, CB, BD, DB, CD, DC

There are 12 possible arrangements.

This can be thought of as having two positions to fill with a choice of four objects:

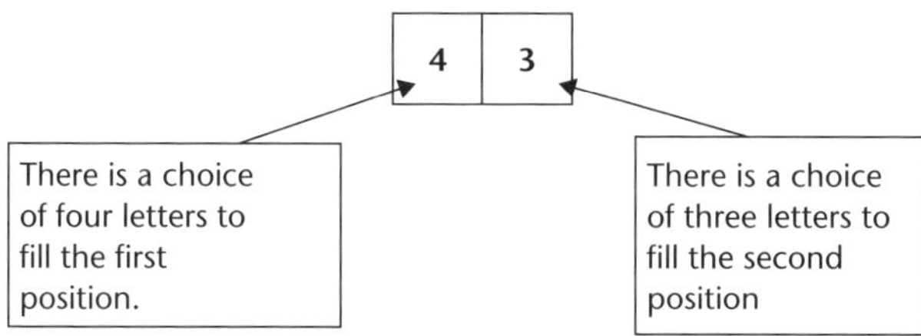

So the number of arrangements of four different objects, taking two at a time, is 4 × 3 = 12.

> There is special notation for the number of ways n different objects can be arranged taking r at a time: ${}^{n}P_{r} = \dfrac{n!}{n-r!}$

Note: the P stands for permutations.

Using this notation for the example above: $n = 4$ and $r = 2$, the number of ways of arranging four different objects taking two at a time is:

$$ {}^{4}P_{2} = \frac{4!}{(4-2)!} = \frac{4!}{2!} = \frac{4\times3\times2\times1}{2\times1} = 4\times3 = 12 $$

Example A

Q. **1.** **a.** Find the value of 8! using your calculator.

b. Simplify, and find the value of, $\frac{9!}{7!}$ without using a calculator.

2. In how many ways can six different books be arranged on a shelf?

3. How many arrangements of the letters MOTHER can be made taking four letters at a time?

A. **1.** **a.** Keying 8! = will give 40320 as the answer.

b. $\frac{9!}{7!} = \frac{9\times8\times7\times6\times5\times4\times3\times2\times1}{7\times6\times5\times4\times3\times2\times1} = 9\times8 = 72;$

All the factors from 7 downwards cancel.

2. All six books are being arranged so there are 6! = 6 × 5 × 4 × 3 × 2 × 1 = 720 ways of doing this.

3. All the letters are different so we are arranging six letters taking four at a time.

This can be done in ${}^{6}P_{4} = \frac{6!}{(6-4)!} = \frac{6!}{2!} = \frac{6\times5\times4\times3\times2\times1}{2\times1} = 6\times5\times4\times3 = 360$ ways.

Note: If we deal with Example 2 as 'How many ways can six different books be arranged taking six at a time?' then we will get:

$$^{6}P_{6} = \frac{6!}{(6-6)!} = \frac{6!}{0!}$$

We know the answer to this question is 6! so this means that 0! must be equal to 1.

$$0! = 1$$

Unit 11.3 Activity 7A: Factorial notation and permutations

1. Use your calculator to find the value of each of the following:

a. $8!$ **b.** $\frac{15!}{9!}$ **c.** $5! \times 8!$

2. Simplify and evaluate each of the following without using your calculator:

a. $\frac{7!}{5!}$ **b.** $\frac{5! \times 6!}{7!}$ **c.** $\frac{5!}{6!}$

d. $\frac{100!}{98!}$ **e.** $\frac{8!}{5! \times 3!}$ **f.** $\frac{0!}{4!}$

g. $\frac{10!}{7! \times 3!}$ **h.** $\frac{11!}{(11-2)! \times 2!}$

3. Evaluate each of the following:

a. $^{7}P_{3}$ **b.** $^{4}P_{4}$ **c.** $^{4}P_{3}$ **d.** $^{7}P_{1}$

4. a. List all the permutations of the letters of the word WORD taken:

i. one at a time

ii. two at a time

b. Write in $^{n}P_{r}$ form and evaluate each of the sets of permutations for part a.

c. How many permutations of the letters of the word WORD are there, taking the letters three at a time?

5. How many ways can five different books be arranged in a row?

6. How many ways can the first, second and third places be awarded for a race that has 10 participants?

7. How many different three-digit numbers can be made from the digits 4, 5, 6, 7, 8 and 9 if numbers can be used only once?

8. A fund-raising competition at school involves matching three different baby photos to three of the seven teachers.

a. How many different ways can the photos be matched to the teachers?

b. If each entry in the competition costs 10t and the prize is K10 is it worth entering every possible arrangement of photo-with-teacher?

Further arrangements (permutations)

Arrangements with identical objects

In general:

> The number of arrangements of n objects, in a row, where p objects are identical to each other, another q objects are identical to each other, another r objects are identical to each other ... etc. is given by $\dfrac{n!}{p!\,q!\,r!\,\ldots}$

Example B

Q. How many ways can the letters of the word MATHEMATICAL be arranged?

A. There are 12 letters but some letters are identical to each other. There are two Ms, three As and two Ts in the word MATHEMATICAL so the number of arrangements of the letters is: $\dfrac{12!}{2!\,3!\,2!} = 19958400$

Arrangements in a circle

In general:

> n different objects can be arranged in a circle in $(n - 1)!$ ways.

Example C

Q. How many ways can 10 people be arranged around a circular table?

A. 10 people can be arranged in a circle in $(10 - 1)! = 9! = 362880$ ways.

Arrangements with restrictions

Sometimes there are restrictions that are placed on arrangements (permutations). The way to deal with these is best illustrated with examples.

Example D

Q. Three children and their parents are to attend a concert. How many ways can this family be seated together, in a row, if:

1. there is no restriction on who they sit next to.
2. the parents are on the ends.
3. the parents sit together.
4. The parents sit together and the children sit together.
5. a parent sits between each of the children.

A. 1. There are five people to be seated in a row.

This can be done in ${}^5P_5 = \dfrac{5!}{0!} = 5! = 120$ ways.

OR using the 'positions to fill' method : Five positions

5	4	3	2	1

So $5 \times 4 \times 3 \times 2 \times 1 = 120$ ways

2. Using the 'positions to fill' method:

   The three positions in the centre can be filled by the children in 3, then 2, then 1 ways.

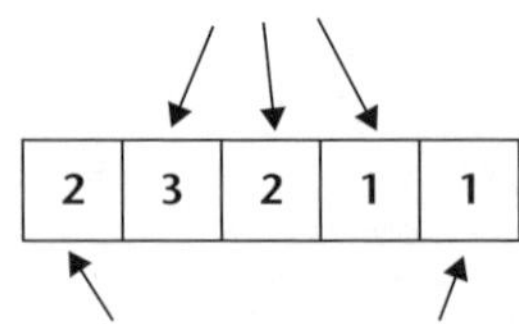

   The parents must occupy the end positions so the first position can be filled in two ways (M or D) and then the last position can be filled in one way.

   So there are 2 × 3 × 2 × 1 × 1 = 24 ways of arranging the family with the parents on the ends.

3. If the parents are to be together then we consider that they are one unit. We now have four units (the parents and each of the 3 children) who can be arranged in 4! ways. However, within the parents unit they can be arranged in 2! ways (MumDad, DadMum) so the number of different arrangements is 4! × 2! = 48 ways

4. Consider the parents as one unit and the children as another unit, these two units can be arranged in 2! ways:

   Within the parents' unit there are two people and these two people can be arranged in 2! ways;

   Within the children's unit there are three children and these children can be arranged in 3! ways.

   So the number of arrangements with the parents together and the children together is: 2! × 2! × 3! = 24.

5. The only possibility for this arrangement is 'child, parent, child, parent, child'.

   Using the 'positions to fill' method:

   Arranging the parents first:

c	p	c	p	c
	2		1	

   The number of ways the parents can sit between the children is 2! × 3! = 24 ways.

   Then arranging the 3 children:

c	p	c	p	c
3	2	2	1	1

Example E

Q. How many different three-digit numbers can be formed from the digits 1, 2, 3, 4, 5, 6, 7, 8 and 9 if:

1. the digits cannot be used more than once?
2. the digits can be used more than once?
3. the numbers must end in 8 and the digits are not used more than once?

4. the numbers are odd numbers and the digits are not used more than once?
5. the numbers are greater than 500 and the digits are not used more than once?

A. 1. There are nine different digits and we are arranging them three at a time so this can be done in ${}^{9}P_{3}=\dfrac{9!}{(9-3)!}=\dfrac{9!}{6!}=9\times8\times7=504$ ways.

Alternatively, using the positions to fill method: Three positions to fill; nine different objects.

9	8	7

So 9 × 8 × 7 = 504 ways.

2. If the digits can be used more than once then there are nine ways of filling the first position, nine ways of filling the second position and nine ways of filling the third position:

9	9	9

So there are $9\times9\times9=9^3=729$ different numbers.

3. If the number must end in 8 then there is only one way of filling the third position, ie with the digit 8. There are now eight remaining digits to fill the first position and then seven remaining digits to fill the second position.

So 8 × 7 × 1 = 56 different numbers ending in 8.

4. If the numbers are to be odd numbers then they must end in an odd digit. There are five odd digits that can fill the last position. The remaining positions can be filled with any of the remaining digits so the first position can be filled in eight ways and the second in seven ways.

8	7	5

So 8 × 7 × 5 = 280 different odd numbers.

5. If the numbers are to be greater than 500 then the first position must be filled with one of the digits 5, 6, 7, 8 or 9, so there are five ways of filling the first position. The remaining positions can be filled with any of the remaining digits, so the second position can be filled in eight ways and then the third in seven ways.

5	8	7

So 5 × 8 × 7 = 280 different numbers greater than 500.

Unit 11.3 Activity 7B: Further arrangements

1. List all the permutations of the letters of the word MOON. Confirm that you have $\dfrac{4!}{2!}$ permutations.
2. How many different ways can the letters of the word LETTER be arranged?
3. Three identical maths textbooks and four identical science textbooks are to be arranged on a shelf. How many ways can these books be arranged on the shelf?
4. List all the arrangements of four people around a circular table. Two of the arrangements are shown below.

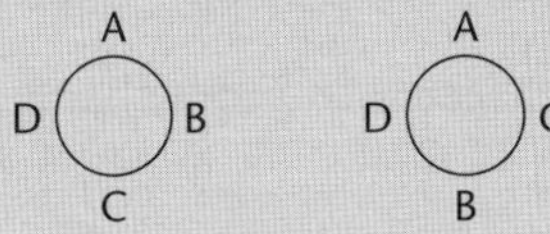

5. How many ways can six people be arranged around a circular table?

6. Six friends go to the movies.

a. How many ways can they be arranged together in a row of seats if:

i. there is no restriction on who they sit next to?

ii. Mani and Joseph are on the ends?

iii. Philip and Nic are to sit together?

b. How many ways can the six friends be arranged together in a row of seats if Philip and Nic are **not** to sit next to each other?

Hint: This is the difference between two of the answers from part a.

7. Four-digit numbers are to be formed from the digits 1, 3, 4, 6, 8 and 9. How many different four-digit numbers can be formed if:

a. the digits cannot be used more than once?

b. the digits can be used more than once?

c. the numbers must end in 3 and the digits are not used more than once?

d. the numbers are even numbers and the digits are not used more than once?

e. the numbers are greater than 8000 and the digits are not used more than once?

8. Three boys and four girls are to be seated in the front row of the school photo. How many ways can they be arranged if:

a. there is no restriction on who they sit next to?

b. the girls are to be together?

c. the girls are to be together and the boys are to be together?

d. the girls and boys are to alternate?

9. Number plates on cars are usually made up of letters (26 letters) and digits (10 digits from 0 to 9).

a. How many different number plates are possible if letters and numbers can be used more than once on a number plate and the number plate is made up of:

i. two letters and then two digits?

ii. three letters and then three digits?

b. It is decided not to use the number 000. How many number plates are possible, using three letters and then three digits, without the number 000?

10. In a game four dice are tossed, one after the other.

a. List three possible outcomes.

b. How many possible outcomes are there if 4-5-3-1 is a different outcome to 4-3-1-5?

c. In how many of these outcomes are all the numbers the same?

Combinations (or selections)

In many situations we need to know the number of ways we can *choose* a sample from a larger number of objects. In these cases *the order of the selection is not important*. For example, if we are choosing two objects from five different objects A, B, C, D, E then the choice AB is the same choice as BA.

If we list all the ways that two objects can be chosen from five different objects A, B, C, D, and E we get:

AB, AC, AD, AE, BC, BD, BE, CD, CE, DE; there are 10 different choices.

The notation for this is ${}^5C_2 = \dfrac{5!}{2!(5-2)!} = \dfrac{5!}{2!\times 3!} = \dfrac{5\times4\times3\times2\times1}{2\times1\times3\times2\times1} = 10.$

In general:

> If we are selecting r objects from a group of n different objects, where order is not taken into account, then the number of ways that this can be done is: ${}^nC_r = \dfrac{n!}{r!(n-r)!}$

Alternative notation: $\dbinom{n}{r} = {}^nC_r = \dfrac{n!}{r!(n-r)!}$

Note: The factors in the calculation are usually cancelled before the numbers are multiplied; this is preferable to using the factorial function on the calculator. Your calculator may have a ${}^nC_r$ function.

Some common simplifications using the ${}^nC_r$ notation

- ${}^nC_1 = \dfrac{n!}{1!(n-1)!} = \dfrac{n\times(n-1)\times(n-2)\times(n-3).....\times2\times1}{1\times(n-1)\times(n-2)\times(n-3).....\times2\times1} = \dfrac{n}{1} = n$

- ${}^nC_r = {}^nC_{n-r}$

$${}^nC_{n-r} = \frac{n!}{(n-r)!\times\big(n-(n-r)\big)!} = \frac{n!}{(n-r)!(n-n+r)!} = \frac{n!}{(n-r)!r!} = {}^nC_r$$

For example ${}^7C_3 = \dfrac{7!}{3\times(7-3)!} = \dfrac{7!}{3!\times4!} = \dfrac{7\times6\times5}{3\times2} = 35$

also ${}^7C_4 = \dfrac{7!}{4!\times(7-4)!} = \dfrac{7!}{4!\times3!} = \dfrac{7\times6\times5}{3\times2} = 35$

- ${}^nC_n = {}^nC_o = \dfrac{n!}{(n!\times0!)} = \dfrac{1}{0!} = \dfrac{1}{1} = 1$

- ${}^nC_2 = {}^nC_{n-2} = \dfrac{n!}{(2!\times(n-2)!)} = \dfrac{(n\times(n-1)\times(n-2)\times(n-3)......\times2\times1)}{(2\times(n-2)\times(n-3).....\times2\times1)} = \dfrac{n(n-1)}{2}$

For example ${}^{12}C_2 = \dfrac{12\times11}{2} = 66$

Example F

Q. **1.** Calculate $^{10}C_4$

2. Calculate:

a. $^{6}C_2$ **b.** $^{6}C_4$

3. How many choices of three books can be made from seven different books?

4. Write down the values for:

a. $^{10}C_1$ **b.** $^{8}C_8$ **c.** $^{9}C_2$ **d.** $^{35}C_{35}$

5. Solve the equation $^{n+2}C_n = 9$ for n.

A. **1.** $^{10}C_4 = \frac{10!}{4!(10-4)!} = \frac{10!}{4!\times 6!} = \frac{10\times 9\times 8\times 7\times 6!}{4\times 3\times 2\times 1\times 6!} = \frac{10\times 9\times 8\times 7}{4\times 3\times 2} = 210$

2. **a.** $^{6}C_2 = \frac{6!}{2!\times 4!} = \frac{6\times 5}{2} = 15$

b. $^{6}C_4 = \frac{6!}{4!\times 2!} = \frac{6\times 5}{2} = 15$

Note: the same answer as part a. because $^{\uparrow n}C_{\downarrow r} = {}^{\uparrow n}C_{\downarrow (n-r)}$

3. There will be $^{7}C_3 = \frac{7!}{3!\times (7-3)!} = \frac{7!}{3!\times 4!} = \frac{7\times 6\times 5}{3\times 2\times 1} = 35$ different choices.

4. **a.** $^{10}C_1 = 10$

b. $^{8}C_8 = 1$

c. $^{9}C_2 = \frac{9\times 8}{2} = 36$

b. $^{35}C_{35} = 1$

5. Simplifying the left-hand side of the equation:

$$^{n+2}C_n = \frac{(n+2)!}{n!(n+2-n)!}$$

$$= \frac{(n+2)(n+1)n(n-1)\ldots\ldots\times 2\times 1}{n(n-1)\ldots\ldots\times 2\times 1\times 2!}$$

$$= \frac{(n+2)(n+1)}{2}$$

We are solving the quadratic equation $\frac{(n+2)(n+1)}{2} = 45$ or $n^2 + 3n + 2 = 90$ which gives $n^2 + 3n - 88 = 0$.

Factorising the left-hand side of this equation: $(n + 11)(n - 8) = 0$ gives the solution $n = 8$. The 'solution' $n = -11$ is not possible as the $^{n}C_r$ notation is only defined for positive numbers.

Unit 11.3 Activity 7C: Combinations

1. Evaluate each of the following without using a calculator:

a. ${}^{10}C_8$ **b.** ${}^{9}C_4$

c. ${}^{6}C_3$ **d.** ${}^{11}C_8$

e. ${}^{14}C_3$ **f.** ${}^{7}C_4$

g. ${}^{12}C_8$ **h.** ${}^{8}C_4$

2. Write down the value of each of the following:

a. ${}^{10}C_{10}$ **b.** ${}^{9}C_1$

c. ${}^{6}C_5$ **d.** ${}^{11}C_9$

e. ${}^{13}C_2$ **f.** ${}^{7}C_5$

g. ${}^{12}C_{11}$ **h.** ${}^{8}C_2$

3. ${}^{9}C_4 = {}^{9}C_{\square}$ Fill in the blank subscript with a number other than 4, so that this equation is true.

4. How many ways can a team of 11 be chosen from a squad of 15 players?

5. How many different triangles can be formed from five randomly placed points (no three points are in a straight line)?

6. To win a lottery you need to choose six correct numbers from the numbers 1 to 45. How many different sets of six numbers are possible?

7. Five playing cards are dealt from a pack of 52 cards. How many different sets of five cards are possible?

8. Solve each of the following equations for n.

a. $\dfrac{(n+2)!}{n!} = 42$

b. ${}^{n+2}C_n = 15$

c. ${}^{n+3}C_{n+1} = 36$

d. ${}^{n}C_{n-2} = 28$

e. ${}^{n+1}C_{n-1} = 21$

Combinations (selections) with restrictions

Dealing with restrictions is best illustrated with an example.

Example G

Q. A committee of four people is to be chosen from three women and four men. How many ways can this be done if:

1. there is no restriction?
2. Pauline, one of the women, must be on the committee?

3. the committee must contain 2 women and 2 men?
4. at least one woman must be on the committee?
5. Sam cannot be on the committee if Mike is on the committee?

A. 1. There is a total of seven people from which to choose the committee of four.

This can be done in $^7C_4 = \frac{7!}{4! \times 3!} = \frac{7 \times 6 \times 5}{3 \times 2} = 35$ ways.

2. If Pauline is on the committee then we need to choose another three people from the remaining six people. This can be done in $^6C_3 = \frac{6!}{3! \times 3!} = \frac{6 \times 5 \times 4}{3 \times 2} = 20$ ways.

3. The committee must contain two women **and** two men. We use the multiplication principle.

To choose two women from three: $^3C_2 = 3$ ways.

To choose two men from four: $^4C_2 = \frac{4 \times 3}{2} = 6$ ways

To choose two women **and** two men: $^3C_2 \times {}^4C_2 = 3 \times 6 = 18$ ways.

4. The number of committees with at least one woman on the committee = (the number of committees with no restriction) – (the number of committees with no women on the committee).

$^7C_4 - {}^3C_0 \times {}^4C_4$

$= 35 - 1 \times 1$

$= 34$

5. We have two mutually exclusive situations to consider:

a. Mike is on the committee (and so Sam is not)

or

b. Mike is not on the committee. (Sam may or may not be on the committee)

We use the addition principle to find the final number of ways.

a. Mike is on the committee (and so Sam is not). Consider Mike on the committee, and we cannot choose Sam, so we can choose the other three people to go on the committee from a total of five people.

This can be done in $^5C_3 = \frac{5 \times 4}{2} = 10$ ways.

b. Mike is not on the committee so Sam can be on the committee so we have a choice of six people from which to choose a committee of four.

This can be done in $^6C_4 = \frac{6 \times 5}{2} = 15$

So the total number of ways a committee can be chosen where 'if Mike is on the committee then Sam cannot be' = 10 + 15 = 25 ways.

Unit 11.3 Activity 7D: Combinations with restrictions

1. A school advisory council is to contain two students, five parents, three teachers and one representative from the school board. If eight students, ten parents, five teachers and three board members nominate for this committee, how many different councils can be formed?
2. For a study course at school the students have a choice of three science subjects, two mathematics subjects, four humanities subjects and two foreign languages. From these subjects a student has to choose one each of the science subjects, mathematics subjects and foreign languages and two humanities subjects. How many different courses of five subjects are possible?
3. Kenime has coordinated her working clothes so that she has five skirts, six blouses, three jackets and five pairs of shoes. Kenime says she has a different outfit for every day of the year. If she wears one of each of the skirts, blouses, jackets and shoes every day, is this true?
4. As a school prize Rose can choose three books from a group of five different fiction books, three different biographies and four history books. How many different choices of these books can Rose make if:
   - **a.** she chooses the books randomly?
   - **b.** she chooses one fiction, one biography and one history book?
   - **c.** she chooses all fiction books?
   - **d.** she chooses all history books?
   - **e.** she chooses two fiction books and one biography?
   - **f.** she chooses at least two fiction books?
5. A team of 15 rugby players are to be chosen from a squad of 10 forwards and 10 backs. How many different teams of eight forwards and seven backs can be chosen if:
   - **a.** there is no restriction?
   - **b.** the captain, a forward, must be on the team?
   - **c.** the captain, a forward, and/or the vice-captain, a back, must be on the team?
6. A 'hand' of five cards is dealt from a pack of 52 playing cards. How many different hands can be dealt:
   - **a.** with no restriction?
   - **b.** that are all black cards?
   - **c.** that are all hearts?
   - **d.** that contain three aces?
   - **e.** that contain two kings and three queens?
   - **f.** that contain at least three jacks?
   - **g.** that all are of the same suit?

- A pack of 52 playing cards contains four 'suits' (of 13 cards each) which are called hearts, diamonds, clubs and spades.
- There are 26 red cards and 26 black cards. The suits hearts and diamonds are red cards and clubs and spades are black cards.
- Each suit contains 13 cards: Ace, two, three, four, five, six, seven, eight, nine, ten, jack, queen and king.

7. A committee of five people is to be chosen from six women and seven men. How many ways can this be done if:

a. there is no restriction?

b. Esther, one of the women, must be on the committee?

c. the committee must contain two women and three men?

d. at least two women must be on the committee?

e. Arnold cannot be on the committee if Alice, his wife, is on the committee?

8. Ari has paint in the basic colours of red, blue, yellow, black and white. If he can make a new colour by mixing equal quantities of at least two of these colours, how many new colours is it possible to make?

Unit 11.3 Managing Data

Topic 8: Probability

The focus of Unit 11.3 is on statistics, probability, permutations and combinations. Topic 8 relates to the content set out under the heading 'Probability' (Syllabus p. 16):

- Calculate simple probability of events.
- Use simulations and theoretical probability techniques to solve problems, including the use of probability trees and informal conditional probability (reduced sample space).

Important property of probabilities

The notation **P(E)** means the **probability** that an **event** E occurs.

If E is any event, then $0 \le P(E) \le 1$.

Proof: P(E) is the limit of the ratio $\frac{\text{number of occurrences}}{\text{number of trials}}$ as N, the number of trials, gets larger. The smallest value, 0, will occur when the number of occurrences is 0. [since $\frac{0}{N} = 0$]

The largest value, 1, will occur when the number of occurrences is N. [since $\frac{N}{N} = 1$]

Thus $0 \le P(E) \le 1$.

Probabilities can be written as fractions, decimals or percentages.

Terminology

A **trial** is one repetition of an experiment. The **sample space** is the set of all possible outcomes of a trial.

Example A

1. In an experiment a coin is to be tossed twice a number of times. The trial consists of tossing the coin twice and recording the outcome. The following are possible sample spaces.
   - **i.** {(H, H), (H, T), (T, H), (T, T)}, where H is a head and T a tail.
   - **ii.** {0, 1, 2} where each number is the possible number of tails.
2. If a netball team plays a series of games, the trial is the playing of a game and the outcome is the result recorded. The following are possible sample spaces:
   - **i.** {L, W, D} where the letters stand for 'Lost', 'Won', 'Drew'.
   - **ii.** {PB, PW} where PB stands for 'played badly', PW stands for 'played well'.

Equiprobable (equally likely) outcomes occur when every element of a sample space has an *equal* likelihood of occurring.

Example B

From Example A:

1. The probability of each element of the sample space {(H, H), (H, T), (T, H), (T, T)} would be found to be the same if two coins were tossed many times.
2. However, {0, 1, 2} is not a sample space of equiprobable outcomes. Probability experiments would show that 1 tended to occur twice as often as 0 or 2.

Mathematically an **event** is a subset of a sample space.

Example C

The event E, 'one or more heads in two tosses of a coin', when written as a subset of the sample space is: {(H, H), (H, T), (T, H)}. This is illustrated by the **Venn diagram** of set theory. S, the sample space, is the universal set (the set of all elements) and E, the event, is the subset.

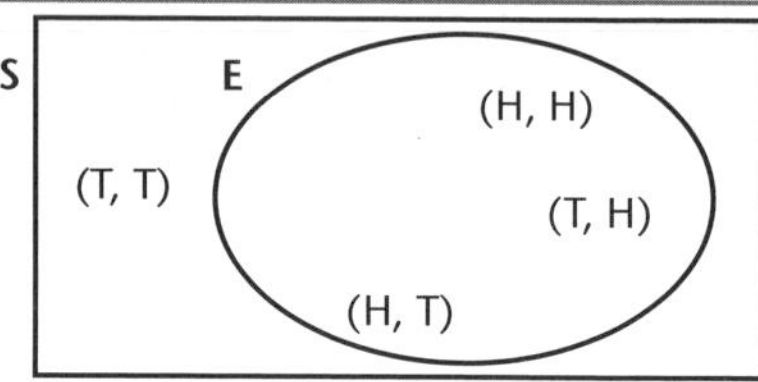

Venn Diagram

Finding probabilities when all outcomes are equiprobable

When all possible outcomes of a trial have an equal chance of occurring (ie they are equiprobable) the probability of an event can be *calculated* rather than determined by performing experiments. The probability of such an event E is given by:

$$P(E) = \frac{\text{number of elements in E}}{\text{number of elements in S}} = \frac{n(E)}{n(S)},$$ where the outcomes in the sample space S are equiprobable

The following problems are solved by establishing a full sample space of **equally likely outcomes** and using the formula for P(E).

Example D

Find the probability that a person chosen at random was born on a day beginning with T.

Solution

Let S be the sample space.

S = {Monday, Tuesday, Wednesday, Thursday, Friday, Saturday, Sunday}

Let E be the event: Born on a day starting with T. E = {Tuesday, Thursday}

$$P(E) = \frac{n(E)}{n(S)}$$

$$= \frac{2}{7}$$

Note: Listing a sample space {M, T, W, F, S} of possible first letters for days of the week gives a sample space whose outcomes are not equally likely (T and S are twice as likely to occur as the other letters) and so the formula $P(E) = \frac{n(E)}{n(S)}$ is not applicable ($P(E) \neq \frac{1}{5}$).

Example E

Find the probability that when a pair of dice is thrown, the two numbers will total four.

Solution

Letting S be sample space, and using ordered pairs for the outcomes,

S = {(6, 6), (6, 5), ..., (6, 1), (5, 6), (5, 5), ..., (1, 1)} which can be written in column form:

6, 6	5, 6	4, 6	3, 6	2, 6	1, 6
6, 5	5, 5	4, 5	3, 5	2, 5	1, 5
6, 4	5, 4	4, 4	3, 4	2, 4	1, 4
6, 3	5, 3	4, 3	3, 3	2, 3	1, 3
6, 2	5, 2	4, 2	3, 2	2, 2	1, 2
6, 1	5, 1	4, 1	3, 1	2, 1	1, 1

Each outcome is equally likely.

The three ordered pairs (3, 1), (2, 2), (1, 3) are the only combinations of numbers on the two dice which give a sum of four, so

E = {(3, 1), (2, 2), (1, 3)}.

$$\therefore \; P(E) = \frac{3}{36} \qquad [n(E) = 3, n(S) = 36 \text{ in formula}]$$

$$= \frac{1}{12}$$

Unit 11.3 Activity 8A: Probability and sample spaces

1. A ball is drawn at random from a bag which contains nine yellow and seven red balls. Find:

a. P(a yellow ball)

b. P(a red ball)

c. P(a yellow or red ball)

d. P(neither a yellow nor a red ball)

2. If, in Question 1, a red ball is drawn and a second draw is made, find:

a. P(a yellow ball if the red one is put back into the bag)

b. P(a yellow ball if the red one is *not* put back into the bag)

c. P(a red ball if the first one is put back into the bag)

d. P(a red ball if the first one is *not* put back into the bag)

3. A ball is drawn at random from a bag which contains eight red, five white and seven blue balls. Find:

a. P(a red ball)

b. P(a white ball)

c. P(a blue ball)

d. P(*not* a red ball)

4. If in Question 3, a blue ball is drawn, and a second draw is made, find:

a. P(a white ball if the blue one is put back into the bag)

b. P(a white ball if the blue one is *not* put back into the bag)

c. P(a blue ball if the first one is put back into the bag)

d. P(a blue ball if the first one is *not* put back into the bag)

5. Find the probability that a day of the week drawn at random will:

a. start with the letter S

b. start with the letter W

c. start with the letters W or T

d. *not* start with the letters W or T

e. end with the letter Y

f. *not* end with the letter Y

6. Find the probability that a month of the year drawn at random will:

a. have more than 30 days

b. have less than 30 days

c. start with the letter M

d. end with the letter Y

e. start with the letter J *and* end with the letter Y

f. start with the letter J *or* end with the letter Y
(**Note:** 'or' includes both, eg January)

7. Find the probability that a number drawn at random from {1, 2, 3, 4, 5, 6, 7, 8, 9} will be:

a. even

b. odd

c. either even or odd

d. neither even nor odd

e. a multiple of 3

f. a factor of 4

g. a multiple of 3 *and* a factor of 3

h. a multiple of 3 *or* a factor of 6

8. Find the probability that a card drawn at random from a pack of 52 cards is:

a. an ace

b. a heart

c. the ace of clubs

d. an ace or a heart

e. neither an ace nor a club

f. an ace but *not* a spade

g. a spade but *not* an ace

9. In a room there are four men and two women.

a. One person is randomly withdrawn and the gender of the person noted.

i. Write down the sample space of outcomes for this experiment.

ii. By using your answer to **i.** find the probability of a woman being withdrawn.

iii. What is the probability of not withdrawing a man?

b. One person is randomly withdrawn. The person returns and another person is randomly withdrawn.

i. Find the probability of getting two people of a different gender.

ii. Find the probability of getting at least one man.

iii. Find the probability of the first person withdrawn being male.

c. One person is randomly withdrawn and the person's gender is noted.
Without the person returning another person is withdrawn, and their gender noted.

i. Find the probability of more than one man leaving.

ii. Find the probability of no women leaving.

iii. Find the probability of the second person who leaves being a woman.

iv. Find the probability of both people leaving being women.

10. A survey about unemployment was taken from the voters in a suburb. The results are shown below.

	Party X	Party Y	Party Z
Employed	47	53	21
Unemployed	13	7	19

a. How many of the voters from Party Y are employed?

b. How many of the unemployed prefer Party X?

c. What is the total number of people surveyed?

d. What is the number of people surveyed who are employed?

e. What is the probability a randomly selected person is employed?

f. What is the probability a randomly selected person prefers Party Y?

g. What is the probability a randomly selected voter prefers Party Z?

h. What is the probability a randomly selected voter is unemployed?

i. What is the probability a randomly selected voter prefers Party Z and is employed?

j. What is the probability a randomly selected voter is employed or prefers Party Y (or both)?

Probability trees

The process of writing out sample spaces as in the previous two examples always works, but becomes time consuming and error prone when the sample spaces become large.

In many circumstances, probability problems are tackled by construction of a branching diagram called a **probability tree**. A probability tree follows an experiment or process through its various stages with probabilities being calculated at each stage.

Example F

A jar contains three red balls and two green balls. One ball is withdrawn and then another ball, without the first one being replaced.

a. Find the probability that both balls are red.

b. Find the probability that the balls are different in colour.

Solution

The probability tree for withdrawing one ball then another without replacing the first is as shown, where R stands for red and G stands for green.

It is drawn as follows:

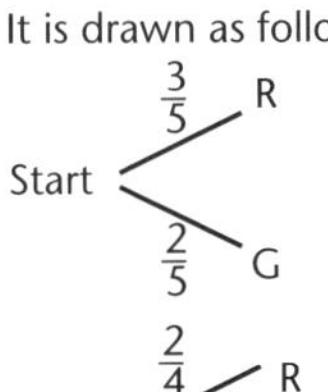

In the first withdrawal there is a $\frac{3}{5}$ chance of removing a red ball and a $\frac{2}{5}$ chance of removing a green ball.

If a red ball had been removed initially there would be two reds and two greens left in the jar. So for the second withdrawal there is a $\frac{2}{4}$ chance of removing a red ball and a $\frac{2}{4}$ chance of removing a green ball.

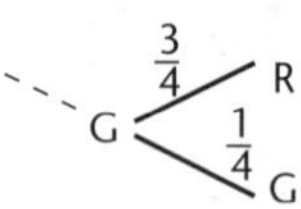

Similarly, if a green ball had been removed initially, there would be three reds and one green left in the jar. So for the second withdrawal there is a $\frac{3}{4}$ chance of removing a red ball and a $\frac{1}{4}$ chance of removing a green ball.

To complete the problem, some simple calculations are done.

1. The probability of two red balls is found by multiplying the probabilities on the branches corresponding to (R, R):
   probability of two red balls = P(R, R)
   $= \frac{3}{5} \times \frac{2}{4}$
   $= \frac{3}{10}$

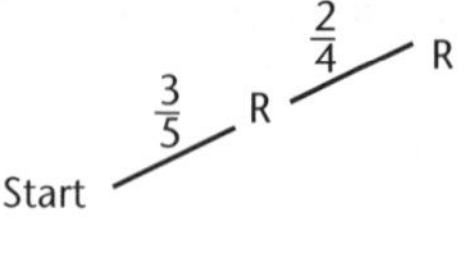

2. Different coloured balls on each draw means either a red followed by a green ball or a green followed by a red ball. The probability of 'different coloured balls on each draw' is:
   $P(R, G) + P(G, R) = \frac{3}{5} \times \frac{2}{4} + \frac{2}{5} \times \frac{3}{4}$
   $= \frac{12}{20}$
   $= \frac{3}{5}$

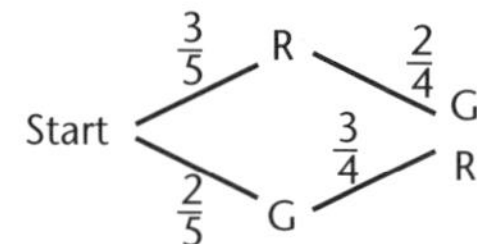

Note:

1. (R, G) and (G, R) are different outcomes and should not be confused.
2. The sum of probabilities for each set of branches from a point is 1.

eg

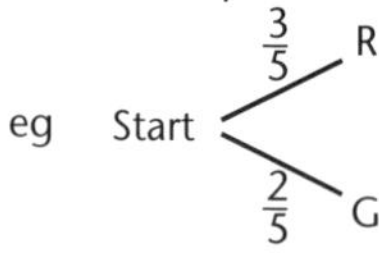

$\frac{3}{5} + \frac{2}{5} = 1$

or

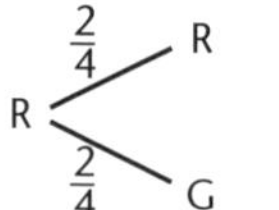

$\frac{2}{4} + \frac{2}{4} = 1$

Example G

A bag contains twelve black cards and ten white cards. A card is withdrawn and then replaced and another card withdrawn. Find the probability, using a probability tree, that both cards are the same colour.

Solution

The probability tree is:

P (two cards the same) = P(B, B) + P(W, W)

$= \frac{12}{22} \times \frac{12}{22} + \frac{10}{22} \times \frac{10}{22}$

$= \frac{244}{484}$

$= \frac{61}{121}$

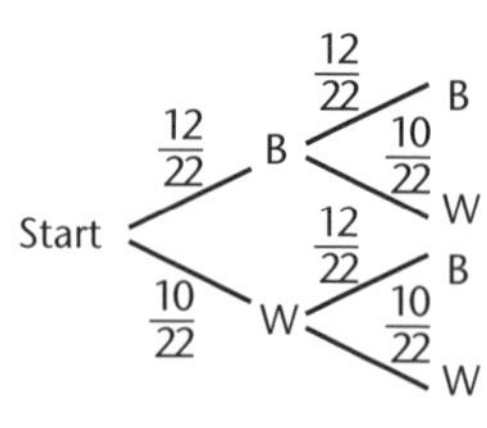

Note: In the second stage (right) branches, the probabilities of getting a black or a white card remain the same as in the first stage. This is because the original card withdrawn was replaced in the bag before the second withdrawal.

Unit 11.3 Activity 8B: Probability trees

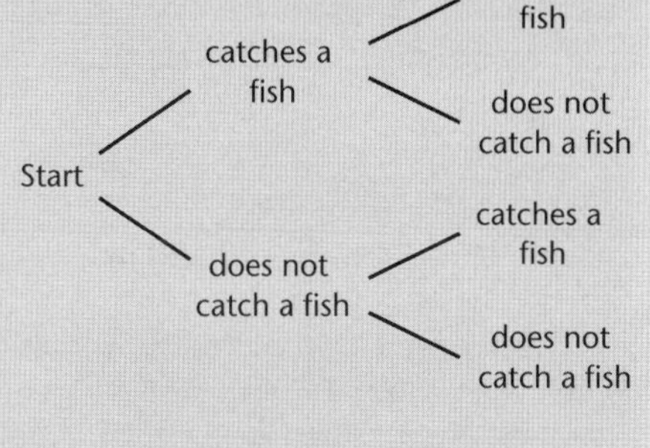

1. The tree diagram alongside describes the possible outcomes for a villager who goes fishing two days in a row.

a. If the probability of catching a fish is always $\frac{1}{3}$ whenever they go fishing, what is the probability that:

i. they don't catch a fish on either day.

ii. they are successful on the first day and not the second day.

iii. they catch a fish on at least one day.

b. Answer the same question as for (a) with the following changes. The probability of success is $\frac{1}{3}$ on the first day. If they are successful on the first day their chance of success on the second day is $\frac{3}{4}$. If they are unsuccessful on the first day their chance for success on the second day reduces to $\frac{1}{4}$.

2. A box contains equal numbers of 10-toea and 20-toea pieces. A coin is randomly withdrawn from the box and replaced. This is done three times.

a. Draw a tree diagram for this.

b. Use your diagram from (a) to answer the following:

i. What is the probability of getting three 10-toea pieces?

ii. What is the probability of two 20-toea pieces and one 10-toea piece in the three draws?

iii. What is the probability of getting at least one 20-toea piece?

iv. What is the probability of getting a 10-toea piece on each of the last two draws?

v. What is the probability of not getting three 20-toea pieces?

3. A lazy Grade 12 student has a three-question test which he has not studied for. Each answer is either true (T) or false (F). He guesses the answer for each question.

a. Draw a tree diagram for this.

b. Use your answer to (a) to answer the following:

i. What is the probability that he got all three questions right?

ii. What is the probability that he got all the questions wrong?

iii. What is the probability, if the correct answers were T, F, F, that he got the last two questions right?

iv. What is the probability that he will fail the test if a fail is fewer than half right?

4. A tourist travels to a tropical country. There are two types of mosquito in this country, type A and type B. Type A is twice as common as type B. The tourist gets bitten twice in one night.

a. What is the probability that she was bitten by type A on each occasion?

b. What is the probability that type B bit her first?

c. What is the probability that she was bitten at least once by type B?

d. What is the probability that she was bitten by type A first?

5. A box contains twelve blue balls, ten red balls and five white balls. A ball is withdrawn, and without replacing it another is withdrawn.

a. What is the probability that at least one of the balls was white?

b. What is the probability that both balls were blue?

c. What is the probability that the balls did not include a red one?

d. What is the probability that at least one of the balls was blue?

e. What is the probability that the second ball withdrawn was blue?

Unit 11.3 Managing Data

Topic 9: Probability, set notation and Venn diagrams

Topic 9 looks at the content under the heading 'Probability' (Syllabus p. 16), in particular:

- The probability of events.
- Mutually exclusive events.

The probability of one event and/or another event

If A and B are two events, then the event 'A **and** B' is the event that both A and B occur. The **intersection** symbol $\cap$ is frequently used to represent 'and', so $A \cap B$ is the event 'A and B'.

In set terms $A \cap B$ is the set of elements that set A and set B have in common.

On a **Venn diagram** $A \cap B$ is the shaded region at right.

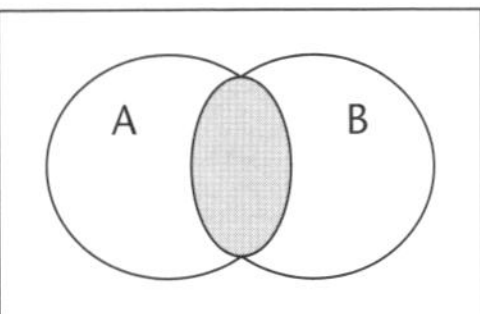

If A and B are two events, then the event 'A **or** B' is the event that either 'A **or** B or **both**' occur. The **union** symbol $\cup$ is frequently used to represent 'or', so $A \cup B$ is the event 'A or B'.

In set terms $(A \cup B)$ is all the elements 'A but not B' plus the elements 'B but not A' plus the elements 'A and B'.

On a Venn diagram $(A \cup B)$ is the shaded region at right.

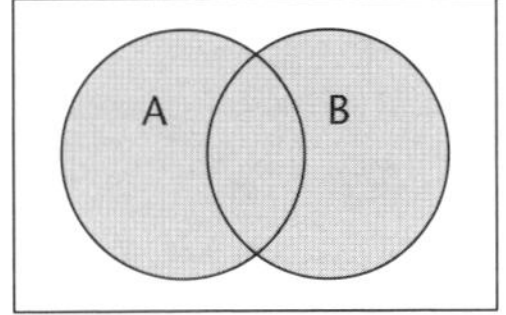

The following formula can be used to calculate the probability of 'A or B':

$$P(A \cup B) = P(A) + P(B) - P(A \cap B)$$

The derivation of this formula can be seen from Venn diagrams:

$A \cup B \quad = \quad A \quad + \quad B \quad - \quad A \cap B$

By adding sets A and B the set $A \cap B$ has been included twice, hence we need to subtract set $A \cap B$.

Example A

Event A has probability 0.5, event B has probability 0.4 and $P(A \cap B) = 0.3$.

Find the probability that A or B occur.

Solution

$P(A \text{ or } B) = P(A \cup B)$

$= P(A) + P(B) - P(A \cap B)$ [by formula]

$= 0.5 + 0.4 - 0.3$ [substituting given values]

$= 0.6$

Example B

In a group of people, the probability of having a deep voice is 0.3 and the probability of being tall is 0.6. The probability of being tall or having a deep voice is 0.7. Find the probability a person is tall and has a deep voice.

Solution

Let T be the event 'being tall' and D be the event 'having a deep voice'.

It is required to find P(D and T) = $P(D \cap T)$. Now

$P(D \cup T) = P(D) + P(T) - P(D \cap T)$	[formula]
$\therefore\ 0.7 = 0.3 + 0.6 - P(D \cap T)$	[substituting given values]
$0.7 = 0.9 - P(D \cap T)$	[simplifying]
$P(D \cap T) = 0.9 - 0.7$	[rearranging]
$\therefore\ P(D \cap T) = 0.2$	

Mutually exclusive events

> Two events are said to be **mutually exclusive** if the fact that one event occurs means that the other event does not (or cannot) occur.

On a Venn diagram two mutually exclusive events, A and B, look like:

The events have no elements in common.

$A \cap B = \varnothing$, the null set.

> For two mutually exclusive events, A and B,
> $P(A \cup B) = P(A) + P(B)$

Example C

Consider rolling a fair die and recording the number on the uppermost side.

Event A is the event 'a number greater than 3' and event B is the event 'a number less than 3'.

A = {4, 5, 6} and B = {1, 2}

Either event A occurs or event B occurs; they cannot occur at the same time. They have no elements in common.

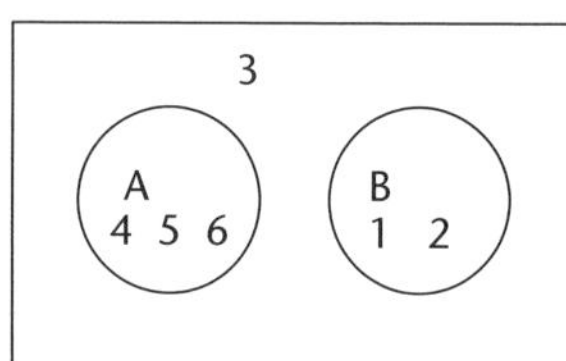

Unit 11.3 Activity 9A: Probabilities of one event and/or another event

1. P(A) = 0.8, P(B) = 0.7 and P(A ∩ B) = 0.58. Find P(A ∪ B).

2. P(C) = 0.68, P(D) = 0.35 and P(C ∩ D) = 0.26. Find P(C ∪ D).

3. In a city, the probability a person drives to work is 0.4 and the probability of living in suburb A is 0.3. The probability of driving to work and living in suburb A is 0.2. Find the probablity of living in suburb A or driving to work.

4. The probability John has toast for breakfast is 0.45 and the probability Janet has toast for breakfast is 0.5. The probability of them both having toast for breakfast is 0.4.

- **a.** Find the probability Janet or John has toast for breakfast.
- **b.** Find the probability of only one of them having toast for breakfast.

5. P(M) = 0.45, P(N) = 0.65 and P(M ∪ N) = 0.73. Find P(M ∩ N).

6. P(K) = 0.77, P(L) = 0.44 and P(K ∪ L) = 0.66. Find P(K ∩ L).

7. P(A) = 0.5, P(A ∪ B) = 0.75, P(A ∩ B) = 0.35. Find P(B).

8. The tree diagram shows the probabilities of events A, B, C and D. Find:

- **a.** P(A)
- **b.** P(B)
- **c.** P(A ∩ C)
- **d.** P(B ∩ C)
- **e.** P(C)

Hint: C is made up of A ∩ C and B ∩ C.

- **f.** P(A ∪ C)
- **g.** P(B ∪ D)

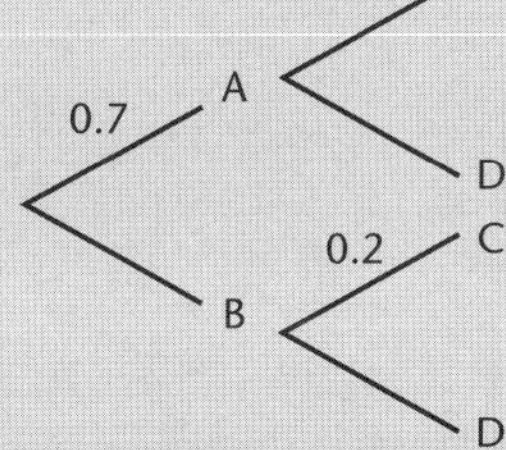

9. A fair dice is rolled.
A is the event 'a number greater than 3'; B is the event 'a number less than 4'.
C is the event 'the number 5'; D is the event 'an even number'.
E is the event 'an odd number'; F is the event 'the number 3'.

- **a.** List the elements for each of the events A to F.
- **b.** State whether or not the following pairs of events are mutually exclusive:
  - **i.** A and B
  - **ii.** A and C
  - **iii.** B and D
  - **iv.** F and B
  - **v.** A and F
  - **vi.** E and F
- **c.** Find:
  - **i.** P(A)
  - **ii.** P(B)

- **iii.** P(C)
- **iv.** P(F)
- **v.** P(A ∪ B)
- **vi.** P(C ∪ E)
- **vii.** P(C ∪ D)
- **viii.** P(E ∪ F)

10. A card is selected at random from a pack of 52 playing cards. S is the event 'the card is a spade'. H is the event 'the card is a heart'. A is the event 'the card is an ace'. R is the event 'the card is red'.

- **a.** State whether the following pairs of events are mutually exclusive:
  - **i.** S and H
  - **ii.** S and A
  - **iii.** H and R
  - **iv.** S and R
- **b.** Find:
  - **i.** P(S)
  - **ii.** P(H)
  - **iii.** P(A)
  - **iv.** P(R)
- **c.** Find:
  - **i.** P(A ∩ R)
  - **ii.** P(S ∪ H)
  - **iii.** P(R ∩ H)
  - **iv.** P(R ∩ S)
  - **v.** P(A ∪ R)
  - **vi.** P(S ∩ A)
  - **vii.** P(R ∪ H)

Unit 11.3 Managing Data
Topic 10: Independent events

Topic 10 looks at the content under the heading 'Probability' (Syllabus p. 16), in particular:

- Classify and calculate events as independent and dependent.

In the context of probability, independence has a precise meaning and needs to be carefully defined: two events, A and B, are **independent** if the occurrence of one has absolutely no effect on the occurrence of the other.

Example A

1. Suppose a dice is being rolled and a coin is being flipped. If event A is 'a number less than 3 occurs' and event B is 'heads', then A and B are clearly independent.
2. Joe lives in the USA; Joan lives in the UK; neither knows the other.
   E is the event: 'Joe eats cornflakes on July 14',
   F is the event: 'Joan eats muesli on July 14'.
   E and F are obviously independent events except in the most contrived circumstances.

An important property of two independent events is that the probability of events both occurring is the product of their probabilities.

$$P(A \cap B) = P(A) \times P(B) \text{ if A and B are independent events}$$

Example B

A is the event that a number less than 3 occurs on the first roll of a dice. B is the event that a number greater than 3 occurs on the second roll of a dice. Find the probability that both A and B occur if a dice is rolled twice.

Solution

$P(A) = \frac{1}{3}$, $P(B) = \frac{1}{2}$. Require $P(A \cap B)$.

A and B are independent (a number less than 3 occurring on the first roll will have absolutely no effect on whether a number greater than 3 occurs on the second roll).

$\therefore$ probability of both A and B occurring is $\frac{1}{3} \times \frac{1}{2} = \frac{1}{6}$ [using the formula]

ie $P(A \cap B) = \frac{1}{6}$

Unit 11.3 Activity 10A: Independent events

1. A bag contains two blue balls and three red balls. A ball is removed, its colour noted then the ball is replaced in the bag. This is repeated once. Which of the following are independent pairs of events?
   - **a.** (First ball is blue), (second is red)
   - **b.** (First ball is blue), (second is blue)
   - **c.** (First ball is blue), (first ball is red)
   - **d.** (First ball is blue, second is red), (second ball is blue)
   - **e.** (First ball is blue, second is red), (second ball is red)

2. Answer the questions from (1) if the first ball is removed and not replaced.
3. For each of the following pairs of events write down a situation in which they would be independent and a situation when they would not be independent.
   a. James is tall; Dorcas is short.
   b. Sabina is happy; Samuel is sad.
   c. The dog wags his tail; David has a bone.
   d. Sila passes her exam; Sarah passes her exam.
   e. It rains on Saturday; it rains on Sunday.
4. **a.** If you select a random two-digit number, which of the following are independent pairs of events?
      i. First digit less than 3; second digit greater than 7.
      ii. First digit is prime; second digit is prime.
      iii. First digit is less than 4; sum of digits is 7.
      iv. First digit is a factor of the second; second digit is greater than 5.
   b. For each pair of events which are independent, find the probability of both occurring.
5. The probability of event A is 0.6. The probability of event B is 0.43. The probability of events A *and* B occurring is 0.37. Explain why A and B are not independent.

Conditional probability

A **conditional probability** is a probability worked out using a restricted sample space. For example, the probability of randomly selecting a person with grey hair from the general population may be as low as 10%. However, suppose the population were restricted to persons over the age of 60. The probability of randomly selecting a person with grey hair from this restricted population would be more likely to be around 90%.

Example C

Assume that 15% of the population is aged 60 or more. Assume that 80% of those over 60 have grey hair. Assume that 5% of those under 60 have grey hair. Represent this on a tree diagram.

Solution

Let the event 'aged 60 or more' be represented by 60^+. The event 'aged less than 60' is represented by 60^-. Let the event 'has grey hair' be G. The event G′ represents 'does not have grey hair'. The tree appears alongside.

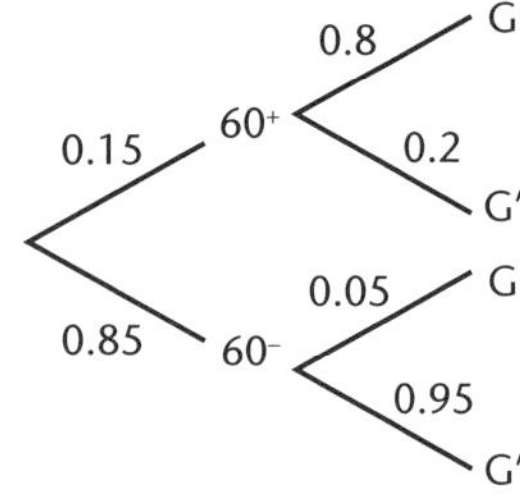

You need to have a good understanding of the meanings of the probabilities on a tree diagram and how to use them. Example D discusses the tree diagram from Example C in more detail.

Example D

Using the tree diagram from Example C, the following aspects should be noted:

1.

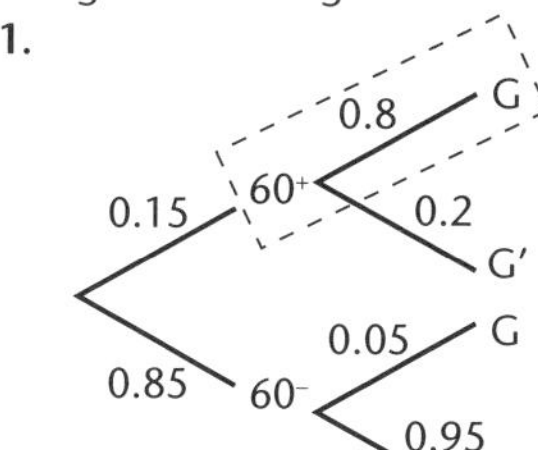

This says the conditional probability of 'someone having grey hair if they are 60 or more' is 0.8.

2.

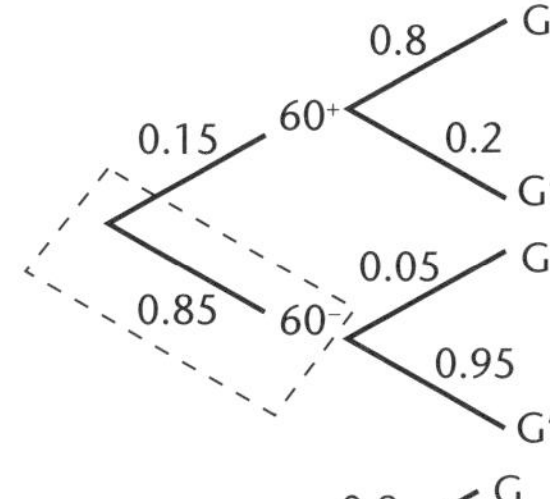

This says the probability of 'someone being under 60' is 0.85.

3.

0.15
60⁺
0.8
G
0.2
G′
0.85
60⁻
0.05
G
0.95
G′

This says the probability of a randomly selected person 'being 60 or more and not having grey hair' is $0.15 \times 0.2 = 0.03$.

Unit 11.3 Activity 10B: Conditional probability

1.

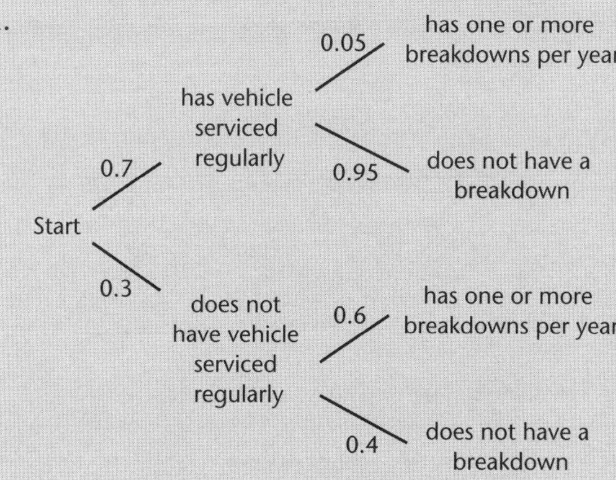

This tree diagram shows the results of a recent survey of users of a certain type of car.

a. What is the probability someone with this sort of car does not have their car serviced regularly?
b. What is the probability someone with this sort of car does not have a breakdown during the year if they have the vehicle serviced regularly?
c. What is the probability someone with this sort of car has one or more breakdowns during the year if they do not have their car serviced regularly?
d. What is the probability someone with this sort of car has their vehicle serviced regularly and has had one or more breakdowns during the year?

2. Copy this tree diagram and put on the six missing pieces of information relating to this diagram. Note A′ is the event 'A does not occur' etc.

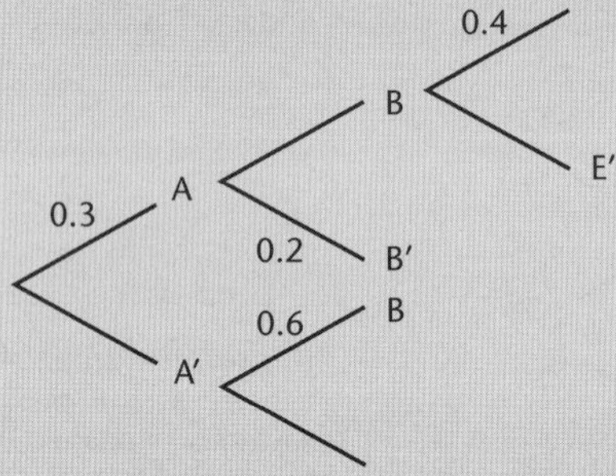

a. What is the probability of B′ occurring if A′ has occurred?
b. What is the probability of E occurring if A and B have occurred?
c. What is the probability of A′ and B both occurring?
d. What is the probability of A, B, and E′ all occurring?

3. In a school all students study maths and physics. The probability of a student passing the maths exam is 0.7. If a student passes the maths exam the probability of passing the physics exam is 0.8. If a student fails the maths exam the probability of passing the physics exam is 0.1.
a. Draw a probability tree showing this information.
b. What is the probability a student does not pass physics if they don't pass maths?
c. What is the probability a randomly selected student passes physics?

4. Use the tree diagram to find the probability of:

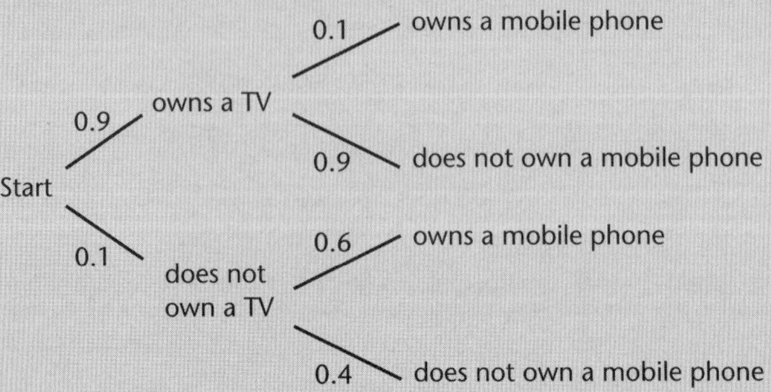

a. owning a mobile phone if you already own a TV.
b. not owning a mobile phone if you do not own a TV.
c. owning a TV.
d. owning a TV and not owning a mobile phone.
e. owning a mobile phone and not owning a TV.

5. The probability of studying maths if you study chemistry is 0.85. The probability of studying chemistry is 0.60.
a. Find the probability of studying chemistry and maths.
b. If the probability of studying maths is 0.67, find the probability of studying maths if you don't study chemistry.

6. The probability of playing soccer and rugby is 0.32. The probability of playing soccer and not playing rugby is 0.13.
a. What is the probability of playing soccer?
b. What is the probability of playing rugby if you play soccer?
c. If the probability of not playing rugby is 0.31, what is the probability of playing neither soccer nor rugby?
d. What is the probability of not playing rugby if you don't play soccer?

Expected number or value

If the probability of an event occuring is p and there are N trials, then the **expected number** of occurrences is Np. This expected number is also called the expected value.

Example E

The probability of a gambler winning on any occasion he plays a poker machine is 0.35. If the gambler plays a machine on 180 occasions, what is the expected number of times he will win?

Solution

Probability of winning, $p = 0.35$
Number of trials, $N = 180$

Expected number of wins $= Np$ [formula]
$= 180 \times 0.35$ [substituting]
$= 63$

Unit 11.3 Activity 10C: Probability and expected values

1. In New Zealand cars driven on the road are required to have a current warrant of fitness and be registered. A recent newspaper article contained the following statement: '80% of cars on the road in New Zealand have a current Warrant of Fitness and 90% are currently registered'.

   A police officer parked on the main street in his town and stopped all passing cars. He issued tickets to motorists found breaching either of these requirements.

   [Assume these statistics are correct, and that a car's having a warrant of fitness is independent of its being registered (an important assumption which needs to be stated).]

   a. What is the probability that a car which is stopped has no warrant of fitness and is not registered?

   b. Calculate the probability that the next driver will be given a ticket.

   c. Of the next 50 cars stopped, what is the expected number of drivers who will be issued a ticket?

2. Mathias is collecting a set of four sports cards from Gutpela cereal packets (there is one card in each packet). His younger sister, Nori, is collecting a set of three plastic dolls from Mobeta cereal packets (one doll per packet).

   Over any three-week period, Nori and Mathias's mother buys Gutpela cereal in two of the weeks and Mobeta cereal in the other week.

   a. Mathias has already collected three different sports cards. What is the probability that when his mother buys the next packet of Gutpela cereal, he obtains the card he needs to complete the set?

   b. Nori also needs one more doll to complete her collection. Find the probability that either Nori or Mathias will complete their set when their mother buys the next cereal packet.

3. The probability of owning an encyclopedia if you own a dictionary is 0.25. The probability of owning a dictionary is 0.93.

   a. Show this information on a tree diagram.

   b. Find the probability of owning a dictionary and an encyclopedia.

   c. If the probability of owning an encyclopedia if you do not own a dictionary is 0.3, find the probability of owning an encyclopedia.

4. The probability of studying history if you study geography is 0.75. The probability of studying geography is 0.5.

   a. Draw a tree diagram for this information.

   b. Find the probability of studying history and geography.

   c. If the probability of not studying history if you do not study geography is 0.6, find the probability of studying neither history nor geography.

Unit 11.4 Geometry

Topic 1: Triangles

Unit 11.4 focuses on mathematics that deals with properties of triangles and circles. This first Topic begins with a brief introduction to polygons and then moves on to triangles (three-sided polygons). It covers:

- Properties of polygons.
- Different types of triangles.

Polygons

A **polygon** is a closed *many-sided* figure whose sides are straight line segments. The names of some important polygons are shown in the table alongside.

Number of sides	Name of Polygon
3	triangle
4	quadrilateral
5	pentagon
6	hexagon
8	octagon
10	decagon
12	dodecagon

A **regular polygon** has all sides of equal length and all angles of equal size.

In a polygon, a **diagonal** is any line segment that joins two **vertices** (corners) and is not a side.

Example A

The pentagon PQRST has 5 diagonals: diagonals PR and QT are shown.

The remaining diagonals are the line segments PS, QS and RT (not shown).

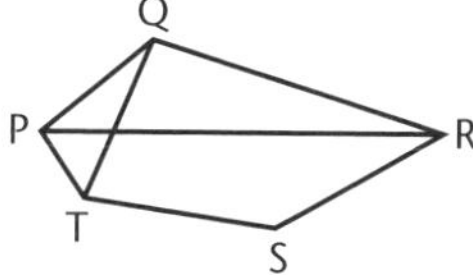

The angles inside a polygon are called **interior angles**. **Exterior angles** are formed by extending the sides of the polygon. An interior angle and an exterior angle are adjacent angles on a straight line (they add to 180°).

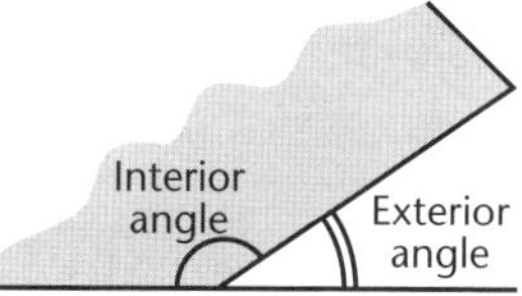

The interior and exterior angles of a **triangle**, **quadrilateral**, and **hexagon** are shown below:

Triangle	Quadrilateral	Hexagon
Interior angles: *a*, *b*, *c* Exterior angles: *p*, *q*, *r*	Interior angles: *a*, *b*, *c*, *d* Exterior angles: *p*, *q*, *r*, *s*	Interior angles: *a*, *b*, *c*, *d*, *e*, *f* Exterior angles: *p*, *q*, *r*, *s*, *t*, *u*

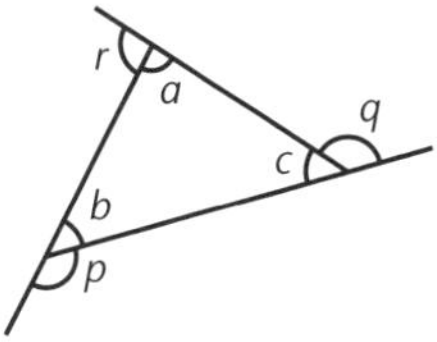

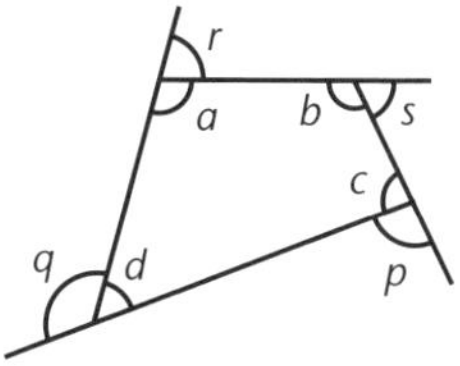

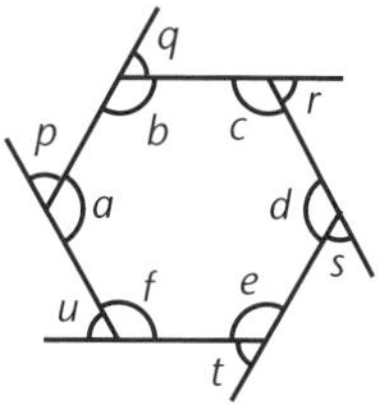

Triangles

A **triangle** is a three-sided polygon. Some of the important geometrical properties of a triangle are discussed below.

The sum of the interior angles of any triangle is 180°. (Abbreviated to: ∠ sum Δ.)

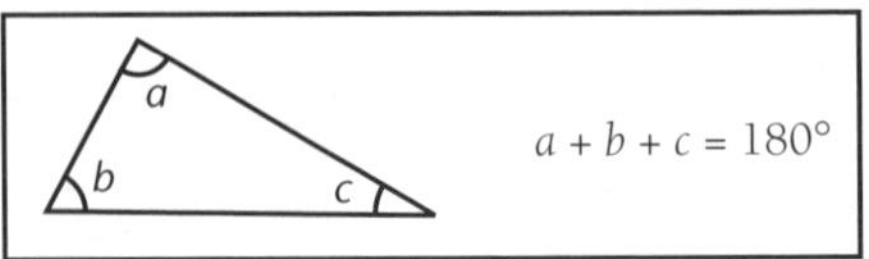

Example B

The size of the angle labelled x is given by:

$x + 70 + 30 = 180$ (∠ sum Δ)

$x + 100 = 180$

$x = 80°$

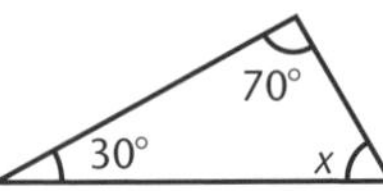

The size of an exterior angle of any triangle is equal to the sum of the two opposite interior angles. (Abbreviated to: ext ∠ Δ.)

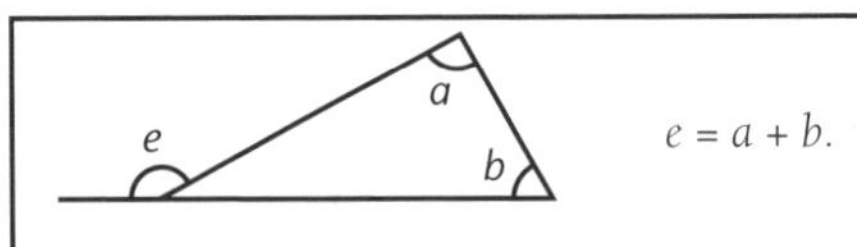

Example C

The angle x shown in the diagram is given by:

$x + 80 = 150$ (ext ∠ Δ)

$x = 70°$

Types of triangle

Triangles can be classified into different types according to the relative lengths of the sides and the sizes of the interior angles. Most of these triangles have different **axes of symmetry**.

Equilateral triangle	Isosceles triangle	Scalene triangle
60° 60° 60°	a b b	b a c
Three sides of equal length. Each interior angle is 60°. Three axes of symmetry.	Two equal sides, other side called **base**. Two equal angles called **base angles**. One axis of symmetry.	No equal sides. No equal angles. No axes of symmetry.

Note: **1.** Matching marks on sides indicate line segments of equal length.

2. The longest side of a triangle is always opposite the largest angle, and the shortest side is always opposite the smallest angle.

Example D

The size of the angle labelled x is given by:

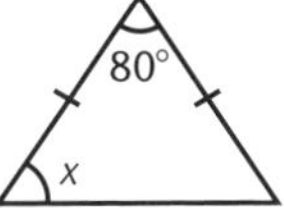

$x + x + 80 = 180$ (sum ∠s isos Δ)

$2x = 100$ [subtracting 80]

$x = 50°$

Note: Stating that the triangle is an 'isos Δ' in the solution shows that the unmarked angle is equal to x (since the base angles of an isosceles triangle are equal).

Triangles can also be classified according to the size of the angles:

Acute-angled triangle	Right-angled triangle	Obtuse-angled triangle
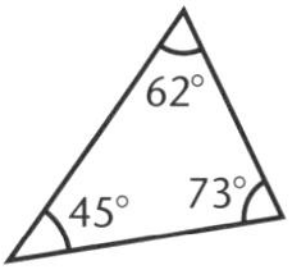 62°, 45°, 73°	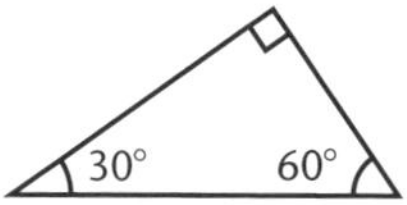 30°, 60°	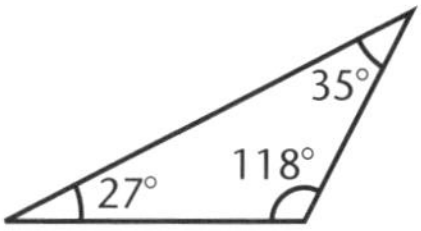 35°, 27°, 118°
Each angle is less than 90°.	One angle is equal to 90°.	One angle is greater than 90°.

Unit 11.4 Activity 1A: Angles and triangles

1. Find, with reasons, the size of the labelled angles.

a. a, 46°, 42°

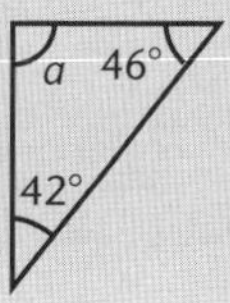

b. 88°, b

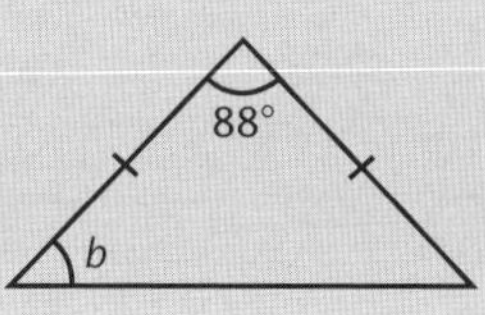

c. c, 54°, 128°

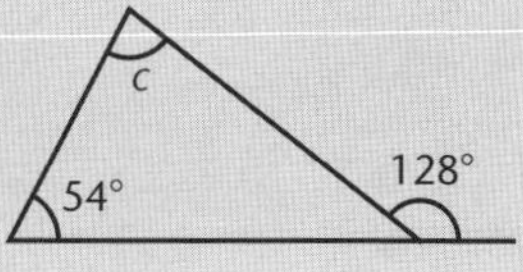

d. e, d, 55°

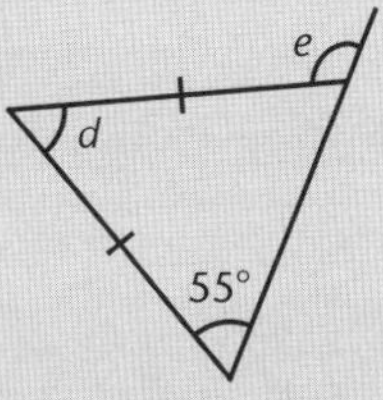

e. 56°, g, f, 43°, 37°, 51°

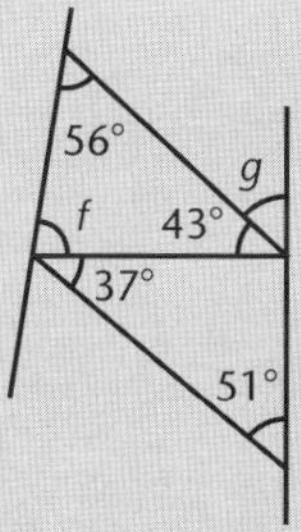

f. 106°, i, h, 84°

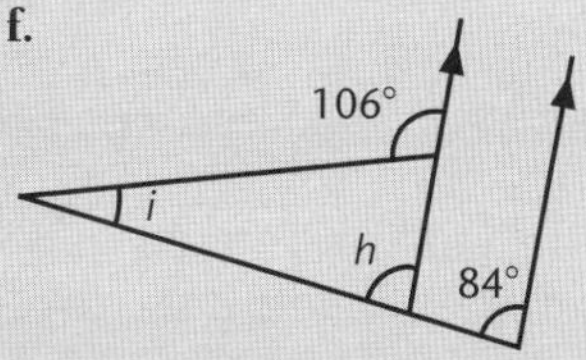

g. d, 40°, e

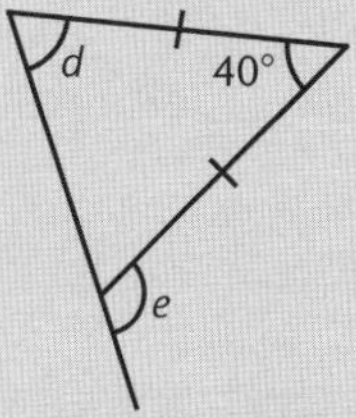

h. f, g, 28°, 81°, 32°, 47°

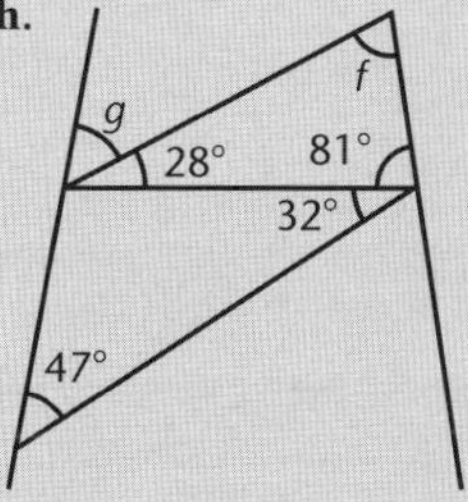

i. 108°, h, i, 40°

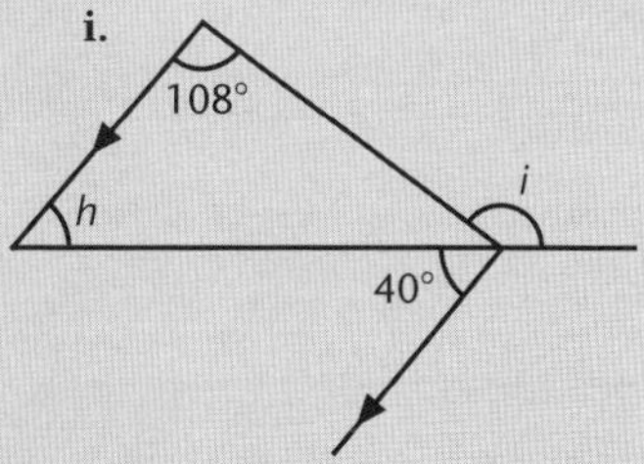

2. Prove that PR is parallel to ST, given that length PQ = length QR and the angle sizes are as marked on the diagram (// means 'is parallel to').

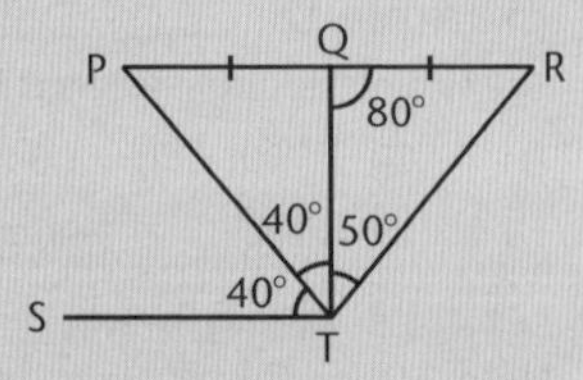

Copy the following and fill in the blanks to complete the proof.

$\angle$QRT = 50° (__________)

∴ length QR = length _____ (ΔQRT is isosceles)

∴ $\angle$QPT = _____° (base $\angle$s isos Δ)

∴ PR // ST (_______________)

3. AB is parallel to CD. Lengths AC, AD and BD are equal.

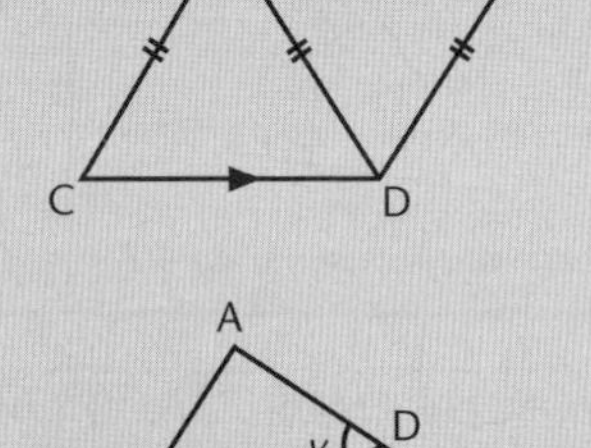

Copy the following and fill in the blanks to prove that $\angle$ACD = $\angle$ABD.

Let $\angle$ACD = x

$\angle$ADC = __ (base $\angle$s isos Δ)

$\angle$BAD = __ (alt $\angle$s // lines)

$\angle$ABD = x (_______________)

4. ABC and ADE are straight lines and $a = b$.

Prove that $x = y$.

Hint: Let $\angle$BFC = c.

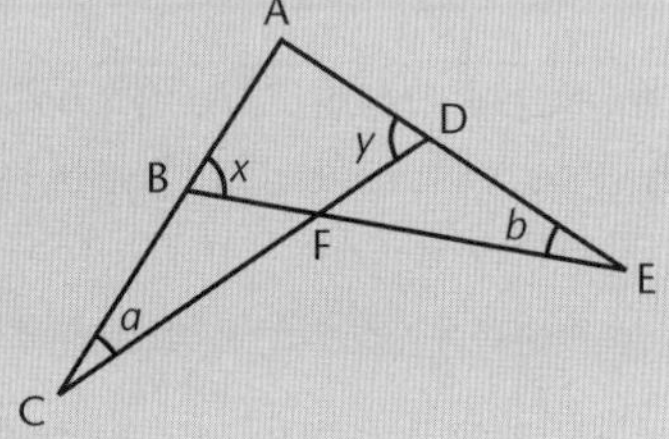

Unit 11.4 Geometry

Topic 2: Similar triangles

Unit 11.4 focuses on mathematics that deals with properties of triangles and circles. Topic 2 looks at similar triangles (Syllabus p. 17):

- Special and similar triangles.

Introduction

Many construction principles are based on similar triangles and recognising that two triangles are similar enables us to use ratio, or **scale factor**, to calculate lengths.

There are three ways that we can establish that a pair of triangles are similar.
Two triangles are similar if:

1. All their corresponding sides are in the same ratio (SSS).

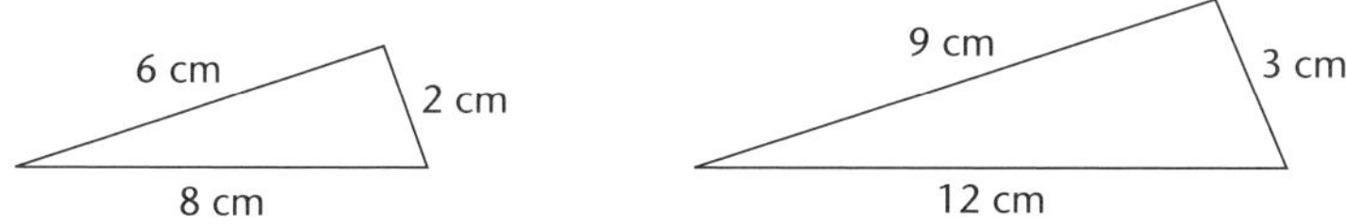

The corresponding sides in this pair of triangles are in the ratio 2 : 3.

2. All the **corresponding angles** are of equal size (AAA).

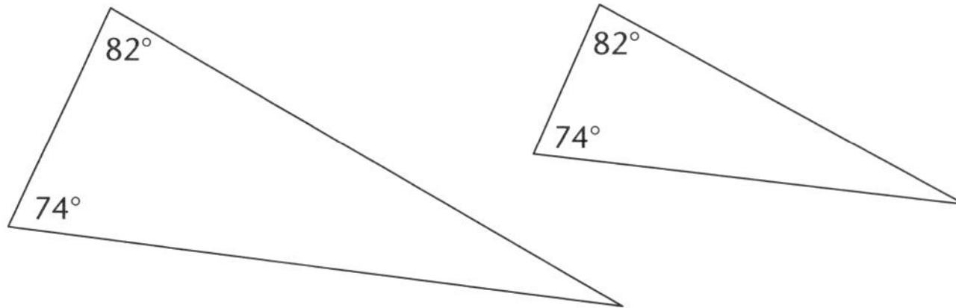

Note: If triangles show two angles that are the same size then the third angles must also be the same size because we can calculate the size of the third angles using the rule:

The sum of the three angles of a triangle is 180°.

For both of the triangles above the unknown angle is $180° - 82° - 74° = 24°$

3. A corresponding pair of **adjacent sides** in the triangles are in the same ratio and the angle between these adjacent sides (the **included angle**) is the same size (SAS).
These triangles have sides in the ratio 4 : 3.

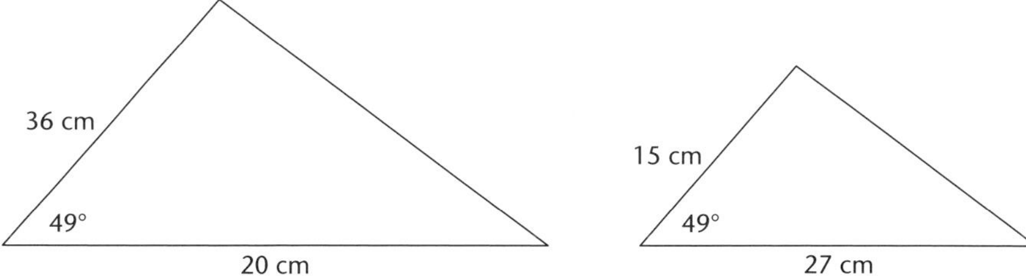

Often we are told, or it is obvious, that a pair of triangles are similar, and in these cases we can use the fact that the corresponding sides are in the same ratio to solve problems.

Example A

Q. Triangles ABC and PQR are similar triangles.

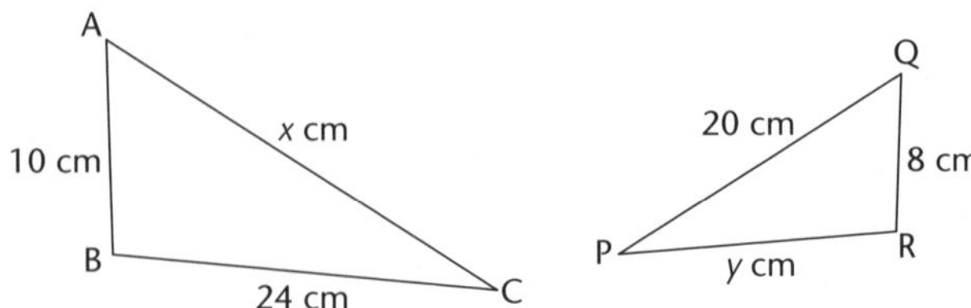

1. Which side in triangle PQR corresponds to side AB in triangle PQR?
2. Find the ratio of the sides of these triangles.
3. Find the value of *x*.
4. Find the value of *y*.

A. **a.** AB is the shortest side in triangle ABC. So AB corresponds to QR, the shortest side in triangle PQR.

b. The ratio of side lengths is AB : QR = 10 : 8 = 5 : 4 (scale factor $\frac{4}{5}$).

c. $x : 20 = 5 : 4$

$$\frac{x}{20} = \frac{5}{4}$$

$$x = \frac{5 \times 20}{4} = 25$$

d. $24 : y = 5 : 4$

$$\frac{y}{24} = \frac{4}{5}$$

$$y = \frac{4 \times 24}{5} = 19.2$$

Example B

Q. Find *x* in the following diagram:

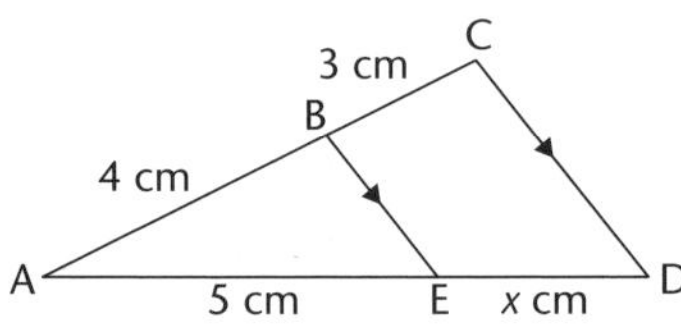

A. The triangles ABE and ACD are similar because their angles are the same; they have an angle at A in common and ∠ABE and ∠ACD are a pair of corresponding angles so are of equal size.

Drawing the triangles separately:

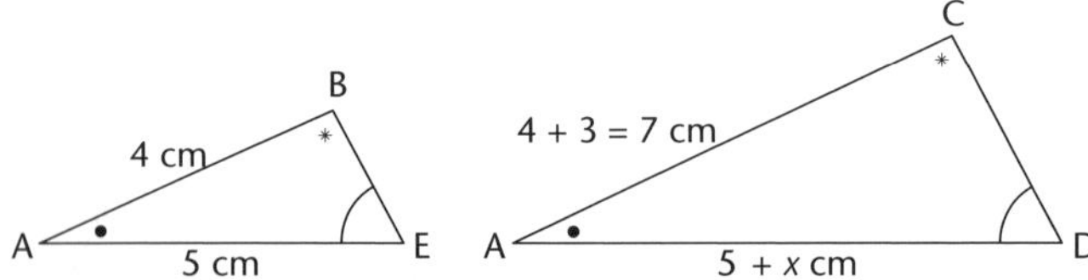

The value of x can be found using the fact that corresponding sides are in the same ratio.

AC and AB are corresponding sides AC : AB = 7:4

AD and AE are corresponding sides AD : AE = (5 + x) : 5

The sides are in the same ratio hence:

$$(5+x):5 = 7:4$$
$$\frac{5+x}{5} = \frac{7}{4}$$
$$4(5+x) = 5\times 7$$
$$20+4x = 35$$
$$4x = 15$$
$$x = 3.75$$

Unit 11.4 Activity 2A: Similar triangles

1. There is a pair of similar triangles in each of the following sets of three triangles (labelled A, B and C). State which pair is similar. The triangles are not drawn to scale.

a.

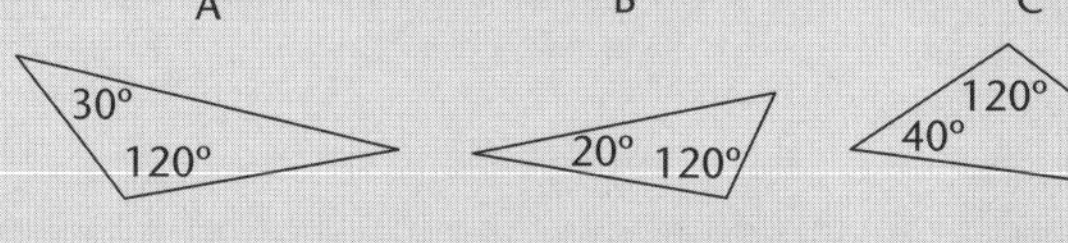

b.

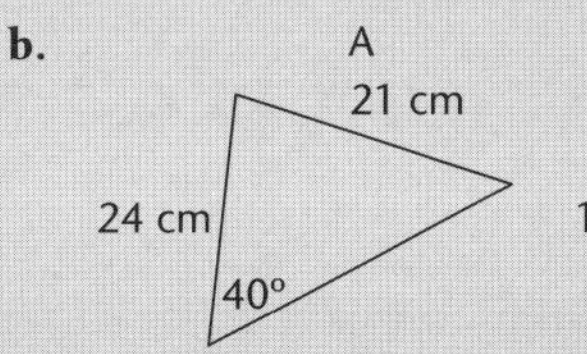

B
12 cm
40°
14 cm

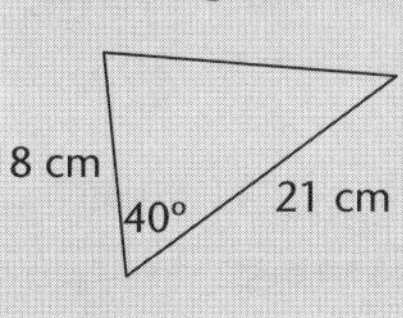

c.

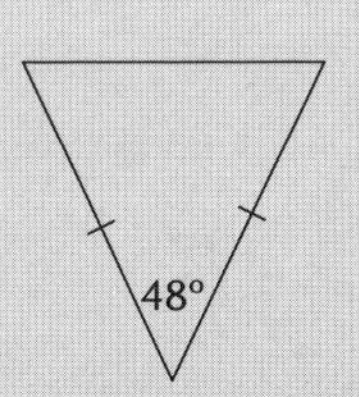

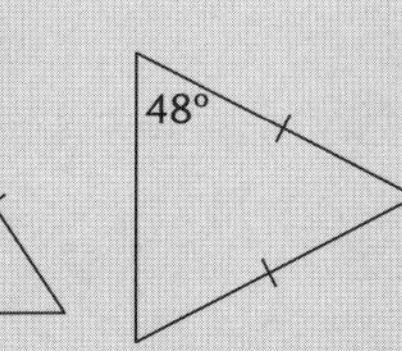

C
48°

d.

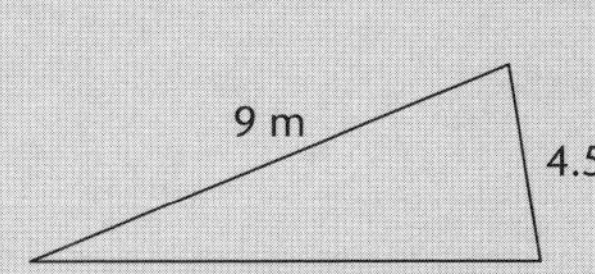

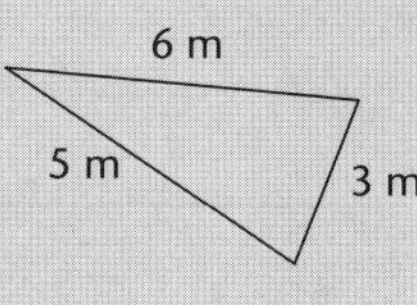

C
4.5 m
10 m
9 m

2. Show that the following pairs of triangles are similar triangles.

a.

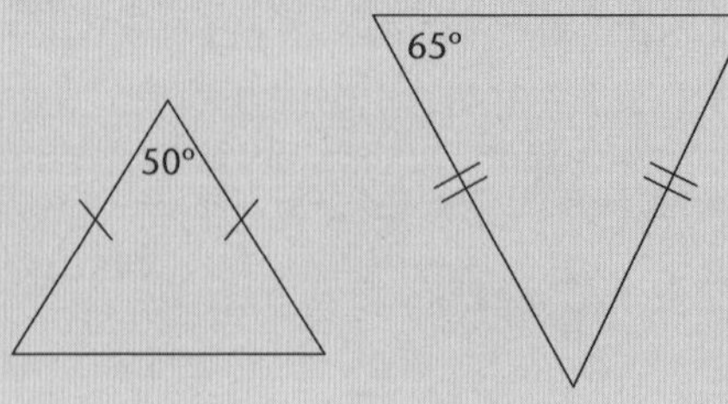

b.

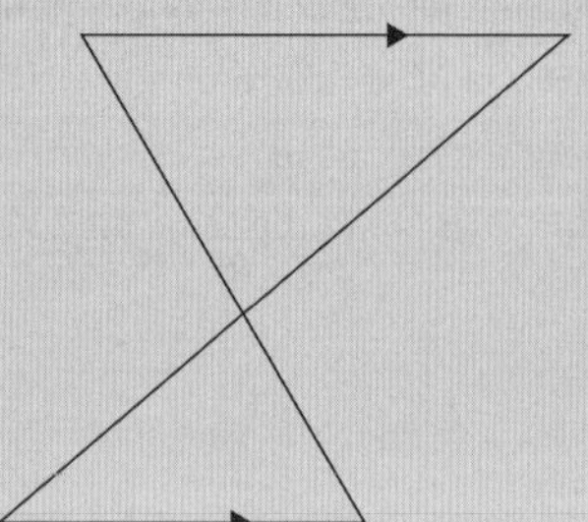

c.

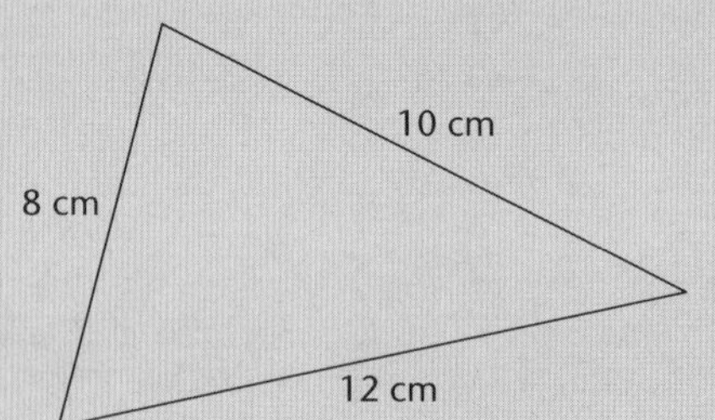

d.

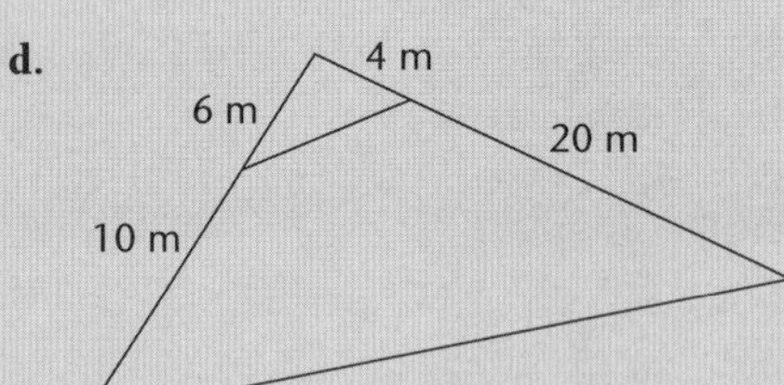

3. For each of the following triangles:

i. Construct a triangle similar to the given triangle. Use a ruler, compass or protractor if necessary.

ii. Mark on your triangle the angles or lengths that show that it is similar to the given triangle.

a.

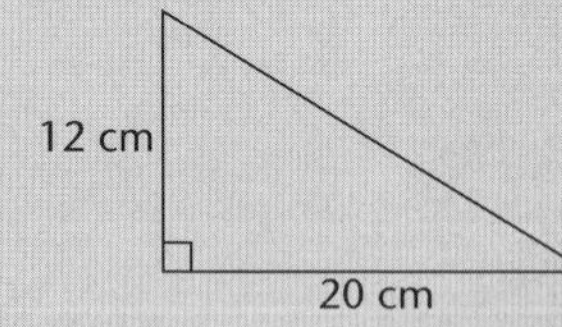

b.

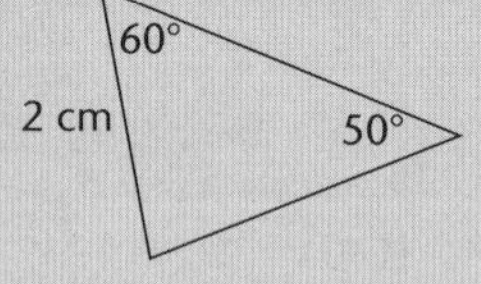

c.

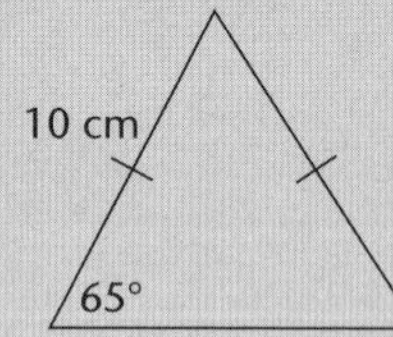

4. Find x and y in the similar triangles below.

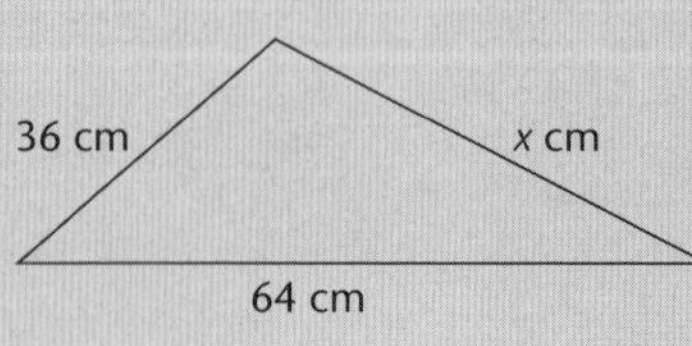

5. Find the length of the sides marked with pronumerals in the following pairs of similar triangles:

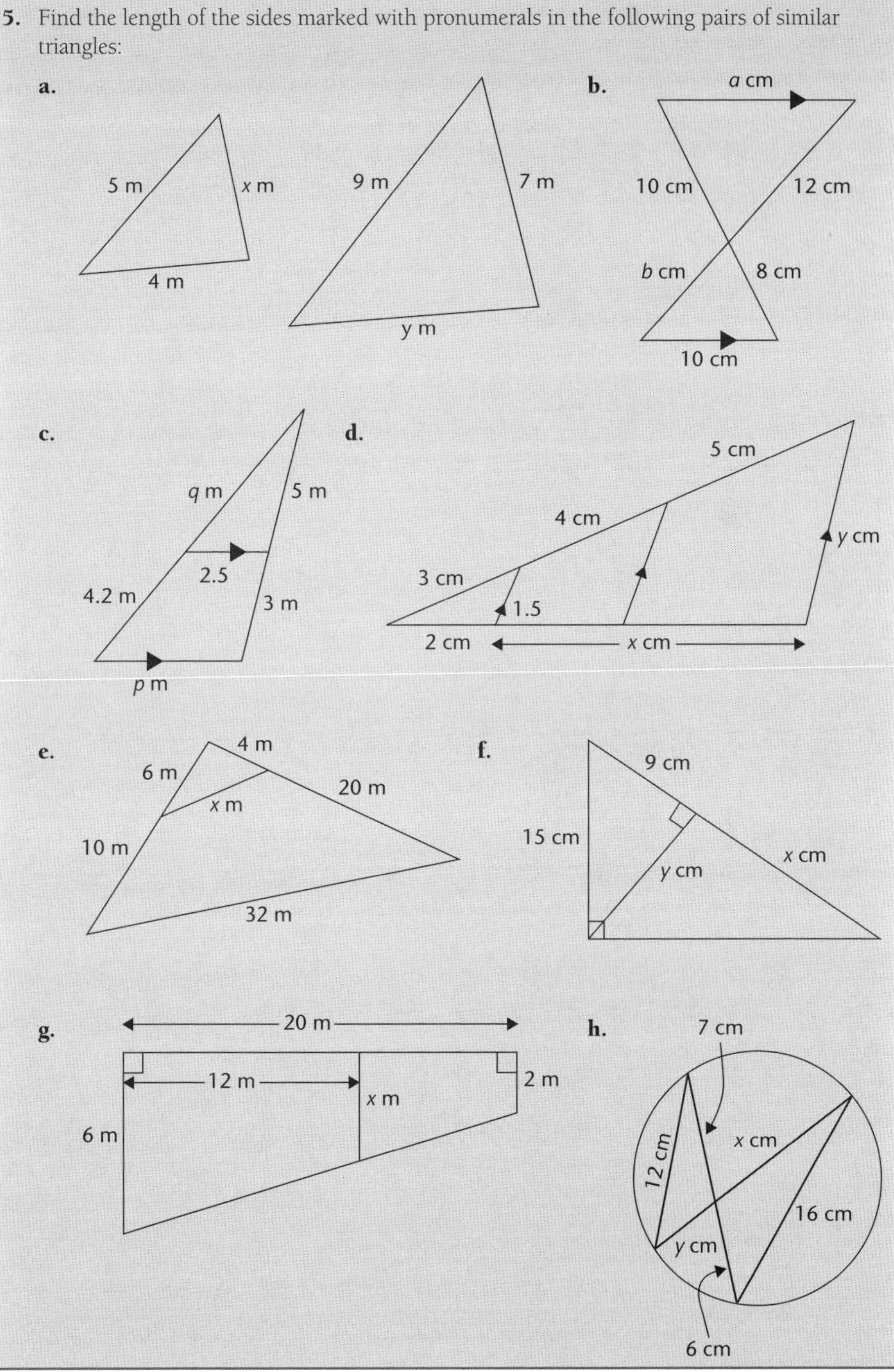

Unit 11.4 Geometry
Topic 3: Congruent triangles

Unit 11.4 focuses on mathematics that deals with properties of triangles and circles. Topic 2 looks at congruent triangles (Syllabus p. 17):

- Applying tests for congruent triangles.
- Constructing, investigating and providing congruency.

Congruent triangles

Congruent triangles are identical in size and shape; their angles are the same size and their corresponding sides are the same length.

Two triangles are congruent if:

- All three corresponding sides are equal; this is called the side-side-side rule (SSS).

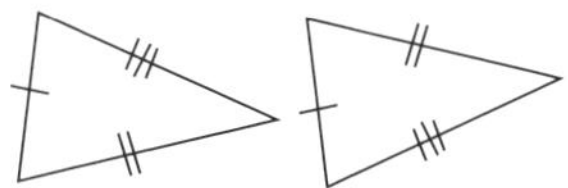

- Two sides and the included angle are equal; this is called the side-angle-side rule (SAS).

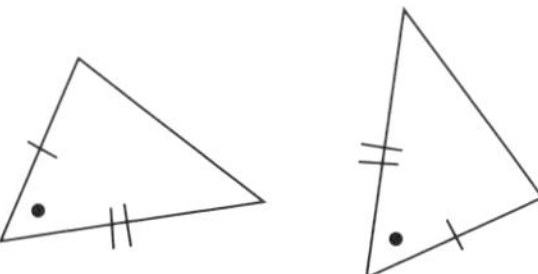

- Two angles and the corresponding side are equal (ASA).

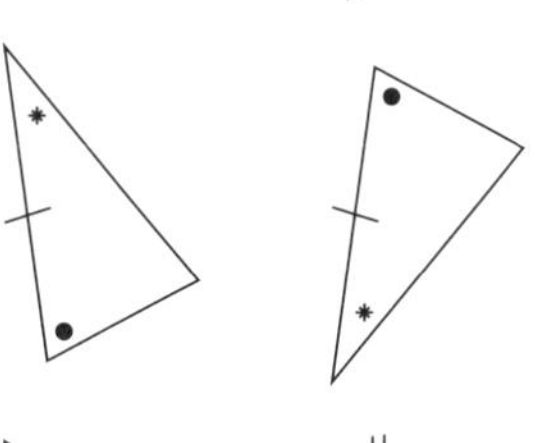

- They are right-angled with equal hypotenuses and one pair of corresponding sides is equal (RHS).

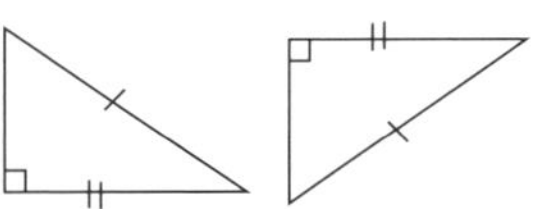

Constructing a triangle that is congruent to a given triangle

1. Where the lengths of the three sides are known:

i. Measure the length of one of the sides (AB) and rule a line of this length.

ii. Use a compass to measure and mark an **arc** for each of the other two sides.

iii. Join the ends of the drawn side to the point where the arcs meet.

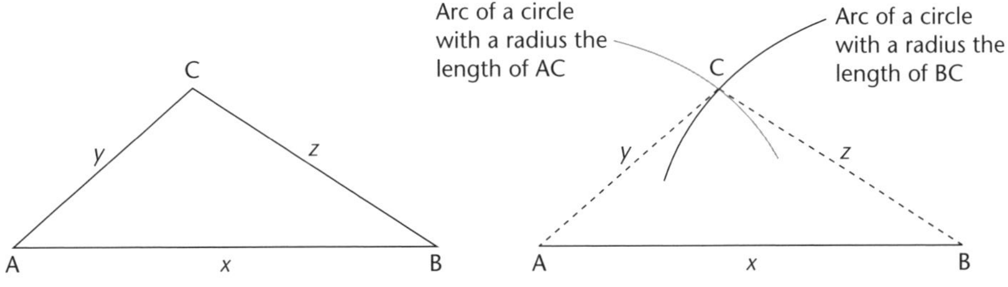

2. Where two angles and one side are known:
    - i. Measure the length of the known side (AB) and rule a line of this length.
    - ii. Use a protractor to find the size of one of the angles (∠CAB). Draw a straight line at this angle to the drawn line (AB).
    - iii. Repeat ii. for the other angle (∠CBA).
    - iv. The **vertex** at C is the intersection of the lines from **ii.** and **iii.**

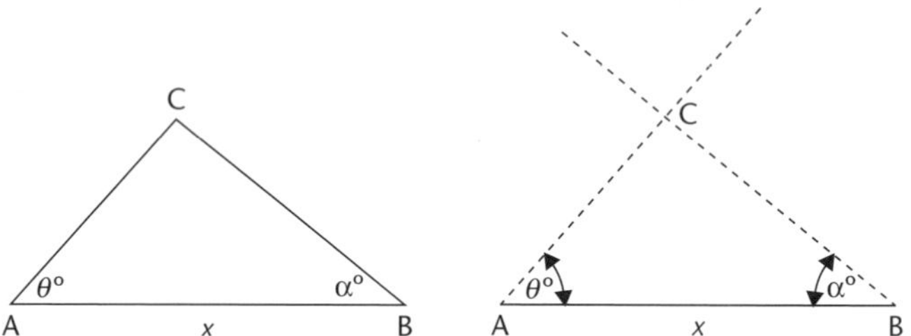

3. Where two sides and the included angle are known:
    - i. Measure the length of one of the known sides (AB) and rule a line of this length (x).
    - ii. Use a protractor to find the size of the included angle (∠CAB). Draw a straight line at this angle to the drawn line (AB). Mark the point that is the length of AC along this line.
    - iii. Join vertex B to the marked point.

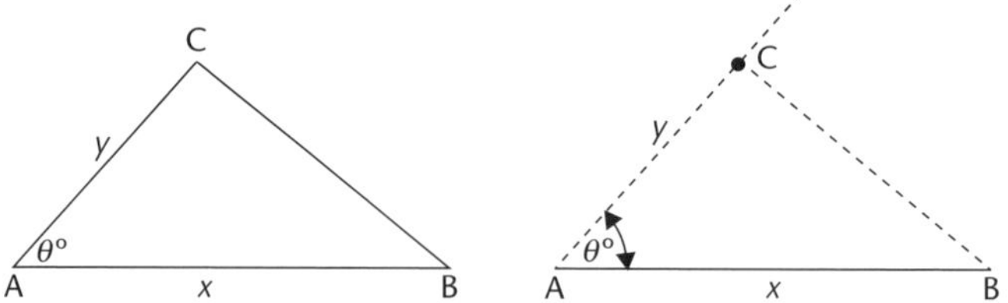

4. Where the hypotenuse and one other side in a right-angled triangle are known:
    - i. Measure the length of the known side (AB) and rule a line of this length (x).
    - ii. Measure an angle of 90° at B. Rule a line from B at this angle.
    - iii. Use a compass (point at A) to draw an arc of a circle with a radius the same length as the hypotenuse.
    - iv. The point where this arc cuts the line drawn from B is the vertex C.

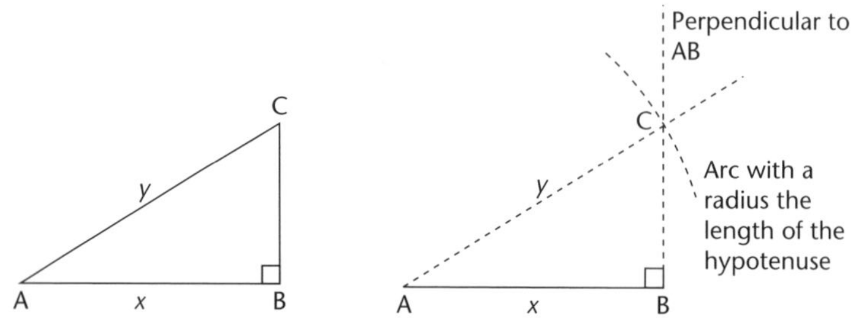

Unit 11.4 Activity 3A: Congruent triangles

1. Which of the triangles given below is congruent to triangle ABC? (Triangles are not drawn to scale.)

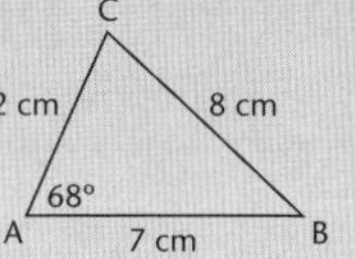

a.

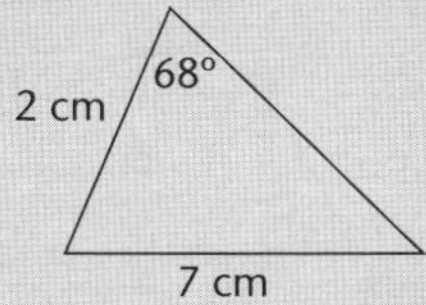

b.

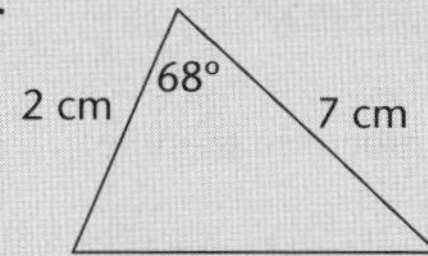

c.

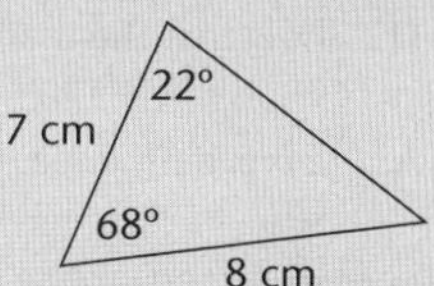

2. Determine the pairs of congruent triangles from those drawn below.

a.

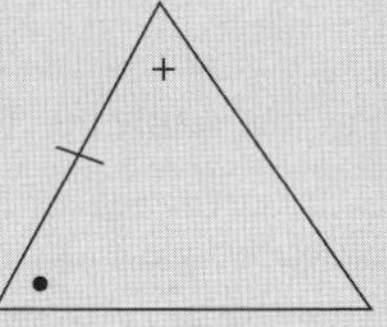

b.

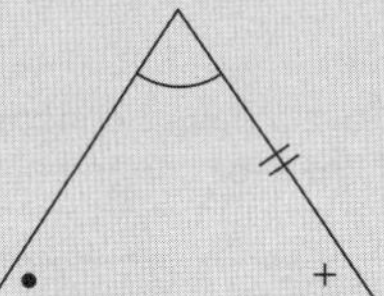

c.

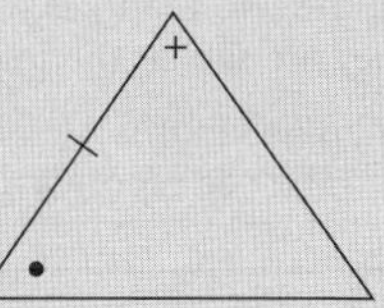

d.

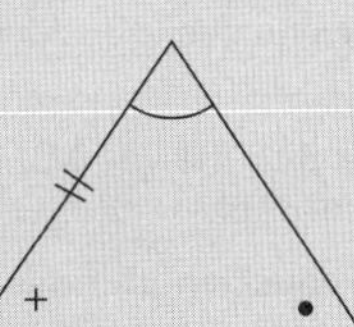

e.

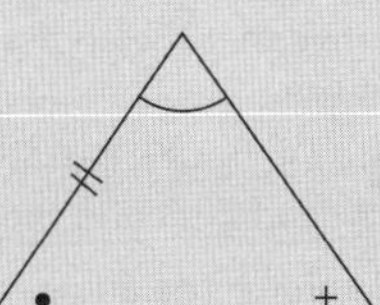

f.

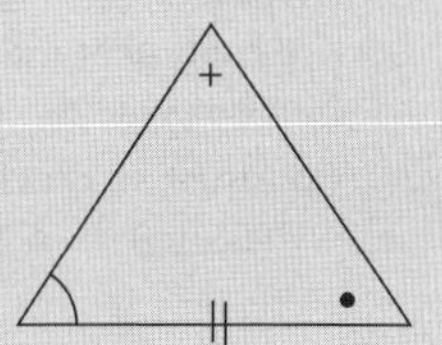

3. Use the construction methods given to construct a triangle that is congruent to each of the following triangles:

a.

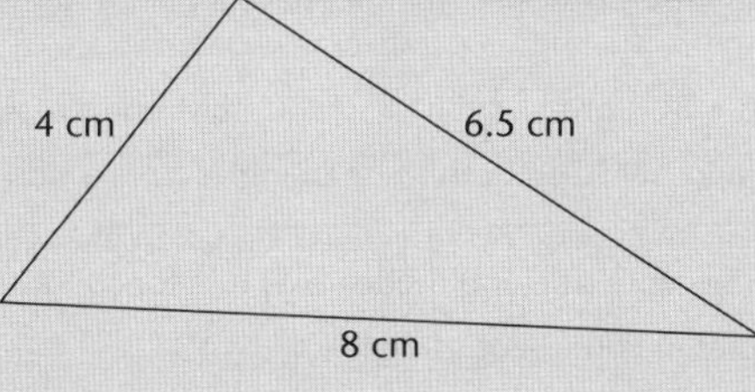

b.

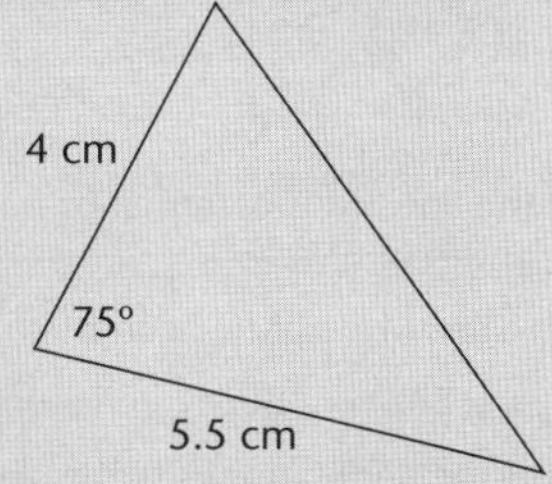

c.

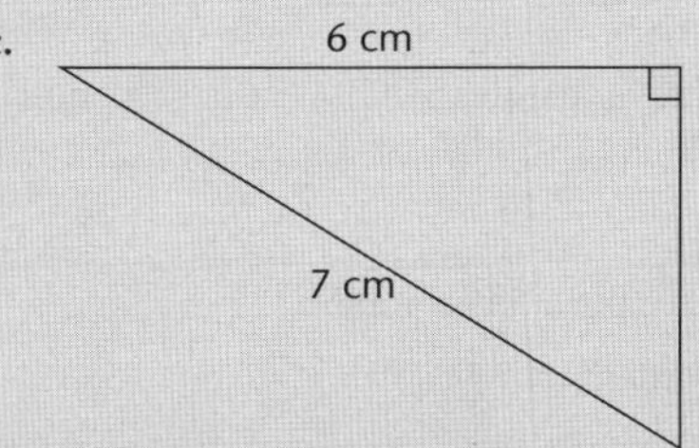

d.

Geometric proofs involving congruent triangles

Many geometric proofs use the fact that two triangles are congruent to demonstrate that two angles are equal or two sides are of equal length.

Example A

Q. Show that the triangles ABC and PQR are congruent triangles.

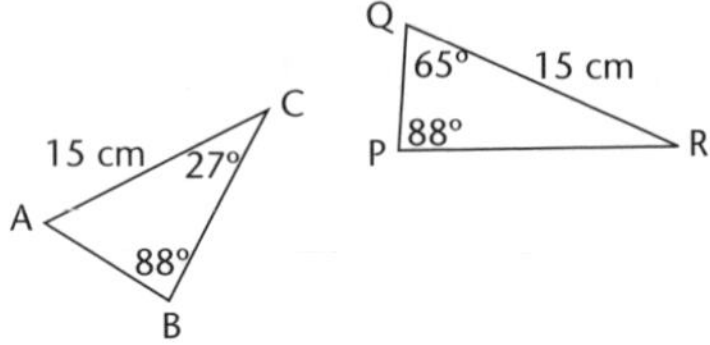

A. *To prove:* Triangle ABC is congruent to triangle PQR.

In triangle ABC, ∠CAB = 180° – 27° – 88° = 65°.

And in triangle PQR, ∠QRP = 180° – 65° – 88° = 27°.

So the two triangles ABC and PQR have angles that are the same size. (This alone means that they are similar triangles.)

The side lengths AC = QR so we have a pair of congruent triangles that satisfy the 'two angles and the corresponding side' (ASA) rule for congruency.

If the triangle ABC is congruent to triangle PQR then we can write this as ΔABC ≡ ΔPQR

Example B

Q. Use congruent triangles to prove that the angles opposite the equal-length sides of an isosceles triangle are equal in size.

A. Construct an isosceles triangle ABC and a line from the vertex B to the line AC, intersecting AC at the point M and bisecting the angle ABC so ∠ABM = ∠MBC.

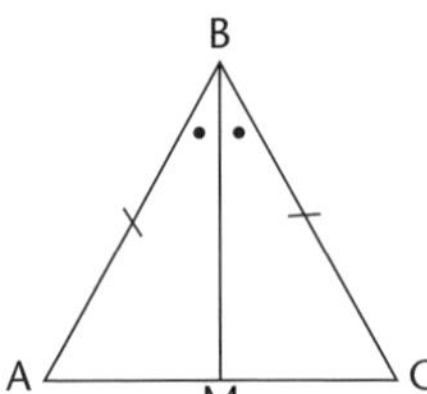

To prove: The angles opposite the equal length sides of an isosceles triangle are equal in size.

This means that on the diagram we must prove that ∠BAM = ∠BCM in the ΔABC.

Given: AB = BC : isosceles triangles have a pair of sides of equal length.

Proof: AB = BC : given ∠ABM = ∠MBC : BM bisects ∠ABC.

The line BM is common to both ΔABM and ΔBMC so both these triangles have a side the length of BM.

⇒ ΔABM ≡ ΔBMC : congruent triangles because side-angle-side (SAS) rule for congruency.

So ∠BAM = ∠BCM : corresponding angles in congruent triangles.

⇒ the angles opposite the sides of equal length, in an isosceles triangle, are equal in size.

Following on from the proof in Example B, we can prove that the line BM is the perpendicular bisector of the line AC:

To prove: AM = MC and ∠BMA = ∠BMC = 90°.

We have already shown that ΔABM ≡ ΔBMC (SAS) (see proof in example 2).

So AM = MC : corresponding sides in these two congruent triangles.

⇒ point M is a bisector of AC .... (1).

Also:

$\angle AMB = \angle BMC$: corresponding angles in a pair of congruent triangles are equal in size.

$\angle AMB + \angle BMC = 180°$: angles on a straight line sum to 180°.

$2 \times \angle BMC = 180°$: Substituting $\angle BMC$ for $\angle AMB$.

$\angle BMC = 90°$: Divide both sides by 2.

$\angle AMB = 90°$: Angles AMB and BMC are equal.

$\Rightarrow$ BM is perpendicular to ($\perp$) AM and BM $\perp$MC ... (2)

(1) and (2) combined $\Rightarrow$ BM is the perpendicular bisector of AC.

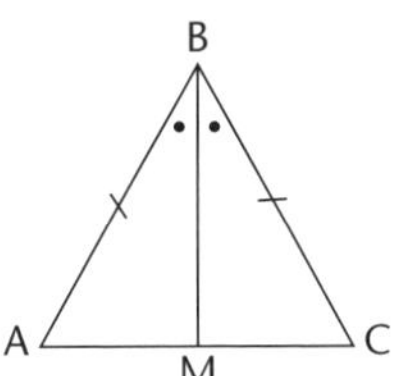

When establishing a geometric proof it is initially important to draw a suitably labelled diagram and express the theorem to be proved in terms of the angles and sides.

Unit 11.4 Activity 3B: Congruent triangles and other figures

1. Show that triangles ABC and ADC are congruent triangles.

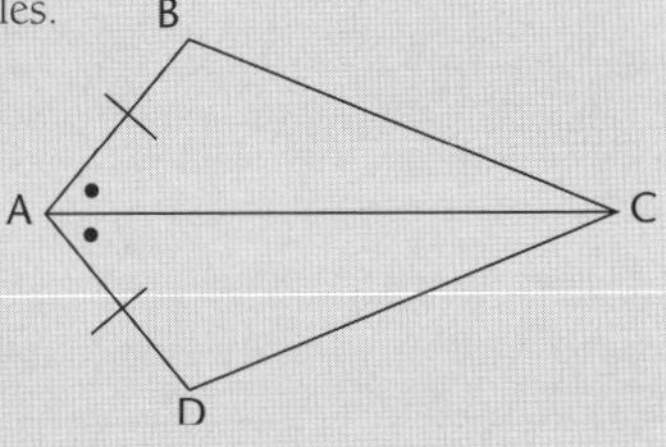

2. Show that triangles OPQ and ORS are congruent triangles.

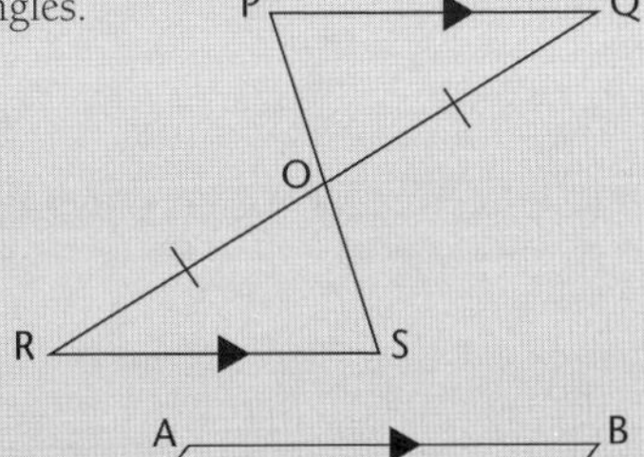

3. ABCD is a **parallelogram**. Prove that the opposite angles are equal in size.

4. Two straight lines are drawn so that they bisect each other, as shown in the diagram.

a. Show that the triangles POS and ROQ are congruent triangles.

b. Prove that PS is parallel to RQ.

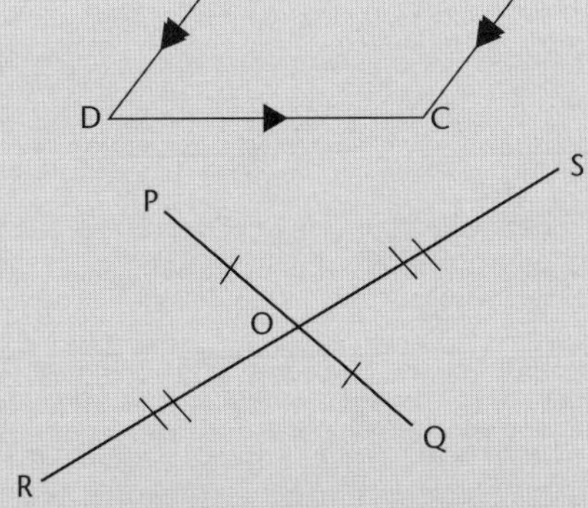

5. ABCD is a **rhombus**.

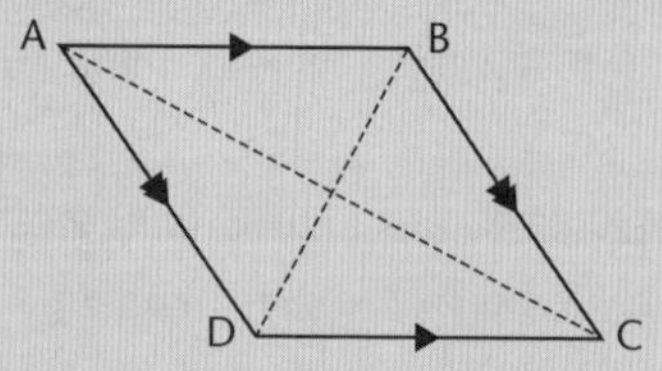

   a. Write down the properties of a rhombus.
   b. Prove that the diagonals of a rhombus bisect each other at right angles.

6. In the diagram the tangents to the circle, centre O, from the points P and Q meet at the point R.

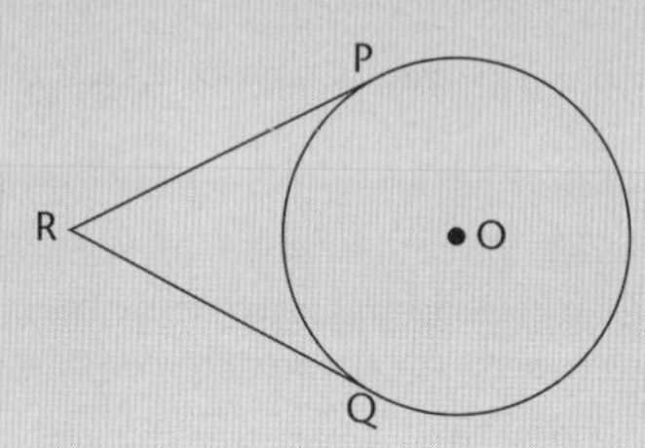

   a. Show that the triangles OPR and OQR are congruent triangles.
   b. Prove that the tangents RP and RQ are equal in length.

7. Use congruent triangles to prove:
   a. The opposite angles of a parallelogram are equal in size.
   b. The diagonals of a kite intersect at right angles.
   c. One pair of opposite angles in a **kite** are equal in size.

Unit 11.4 Geometry

Topic 4: Pythagoras' Theorem—revision

A revision of Pythagoras' Theorem is included because of its importance in the study of Geometry. This Topic covers:

- The use of Pythagoras' Theorem to find unknown sides in right-angled triangles.
- Finding unknown sides in practical situations involving right-angled triangles (the right-angled triangle will either be given or easily identified and extracted from a diagram).
- Finding unknown sides in word problems (student to extract right-angled triangles implicit in a context and use them to solve the problem in a well-reasoned and logical solution).
- Using appropriate rounding, units and mathematical statements.

Introduction

In a **right-angled triangle**, one of the angles of the triangle is 90° (on a diagram a small square is used to mark the right angle). The side opposite the right angle is called the **hypotenuse** and is always the longest side of the triangle.

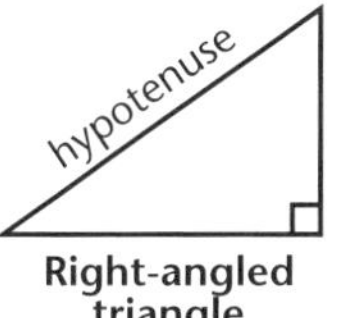

Right-angled triangle

This Achievement Standard focuses on finding unknowns in right-angled triangles.

Pythagoras' Theorem

There is an important relationship between the lengths of the three sides of a right-angled triangle. This relationship is called **Pythagoras' Theorem**, and states:

> For any right-angled triangle, the square on the hypotenuse is equal to the sum of the squares on the other two sides.

Pythagoras' Theorem is more useful when written as a mathematical formula:

$$a^2 = b^2 + c^2$$

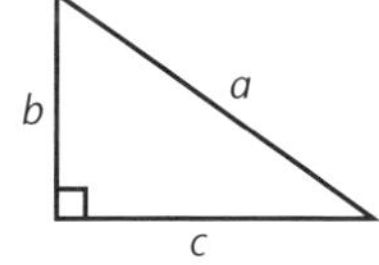

where a is the length of the hypotenuse, and b and c are the lengths of the other two sides.

The following example illustrates Pythagoras' Theorem.

Example A

A right-angled triangle with sides of length 3, 4, and 5 is shown alongside.

A square is drawn on each side of the triangle. Each of these squares is divided into square units labelled 1–9 and a–p. The 'square on each side' is the number of square units on that side. The number of square units on the hypotenuse is compared with the total number of square units on the other two sides.

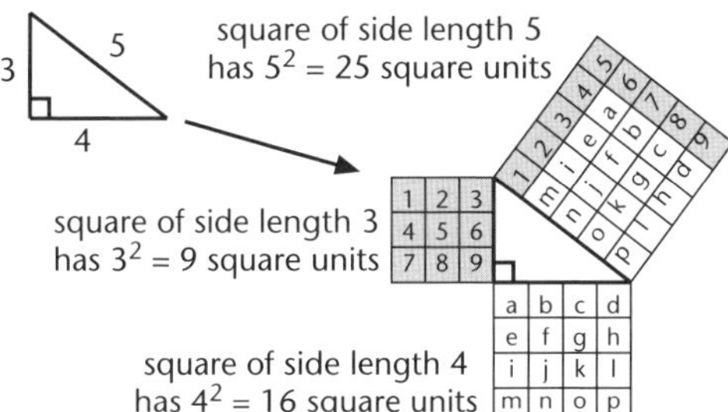

The number of square units on the hypotenuse (25) is equal to the sum of the numbers of square units on the other two sides (9 + 16).

Conversely, if the three sides of a triangle obey Pythagoras' Theorem, then the triangle is right-angled. For example, if the three sides of a triangle are 5 cm, 12 cm and 13 cm, then the relationship $a^2 = b^2 + c^2$ holds , where $a = 13$, $b = 5$ and $c = 12$ [since $13^2 = 5^2 + 12^2$] and the triangle is right-angled.

Finding unknown sides using Pythagoras' Theorem

Pythagoras' Theorem is used to find an unknown side of a right-angled triangle when two sides of the triangle are known. Be sure to set out each step of the working carefully and use appropriate rounding, including units with your answer where required.

Example B

Q. 1. Find x in the triangle shown.

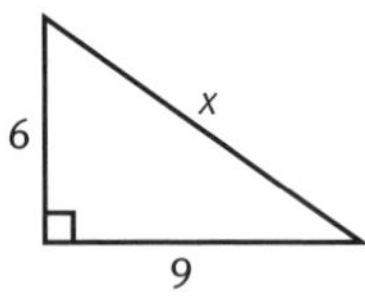

2. Find y in the triangle shown.

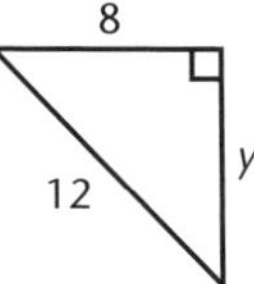

A. 1. $x^2 = 6^2 + 9^2$ [Pythagoras' Theorem]

$= 36 + 81$

$= 117$

$x = \sqrt{117}$ [taking square roots]

$= 10.8$ (1 dp)

2. $12^2 = y^2 + 8^2$ [Pythagoras' Theorem]

$144 = y^2 + 64$

$80 = y^2$ [subtracting 64]

$y = \sqrt{80}$ [taking square roots]

$= 8.9$ (1 dp)

Note: Both parts of Example B begin with 'hypotenuse squared ='. This is the 'safest' way of writing an equation involving Pythagoras' Theorem.

In a word problem, no diagram may be provided. A right-angled triangle must be drawn from the information given.

Example C

Q. In a triangle ABC, angle A = 90°, the length of AB is 2.7 cm and the length of BC is 4.9 cm. Find the length of AC.

A. Draw a triangle, right-angled at A, hypotenuse BC, using the information given.

$4.9^2 = x^2 + 2.7^2$ [Pythagoras]

$24.01 = x^2 + 7.29$ [calculator]

$16.72 = x^2$ [subtracting 7.29]

$x = \sqrt{16.72}$ [taking square roots]

$x = 4.1$ (1 dp) [rounding sensibly]

The length of AC is 4.1 cm [answer in terms of given labels, not x]

C
4.9 cm
x
A
B
2.7 cm

Note: Never round an answer to the nearest whole number when using Pythagoras' Theorem. Doing so may imply that all the lengths of the sides are whole numbers, and this is seldom so.

Unit 11.4 Activity 4A: Pythagoras' Theorem

1. Find the lengths of the marked sides, rounding sensibly when required.

a.
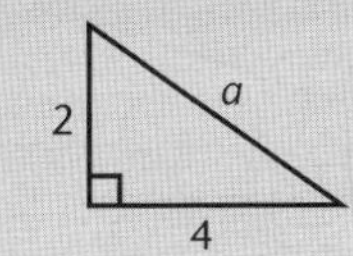

b.
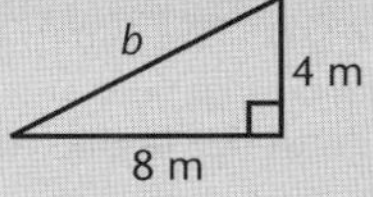

c.
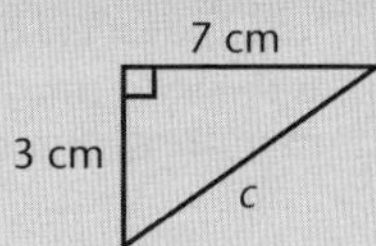

d.
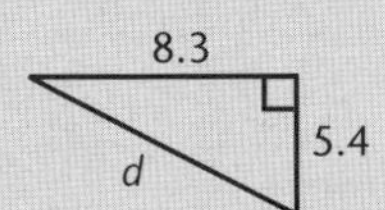

e.
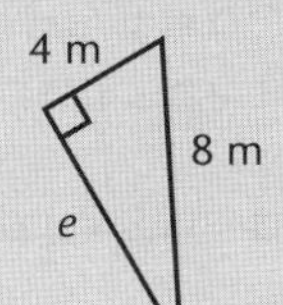

f.
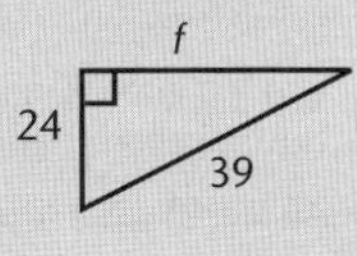

g.
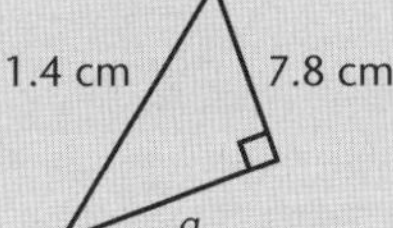

h.
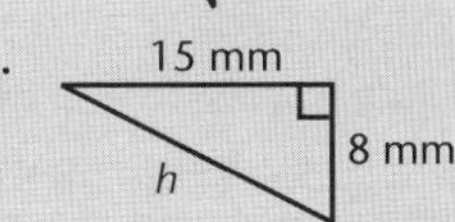

i.
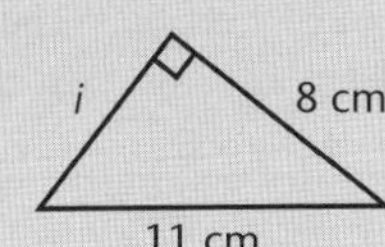

2. A right-angled triangle has a hypotenuse of unknown length. Its other two sides are of length 4.7 m and 6.3 m. Find the length of the hypotenuse.

3. A right-angled triangle has a hypotenuse of length 154 cm and another side of length 65 cm. Find the length of the third side.

4. The longest side of a right-angled triangle is 16.5 mm. If another side is 9.8 mm find the length of the third side.

5. If a triangle is right-angled then its sides obey Pythagoras' Theorem. Which of the following triangles, whose side lengths are given, are right-angled? (Remember the hypotenuse is always the longest side.)

a. 2.4 cm, 3.2 cm, 4 cm **b.** 6 cm, 9 cm, 12 cm **c.** 3.5 cm, 12 cm, 12.5 cm

Solving problems using Pythagoras' Theorem

Many practical problems involve finding side lengths of right-angled triangles. A clear diagram, including all information given, is the starting point for solving such problems.

Example D

Q. A 4 m ladder is leaning against a wall. The foot of the ladder is 1 m out from the wall, as the diagram shows. How far up the wall does the ladder reach?

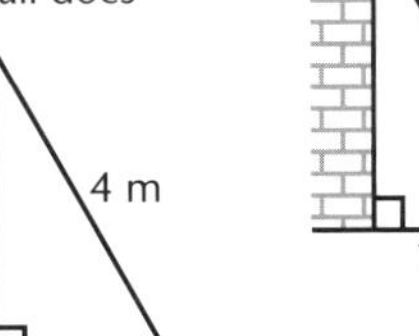

A. A right-angled triangle is drawn with *h* being the height the ladder reaches up the wall.

Using Pythagoras' Theorem:

$4^2 = h^2 + 1^2$

$16 = h^2 + 1$

$h^2 = 15$ [subtracting 1 and rearranging]

$h = 3.87$ m (2 dp) [taking square roots]

Include extra lines where necessary to complete a right-angled triangle in your diagram.

Example E

Q. Two parallel sides of a field are 80 m and 140 m long, as shown in the diagram. If the parallel sides are 68 m apart, how long is the fourth side of the field?

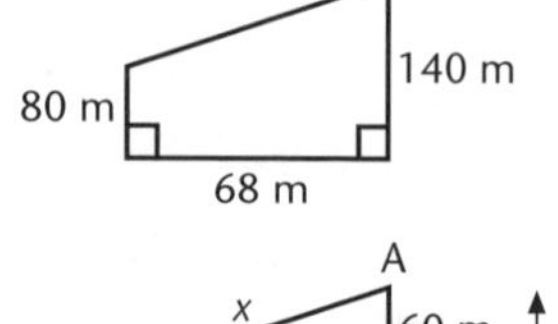

A. Label the corners of the field A, B, C, D as shown.

Draw a line DE parallel to CB so that AED is a right-angled triangle.

AE = 140 – 80 = 60 m [subtracting CD from AB]

DE = 68 m

By Pythagoras' Theorem $x^2 = 68^2 + 60^2$

$= 8\,224$

$x = 90.7$ m (1 dp) [taking square root]

Drawing the axis of symmetry of an isosceles triangle creates two right-angled triangles of the same size and shape. Pythagoras' Theorem can then be applied.

Example F

Q. A flagpole is held upright by two wires both of length 25 m. The poles are 16 m apart. How tall is the flagpole?

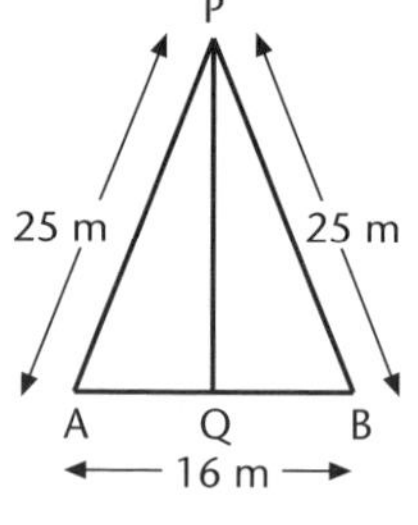

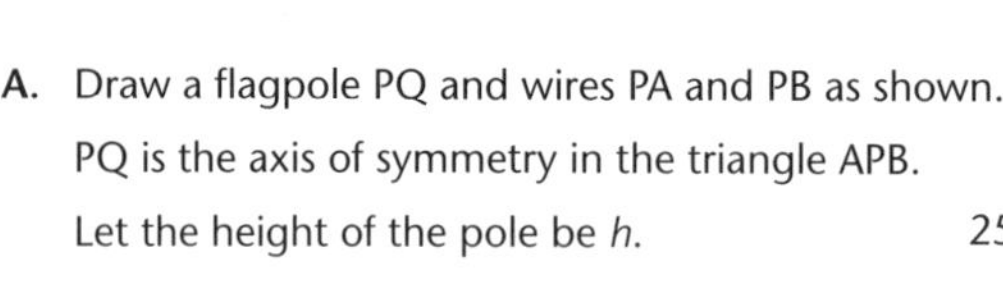

A. Draw a flagpole PQ and wires PA and PB as shown.

PQ is the axis of symmetry in the triangle APB.

Let the height of the pole be h.

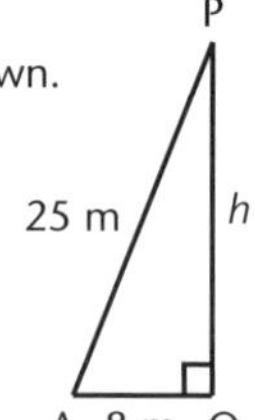

In triangle APQ, AQ = 8 m [half of AB by symmetry]

$25^2 = h^2 + 8^2$ [Pythagoras' Theorem]

$625 = h^2 + 64$

$h^2 = 561$ [subtracting 8^2 and swapping sides]

$h = 23.7$ m [taking square root]

The flagpole is 23.7 m tall.

Unit 11.4 Activity 4B: Solving problems using Pythagoras' Theorem

1. In an orienteering event, Simon ran from point A to point B along two paths at right angles to each other. He ran 460 m along one path and 372 m along the other.

   Mary ran from point A to point B in a direct line.

   How much shorter was Mary's course, to the nearest metre?

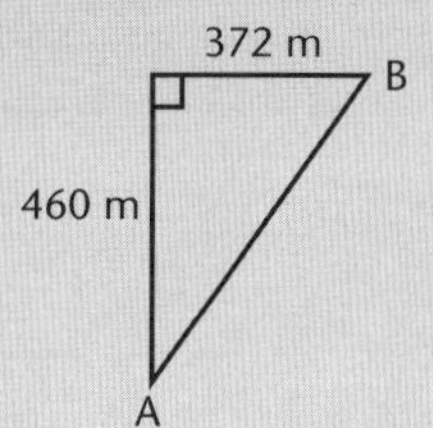

2. A flagpole is held upright by four wires. Each wire is fastened to the top of the pole, 24 m above ground, and to pegs in the ground 16 m from the base of the pole. How long is each wire?

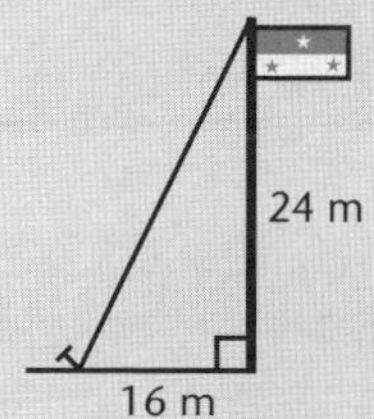

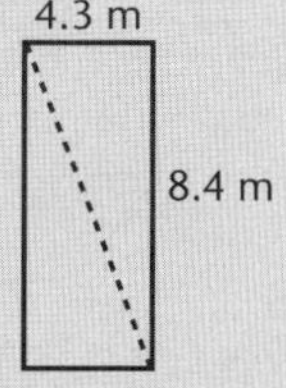

3. A builder laying the foundations of a house wants to be sure she has the walls of the house at right angles.

   She knows the length of the house should be 8.4 m and the width 4.3 m. To check this, she will measure the length of the diagonal.

   a. Explain how measuring the length of the diagonal will check that the walls of the house meet at right angles.

   b. How long should the diagonal be so that the walls are at right angles?

4. In the diagram, AB is the sloping roof of a shed. The front A is 3 m high, the back B is 6 m high. The width of the shed is 6.5 m.

   What length (AB) of roofing iron, to the nearest centimetre, is needed for the roof?

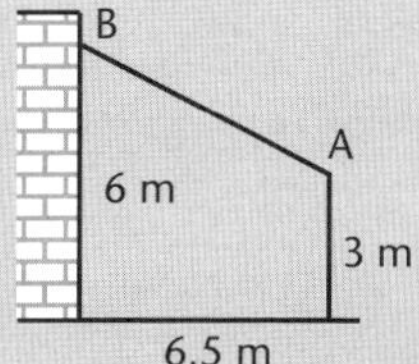

5. The dimensions of the roof and ceiling in a house are shown. Calculate the distance between the top of the roof and the ceiling, h.

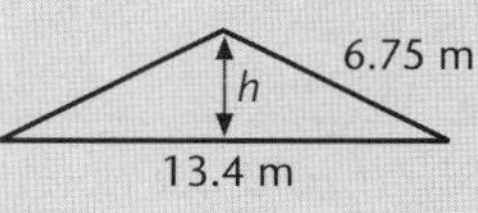

6. Susan has a suitcase with dimensions 90 cm by 60 cm by 30 cm. She has an umbrella 1.05 m long.

   Can she fit her umbrella in the bottom of her suitcase?

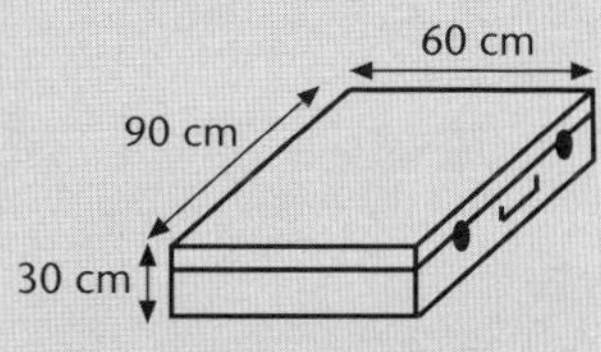

7. A rugby goal kicker places the ball, 18 m to the right of the centre of the goal posts and 34 m out from the goal-line.

   How far is it from the centre of the goal posts to the ball?

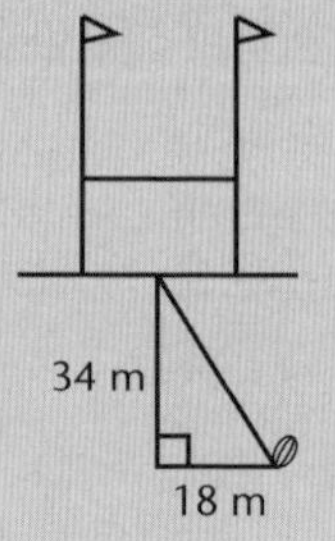

8. James and Rachel were laying a footpath for their new home, and they wanted to make sure that they had a right angle at point A.

   James placed a peg in the ground at B, 4 m from A. He tied a 10 m length of twine to the peg. Rachel held a tape at point A and James held the free end of the twine.

   At what point on the tape should the twine meet the tape to get a right angle at A?

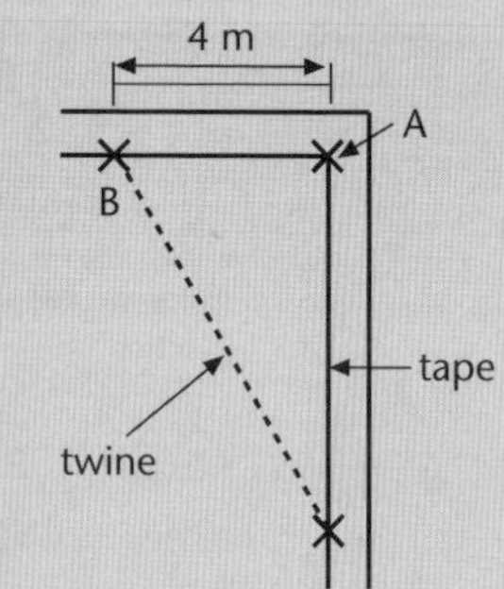

9. A triangle ABC has side lengths AB = 31 m, BC = 42 m and the distance between the vertex B and the side AC is 15 m. Find the length of AC.

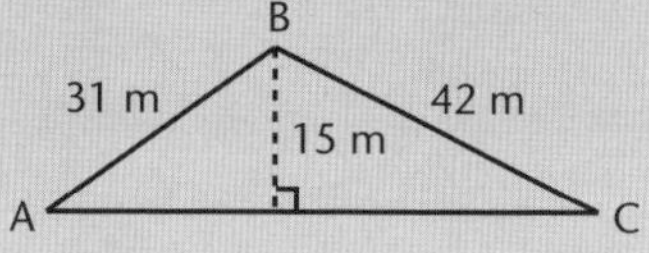

10. Ming is designing a 3 cm wide headband. She draws lines 4 cm long making a zig-zag pattern.

    a. How far apart are the points P and Q marked on Ming's headband?

    b. Ripeka wants to make a similar headband. It is to be the same width as Ming's band, but Ripeka wants the points P and Q to be exactly 4 cm apart. How long should each of the zig-zag lines be?

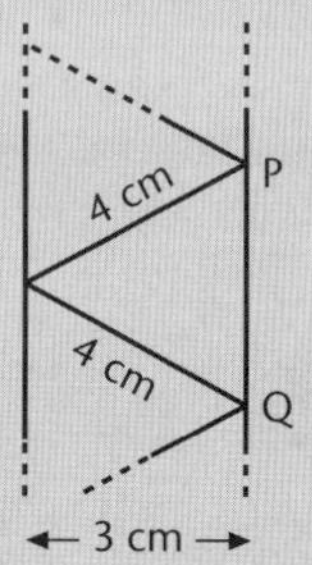

Unit 11.4 Geometry

Topic 5: Trigonometric ratios

A revision of trigonometric ratios is included because of its relevance in solving application problems on right-angled triangles. Topic 5 covers:

- The use of trigonometric ratios to find unknowns in right-angled triangles.
- Using trigonometric ratios to find and interpret unknowns in practical situations involving right-angled triangles (the right-angled triangle will either be given or easily identified and extracted from a diagram).
- Using trigonometric ratios to find unknowns in word problems (student to extract right-angled triangles implicit in a context and use them to solve the problem in a well-reasoned and logical solution).
- Using appropriate rounding, units and mathematical statements.

Introduction

Trigonometric ratios provide a relationship between angle size and side lengths in right-angled triangles. In this chapter trigonometric ratios are used with right-angled triangles:

- To find an unknown side length when an angle and side are known, or
- To find the size of an unknown angle when two side lengths of the triangle are known.

Ratios of side lengths in right-angled triangles

Any angle between 0° and 90° can be drawn as part of a right-angled triangle. No matter what *size* right-angled triangle is drawn for a particular angle size, the *ratio* of one side of the triangle compared with another side of the same triangle *will always be equal*, as shown below.

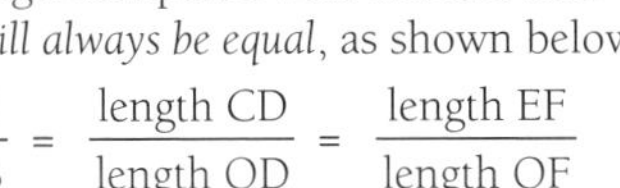

$$\frac{\text{length AB}}{\text{length OB}} = \frac{\text{length CD}}{\text{length OD}} = \frac{\text{length EF}}{\text{length OF}}$$

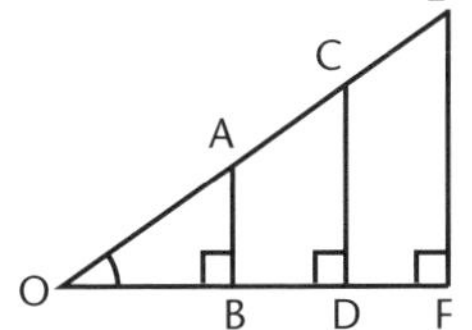

This relationship is the basis of the **trigonometric ratios** called the **tangent**.

Similar equalities give rise to other trigonometric ratios called the **sine** and **cosine**.

Labelling the sides of right-angled triangles

The three differe.nt sides of a right-angled triangle are named the **hypotenuse**, **opposite side** and **adjacent side**. The labelling of the adjacent and opposite side is relative to the angle chosen in the triangle.

h	**h**ypotenuse	The side *opposite the right angle*, as shown by the arrow pointing from the right angle to *h* in the diagram.
o	**o**pposite side	The side *opposite* the angle referred to (in this case θ), as shown by the arrow pointing from θ to *o* in the diagram.
a	**a**djacent side	The side *adjacent* (next) to the angle referred to (in this case θ).

h
o
θ
a

The trigonometric ratios

In a right-angled triangle each trigonometric ratio compares the length of one side with another. These rules are shown in full and in abbreviated form in the box below.

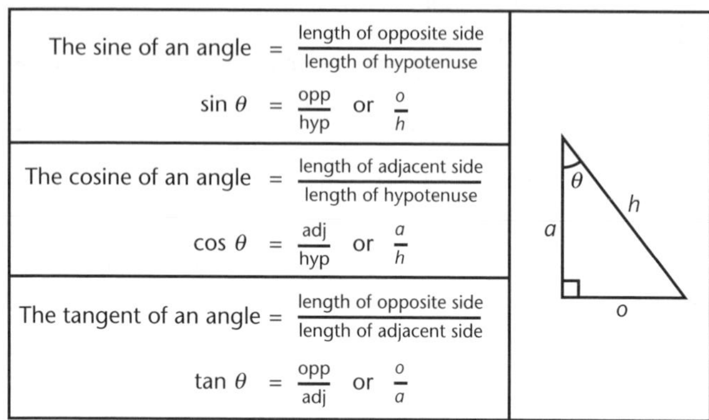

The sine of an angle $= \dfrac{\text{length of opposite side}}{\text{length of hypotenuse}}$

$\sin\theta = \dfrac{\text{opp}}{\text{hyp}}$ or $\dfrac{o}{h}$

The cosine of an angle $= \dfrac{\text{length of adjacent side}}{\text{length of hypotenuse}}$

$\cos\theta = \dfrac{\text{adj}}{\text{hyp}}$ or $\dfrac{a}{h}$

The tangent of an angle $= \dfrac{\text{length of opposite side}}{\text{length of adjacent side}}$

$\tan\theta = \dfrac{\text{opp}}{\text{adj}}$ or $\dfrac{o}{a}$

These three trigonometric ratios must be learnt. **SOH CAH TOA** is a useful mnemonic for doing this:

S	**S**ine of angle	**C**	**C**osine of angle	**T**	**T**angent of angle
O	**O**pposite over	**A**	**A**djacent over	**O**	**O**pposite over
H	**H**ypotenuse	**H**	**H**ypotenuse	**A**	**A**djacent

Example A

Using ΔUVW shown, the hypotenuse is UV (5.8 cm). Relative to angle V, the adjacent is VW (5 cm) and the opposite is UW (3 cm).

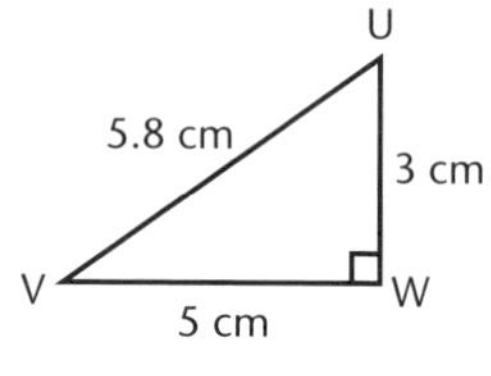

$\sin V = \dfrac{3}{5.8} = 0.52$ (2 dp) $\quad [\sin = \dfrac{\text{opp}}{\text{hyp}}]$

$\cos V = \dfrac{5}{5.8} = 0.86$ (2 dp) $\quad [\cos = \dfrac{\text{adj}}{\text{hyp}}]$

$\tan V = \dfrac{3}{5} = 0.6 \quad [\tan = \dfrac{\text{opp}}{\text{adj}}]$

Relative to angle U, the opposite is VW (5 cm) and the adjacent is UW (3 cm).

$\tan U = \dfrac{5}{3} = 1.67$ (2 dp) $\quad [\tan = \dfrac{\text{opp}}{\text{adj}}]$

Unit 11.4 Activity 5A: Trigonometric ratios

1. Use the diagram of ΔABC to state:

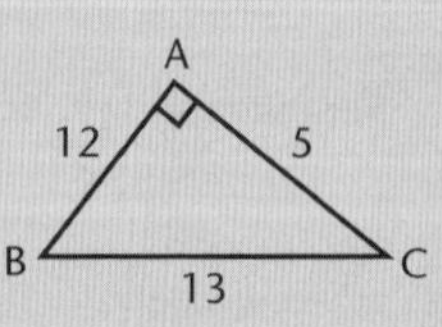

a. The length of the hypotenuse.

b. The length of the side adjacent to angle B.

c. The length of the side opposite angle C.

d. sin C **e.** tan B **f.** cos B **g.** sin (90° – C)

2. In ΔPQR, state the values of:

a. The length of the hypotenuse.

b. The length of the side opposite angle P.

c. The length of the side adjacent to angle P.

d. tan P **e.** sin R **f.** cos (90° – P) **g.** sin (90° – R)

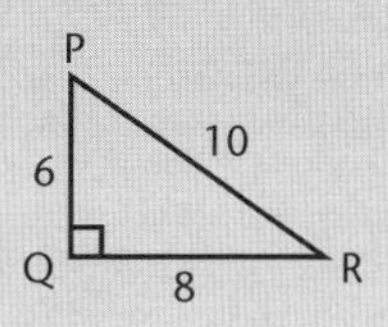

Finding the value of a trigonometric ratio

The procedure for finding the value of a trigonometric ratio with a calculator is shown in the following example (in this book, access to a calculator with trigonometric function keys is assumed).

Example B

All answers are rounded to four decimal places.

sin 60° = 0.8660

tan 49.2° = 1.159

Make sure your calculator is in the correct MODE, ie DEG (degrees).

Finding angle sizes from the value of a trigonometric ratio

The inverse [inv] or second function key [Shift] or [2nd F] is used to find an angle size when a trigonometric ratio is known.

Example C

Q. Find θ if **1.** sin θ = 0.766 **2.** cos θ = 0.3420 **3.** tan θ = 0.7266

A. **1.** If sin θ = 0.7660, θ = 50.0° [Shift] [sin] [•] [7] [6] [6] [=]

2. If cos θ = 0.3420, θ = 70.0° [Shift] [cos] [•] [3] [4] [2] [=]

3. If tan θ = 0.7266, θ = 36.0° [Shift] [tan] [•] [7] [2] [6] [6] [=]

Note: It is common practice to round angle sizes to 1 dp, unless an alternative accuracy (rounding) is suggested.

The notation '**cos^{-1}**' means 'the angle whose cosine is'. For example, 'cos^{-1} (0.3420)' means 'the angle whose cosine is 0.3420'. From Example C it can be seen cos^{-1} (0.3420) = 70.0°, ie cos 70.0° = 0.3420.

Similarly, the notation 'sin^{-1}' means 'the angle whose sine is' and the notation 'tan^{-1}' means 'the angle whose tangent is'.

Unit 11.4 Activity 5B: Using your calculator

1. Use your calculator to find the value of each of the following to 4 decimal places:

a. sin 20° **b.** cos 10° **c.** tan 70° **d.** cos 38° **e.** tan 48°

f. cos 64° **g.** sin 56.3° **h.** tan 78.4° **i.** sin 12.4°

2. Find the value of θ in each of the following, giving answers to 1 decimal place.

a. $\sin\theta = 0.6428$ **b.** $\cos\theta = 0.7660$ **c.** $\tan\theta = 1.7324$

d. $\cos\theta = 0.6428$ **e.** $\tan\theta = 0.7931$ **f.** $\sin^{-1}(0.4532) = \theta$

g. $\theta = \cos^{-1}(0.8751)$ **h.** $\theta = \tan^{-1}(1.234)$ **i.** $\theta = \sin^{-1}(0.1159)$

Finding unknown side lengths (unknown in numerator)

For any right-angled triangle, the length of an unknown side can be found if the length of another side *and* the size of another angle (other than the right angle) are known.

Example D

Q. Find the length of the unknown side (a, b or c) in each of the following triangles:

1.

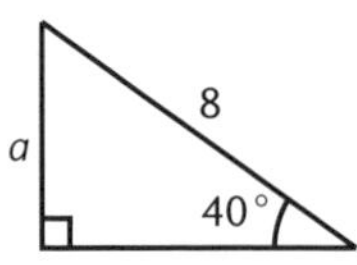

2.

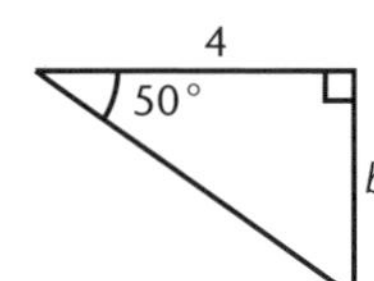

3.

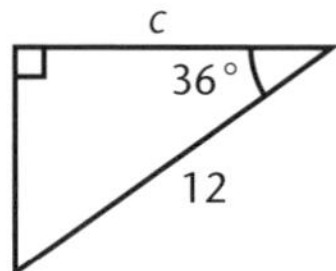

A. 1. In the triangle, the hypotenuse is 8 and (relative to the given angle) the opposite side is a. The rule linking opposite and hypotenuse is sine [SOH].

$\sin 40° = \frac{a}{8}$ $\quad$ [$\sin = \frac{o}{h}$]

$8 \times \sin 40° = a$ $\quad$ [multiplying by 8]

$a = 5.14$ (2 dp) $\quad$ [using a calculator]

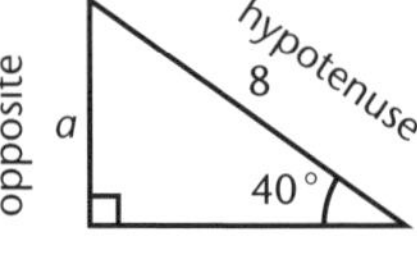

To calculate 8 sin 40° using a scientific calculator:

or
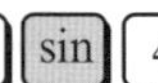

Note: Since the horizontal side is neither given, nor required, it is left unlabelled (this makes the identification of the correct trigonometric ratio easier).

2. $\tan 50° = \frac{b}{4}$ $\quad$ [$\tan = \frac{o}{a}$]

$b = 4 \times \tan 50°$ $\quad$ [multiplying by 4]

$= 4.77$ (2 dp) $\quad$ [using a calculator]

3. $\cos 36° = \frac{c}{12}$

$c = 12 \cos 36°$

$= 9.71$ (2 dp)

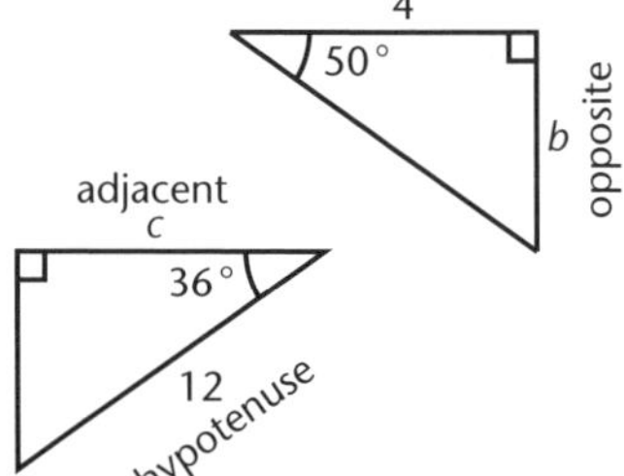

Finding unknown side lengths (unknown in denominator)

Sometimes, the unknown side appears in the denominator of the trigonometric ratio. An extra line of working will be required to rearrange the equation in these types of problem.

Example E

Q. Find d in the right-angled triangle shown.

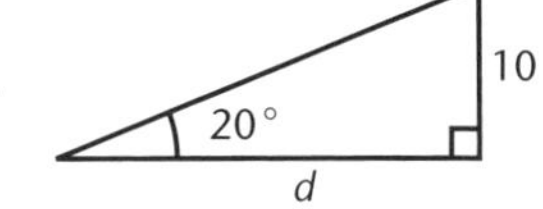

A. Relative to the angle 20°, d is 'adjacent' and 10 is 'opposite', so tangent is used (TOA).

$\tan 20° = \dfrac{10}{d}$ [$\tan = \dfrac{o}{a}$]

$10 = d \tan 20°$ [multiplying by d (to bring d into the numerator)]

$\dfrac{10}{\tan 20°} = d$ [dividing by tan 20° (the coefficient of d)]

$d = 27.47$ (2 dp) [using a calculator]

To calculate $\dfrac{10}{\tan 20°}$ using a calculator:

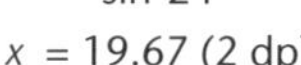

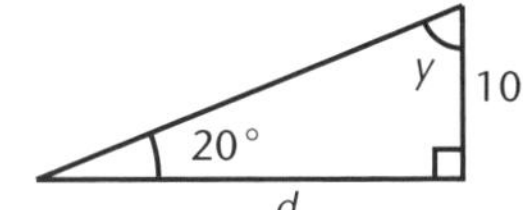

Note: d can be found using the other angle in the triangle, labelled y. Investigate this for yourself.

Example F

Q. Find x in the right-angled triangle shown.

A. $\sin 24° = \dfrac{8}{x}$ [using SOH]

$x \sin 24° = 8$ [multiplying by x]

$x = \dfrac{8}{\sin 24°}$ [dividing by sin 24°]

$x = 19.67$ (2 dp) [using a calculator]

Unit 11.4 Activity 5C: Finding unknown sides

1. Find the lengths of the lettered sides for each of the triangles. When required, round your answers sensibly.

a.

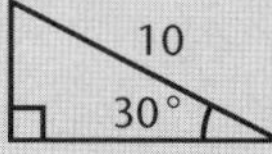

b.

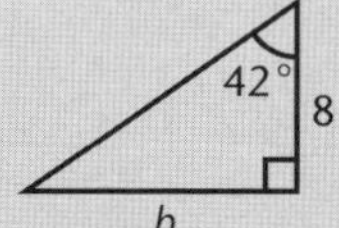

c.

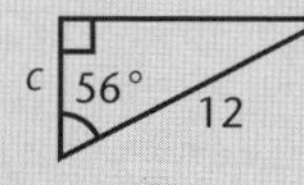

d.

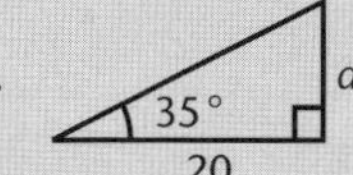

e.

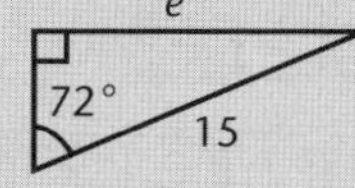

f. (15, 26°, f)

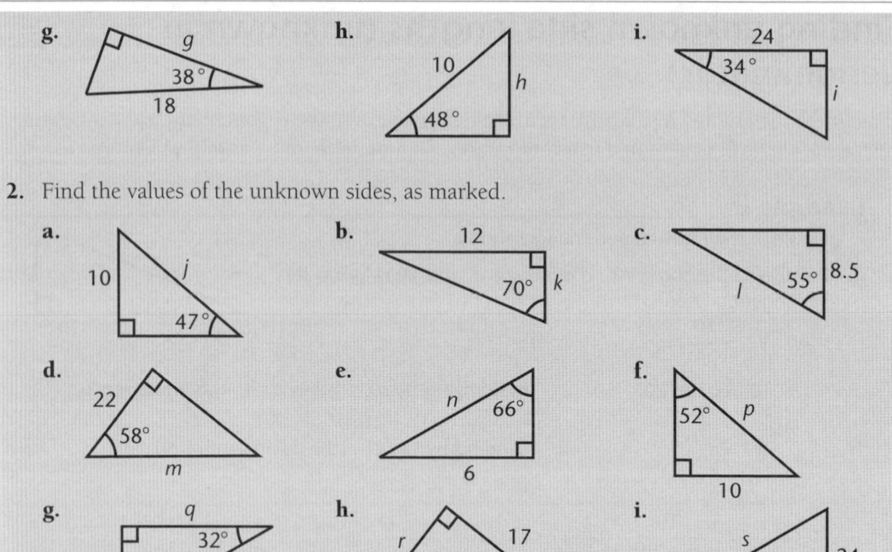

2. Find the values of the unknown sides, as marked.

Finding unknown angles

If two sides of a right-angled triangle are known, the size of an unknown angle can be found.

Label the known sides relative to the angle required and choose the relevant trigonometric ratio for this angle.

Example G

Q. Find the size of the angles marked in each of the following:

1.

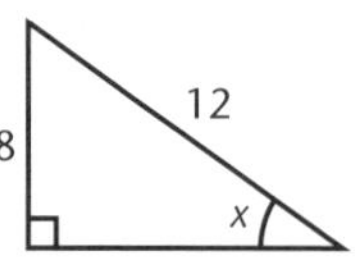

2.

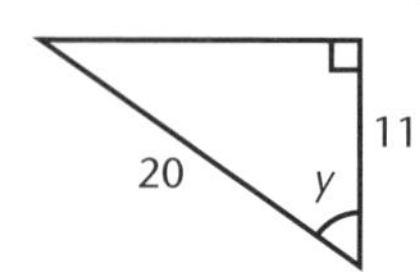

3.

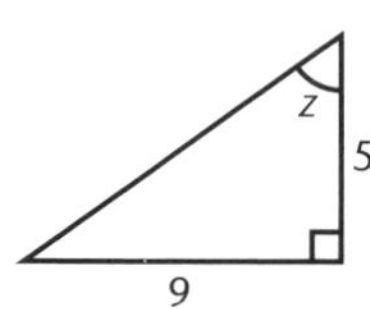

A. **1.** $\sin x = \frac{8}{12}$ $\quad$ $[\sin = \frac{o}{h}]$

$x = \sin^{-1}\left(\frac{8}{12}\right)$

$x = 41.8°$ (1 dp)

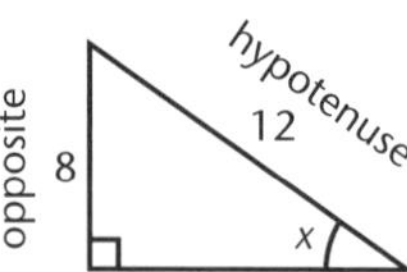

To find the value of x using a calculator:

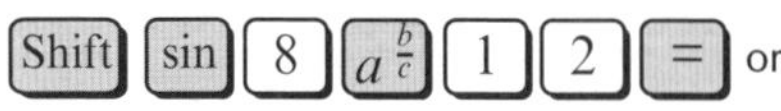

or

Note: No information is given about the adjacent side, which means that the trigonometric ratios cos and tan cannot be used directly to solve this problem.

2. $\cos y = \dfrac{11}{20}$ $\quad [\cos = \dfrac{a}{h}]$

$y = \cos^{-1}\left(\dfrac{11}{20}\right)$

$y = 56.6°$ (1 dp)

3. $\tan z = \dfrac{9}{5}$ $\quad [\tan = \dfrac{o}{a}]$

$z = \tan^{-1}\left(\dfrac{9}{5}\right)$

$z = 60.9°$ [rounding to one decimal place]

Note: Once one unknown angle in a right-angled triangle is found, the third angle can be found using the angle sum of a triangle rule. For example, in part **3** of Example G, the third angle in the triangle is $90° - z = 90° - 60.9° = 29.1°$.

Unit 11.4 Activity 5D: Finding unknown angles

1. For each of the triangles find the size (to one decimal place) of the angle labelled.

a. 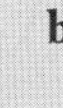

b. 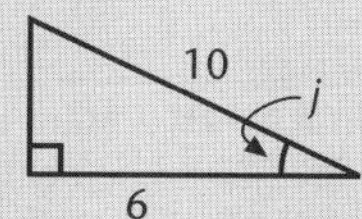

c. 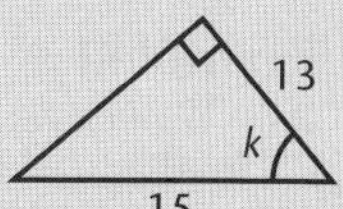

d. 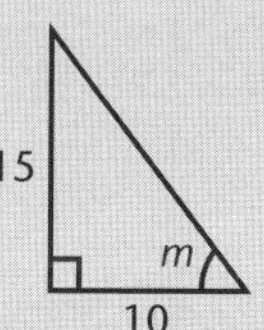

e. 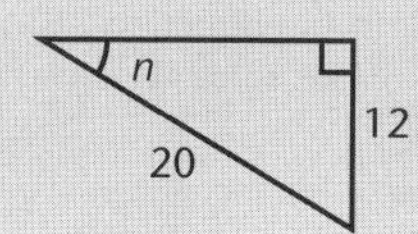

f. 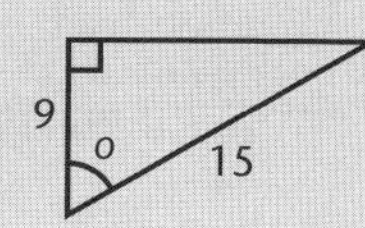

g. 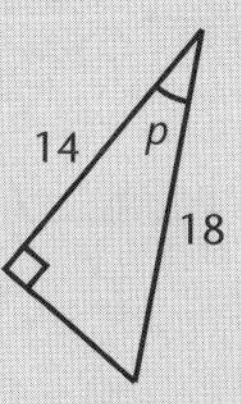

h. 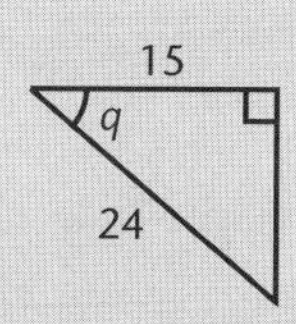

i. 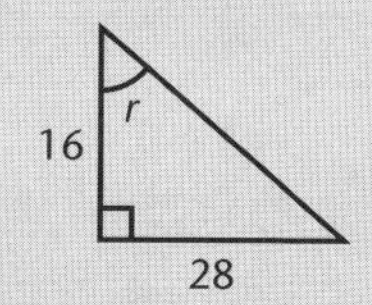

2. Find the size (to one decimal place) of the angles labelled.

a. 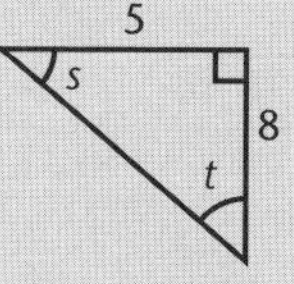

b. 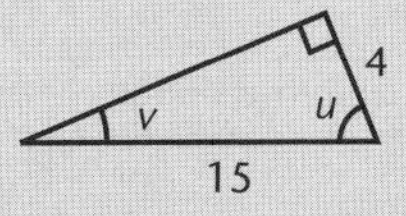

c. 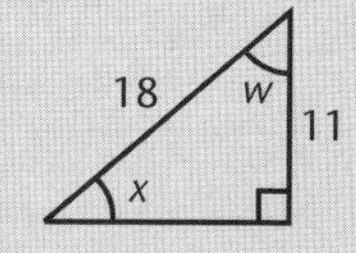

d. 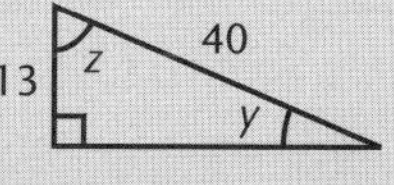

e. 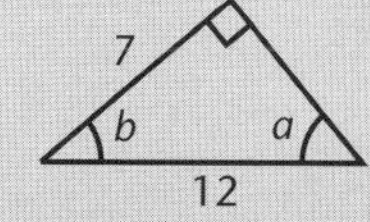

f. 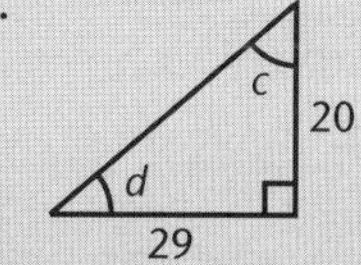

Solving problems using trigonometry

To solve practical problems involving right-angled triangles, draw a clear diagram from the given information.

Example H

Q. A ladder 3.5 m long is leaning against a wall. If the foot of the ladder is 1 m out from the base of the wall, what angle does the ladder make with the wall?

A. $\sin x = \dfrac{1}{3.5}$ $\quad$ $[\sin = \dfrac{o}{h}]$

$x = 16.6°$ $\quad$ [taking $\sin^{-1}\left(\dfrac{1}{3.5}\right)$]

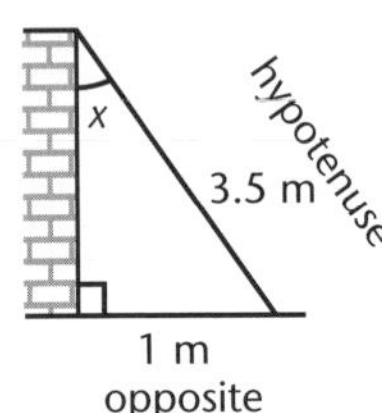

Note: Once the equation has been established to find the missing angle, the entire calculation may be done on the calculator:

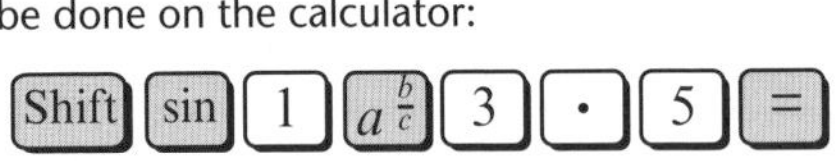

If your diagram of a practical problem does not contain a right-angled triangle, you may need to add at least one extra line perpendicular to an existing line of your diagram.

Example I

Q. A boat leaves from point A which is 10 m downstream from a 12.5 m long bridge (CD). The boat is aimed directly at the opposite side of the river, but is pushed downstream, so that it travels in the direction AB, making a 73° angle with the opposite bank. How far from the bridge is the point B?

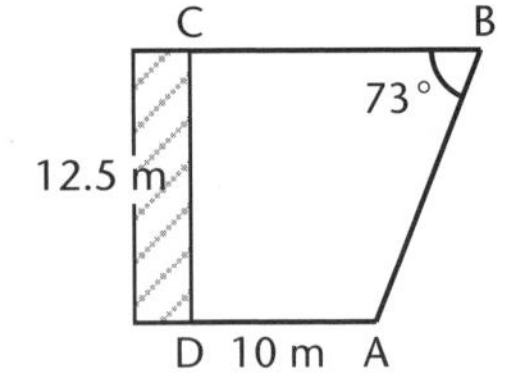

A. In the diagram, construct AP, so that AP is perpendicular to BC and ΔAPB is right-angled, AP = 12.5 m $\quad$ [same as CD]

Require to find x (the length PB).

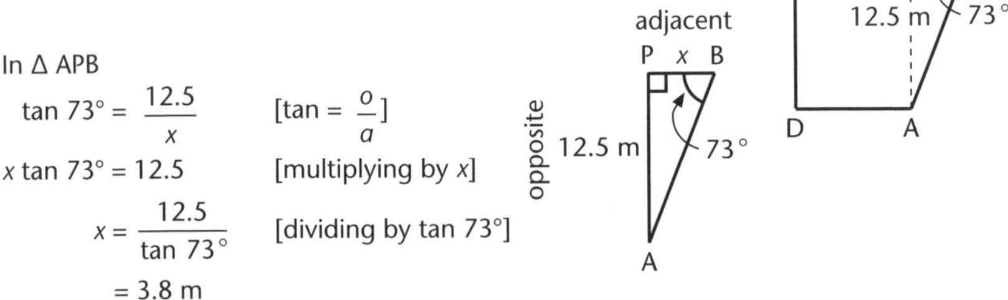

In Δ APB

$\tan 73° = \dfrac{12.5}{x}$ $\quad$ $[\tan = \dfrac{o}{a}]$

$x \tan 73° = 12.5$ $\quad$ [multiplying by x]

$x = \dfrac{12.5}{\tan 73°}$ $\quad$ [dividing by tan 73°]

$= 3.8$ m

Thus B is 10 + 3.8 = 13.8 m from the bridge $\quad$ [since CP = DA = 10 m]

Angles of elevation and depression

Two expressions in common usage when describing the position of an object are **angle of elevation** and **angle of depression**.

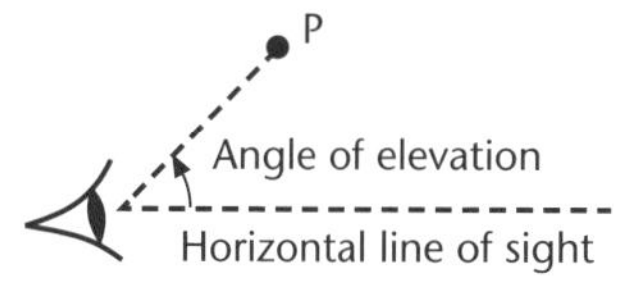

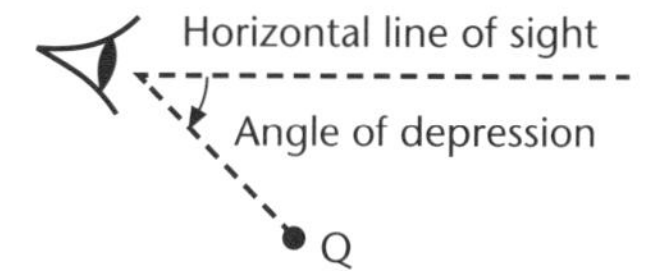

If you look up to a position P, ie P is above the horizontal line of sight, then the angle of elevation is measured up from the horizontal.	If you look *down* to a position Q, ie Q is below the horizontal line of sight, then the angle of depression is measured down from the horizontal.

Example J

Q. A kite is flying at the end of a 60 m length of string. How high above the ground is the kite if the angle of elevation to the kite is 55°?

A. The information can be shown on a diagram. K is the position of the kite and S is the other end of the string. The height of the kite above the ground is *d*.

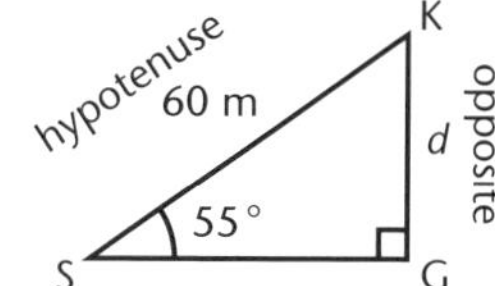

$\sin 55° = \dfrac{d}{60}$ [$\sin = \dfrac{o}{h}$]

$d = 60 \sin 55°$ [multiplying by 60 and swapping sides]

$= 49.1$ m (1 dp)

Unit 11.4 Activity 5E: Problem solving using trigonometry

1. A ladder 2.4 m long is leaning against a wall.

If the ladder reaches 2.0 m up the wall, what angle does the ladder make with the ground?

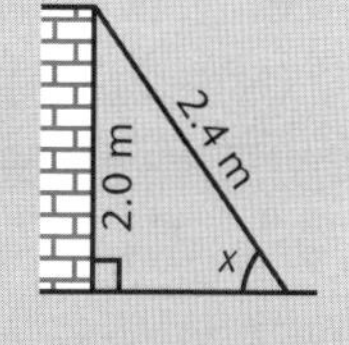

2. A farmer has to build a ramp to load cattle onto a truck.

The ramp is to make an angle of 30° with the ground. The horizontal distance between the beginning and end of the ramp is 2.3 m.

Calculate the length *h* to find the height of the end of the ramp.

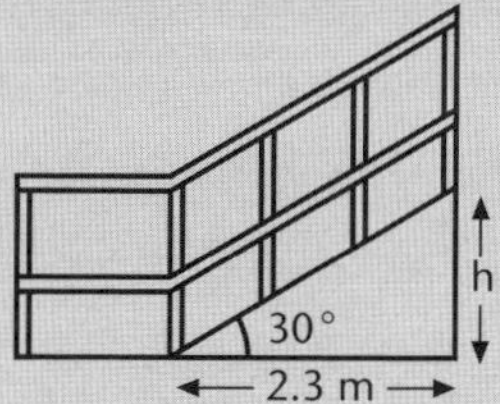

3. In a calm sea, a sailor measures the angle of elevation of a ship's telescope to the top of a lighthouse on a cliff to be 21°. The telescope is 28.4 metres above sea level, and radar places the ship 1 250 metres from the foot of the cliff.

Calculate the height of the top of the lighthouse above sea level, to the nearest metre.

4. A sailor, 1 km from a vertical cliff, notes that the angle of elevation of the top of the cliff is 10°.

   a. How high is the cliff? Give answer to the nearest metre.

   b. An observer on the top of the cliff later notices that a boat is passing a marker 500 m from the base of the cliff. Calculate the angle of depression of the observer's line of vision.

5. A tent has a cross-section which is an isosceles triangle. The base of the triangle is of length 3.5 m and the base angles are of size 51°.

   What is the height of the tent?

6. A kite is flown on a cord 125 m long. The cord, held taut with its end at a height of 1.5 m above the ground, makes an angle of 70° with the horizontal.

   How high above the ground is the kite?

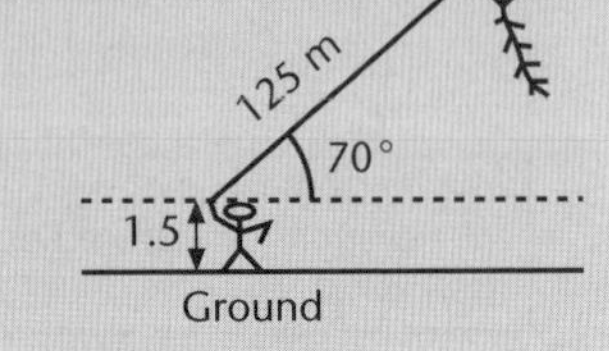

7. Anna was given the task of calculating the height of the school hall. She stood 6 metres from the wall and, using a clinometer, measured the angle of elevation to the topmost point of the hall.

   The angle of elevation was measured to be 56° and Anna's eye level was 1.5 metres above the ground.

   a. Calculate the height, h, of the school hall. Give your answer to 1 dp.

   b. Anna's answer to the height of the school hall was wrong by several metres. Give a possible source for this error.

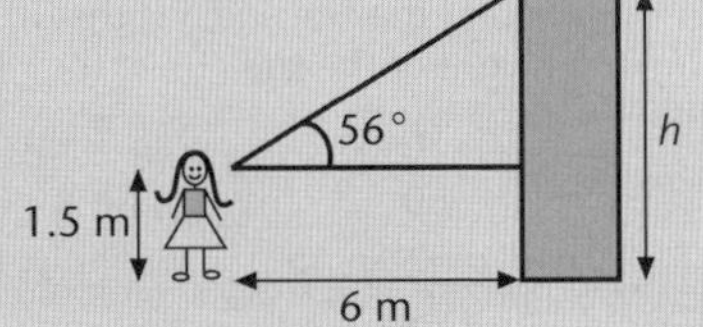

8. From a position 3.5 m above ground level in a building, an an observer measures the angle of elevation of the top of a flagpole to be 48°, and the angle of depression of the foot of the flagpole to be 35°.

   a. How far away from the flagpole is the building?

   b. How tall is the flagpole?

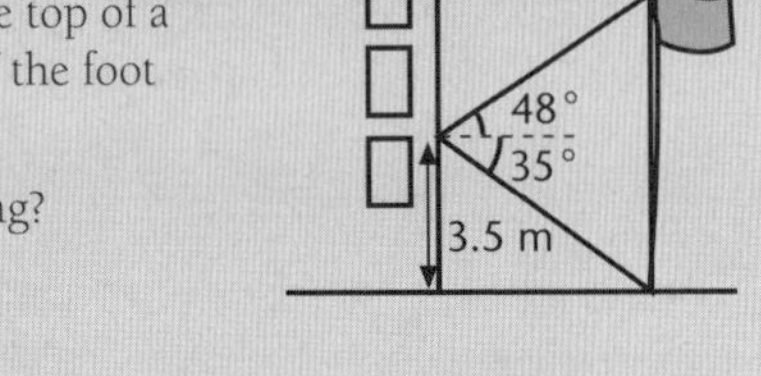

9. A boat sets off from a straight shore at an angle of 67° to the shore and travels in a straight line for 14.5 km.

   The boat then turns and travels in a straight line for home as shown in the diagram.

   If the homeward journey is 17.6 km, find θ, the acute angle the line of the homeward journey makes with the shore.

10. A beam is supported by two ropes both attached to a horizontal rod.

    One rope is 3 m long and makes an angle of 120° with the beam. The other rope is 4 m long.

    What angle does the 4 m rope make with the beam?

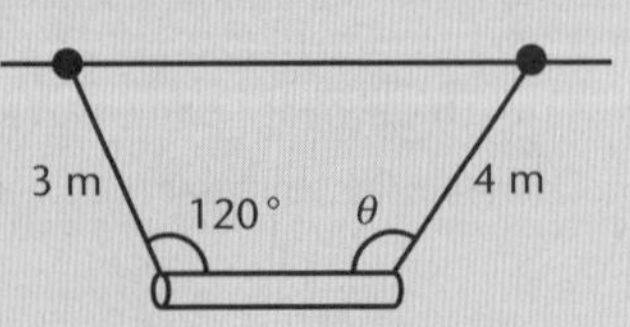

Unit 11.4 Geometry
Topic 6: Three-dimensional trigonometry

This Topic continues the coverage of solving right-angled triangle problems, in particular learning outcome 5 (Syllabus p. 17): 'Interpret, describe and represent properties of and relationships between 2-dimensional shapes and 3-dimensional objects in a variety of situations, orientation and positions.' It covers:

- The use of trigonometric ratios to find and interpret unknowns in straightforward 3-D situations.
- Using trigonometric ratios to find unknowns in word problems (student to extract right-angled triangles implicit in a context and use them to solve the problem in a well-reasoned and logical solution).
- Using appropriate rounding, units and mathematical statements.

Introduction

In order to find the sizes of angles and lengths of lines in three-dimensional figures, right-angled triangles must be identified within the figure. Trigonometric ratios and/or **Pythagoras' Theorem** may then be applied to find the unknowns.

Finding unknown lengths in 3-D figures

If a length is to be found in a 3-D figure, look for a right-angled triangle in which the unknown length is a side. If the other two side lengths are known, Pythagoras' Theorem may then be applied to find the unknown side. (It is best to redraw these right-angled triangles.)

Example A

Q. Rebecca's bag is 75 cm by 40 cm by 16 cm. Rebecca wants to buy an umbrella which will fit in her bag. What is the longest umbrella she could buy?

16 cm

40 cm

75 cm

A. Labelling the **cuboid** representing Rebecca's bag as shown, the length of AG (or HB) is the length of the longest umbrella Rebecca could buy.

AG is part of the right-angled triangle AEG drawn alongside.

The length EG is the hypotenuse of the right-angled triangle EHG (as shown). Therefore,

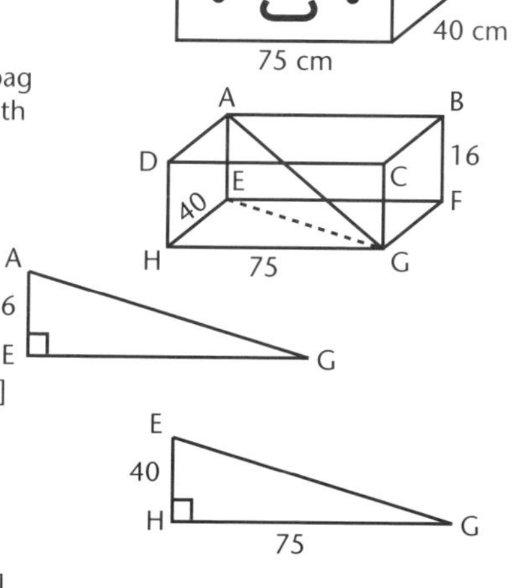

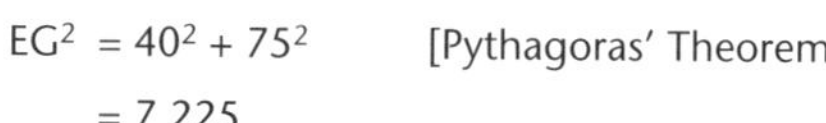

$EG^2 = 40^2 + 75^2$ [Pythagoras' Theorem]

$= 7\,225$

$\therefore$ EG = 85 cm [taking square root]

In triangle AEG

$AG^2 = 16^2 + 85^2$ [Pythagoras' Theorem]

$= 7\,481$

$\therefore$ AG = 86.5 cm (3 sf)

The longest umbrella Rebecca could buy is 86.5 cm.

Angle between a line and a plane

Find the angle between a line and a plane

The line AB meets the shaded **plane** at P.

Draw a perpendicular from A to meet the plane at C.

Join PC.

Angle APC is the angle between the line AB and the shaded plane.

Example B

Q. Name the angle between the line EC and the plane EFGH in the cuboid shown.

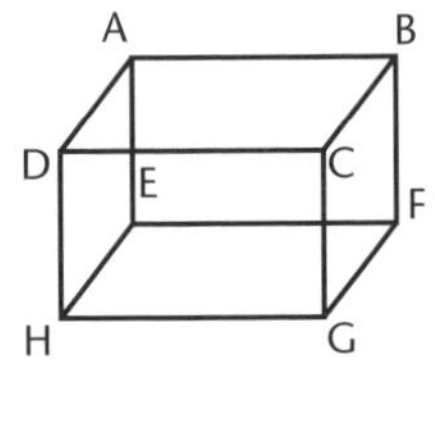

A. Join EC and shade the plane EFGH.

The line EC meets the plane at E.

The perpendicular (CG) from C meets the plane at G.

The required angle is $\angle CEG$ as shown on the diagram.

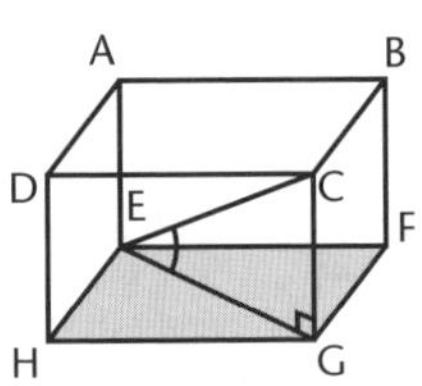

When measurements are given on the figure Pythagoras' Theorem and/or trigonometry can be used to calculate the sizes of angles.

Example C

Q. The cuboid shown has length 4, width 2 and height 3.

Calculate the size of the angle between the line HB and the plane ABFE.

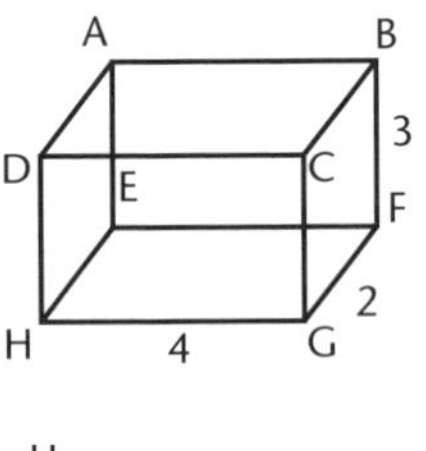

A. HB meets the plane (shown shaded) at B.

The perpendicular from H meets the plane at E.

The required angle is $\angle HBE$.

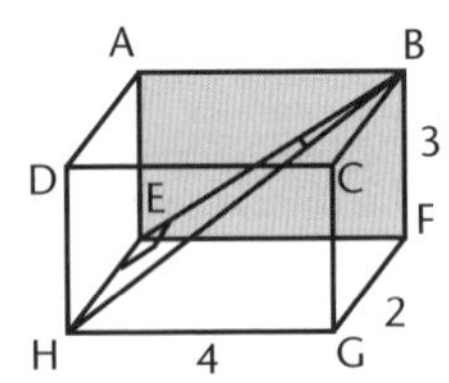

Triangle HBE is drawn alongside.

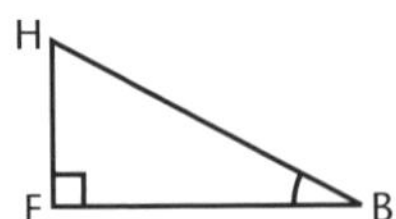

The length EB is the hypotenuse of the right-angled triangle EFB.

Thus $EB^2 = EF^2 + FB^2$ [Pythagoras' Theorem]

$= 4^2 + 3^2$

$= 25$

$\therefore EB = 5$ [taking square root]

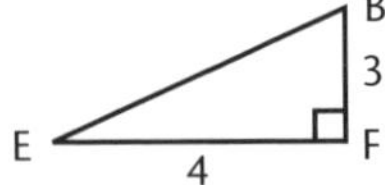

In triangle HEB, let $\angle HBE = \theta$

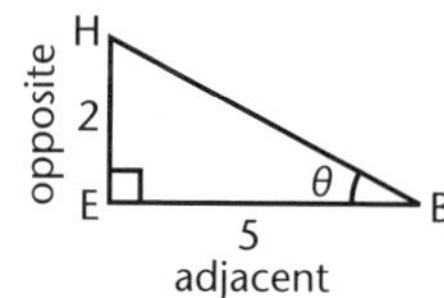

$$\tan\theta = \frac{2}{5}$$
$$\theta = \tan^{-1}\left(\frac{2}{5}\right)$$
$$= 21.8^\circ$$

The angle between the line HB and the plane ABFE is 21.8°.

When working with **pyramids**, use the symmetry of the figure to find unknown lengths. The **apex** (top point of the pyramid) sits over the 'centre' of the base. In a square base the 'centre' is where the diagonals intersect.

Example D

Q. In the pyramid shown, ABCD is a square with sides of 4 cm. The length of EB is 8 cm.

Calculate the size of the angle between line EB and the plane ABCD.

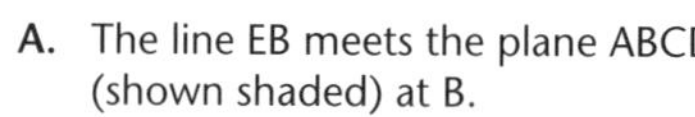

A. The line EB meets the plane ABCD (shown shaded) at B.

The perpendicular from E meets the plane at F, the 'centre' of the square (where its diagonals meet).

The required angle, labelled θ, is $\angle EBF$.

In triangle EBF, the length FB is required.

FB is half the length of the hypotenuse DB of the right-angled triangle DCB (shown alongside).

Thus $DB^2 = 4^2 + 4^2$ [Pythagoras' Theorem]

$= 32$

$\therefore DB = 5.6569$ [taking square root]

$FB = 2.8284$ [since FB is half DB]

Triangle EFB is drawn alongside.

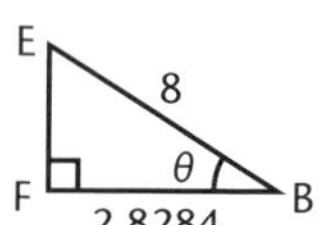

$$\cos\theta = \frac{2.8284}{8}$$
$$= 0.3536$$
$$\theta = \cos^{-1}(0.3536)$$
$$= 69.3^\circ \text{ (1 dp)}$$

The angle between EB and the plane ABCD is 69.3°.

Note: Work to several decimal places and round at the end to avoid rounding error.

Unit 11.4 Activity 6A: 3-D trigonometry

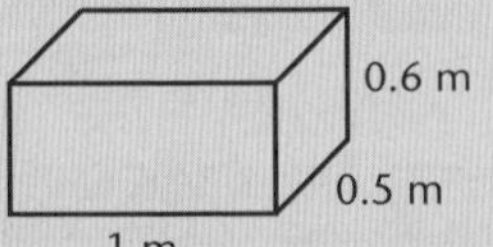

1. Giwi is packing tent poles in a box. He notices that the longest pole just fits in the box. If the box is 1 m by 0.5 m by 0.6 m, how long is the longest pole?
2. The diagram shows a cuboid. Name the following angles:
   a. The angle between line AH and plane EFGH.
   b. The angle between line CE and plane EFGH.
   c. The angle between line BH and plane ABCD.
   d. The angle between line DF and plane BCGF.

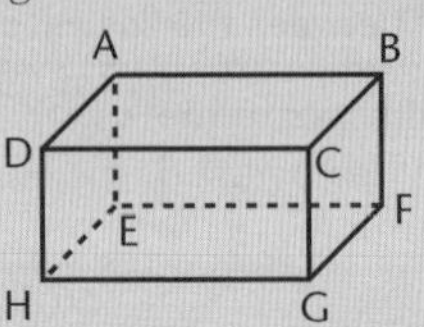

3. The diagram shows a wedge in which ABCD, AEFD and BCFE are rectangles.
   a. Calculate the length of CD.
   b. Name and calculate the size of the angle between the planes ABCD and AEFD.

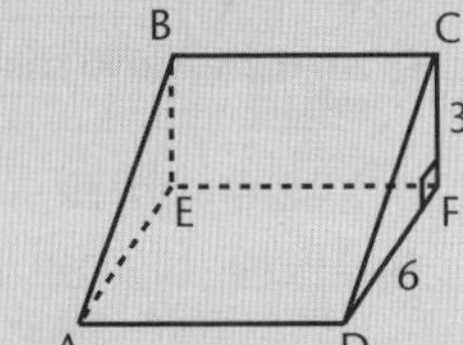

4. ABCDEFGH is a cuboid.
   a. Calculate the length of EG.
   b. Calculate the length of EC.
   c. Name and calculate the angle between CE and the plane EFGH.

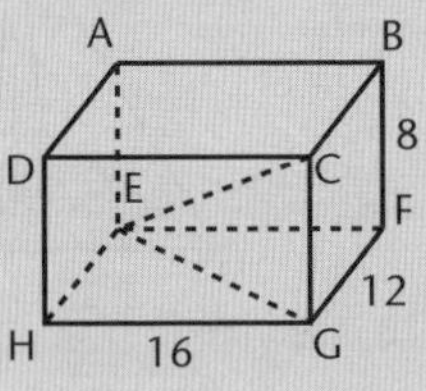

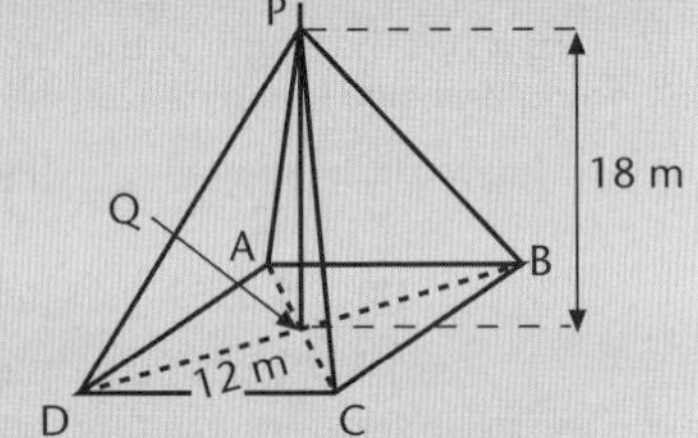

5. A pole QP is held upright by four equal ropes PA, PB, PC, PD. The ropes are fastened to the pole at 18 m above the ground and to pegs at A, B, C and D, each 12 m from the base of the pole Q.
   a. Calculate the size of the angle each rope makes with the ground.
   b. The four pegs form a square ABCD. What is the length of each side of this square, to the nearest centimetre?
6. A pencil box in the shape of a cuboid is 16 cm long and 6 cm wide. If the longest pencil that fits in the box is 18 cm long, how deep is the box?
7. The angle between AH and the plane EFGH in the cuboid shown is 40°. Find the height of the cuboid.

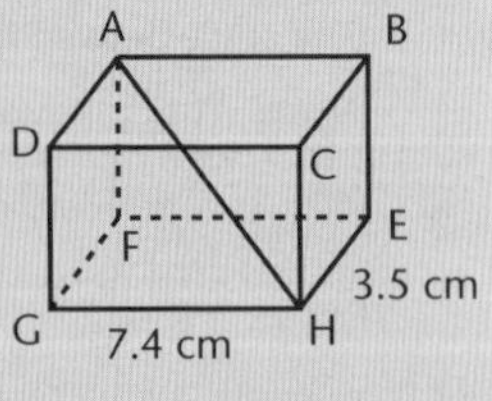

8. A 12 cm by 10 cm rectangle-based pyramid ABCDE is shown. Its sides are isosceles triangles of sloping side length 9 cm.

   Find the angle between the line AE and the base plane BCDE.

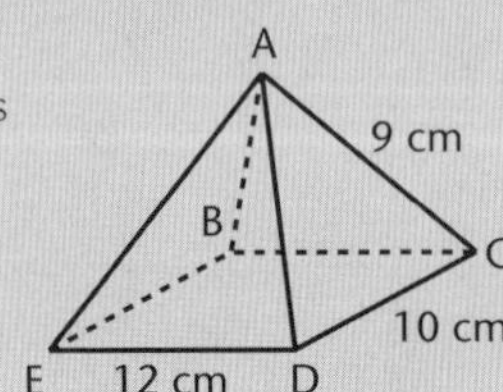

The angle between two planes

Find the angle between two planes

Find the line of intersection: planes ABCD and CDEF meet on the line CD.

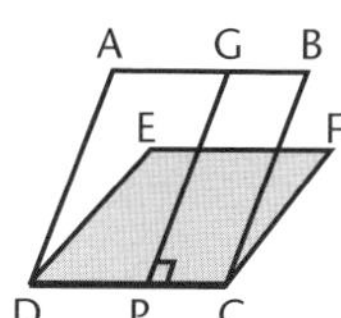

On the plane ABCD draw a line GP perpendicular to CD, meeting CD at P.

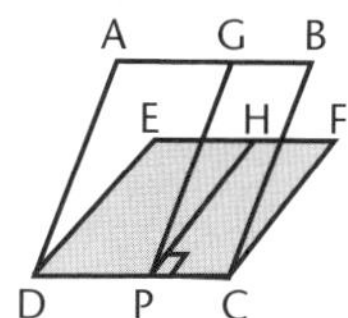

On plane CDEF draw a line HP perpendicular to CD meeting CD at P.

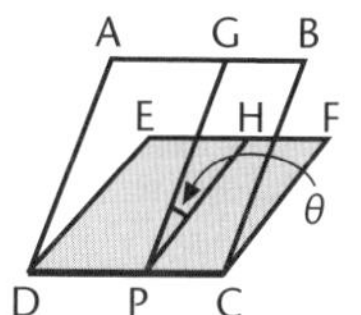

The angle GPH (marked θ) is the angle between the plane ABCD and the plane CDEF.

Example E

Q. In the cuboid shown, find the size of the angle between planes AFGD and EFGH.

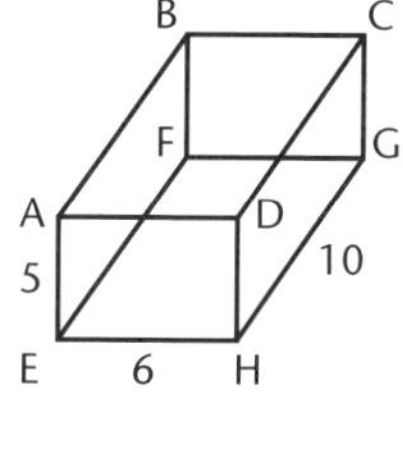

A. The two planes meet on line FG.

- On the plane AFGD, DG is perpendicular to FG.
- On the plane EFGH, HG is perpendicular to FG.

Since DG and HG are perpendicular to FG, the required angle is $\angle DGH$.

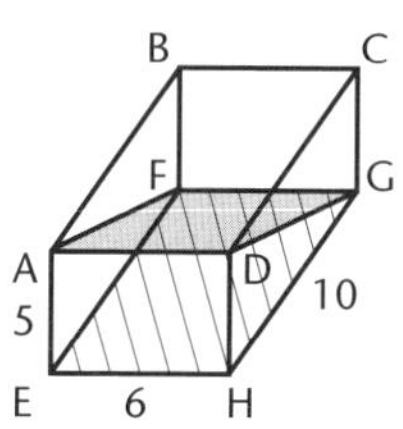

Triangle DGH is shown alongside.

$$\tan \angle DGH = \frac{5}{10}$$

$$\angle DGH = \tan^{-1} 0.5$$

$$= 26.6^\circ$$

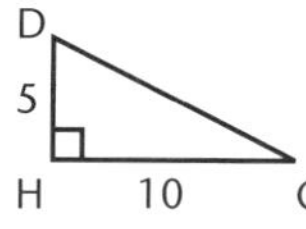

Note: $\angle AFE$ (= $\angle DGH$) is also the angle between the two planes.

In pyramids, take care to *identify perpendiculars correctly* when working with triangular planes.

Example F

Q. In the square-based pyramid shown, the sloping faces are isosceles triangles. X is the midpoint of BC and Y is the midpoint of AD. Name the angle between the planes ECB and ABCD.

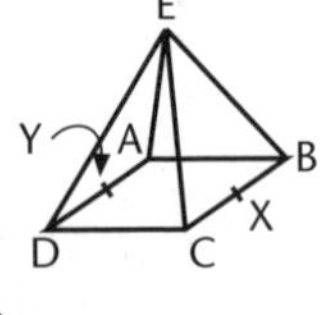

A. The diagrams below show how to identify the required angle, EXY.

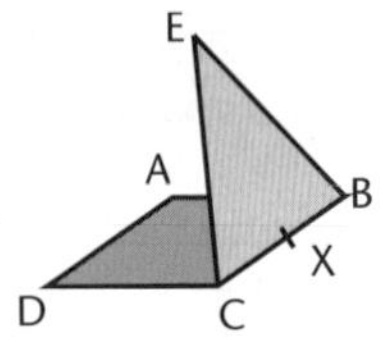

BC is the line of intersection of planes EBC and ABCD.

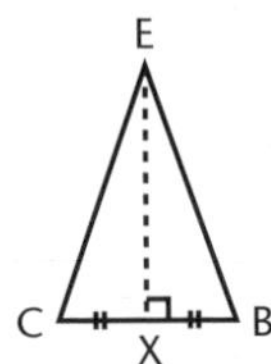

EX is at right angles to BC (ΔEBC isosceles).

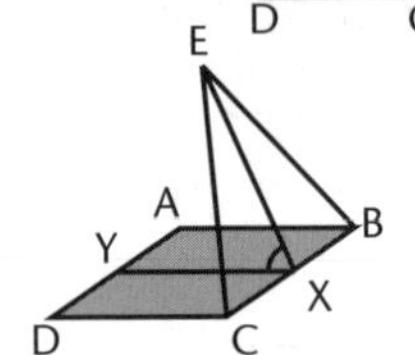

On the plane ABCD, the line XY is perpendicular to BC.

Shortest distance along the faces of a solid

You may be required to find the shortest distance between two points on a 3-D object when distances are measured along the *faces* of the object. To do this, draw a **net** of the object. Join the points concerned with a straight line and use trigonometry and/or **Pythagoras' Theorem** to find the length of the line joining the two points. There may be more than one diagram possible.

Example G

Q. Find the shortest distance between A and B over the faces of the cuboid shown.

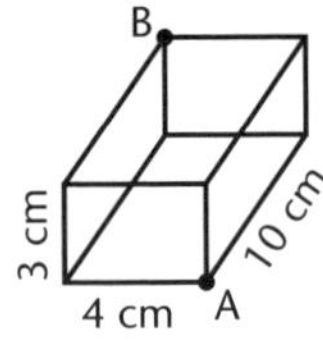

A. The relevant parts of the net of the cuboid are drawn, and the points A and B are joined with a straight line.

There are two cases.
A right-angled triangle is formed in each case.

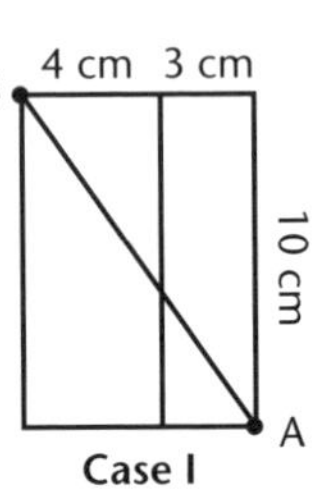

Case I

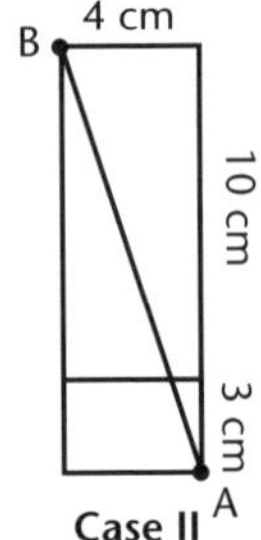

Case II

Case I

Using Pythagoras' theorem:

$AB^2 = 7^2 + 10^2$

$= 149$

$AB = \sqrt{149}$

$= 12.2$ cm (1 dp)

B 7 cm 10 cm A

Thus the shortest distance from A to B is 12.2 cm.

Case II

Using Pythagoras' theorem:

$AB^2 = 4^2 + 13^2$

$= 185$

$AB = \sqrt{185}$

$= 13.6$ cm (1 dp)

B 4 cm 13 cm A

Note: The shortest distance between A and B is still through the interior of the object (check this distance is 11.18 cm) but here distances are being calculated along the faces.

There are many practical applications for this type of problem.

Example H

Q. Sam wants to run a cord from one end of the top of the roof of the village Haus Bung (D) to the bottom corner at the opposite end (C).
What is the shortest length of cord Sam will need?

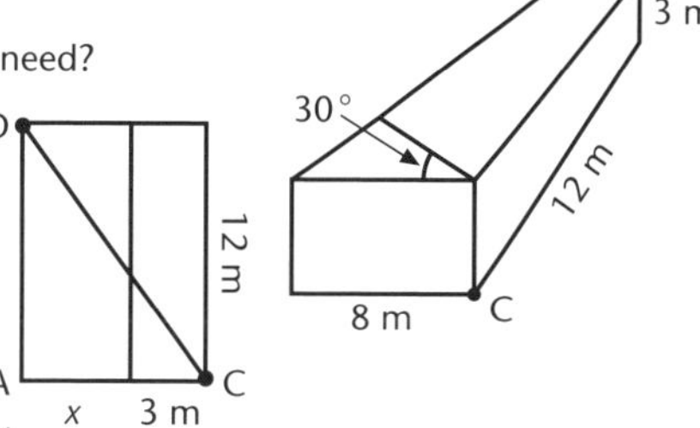

A. Part of the *net* of the Haus Bung is drawn. The roof peak opposite D is labelled A. The length x is the length of the roof from the top to the side wall.

In the triangle of the cross-section of the roof

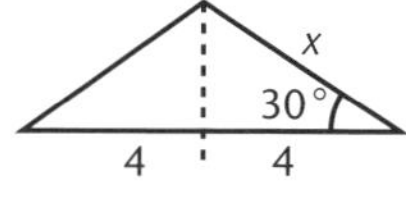

$$\cos 30° = \frac{4}{x}$$

$$\therefore x = \frac{4}{\cos 30°} \quad \text{[rearranging]}$$

$$= 4.619 \text{ m}$$

In triangle DAC drawn alongside,

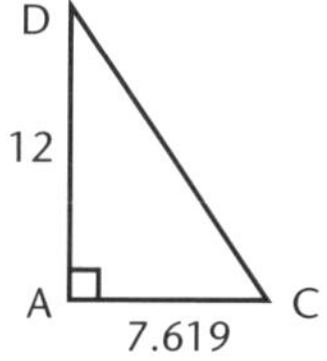

$$AC = x + 3$$

$$= 7.619 \text{ m}$$

$$CD^2 = 12^2 + 7.619^2 \quad \text{[Pythagoras' Theorem]}$$

$$= 202.049$$

$$CD = \sqrt{202.049}$$

$$= 14.2 \text{ m (1 dp)}$$

The shortest length of cord Sam will need is 14.2 m.

Unit 11.4 Activity 6B: Further 3-D trigonometry

1. The diagram shows a cuboid. Name the following angles.

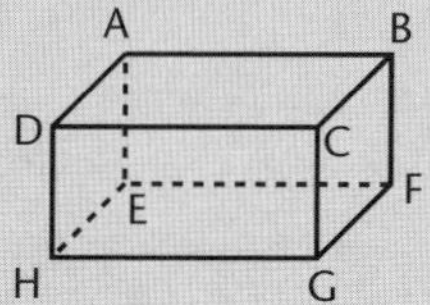

a. The angle between planes ABGH and EFGH.

b. The angle between planes CDEF and ABCD.

c. The angle between planes EFGH and DCFE.

d. The angle between planes ACGE and BCGF.

2. A cuboid ABCDEFGH has dimensions 12 cm by 5 cm by 8 cm.

Find the size of the angle between the plane AFGD and the plane BCGF.

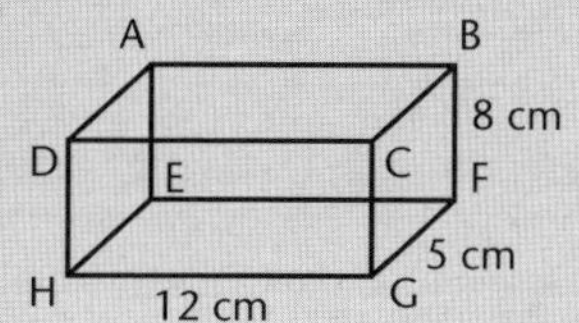

3. ABCDEFGH is a cube of side length 12 cm.

   Find the size of the angle between the planes ACGE and BDHF.

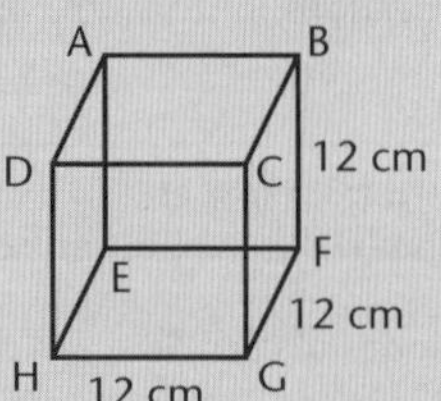

4. The diagram shows a triangular-based **prism**.

   Triangles ABC and DEF are right-angled, as shown.

   The length of AB is 5 cm. The length of BC is 8 cm and the length of CF is 18 cm.

   Calculate the size of the angle between planes BCFE and ADFC.

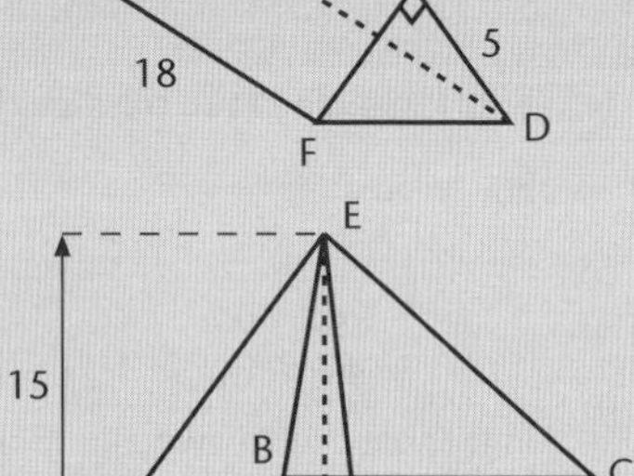

5. ABCDE is a right pyramid with a square base.

   Length of AD is 10 cm. Length of EH is 15 cm. X is the midpoint of CD.

   Name and calculate the angle between planes ECD and ABCD.

6. A rectangle-based pyramid is shown below.

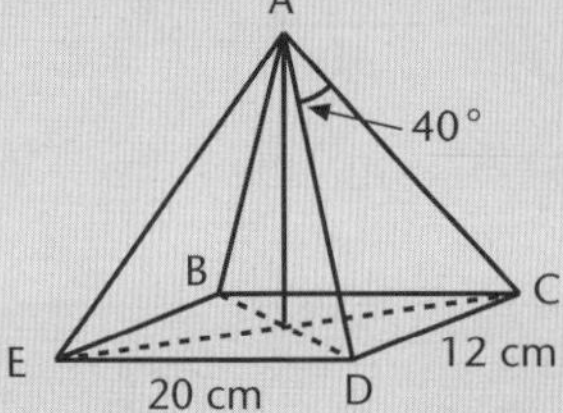

   Find the angle between the planes ADC and BCDE.

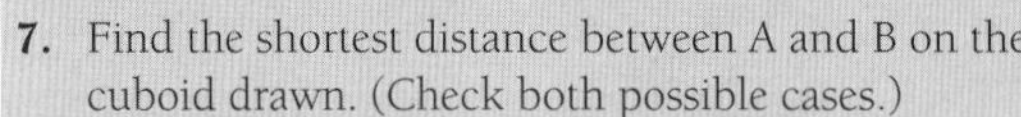

7. Find the shortest distance between A and B on the cuboid drawn. (Check both possible cases.)

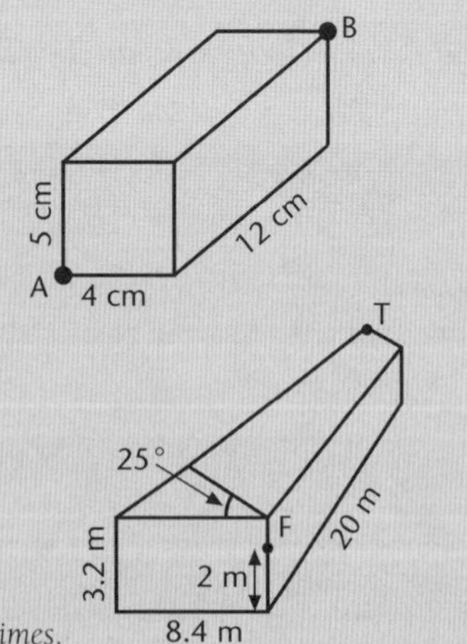

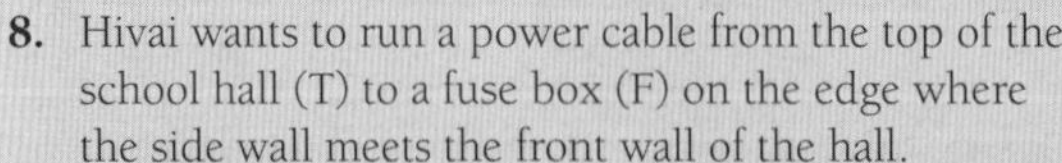

8. Hivai wants to run a power cable from the top of the school hall (T) to a fuse box (F) on the edge where the side wall meets the front wall of the hall.

   The fuse box is 2 m above the ground, and the walls of the hall are 3.2 m high. The hall is 20 m long and 8.4 m wide. The pitch of the roof is 25°.

   The cable is to be attached to the roof and wall *at all times*.

   a. What is the shortest length of cable Hivai needs?

   b. Where along the roof/wall line will the cable cross?

Unit 11.4 Geometry

Topic 7: The sine and cosine rules

The material in this Topic deals with the sine and cosine rule and the solving of practical problems using those rules. This Topic covers:

- Trigonometric modelling of practical problems, using non right-angled triangles.
- Using the sine and cosine rules to find lengths and angles.
- Finding the area of a triangle.
- Solving multi-step trigonometry problems, including 2-D representations of a 3-D situation.

Introduction

Trigonometry involving right-angled triangles (and the mnemonics SOH, CAH, TOA) should by now be familiar to the reader.

In this chapter, trigonometry is extended to include triangles which are not right-angled.

Conventions

An important **convention** in this section is the method used to name the angles and sides of a triangle.

The diagram shows how capital letters are used to show the angles at the corners of the triangle. The length of the side *opposite* each corner is denoted by the same letter in the lower case (side *a* is opposite angle A, etc).

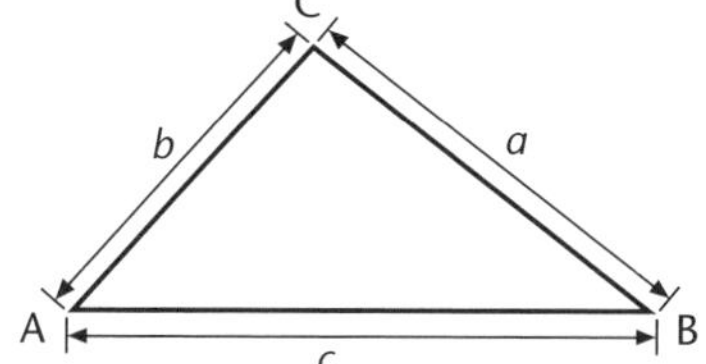

The cosine rule

For any triangle ABC, the **cosine rule** states:

$$a^2 = b^2 + c^2 - 2bc \cos A$$

Making cos A the subject of $a^2 = b^2 + c^2 - 2bc \cos A$ gives another form of the cosine rule, useful for calculating unknown angles:

$$\cos A = \frac{b^2 + c^2 - a^2}{2bc}$$

Note: Different letters can be made subject of these formulae.
By symmetry, $b^2 = a^2 + c^2 - 2ac \cos B$ and $c^2 = a^2 + b^2 - 2ab \cos C$.
Also $\cos B = \dfrac{a^2 + c^2 - b^2}{2ac}$ and $\cos C = \dfrac{a^2 + b^2 - c^2}{2ab}$

Example A

Q. A fishing vessel sets out to sea. It is 5.2 km from point P on the shore and 4.7 km from another point, Q, on the shore. The angle between the boat and these two points is 75°.

1. How far apart are the two points?
2. Failing to catch any fish, the captain sails the boat to a point which is now 5.4 km from point P and 6.3 km from Q. What is the angle between P, the boat, and Q?

A. 1. The diagram shows the points P and Q.

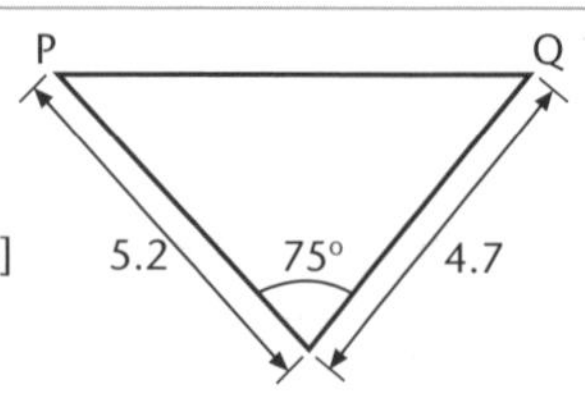

The length PQ is required. Using the cosine rule,

$PQ^2 = 5.2^2 + 4.7^2 - 2 \times 5.2 \times 4.7 \cos 75°$

[substituting into $a^2 = b^2 + c^2 - 2bc \cos A$]

$= 36.48$

$\therefore$ PQ = 6.0 km (1 dp) [taking square roots]

2. The required angle, x, is shown below.

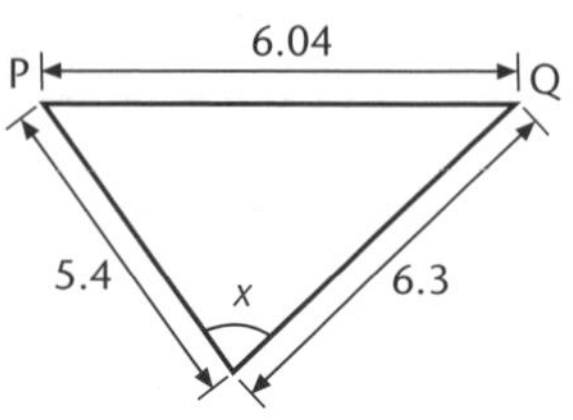

$$\cos x = \frac{6.3^2 + 5.4^2 - 6.04^2}{2 \times 6.3 \times 5.4}$$

[substituting into $\cos A = \dfrac{b^2 + c^2 - a^2}{2bc}$]

$\therefore \cos x = 0.4757$

$\therefore x = 61.6°$ (1 dp) [taking $\cos^{-1}$]

Unit 11.4 Activity 7A: The cosine rule and applications

1. In the diagram shown, find:

a. The length x.

b. The size of $A\hat{B}C$.

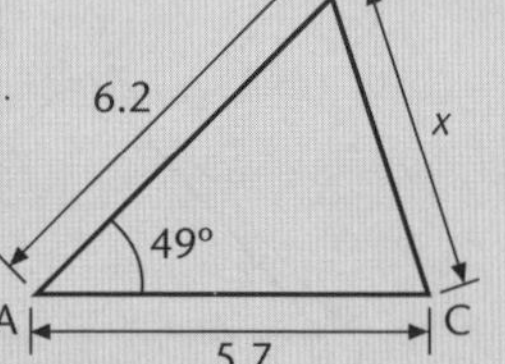

2. In this diagram find:

a. The length x.

b. The angle y.

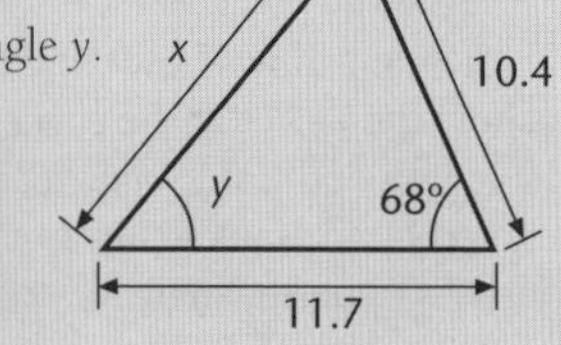

3. In this diagram find:

a. The length of QR.

b. The size of $P\hat{Q}R$.

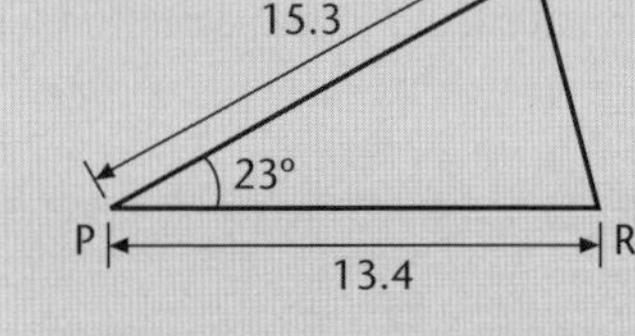

4. In the triangle shown, calculate

a. The length of BC.

b. The size of angle ABC.

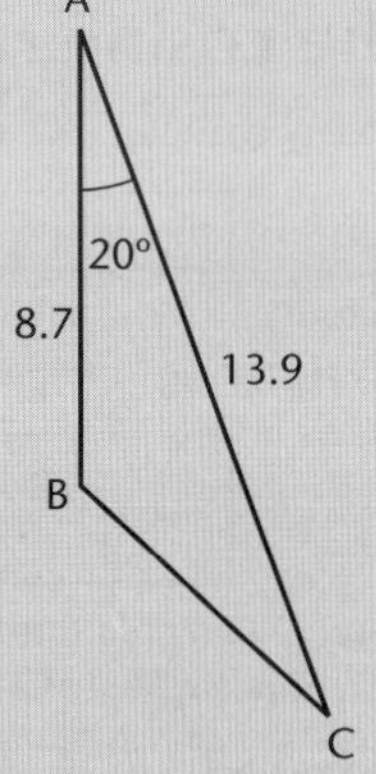

5. A bar of length 2 m is suspended by two strings of 1.3 m and 1.8 m long, attached to a nail.

a. Find the angle between the two pieces of string.

b. If the bar is now shortened so that the angle between the two strings is 75°, find the new length of the bar.

6. A triangular shaped field has one boundary extending 350 m from a gate in a direction N35°E. The other boundary extends N32°W from the gate for 200 m.

a. Find the length of the third side of the field to the nearest metre.

b. Find the largest angle of the field.

The sine and area rules

Using the same triangle labelling conventions as before, the **sine rule** states:

$$\frac{\sin A}{a} = \frac{\sin B}{b} = \frac{\sin C}{c} \quad \text{or} \quad \frac{a}{\sin A} = \frac{b}{\sin B} = \frac{c}{\sin C}$$

Note: The sine rule is a short way of expressing three separate relationships:

$\frac{a}{\sin A} = \frac{b}{\sin B}$, $\frac{a}{\sin A} = \frac{c}{\sin C}$ and $\frac{b}{\sin B} = \frac{c}{\sin C}$ (and their reciprocal relationships).

The **area rule** states that the area of a triangle is:

$$\text{Area} = \frac{1}{2}ab \sin C$$

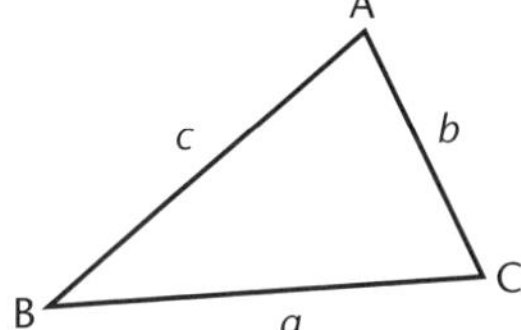

Note: Alternatively, the area is equal to $\frac{1}{2}bc \sin A$ or $\frac{1}{2}ac \sin B$.

Example B

Q. Peter and Moreni tie two ropes to the top of a pole. They stake the other end of each rope into the ground so they are tight and in the same plane with the pole. Peter's rope is 5.6 m long and makes an angle of 26° with the ground. Moreni's rope makes an angle of 32° with the ground.

1. Find the length of Moreni's rope.
2. Find the distance between the two stakes.
3. Find the area of the triangle formed by the two ropes and the line along the ground which joins the two stakes.

A. The following diagram shows Moreni's rope with length L, and the distance between the stakes d. Let the area required be A.

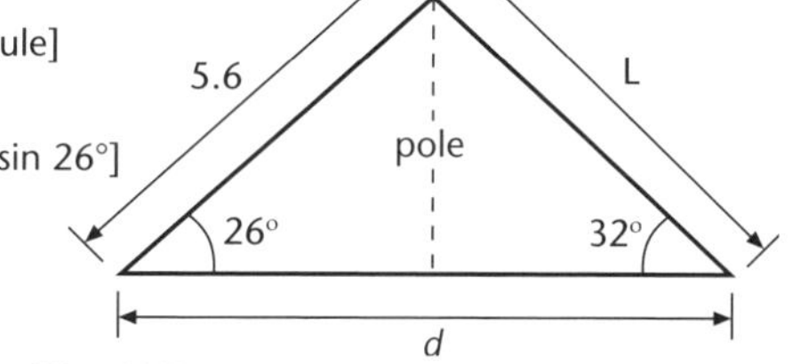

1. $\frac{L}{\sin 26°} = \frac{5.6}{\sin 32°}$ [using the sine rule]

$\therefore L = \frac{5.6}{\sin 32°} \sin 26°$ [multiplying by sin 26°]

$= 4.63255\ldots$

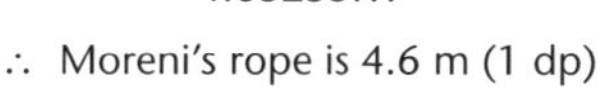

$\therefore$ Moreni's rope is 4.6 m (1 dp)

2. The third angle of the triangle is 180 – 26 – 32 = 122°

$\frac{d}{\sin 122°} = \frac{5.6}{\sin 32°}$ [using the sine rule]

$\therefore d = \frac{5.6 \times \sin 122°}{\sin 32°}$ [multiplying by sin 122°]

$= 8.96187\ldots$

$\therefore$ the distance between the stakes is 9.0 m (1 dp)

3. $A = \frac{1}{2} \times 5.6 \times 8.962 \times \sin 26°$ [using area $= \frac{1}{2}ab \sin C$]

$= 11.000\ldots$

$\therefore$ the area of the triangle is 11.0 m^2 (1 dp)

Unit 11.4 Activity 7B: The sine and area rules and applications

1. In this diagram find:

a. The length AC. **b.** The length BC. **c.** The area of ΔABC.

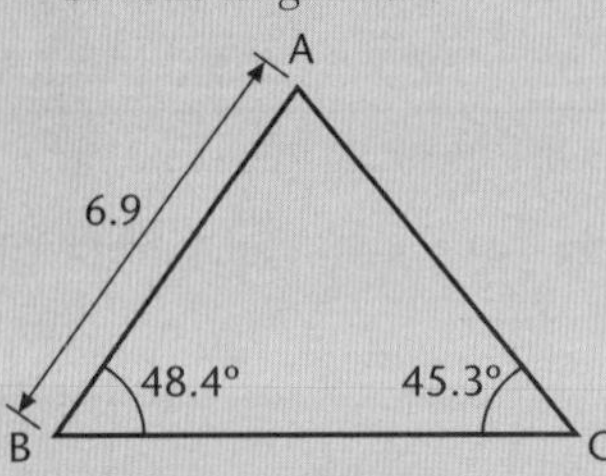

2. In this diagram find the length of: **a.** AB. **b.** AC.

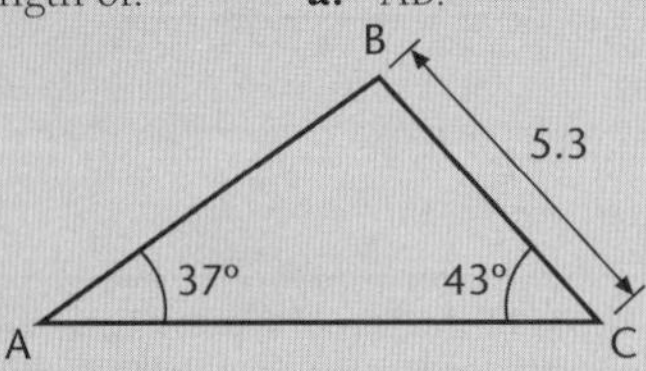

3. In this diagram find:

a. The length of AC. **b.** The length of CD. **c.** The area of ABCD.

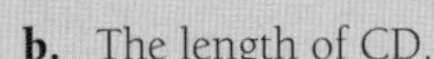

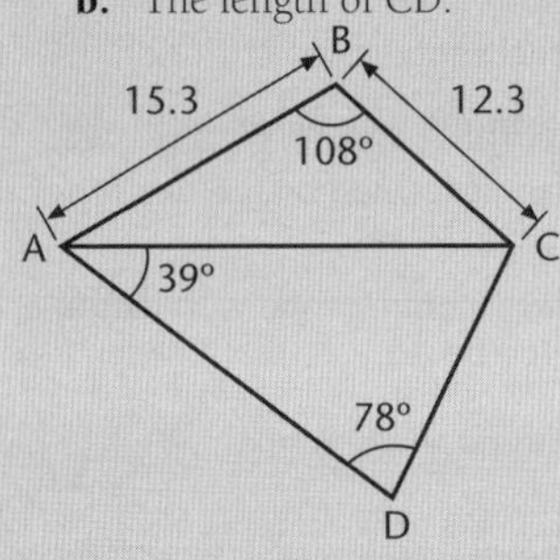

4. A ship sails for 105 km in a direction E32°S. It then sails for another 200 km in a direction N10°E.

a. How far is it from its original position?

b. What is its **compass bearing** from the original position?

5. A father looks down at his son's eyes. The man's eyes are tilted at 38° to the vertical. His son looks at his father's feet. The boy's eyes are tilted at 42° to the vertical. The son's eyes are 1.1 m from his father's eyes. Find the height of the father's eyes from the ground.

6. A triangular wedge (prism) has a base with an angle of 43° enclosed between sides of 4.3 cm and 5 cm. Find:

a. the length of the third side of the triangle.

b. the area of the cross-section.

c. the volume of the wedge if the length is 51 cm.

7. A section of land is of triangular shape. One corner of the land has fences of lengths 52 m and 54 m enclosing an angle of 66°.

a. Find the cost to the nearest dollar of fencing the land at $6.80 a metre.

b. Find the area of the section.

8. A man walks for 5.6 km on a bearing of N33°E from O. He stops and finds that he is now at a bearing of N27°W from P which is due East of O.

a. How far is he from P?

b. How far is P from O?

9. A cyclist rides for 11 km in a direction of N38°E from O. At this time she dismounts and finds she is 12 km away from P which is due East of O.

a. What is her compass bearing from P?

b. What distance is P from O?

10. ABC is an equilateral triangle with sides of length 2. AE, BE, CE are of equal length. Find that length.

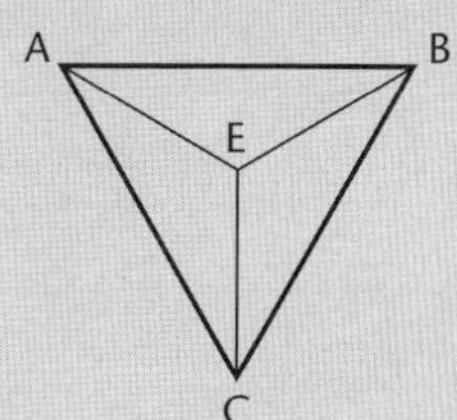

11. A student placed a piece of clean paper on a table in the middle of a field ABCD and by use of a device called an alidade was able to construct the following diagram (not to scale). The lines intersect at a point E on the paper. AE is 30 m long, BE is 40 m long, DE is 40 m long and CE is 35 m long.

a. Find angle BAE.

b. Find the perimeter of the field ABCD.

c. Find the area of the field.

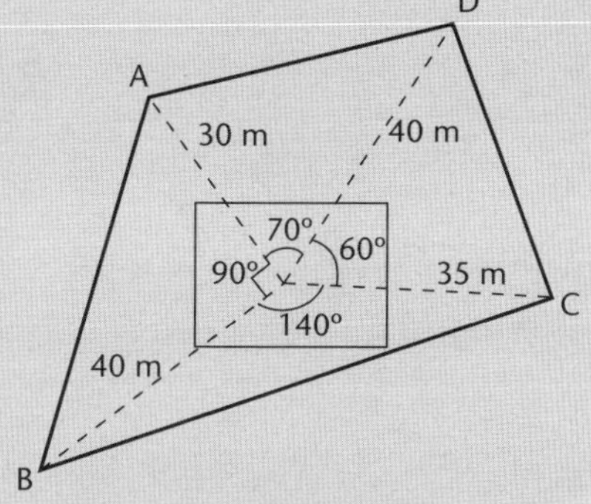

12. A small boat, which is experiencing difficulties, sets off an alarm. This alarm is detected by two aeroplanes which are 8 km apart in a North-South direction. The northernmost plane detects the alarm at a bearing of 120° and the other plane detects the alarm at a bearing of 040°. How far is the northernmost plane from the boat?

Unit 11.4 Geometry
Topic 8: Circle geometry

The material in this topic relates to circle properties (Syllabus p.17) and the use of logical argument in the resolution of mathematical problems. It covers:

- Circle geometry (limited to angle geometry).
- Finding unknowns in circles using two-step processes.
- Investigating a conjecture, presenting a proof or solving a problem using at least three steps of reasoning.

The circle

A **circle** is the set of all points in a plane a fixed distance (the length of the **radius**) from a particular point (the **centre**).

Lines and areas relating to circles are described in the following diagrams.

Centre	Circumference	Diameter	Radius	Tangent
The point about which the circle is formed. 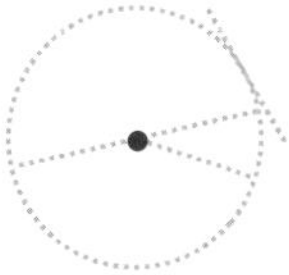	The line that forms the circle. 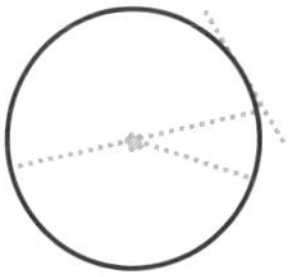	A straight line segment passing through the centre of a circle and joining two opposite points on the circumference. 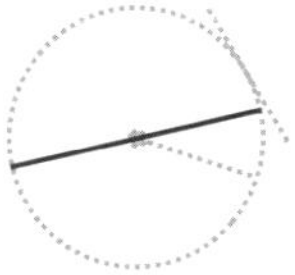	A straight line segment from the centre of a circle to a point on the circumference. 	A straight line touching one point on the circumference of a circle. 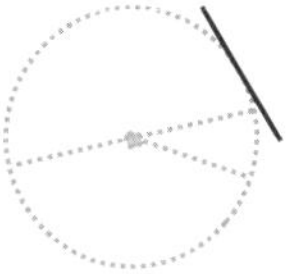

An **arc** is part of the **circumference**. There are always two arcs between any two points P, Q on the circumference, a **minor arc** and a **major arc**. When the length of the minor arc is the same as the length of the major arc, the arcs are called **semi-circles**. Arcs can be drawn using a compass, in the same manner as for drawing a full circle.

Minor Arc	Major Arc	Semi-circle
The shorter arc between any two points on the circumference of a circle. P minor arc Q 	The longer arc between any two points on the circumference of a circle. P major arc Q 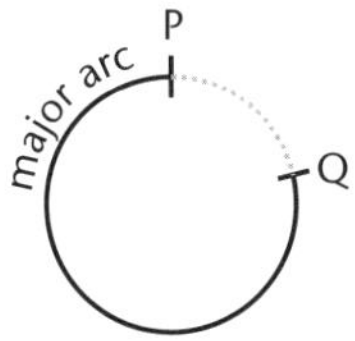	Occurs when the length of the major arc and minor arc are the same. semi-circle P Q 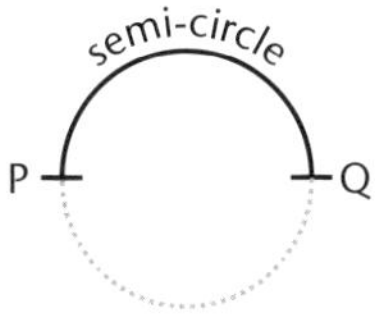

In general, the arc PQ refers to the minor arc PQ (of length less than the semicircle).

Sector	Chord	Segment
A region inside a circle enclosed by an arc and two radii.	A straight line segment joining any two points on the circumference.	A region inside a circle enclosed by a chord and an arc.

The diagrams show angles which the arc AB **subtends** (lies under).

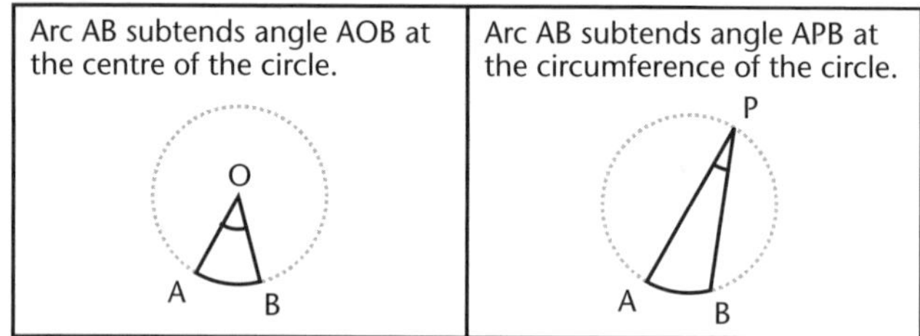

Angles on the same arc

Angles on the circumference of a circle which are subtended by the *same* arc (or **chord**) are equal in size. (Abbreviated to: ∠s same arc or ∠s same chord.)

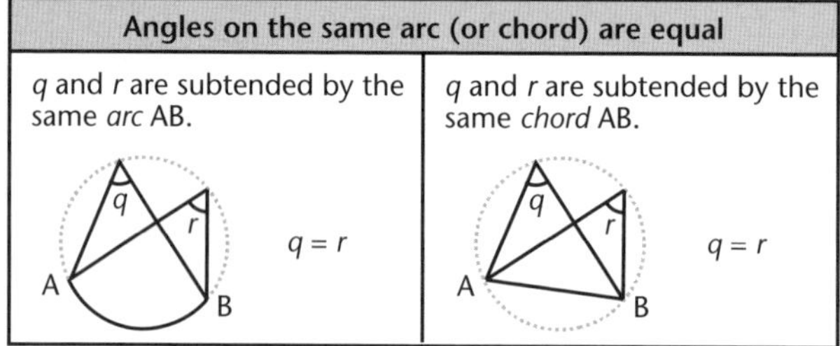

Note: In the diagram shown opposite:

$x \neq y$ because x and y are subtended by different arcs.

$x \neq z$ because, although these two angles are subtended by the same arc, z is not an angle at the circumference.

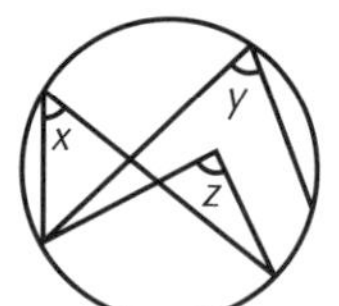

Example A

In the diagram:

$x = 35°$ (∠s same arc)

$y = 42°$ (∠s same arc)

$y + z + 35 = 180$ (sum ∠s triangle)

$42 + z + 35 = 180$ [$y = 42°$]

$\therefore\ z = 103°$ [simplifying]

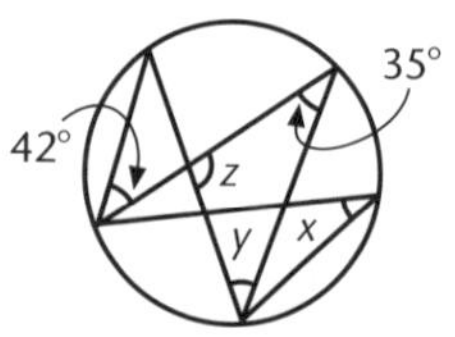

Angles at the centre and circumference

If an angle at the circumference of a circle and an angle at the centre are subtended by the same arc (or chord), then the angle at the centre is twice the angle at the circumference. (Abbreviation: ∠ at centre.)

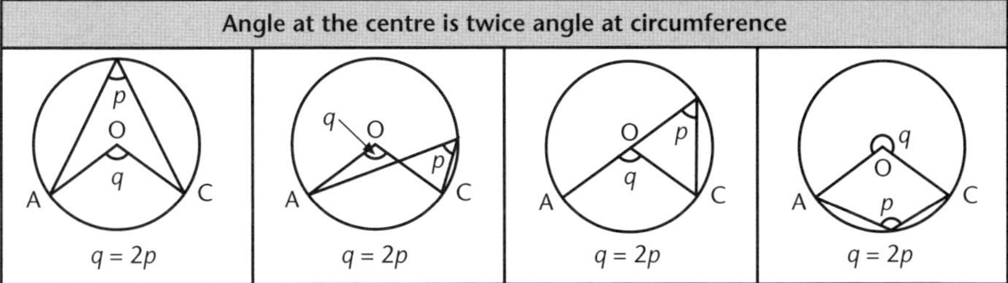

Note: In the diagram shown opposite, $q \neq 2r$, because the angle q is subtended by the arc ABC and the angle r is subtended by the arc ADC.

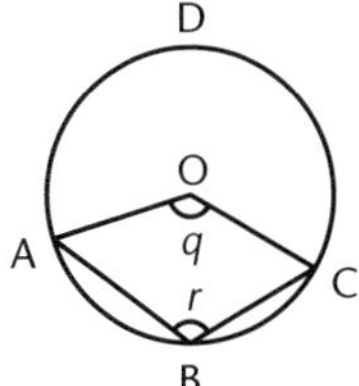

Example B

In the diagram:

$$a = \frac{100}{2} \quad (\angle \text{ at centre})$$
$$= 50°$$

Reflex ∠AOC = 260° (∠s at a point)

$$\angle b = \frac{1}{2} \times 260 \quad (\angle \text{ at centre})$$
$$= 130°$$

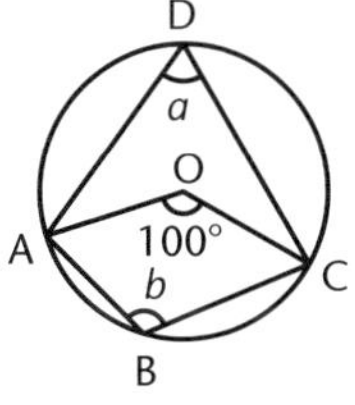

Angle in a semi-circle

The following result is a special case of the rule for the angle at the centre.

When the angle at the centre is 180°, the angle at the circumference is a right angle. (Abbreviation: ∠ in semi.)

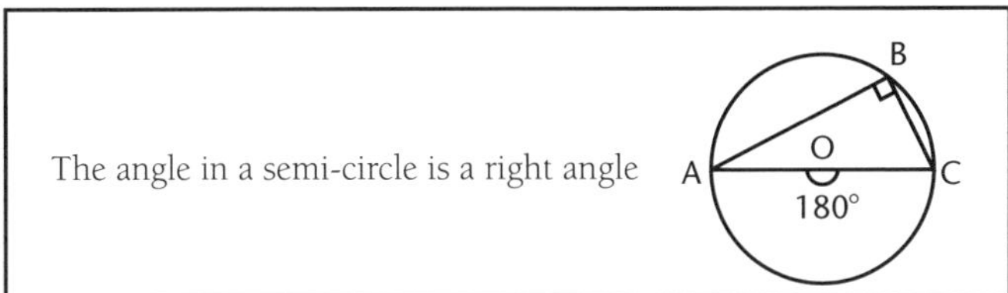

Note: When solving problems involving triangles inside circles, be on the lookout for isosceles triangles, formed when two of the sides of the triangle are radii of the circle.

Unit 11.4 Activity 8A: Circle geometry

1. Find the size of the angles labelled with letters. Give a brief reason for each answer.

a.

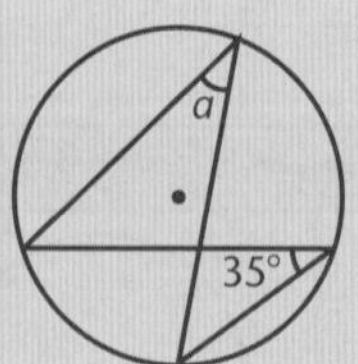

b.

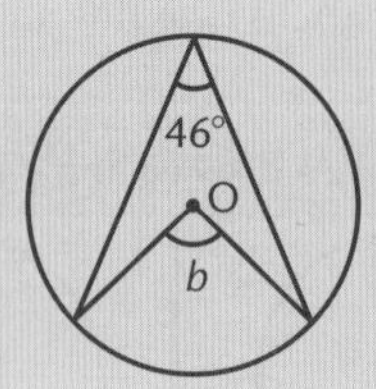

c.

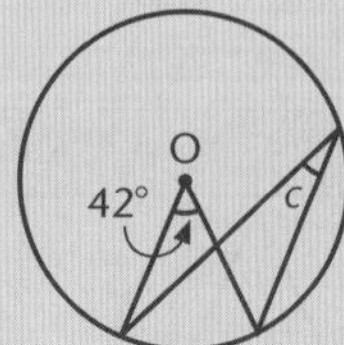

d.

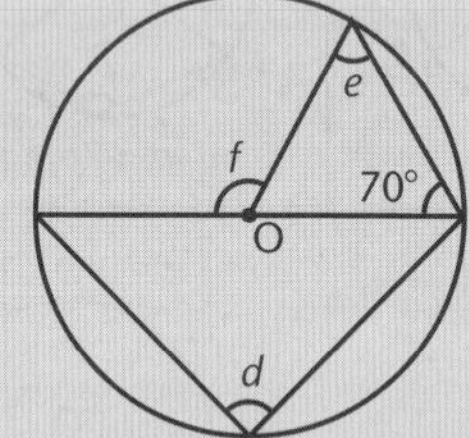

e.

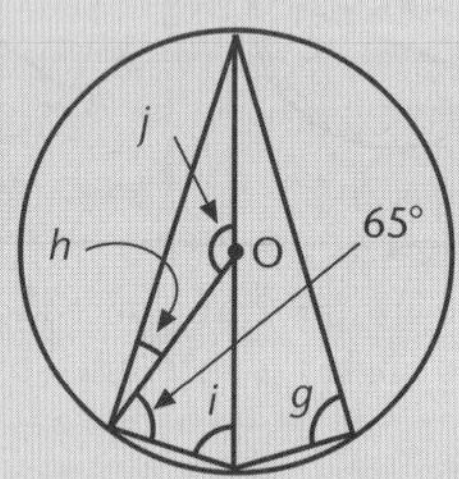

f.

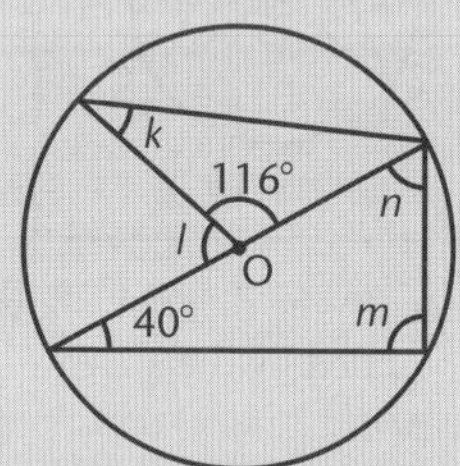

2. Find, with reasons, the size of the angles marked.

a.

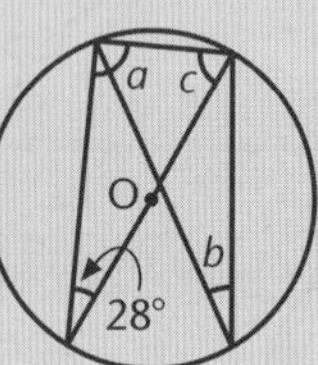

b.

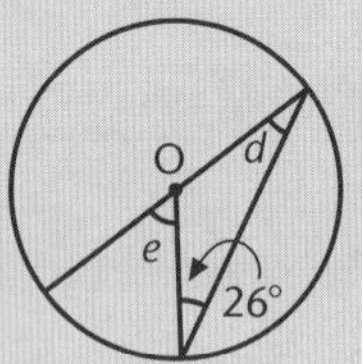

c.

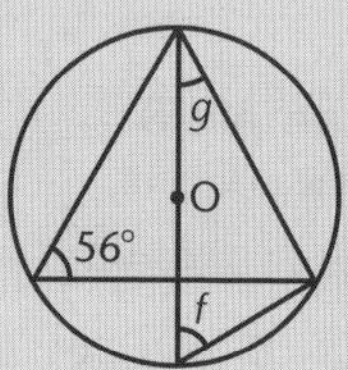

d.

e.

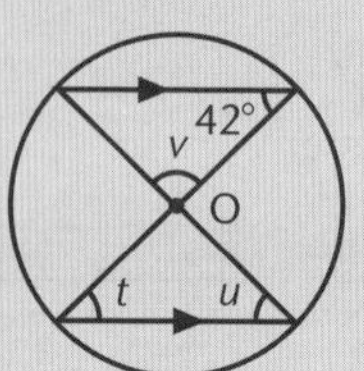

f.

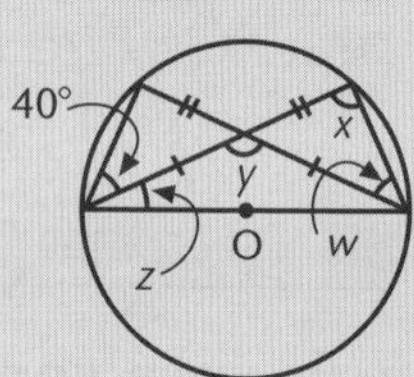

3. In the semi-circle shown O is the centre, ABCD is an **isosceles trapezium** with ∠BAD = 64°. Find, with reasons, the size of ∠DOC.

4. A logo for a sail-maker is shown, where O is the centre of the circle and PQS is an isosceles triangle. If ∠OSQ = 35°, find, with reasons, the size of QPS.

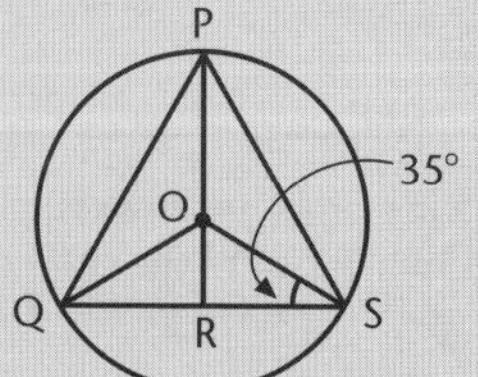

5. A circle has a centre O, and the points A, B, C, D lie on its circumference. If ∠AOB = 54° find:

a. ∠OBC

b. ∠ODA

c. Why is AD parallel to BC?

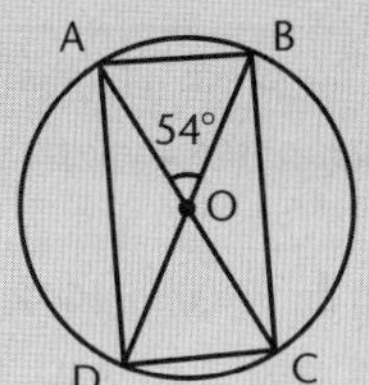

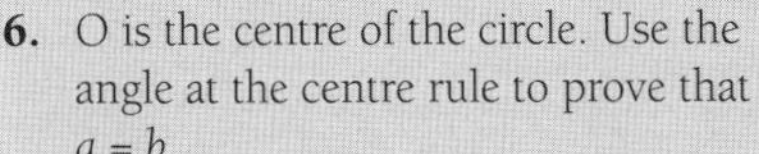

6. O is the centre of the circle. Use the angle at the centre rule to prove that $a = b$.
Hint: Consider the relationship between x and a and x and b.

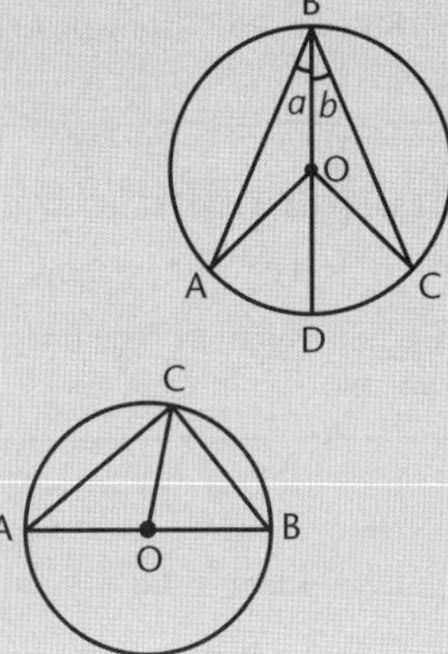

7. O is the centre of the circle ABCD. Prove that ∠AOC = 2 ∠ABC.
Hint: Find the size of ∠BAO in terms of a, then find the size of ∠AOD in terms of a.

8. AB is the diameter of circle centre O. Prove that ∠BCA = 90°.

Hint: Extend CO to D on the circumference and let ∠AOD = x, ∠BOD = y.]

Further properties of the circle

Radius and tangent

A right angle is formed at the point where the radius of a circle meets the tangent to the circle at that point. (Abbreviation: rad ⊥ tan.)

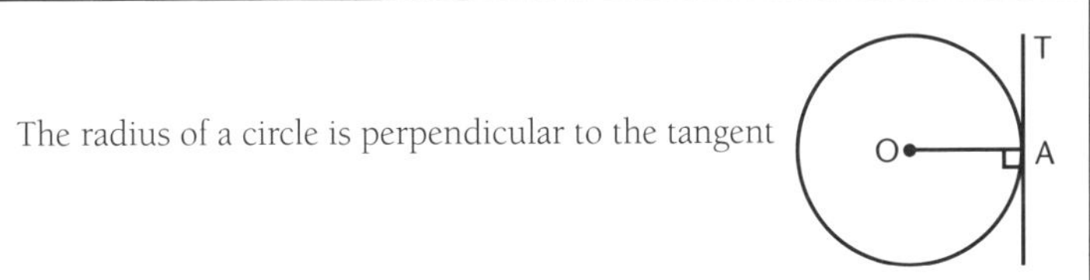
The radius of a circle is perpendicular to the tangent

Example C

Q. A circle centre O has a tangent which touches it at T.

Find the sizes of the angles a, b, c, d. Give reasons.

A. $a = 90°$ (rad $\perp$ tan)

OT and OR are radii, so ΔOTR is isosceles.

$\angle$OTR $= 35°$ (base $\angle$ isos Δ)

$b + 35 = 90$ (rad $\perp$ tan)

$b = 55°$

$c + 35 = 90$ ($\angle$TRU $= 90°$, angle in semicircle)

$c = 55°$

ΔORU is also isosceles, with base angles c and $\angle$OUR. [since OR and OU are radii]

So $d = 180 - 55 - 55$ (sum $\angle$s Δ)

$= 70°$

Tangents from a point

From a point T outside a circle, two tangents to a circle, TA and TB, can be drawn.

The tangents are equal in length. The line, TO, drawn from the point outside, T, to the centre of the circle, O, is an axis of symmetry.

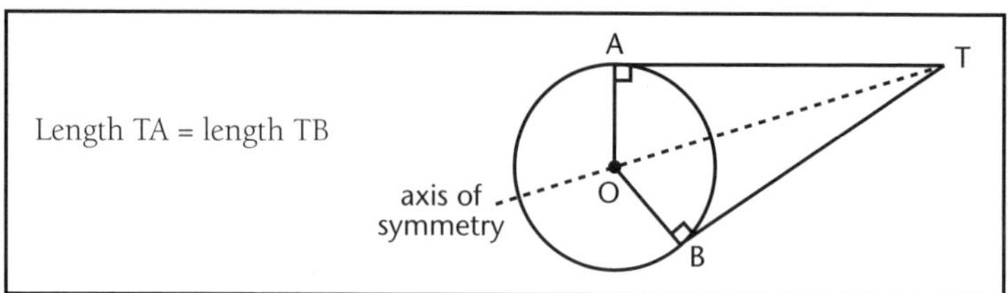

Chords and mediators

The **mediator** (perpendicular bisector) of a chord passes through the centre of the circle, O.

Likewise, the perpendicular from the centre of a circle to a chord will bisect the chord.

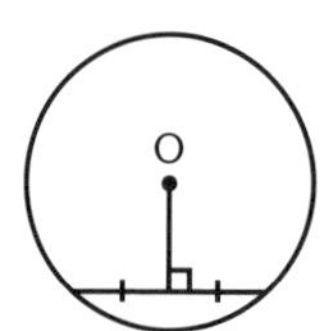

Cyclic quadrilaterals

A **cyclic quadrilateral** is a quadrilateral with all four vertices (corners) on the circumference of the same circle.

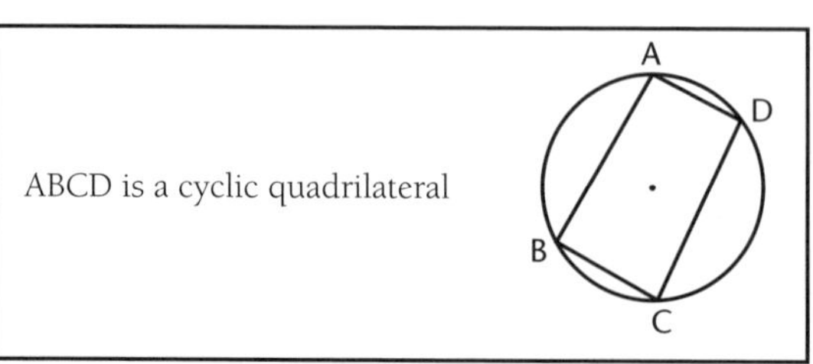

Note: OPQR is not a cyclic quadrilateral, because point O does not lie on the circumference.

Properties of cyclic quadrilaterals

The *opposite* angles of a cyclic quadrilateral are **supplementary** (add to 180°). (Abbreviation: opp ∠s cyc quad.)

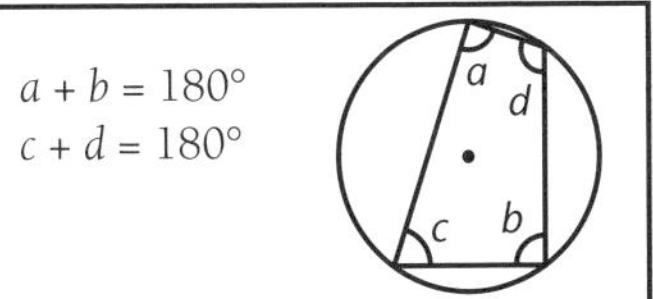

The *exterior* angle of a cyclic quadrilateral is equal to the interior opposite angle. (Abbreviation: ext ∠ cyc quad.)

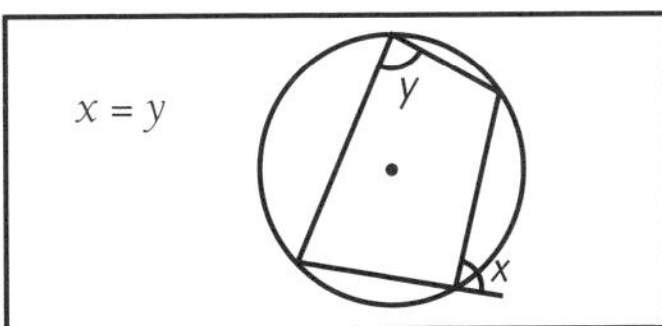

Example D

In the diagram:

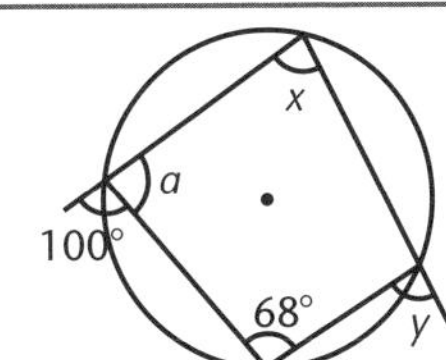

$x + 68 = 180$ (opp ∠s cyc. quad)

$x = 112°$

$a = 80°$ (∠s on line)

$y = 80°$ (ext ∠ cyc quad)

Concyclic points

Concyclic points are points which lie on the circumference of the same circle.

Four points will be concyclic if, in the quadrilateral they form:

1. the opposite angles add to 180°, *or*
2. an exterior angle equals the interior opposite angle, *or*
3. the same straight line subtends two equal angles.

If four points are concyclic, then they form a cyclic quadrilateral.

Example E

Q. Test the following quadrilaterals to see if they are concyclic:

1.

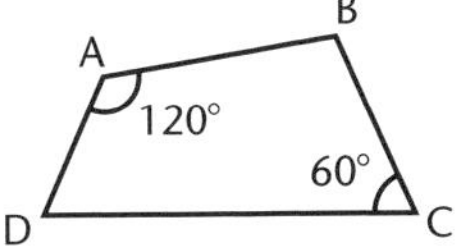

2.

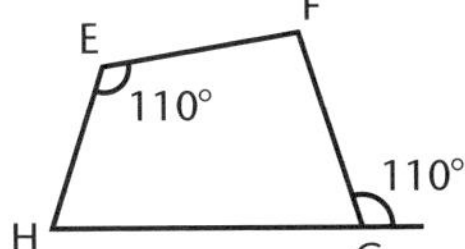

3.

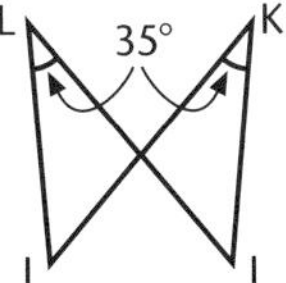

A. **1.** ∠A + ∠C = 180°, so ABCD is a cyclic quadrilateral since opposite angles add to 180°. (A, B, C, and D are concyclic points.)

2. EFGH is a cyclic quadrilateral since the exterior angle (at G) is equal to the interior opposite angle (at E). (E, F, G, and H are concyclic points.)

3. I, J, K, and L are concyclic points since IJ subtends equal angles at K and L.

Unit 11.4 Activity 8B: Cyclic quadrilaterals

1. Find the size of the angles labelled with letters. In each case O is the centre of the circle. Give a brief reason for each answer.

a.

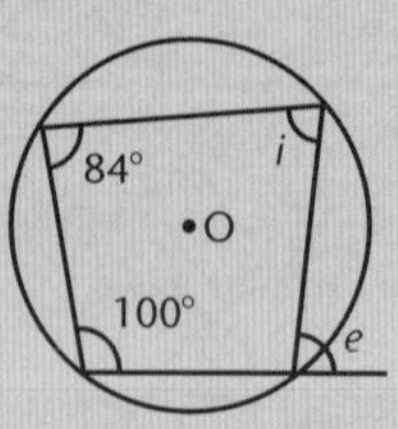

b.

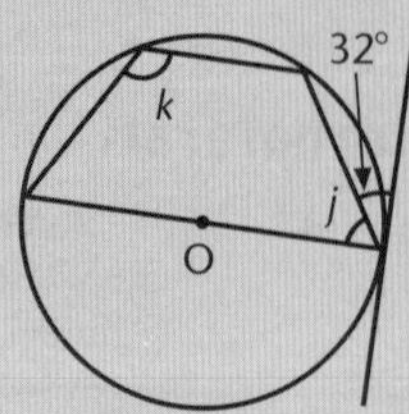

c.

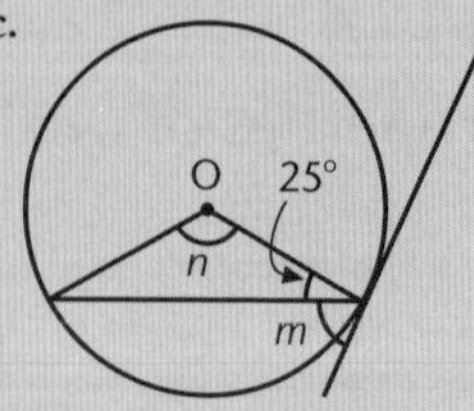

d.

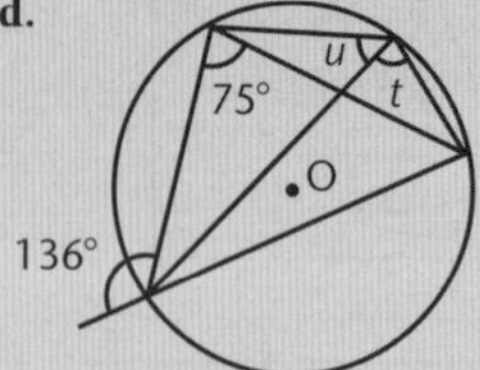

e.

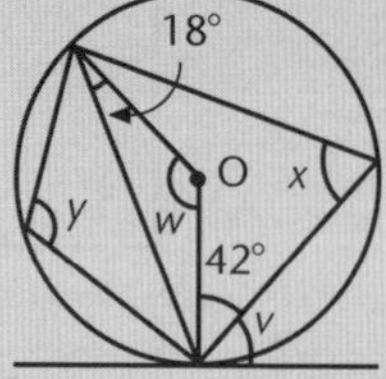

f.

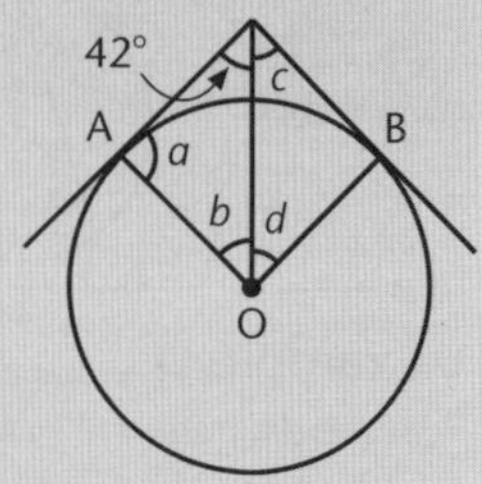

2. Find, with reasons, the sizes of the angles marked. In each case O is the centre of the circle.

a. Circle has tangent at D.

b. AB and AC are tangents to the circle.

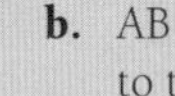

c. TA and TB are tangents to the circle.

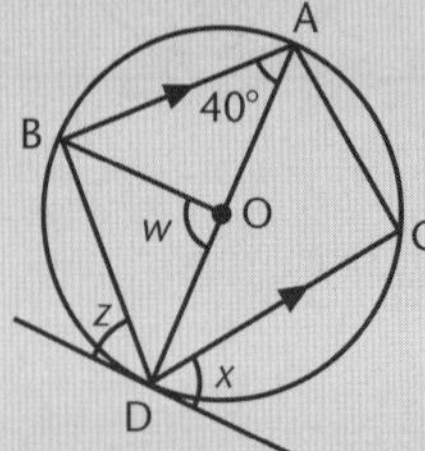

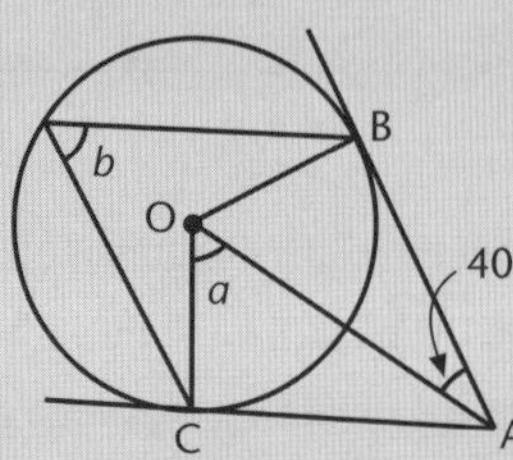

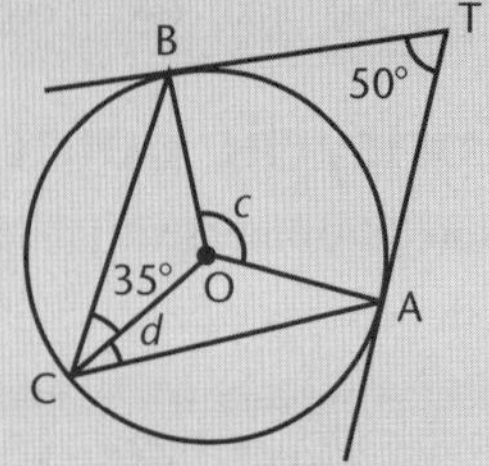

d.

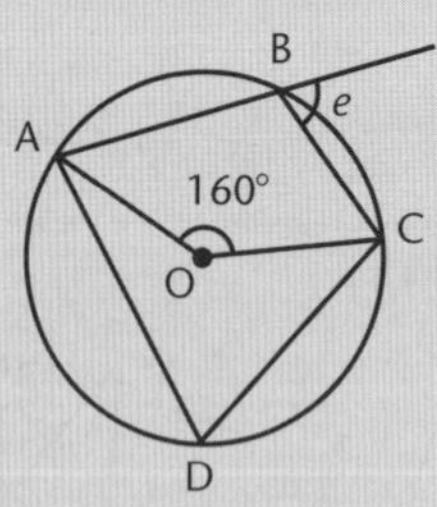

e.

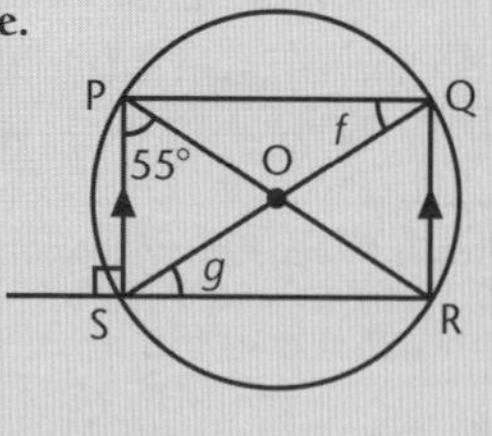

f. Circle has a tangent at T.

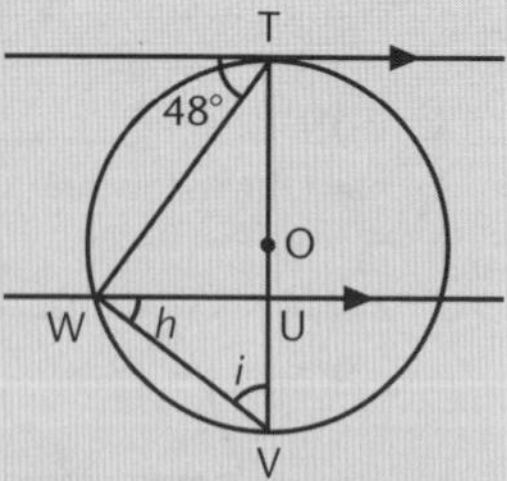

3. In the diagram shown, find the sizes of angles a and b. The circle has centre O and OB = AB. Give reasons.

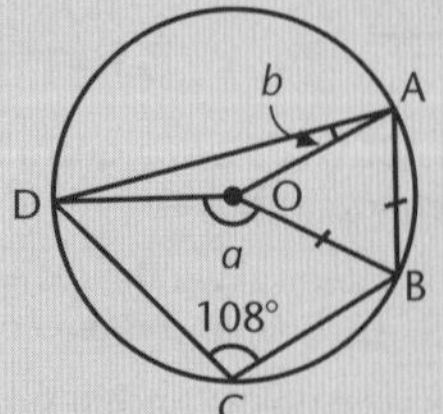

4. Name a set of four concyclic points on each diagram.

a.

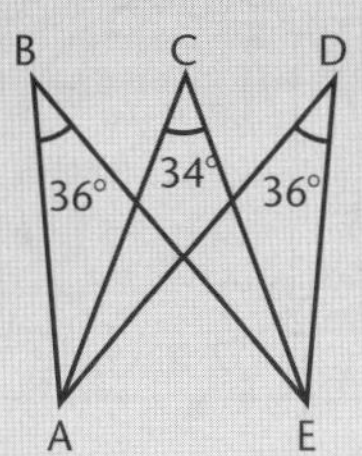

b.

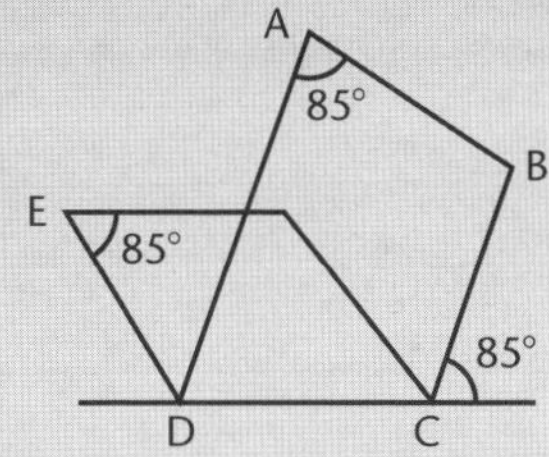

c.

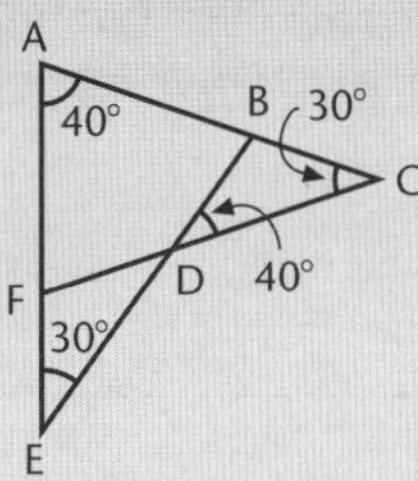

d.

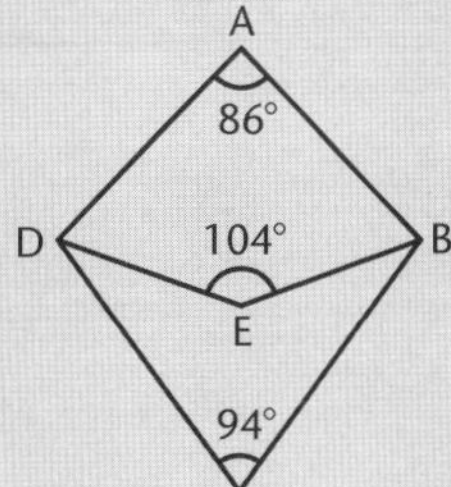

e.

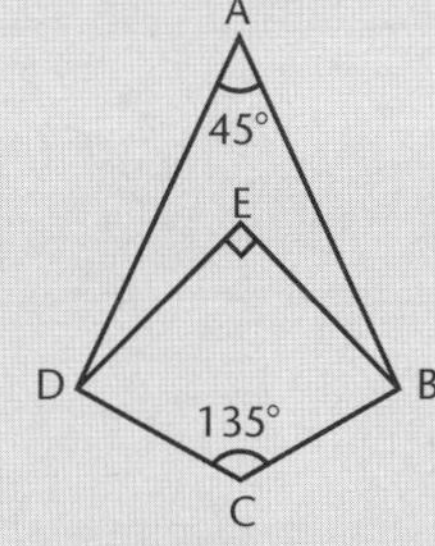

f.

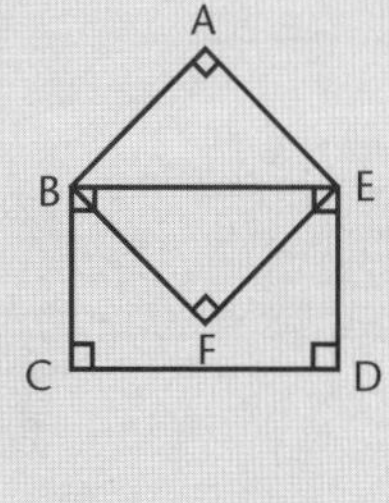

5. A semicircle, centre O, is drawn with ∠BAO = 60°.

Copy the figure and:

a. Mark all sides of equal length on the diagram.

b. Draw the tangent at B. Mark a point on the tangent and name it P. Join P to O.

c. What type of triangle is?

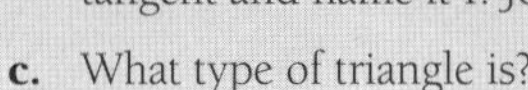

i. ABC **ii.** OBC **iii.** OBP **iv.** AOB

d. Would it be possible to choose a position P on the tangent so that ΔOBP was obtuse-angled? Explain.

6. ABC and PQR are straight lines. Prove that AP is parallel to RC.

Hint: Consider ∠s RCB, BQR and PAB.

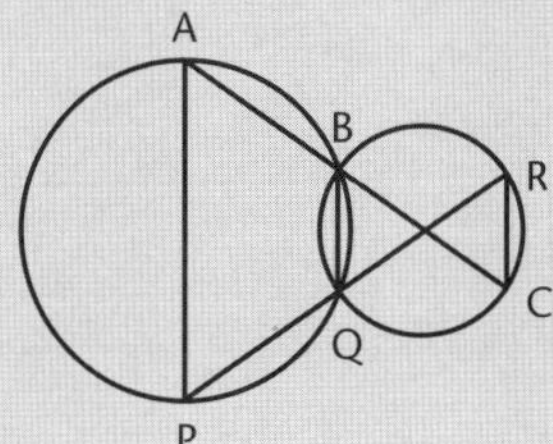

7. TA is a tangent to the circle centre O. Prove that $x = y$.

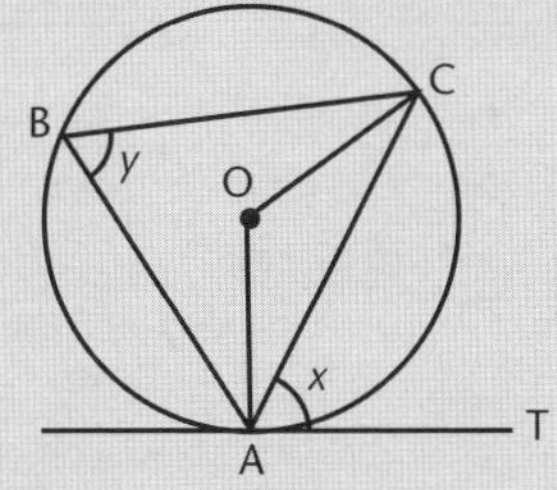

8. Prove that the opposite angles of a cyclic quadrilateral add to 180° (ie A + C = 180° or B + D = 180°).

Hint: Join BO and OD and consider ∠ at centre rule.

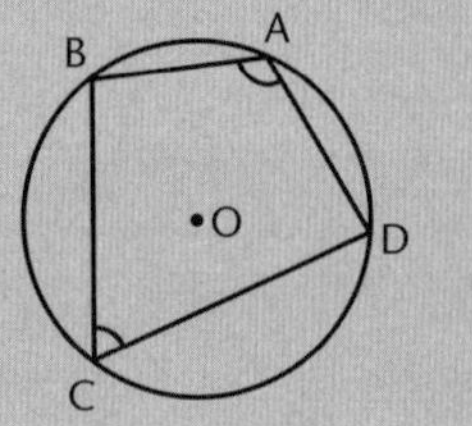

Unit 11.4 Geometry
Topic 9: Equations of a circle

The material in this Topic continues the focus on circle properties and deals with straightforward non-linear graphs. It covers:

- Drawing circles with centre at a point (a, b).
- Drawing circles of $(x - a)^2 + (y - b)^2 = r^2$.
- Identifying and interpreting features of graphs.
- Writing equations of graphs.
- Finding points of intersection.

Circles centred at a point (a, b)

The general equation of a circle with radius r and centre (a, b) can be found from the diagram in a similar way to that above. The equation is:

$$(x - a)^2 + (y - b)^2 = r^2$$

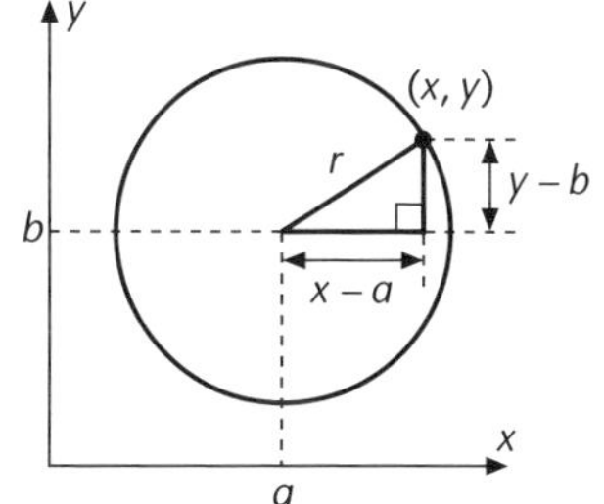

Example A

Find the equation of the circle centred at (1, 2) and with radius 4.

Solution

The equation of a circle centred at (1, 2) and with radius 4 is found by substituting $a = 1$, $b = 2$ and $r = 4$ into $(x - a)^2 + (y - b)^2 = r^2$ giving:

$$(x - 1)^2 + (y - 2)^2 = 4^2$$

$$\therefore\ (x - 1)^2 + (y - 2)^2 = 16$$

$$\therefore\ x^2 + y^2 - 2x - 4y + 1 + 4 = 16 \quad \text{[expanding]}$$

$$\therefore\ x^2 + y^2 - 2x - 4y - 11 = 0 \quad \text{[simplifying]}$$

Example B

Find the equation of the circle centred at (–3, 4) and with radius 2.

Solution

The equation of a circle centred at (–3, 4) with radius 2 is:

$$(x - (-3))^2 + (y - 4)^2 = 2^2 \quad \text{[substituting } a = -3,\ b = 4,\ r = 2\text{]}$$

$$(x + 3)^2 + (y - 4)^2 = 4$$

$$x^2 + y^2 + 6x - 8y + 21 = 0 \quad \text{[expanding and simplifying]}$$

Unit 11.4 Activity 9A: Circles centred at (a, b)

1. Sketch the circle $x^2 + (y - 2)^2 = 1$.
2. Sketch the circle $(x + 3)^2 + y^2 = 3$.
3. Sketch the circle $(x - 1)^2 + (y - 1)^2 = 1$.
4. Sketch the circle $(x + 2)^2 + (y - 2)^2 = 4$.
5. Sketch the circle $(x - 3)^2 + (y - 4)^2 = 25$.

6. Find the equations of the following circles:

a.

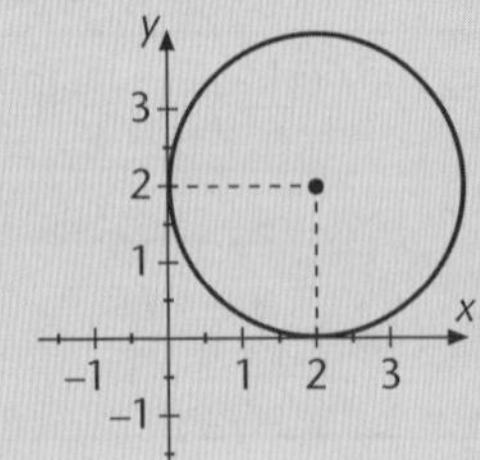

b.

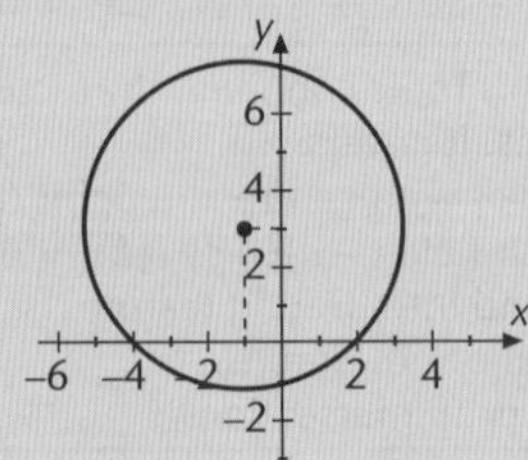

c.

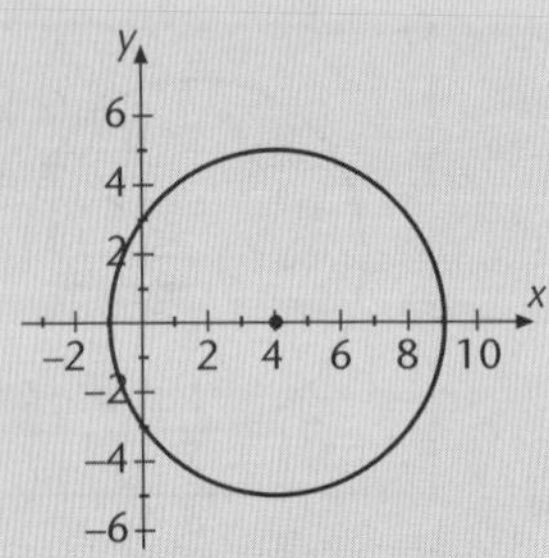

7. Find the equations of the circles with the following features:

a. centre (2, 3), radius 5.

b. centre (–3, 1), radius 4.

c. centre (–2, –3), radius 2.

d. centre (1, 4), radius 3.

e. passing through (–3, 3) and centred at (1, 2).

f. passing through (2, 5) and centred at (4, 2).

g. diameter with end points (–3, 2), (4, 1).

h. diameter with end points (–1, –2), (3, 2).

8. Find the coordinates of the centre and the radius of each of the following circles:

a. $(x + 1)^2 + (y - 3)^2 = 16$

b. $(x - 2)^2 + (y - 4)^2 = 25$

c. $(x - 4)^2 + (y + 1)^2 = 36$

d. $(x - 3)^2 + (y - 4)^2 = 49$

9. Find the equations of the circles on which the following sets of points lie:

a. (4, 6), (4, –2), (–3, 5), (5, –1)

b. (9, 15), (13, 7), (3, –3), (11, 1)

c. (0, 14), (7, –3), (8, 2), (–5, –11)

d. (1, 2), (3, 2), (1, 4), (3, 4)

10. An outpost is surrounded by a circular fence 500 m from the outpost. Any person coming into contact with the fence causes an alarm to ring. Silas was 230 m east and 340 m north of the outpost, Daniel was 330 m west and 215 m south, Saya was 480 m west and 140 m south. Which one set off the alarm?

Intersection of a circle with a straight line

A simple **graphical method** can be used where a graph is drawn and points of intersection read off. For a more accurate solution an **algebraic method** is used, where the equations of the circle and line are solved **simultaneously**. (Using a graphical calculator, great accuracy can be obtained.)

Example C

Find the coordinates of the point of intersection of the line $y = x + 1$ with the circle $x^2 + y^2 = 25$.

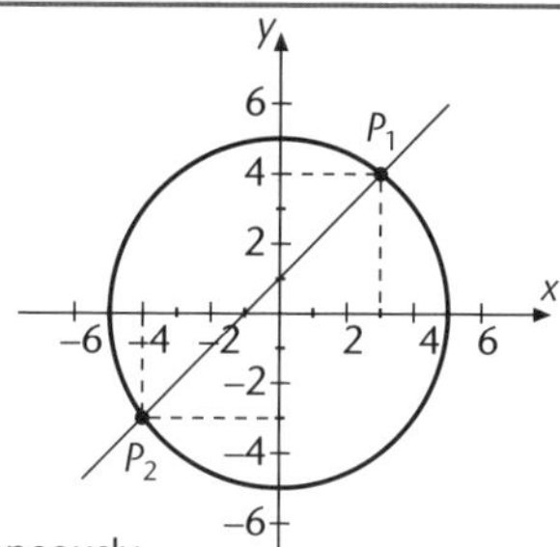

Solution

1. **Graphical:**

There are two points of intersection (P_1 and P_2) which can be read off from the above graph, $P_1 = (3, 4)$ and $P_2 = (-4, -3)$.

2. **Algebraic:**

The equations $y = x + 1$ and $x^2 + y^2 = 25$ are solved simultaneously.

$x^2 + (x + 1)^2 = 25$	[substituting $y = x + 1$ in $x^2 + y^2 = 25$]
$\therefore\ x^2 + x^2 + 2x + 1 = 25$	[expanding]
$\therefore\ 2x^2 + 2x - 24 = 0$	[subtracting 25 and simplifying]
$\therefore\ x^2 + x - 12 = 0$	[dividing by 2]
$(x + 4)(x - 3) = 0$	[factorising]
$x = 3$ or -4	

Substituting $x = 3$ and $x = -4$ into $y = x + 1$ gives the points of intersection (3, 4) and (–4, –3).

When the substitution method gives only one root to the quadratic equation, the line is a **tangent** to the circle (ie the line touches the circle at one point only).

Example D

Prove that the line $y = 2 - x$ is a tangent to $x^2 + y^2 = 2$ at the point (1, 1).

Solution

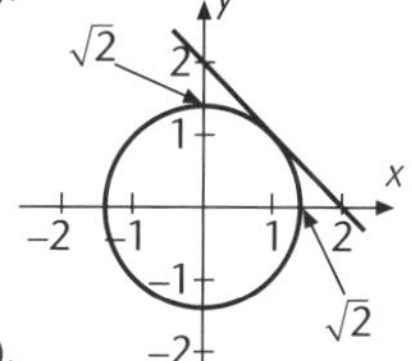

1. **Graphical:**

$x^2 + y^2 = 2$ is a circle centre (0, 0) radius $\sqrt{2}$.

$y = 2 - x$ is a straight line, y intercept 2, gradient –1.

The graph shows the line is a tangent to the circle at the point (1, 1).

2. **Algebraic:**

Substituting for y and solving gives 1 root:

$x^2 + (2 - x)^2 = 2$	[substituting]
$\therefore\ x^2 + 4 - 4x + x^2 = 2$	[expanding]
$\therefore\ 2x^2 - 4x + 2 = 0$	[simplifying]
$\therefore\ x^2 - 2x + 1 = 0$	[dividing by 2]
$\therefore\ (x - 1)^2 = 0$	[factorising]
$\therefore\ x = 1$, confirming *one* root.	

Then substitute $x = 1$ into $y = 2 - x$ to confirm that the point of intersection is (1, 1).

Where a straight line does not cut the circle, the equation resulting from solving simultaneously has *no real roots*.

Example E

Prove that the line $y = 4 - x$ does not cut the circle $x^2 + y^2 = 4$.

Solution

Attempting to solve the equations simultaneously:

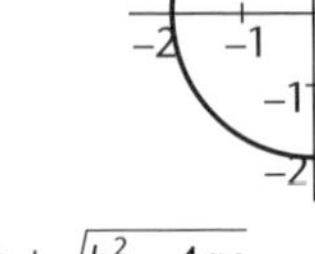

$$x^2 + (4 - x)^2 = 4 \quad \text{[substituting]}$$

$$\therefore\ x^2 + 16 - 8x + x^2 = 4 \quad \text{[expanding]}$$

$$\therefore\ 2x^2 - 8x + 12 = 0 \quad \text{[simplifying]}$$

$$\therefore\ x^2 - 4x + 6 = 0 \quad \text{[dividing by 2]}$$

$$x = \frac{4 \pm \sqrt{-8}}{2} \quad \text{[substituting into } x = \frac{-b \pm \sqrt{b^2 - 4ac}}{2a}\text{]}$$

$x^2 - 4x + 6$ has no real roots as the square root of a negative number cannot be found. Therefore the line does not cut the circle (see the diagram).

Unit 11.4 Activity 9B: Intersection of a circle and a line

1. Find the coordinates of the points of intersection, if any, of the circle $x^2 + y^2 = 4$ with the following straight lines:

a. $y = 2$ **b.** $y = 1$ **c.** $x = 2$ **d.** $x = -1$

e. $y = 1.5$ **f.** $y = 3$ **g.** $y = x$ **h.** $y = 2x$

i. $y = 3x$ **j.** $x = 2y$

2. Find the points of intersection of the circle $x^2 + y^2 = 4$ with the straight lines:

a. $y = x + 1$ **b.** $y = x - 1$ **c.** $y = 2x + 1$ **d.** $y = x + 2$

e. $y = 2x - 2$

3. Investigate the claim that each of the following lines is a tangent to the circle indicated.

a. $3x + 4y = 25,\ x^2 + y^2 = 25$ **b.** $x + 2y = 5,\ x^2 + y^2 = 5$

c. $2x - y = 5,\ x^2 + y^2 = 5$

4. Find k so that the line $x + y = k$ is a tangent to the circle $x^2 + y^2 = 9$.

5. Determine if the line $y = x + 5$ has any intersection with the circle $x^2 + y^2 = 9$.

6. Find the points of intersection of the line $4x - 3y + 3 = 0$ and the circle $x^2 + y^2 - 2y = 24$.

Unit 11.4 Geometry
Topic 10: Arc length and area of circles

The material in this Topic continues the coverage of circle properties and deals with arc length, sectors and segments.

The length of an arc of a circle

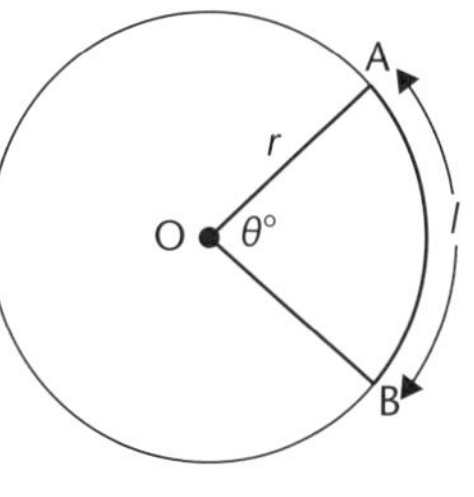

In the diagram at right the arc AB is subtended by the angle $\theta°$ at the centre, O, of the circle.

We use equivalent ratios to find a formula for the length of the arc:

Arc length (l) : Circumference of the circle ($2\pi r$) = the angle subtending the arc (θ) : angle at the centre of a complete circle (360°).

The ratio equation $l : 2\pi r = \theta° : 360°$

gives $$\frac{l}{2\pi r} = \frac{\theta}{360}$$

So the formula is:

$$l = \left(\frac{\theta}{360}\right) \times 2\pi r$$

Example A

Q. Find the length of the arc subtended at the centre of a circle, radius 18 cm, by an angle of 40°.

A. Substituting $\theta = 40$ and $r = 18$ into the formula gives:

Arc length $= \frac{40}{360} \times 2 \times \pi \times 18 = 4\pi = 12.57$ cm (to 2 decimal places).

The area of a sector of a circle

Similarly, we can find a formula for the area of a sector of a circle. In the diagram at right the sector OAB (shaded) contains the angle θ at the centre of the circle.

Using equivalent ratios:

Area of the sector (A) : Area of a whole circle = $\theta°$: 360°

The ratio equation $A : \pi r^2 = \theta : 360$

gives $$\frac{A}{\pi r^2} = \frac{\theta}{360}$$

So the formula is:

$$A = \left(\frac{\theta}{360}\right) \times \pi r^2$$

Example B

Q. Find the area of the following sector of a circle, radius 6 metres.

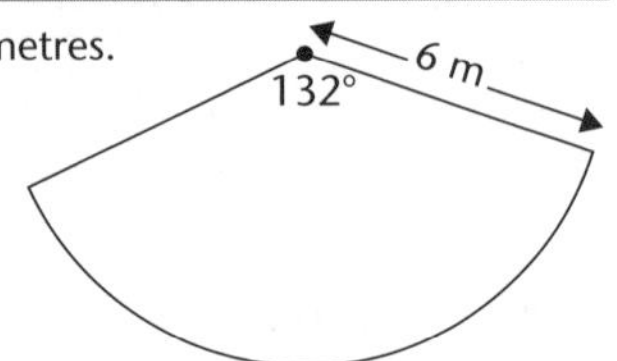

A. Substituting $\theta = 132$ and $r = 6$ in the formula:

$$A = \left(\frac{132}{360}\right) \times \pi \times 6^2$$
$$= 41.47 \text{ m}^2 \text{ (to 2 d.p.)}$$

The area of a segment of a circle

The minor segment, shaded on the diagram at right, is cut off by the chord AB.

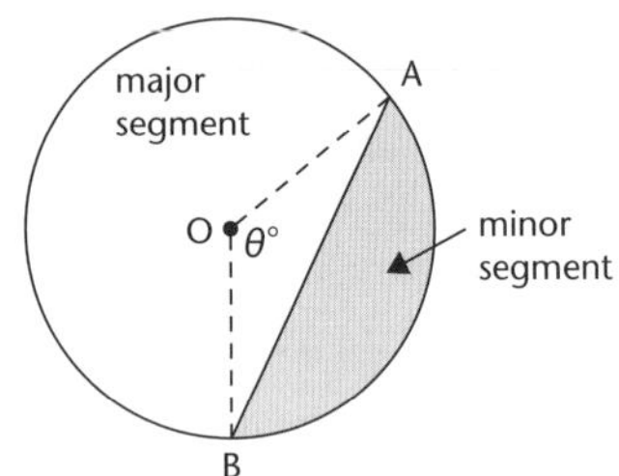

The area of a segment of a circle = the area of the sector OAB – the area of the triangle OAB.

$$A = \left(\frac{\theta}{360}\right) \times \pi r^2 - \frac{1}{2} r^2 \sin \theta$$

Example C

Q. Find the area of:

i. the minor segment

ii. the major segment

cut off by a chord subtended at the centre by an angle of 54°, of a circle of radius 15 cm.

A. i. Substituting $\theta = 54°$ and $r = 15$ into the formula gives:

$$A = \left(\frac{54}{360}\right) \times \pi \times (15)^2 - \frac{1}{2}(15)^2 \times \sin 54$$
$$= 15.01 \text{ m}^2 \text{ (to 2 decimal places)}$$

ii. The area of the major segment

= the area of the whole circle – the area of the minor segment.

$= \pi \times 15^2 - 15.014$

$= 691.84 \text{ m}^2$ (to 2 d.p.)

Unit 11.4 Activity 10A: Arc length and area of circles

Give your answers correct to two decimal places.

1. On a circle of radius 24 centimetres find the arc length subtended by an angle of:

a. 10°

b. 100°

c. 215°

2. On a circle of radius 20 metres find the area of a segment that has the angle:

a. 35°

b. 200°

at the centre of the circle.

3. A chord is subtended by an angle of 98° at the centre of a circle, radius 40 centimetres. Find the area of:

a. the minor segment

b. the major segment

of this circle.

4. Find the perimeter and the area of each of the following:

a.

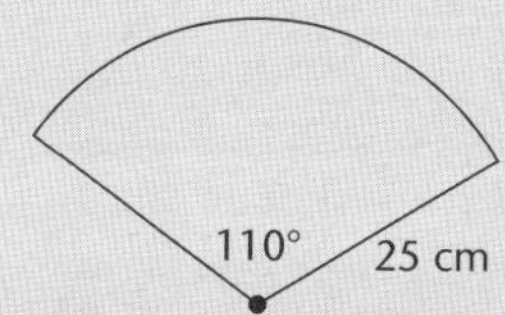

b.

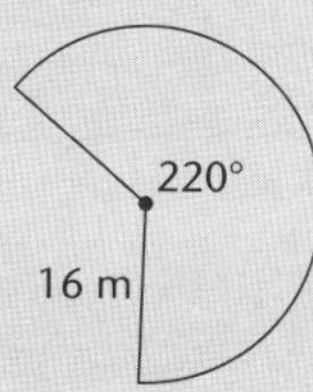

5. The sector of the circle is to be made into a cone by joining the two straight edges (no overlap) as shown in the diagram below. Find the radius of the base of the cone.

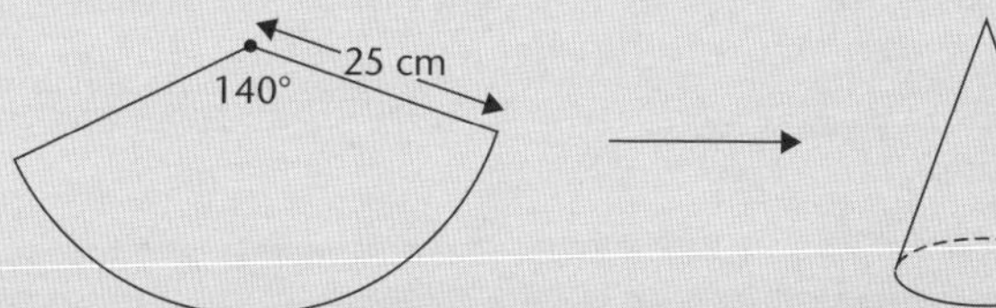

6. A garden has the shape shown on the diagram at right. The boundary of the garden is the arc PQ of a circle of radius 3.6 metres, and the straight edge joining the points P and Q. The arc PQ subtends an angle of 112° at the centre of the circle.
Find: **a.** the area of the garden.

b. the length of the boundary of the garden.

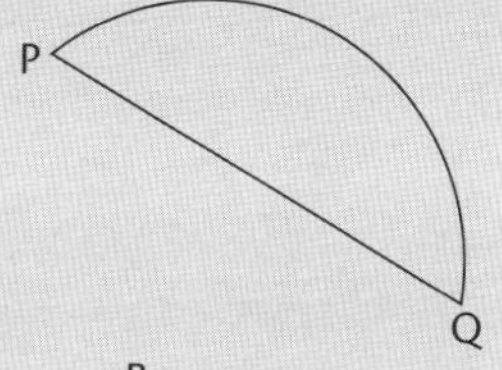

7. The circle at right has a radius of 30 cm and the triangle ABC is an equilateral triangle. Find the shaded area in this diagram.

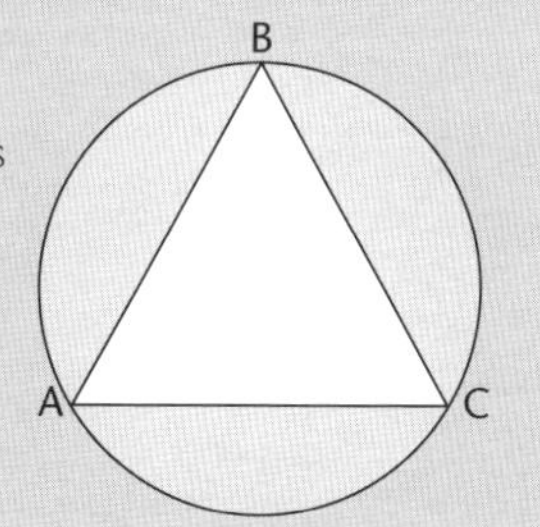

8. On the diagram at right AB and CB are tangents to the circle, centre O and radius 6 metres. Find the shaded area.

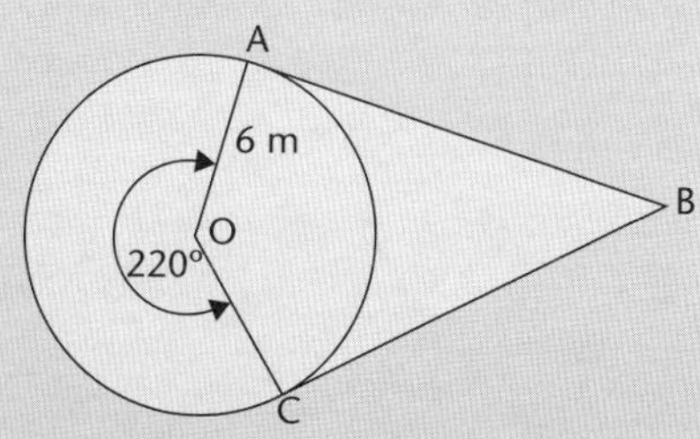

Unit 11.4 Geometry

Topic 11: Spheres, latitude and longitude

Circle properties concludes with an examination of spheres and the general equation of circles. The Topic covers:

- Cross-sections of a sphere.
- Position on a sphere.
- Latitude and longitude.

Spheres

A **sphere** is a three-dimensional figure where every point on the surface of the sphere is the same distance from the centre of the sphere. Half of a sphere is called a **hemisphere**.

For a sphere of radius r:

$$\text{The surface area} = 4\pi r^2$$

$$\text{The volume} = \frac{4}{3}\pi r^3$$

To draw a sphere or hemisphere:

1. Draw a circle.
2. Connect two points on the circumference of the circle with an ellipse.

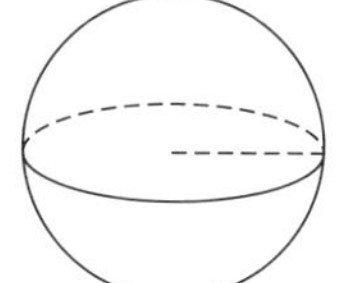

A see-through sphere

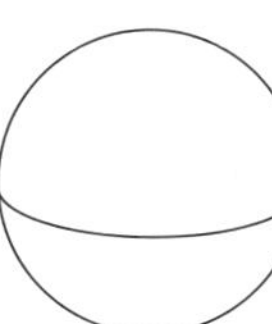

A solid sphere

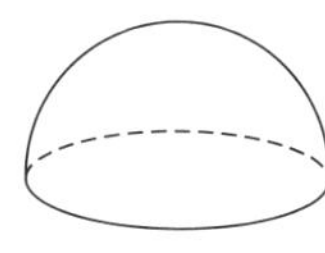

A see-through hemisphere

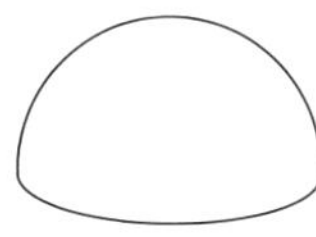

A solid hemisphere

Cross-sections of a sphere—intersections of a plane and a sphere

If a plane intersects with a sphere then the cut surface is always in the shape of a circle. The size of the circle depends on where the plane intersects with the sphere:

If the plane passes through the centre of the sphere then the circle is called a **great circle**. If the plane does not go through the centre then the circle is called a **small circle**.

Finding the radius of a small circle

A sphere of radius R is truncated so that a circular surface of radius r is visible, as shown in the diagram at right.

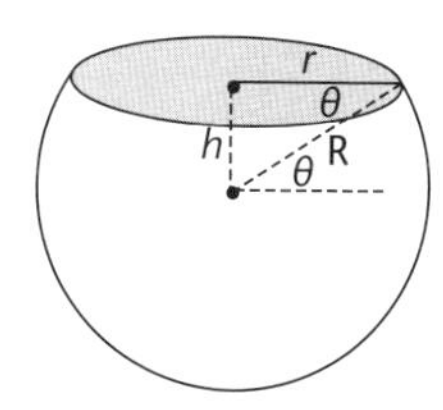

If the distance h is known then we can use Pythagoras' Theorem to give

$R^2 = h^2 + r^2$

and so

$r = \sqrt{R^2 - h^2}$

If the angle θ is known then we can use:

$$\cos\theta = \frac{r}{R}$$

and so

$$r = R\cos\theta$$

Example A

Q. A hemispherical bowl, of radius 12 centimetres, is filled with liquid up to a depth of 5 centimetres. Find the radius of the liquid surface.

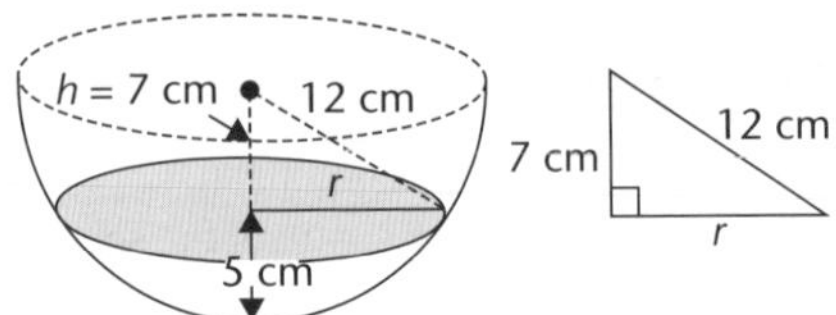

A. If the depth of the liquid is 5 centimetres then the height, h, from the liquid surface to the centre of the hemisphere is $12 - 5 = 7$ cm.

Substituting in $r = \sqrt{R^2 - h^2}$ gives

$r = \sqrt{12^2 - 7^2} = 9.75$ cm (to 2 d.p.)

Position on the Earth's surface

The shape of the Earth is very close to a sphere, with a radius of approximately 6400 kilometres. The earth rotates about an axis that goes through the North Pole and the South Pole.

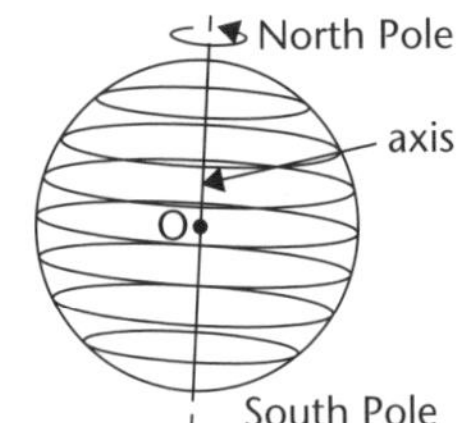

If the Earth intersects with a plane then the planes that are perpendicular to the Earth's axis are called **parallels of latitude**. All parallels of latitude are **small circles** except the parallel of latitude that goes through the centre of the Earth, called the **equator**. The equator is a **great circle**.

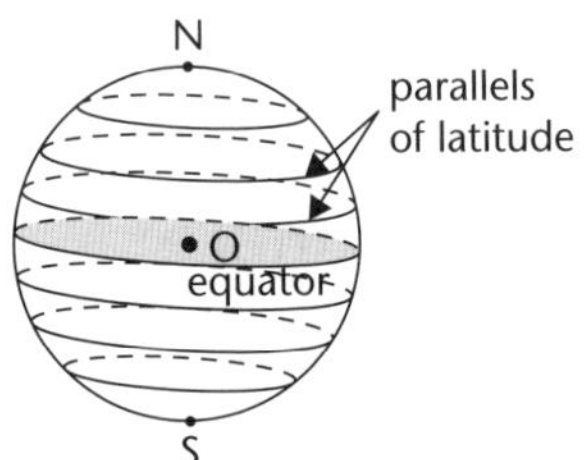

Great circles that pass through the North and South Poles are cut in half at the poles and these semi-circles form the **meridians of longitude**.

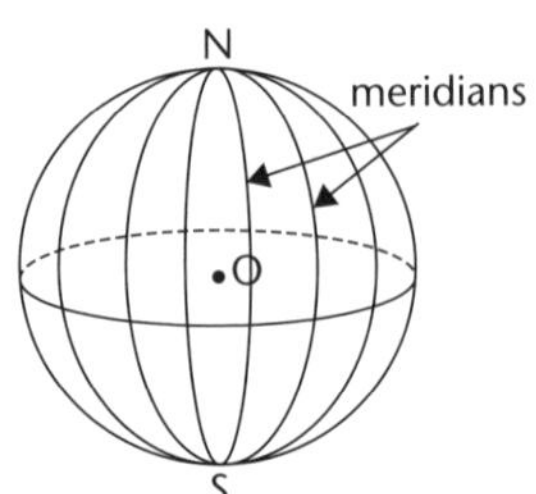

The parallels of latitude and the meridians of **longitude** form a grid on the Earth's surface which means that the position of any point on the Earth's surface can be determined.

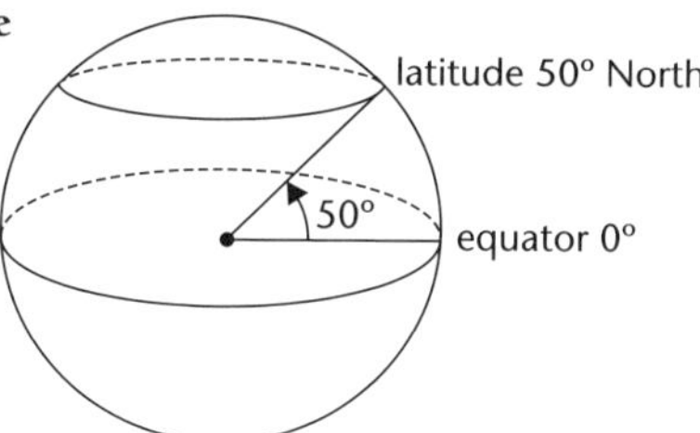

The latitude of every point on the equator is 0° and the latitudes of other points are given as **degrees**, up to 90°, North or South of the equator. All points on the same parallel of latitude have the same latitude.

The Prime Meridian is the meridian that goes through the observatory at Greenwich in England. The Prime Meridian is given the longitude 0° and all other points are given a longitude up to 180° West or 180° East. All points on the same meridian have the same longitude.

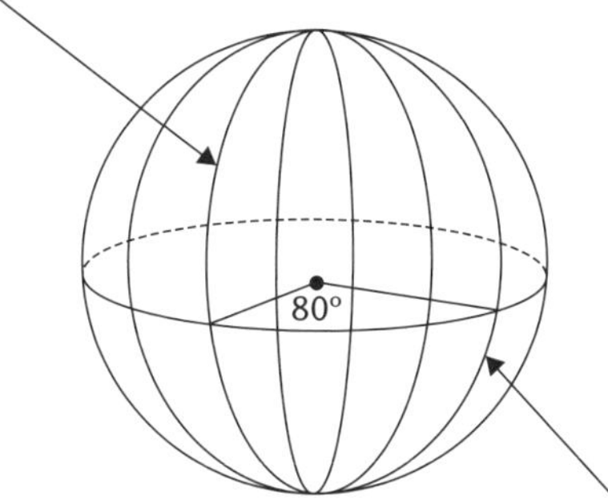

The position of a point on the Earth's surface is given as (latitude, longitude). Angles are measured in degrees, minutes and seconds (60 minutes in a degree, 60 seconds in a minute). It is convention now to use decimal minutes rather than seconds. eg 32° 15 '12" would be written as 32° 15.2'.

Example B

Q. The cities of Saipan, in the North Mariana islands of the United States, and Cairns in Australia are on the same meridian of longitude (145°45'E). The latitude of Saipan is 15°11'N and the latitude of Cairns is 16°55'S. Find the distance between these two cities along the meridian. Assume the radius of the earth is 6400 km.

A. The angle subtended at the centre of the earth is

$15°11' + 16°55' = 32°06' = \left(32 + \frac{6}{60}\right)^{\circ} = 32.1°$.

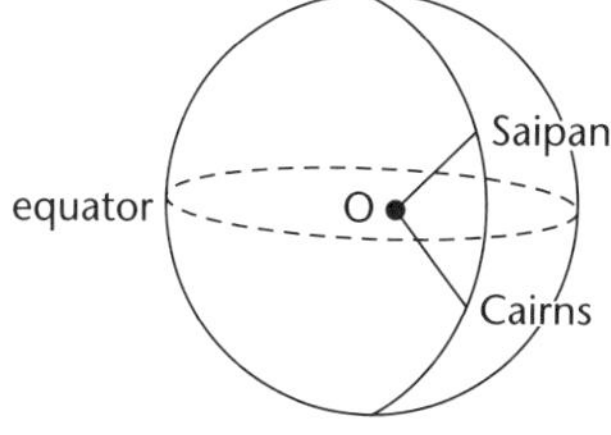

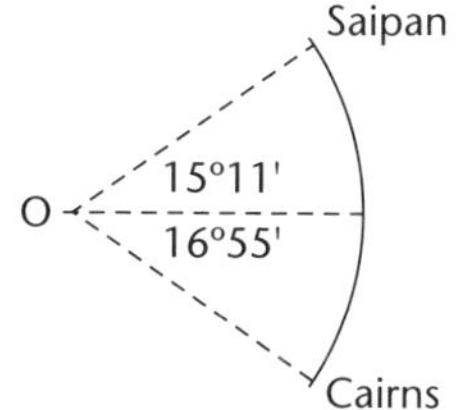

The distance between Saipan and Cairns along the meridian $= \frac{32.1}{360} \times 2 \times \pi \times 6400 \approx 3586$ km.

Example C

Q. The cities of Santa Cruz, in Bolivia, and Port Vila in Vanuatu are at the same latitude (17°45'S). The longitude of Santa Cruz is 63°14'W and the longitude of Port Vila is 168°18'E. Find the shortest distance along the parallel of latitude between these two cities.

A. The first step is to find the radius, r, of the parallel of latitude 17°45' S (= 17.75° S).

$\cos 17.75° = \dfrac{r}{6400}$ so $r = 6400 \cos 17.75 \approx 6095$ km.

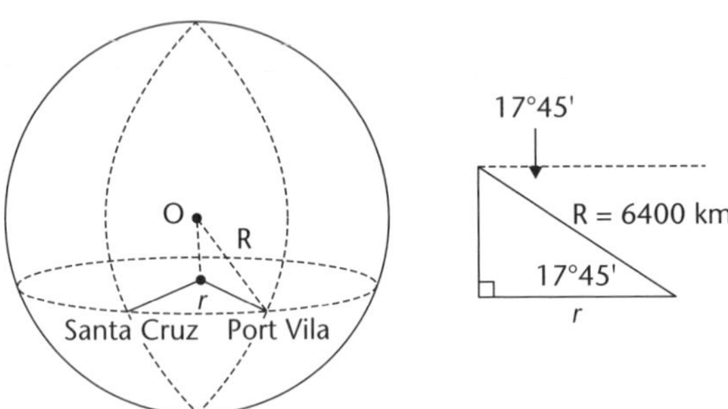

The angle along the parallel of latitude between the two cities is 168°18' + 63°14' = 231°32'.

This is obviously more than 180° so the cities are closer if we go in the other direction. The angle will be 360° – 231°32' = 128°28' ≈ 128°

The closest distance along the parallel of latitude between Santa Cruz and Port Vila

$= \dfrac{128}{360} \times 2 \times \pi \times 6095 \approx 13616$ km.

The shortest route between two locations on the Earth

Mariners have known for a long time that the **shortest route between two locations on the Earth is along the great circle** that goes through the two places. If the two places are on the same meridian of longitude then they are already along a great circle and the distance between them can be calculated as in Example B above.

The situation is more complex if the two locations are on the same latitude (not the equator). Consider Example C above:

The circle of the parallel of latitude for Santa Cruz and Port Vila has a radius of 6095 km.

The great circle going through Santa Cruz and Port Vila has a radius of 6400 km.

On the diagram below you can see that, when the small circle is imposed on the diagram of the great circle, the distance is less along the great circle.

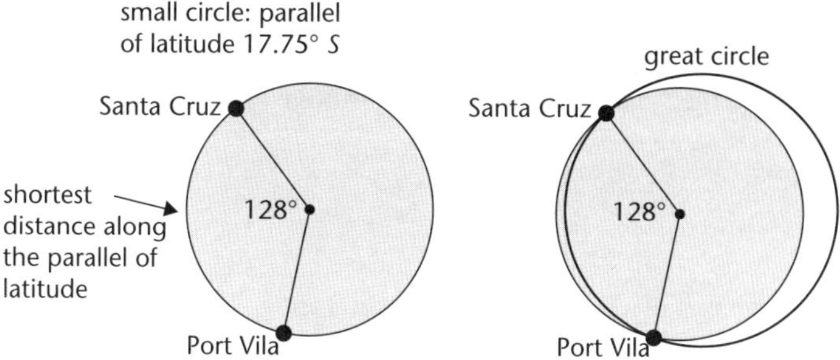

Finding the great circle distance between the two cities

We need to find the angle subtended at the centre of the Earth by the great circle. This is the angle θ on the diagram below right.

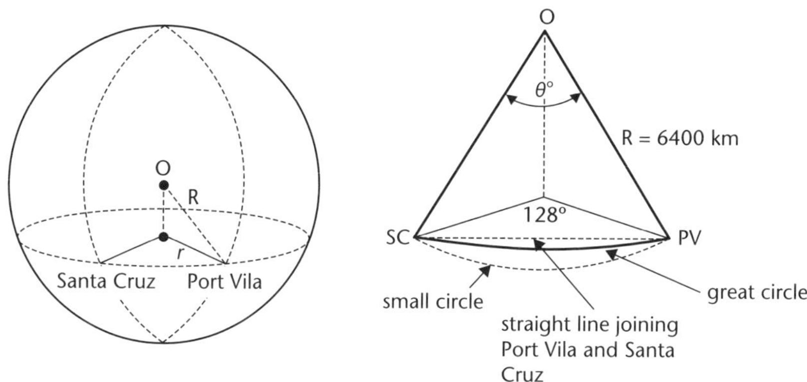

We use the small circle angle, 128°, and radius 6095 km to find the length of the straight line joining Port Vila and Santa Cruz:

Using the cosine rule:

Length of straight line = $\sqrt{6095^2 + 6095^2 - 2 \times 6095 \times 6095 \cos 128°} = 10956.3$ km

Using the cosine rule to find the angle in the great circle:

$$\cos\theta = \frac{6400^2 + 6400^2 - 10956.3^2}{2 \times 6400 \times 6400} = -0.46534$$

$$\theta = 117.73°$$

Length along the great circle = $\frac{117.73}{360} \times 2 \times \pi \times 6400 \approx 13151$ km

Note: The distance along the great circle connecting Santa Cruz and Port Vila is 13616 – 13151 = 465 km shorter than the distance along the parallel of latitude.

Unit 11.4 Activity 11A: Spheres, latitude and longitude

1. A circle of radius 20 cm is truncated so that the height through the centre is 35 cm. Find the radius of the cut surface, shaded on the diagram at right.

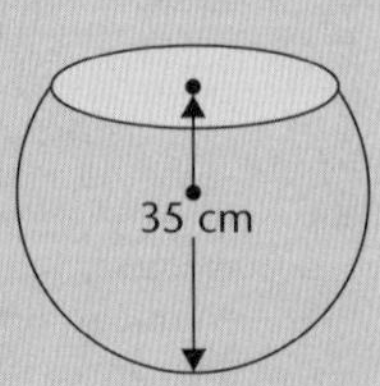

2. A bowl is made from part of a sphere of radius 12 cm. Liquid fills the bowl to a height of 15 cm. Find the radius of the surface of the liquid.

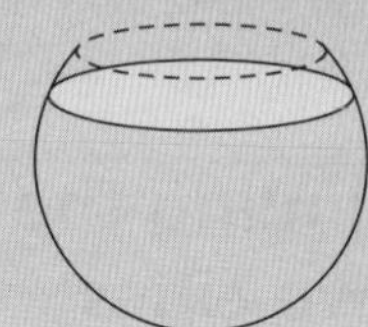

3. A circular mound has a height of 1.2 m at its highest point. If the mound was a hemisphere then it would have a diameter of 7 m. What is the diameter of this mound?

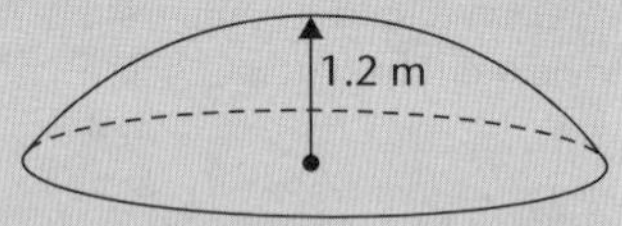

4. Each of the following pairs of cities is on the same meridian of longitude. Find the shortest distance between the cities.

a. Willemstad, Netherlands (12°07'N, 68°56'W)

Calama, Chile (22°28'S, 68°56'W)

b. Saint Pierre, France (46°46'N, 56°10'W)

Montevideo, Uraguay (34°55'S, 56°10'W)

c. Leeds, United Kingdom (53°48'N, 1°33'W)

Nantes, France (47°13'N, 1°33'W)

d. Port Vila, Vanuatu (17°45'S, 168°18'E)

Invercargill, New Zealand (46°25'S, 168°18'E)

5. The following pairs of cities are on the same parallel of latitude. For each pair:

i. Find the shortest distance along the parallel of latitude.

ii. Find the shortest path along the great circle going through the two cities.

a. Port Moresby, PNG (9°28'S, 147°12'E)

Honiara, Solomon Islands (9°28'S, 159°49'E)

b. Hamburg, Germany (53°34'N, 10°E)

Edmonton, Canada (53°34'N, 113.5°W)

c. New Delhi, India (28°38'N, 77°13'E)

Chihuahua, Mexico (28°38'N, 106°5'W)

d. Bhisto, South Africa (32°53'S, 27°23'E)

Mendoza, Argentina (32°53'S, 68°50'W)

Appendix

Mathematical symbols

Sets

$b \in A$	b is an element of A
$c \notin A$	c is not an element of A
U or ε	the universal set
ϕ or { }	the empty or null set
A´	the complement of A
$A \subset B$	A is a subset of B
$A \not\subset B$	A is not a subset of B
$A \cup B$	the union of A and B, ie, the set containing elements in A or B (or in both)
$A \cap B$	the intersection of A and B, ie, the set containing elements in both A and B
$n(A)$	the number of elements in set A

Number Sets

N	Natural numbers
W	Whole numbers
J	Integer numbers
Q	Rational numbers
Q´	Irrational numbers
R	Real numbers

Language

:	such that
...	and so on
$\therefore$	therefore
=	is equal to
$\neq$	is not equal to
$\approx$	is approximately equal to
<	is less than
$\leq$	is less than or equal to
>	is greater than
$\geq$	is greater than or equal to
$\Rightarrow$	implies
Σ	sum
$\sqrt{x}$	the positive square root of x
$\pm x$	plus or minus x
$\lvert x \rvert$	the absolute value of x

Geometry

$\angle$	angle
$\angle ABC$ or $A\hat{B}C$	the angle at B defined by the points A, B and C.
$\overline{AB}$	the line segment A to B.
//	parallel to
$\perp$	perpendicular to
Δ	triangle
$^\circ$	degree
π	Pi = 3.14159 (5 dp)

Probability

P(A)	the probability of event A occurring
$\overline{x}$	the mean of a sample
s.d. or s	the standard deviation of a sample
μ	population mean
σ	population standard deviation

Mathematical formulae

Area

$\frac{1}{2}bh$	triangle	a^2	square (side a)
bh	rectangle or parallelogram	$\frac{1}{2}(a+b)h$	trapezium
$\frac{1}{2}d_1d_2$	rhombus or kite	πr^2	circle
$\frac{\theta}{360} \times \pi r^2$	sector (θ in degrees)	$4\pi r^2$	sphere (surface area)
$\frac{1}{2}r^2\theta$	sector (θ in radians)	$2\pi r(r+h)$	closed cylinder (surface area)

Volume

l^3	cube (side l)	Ah	prism (A = base area)
lbh	cuboid	$\frac{1}{3}Ah$	pyramid (A = base area)
$\frac{1}{2}bhl$	triangular prism	$\frac{4}{3}\pi r^3$	sphere
πr^2h	cylinder	$\frac{1}{3}\pi r^2h$	cone

Logarithms and indices

Logarithms

If $y = b^x$ then $\log_b y = x$

$\log_b x + \log_b y = \log_b xy$

$\log_b x - \log_b y = \log_b \frac{x}{y}$

$\log_b x^n = n\log_b x$

$\log_b x = \frac{\log_a x}{\log_a b}$

Indices

$a^m \times a^n = a^{m+n}$ $\quad$ $a^m \div a^n = a^{m-n}$

$(a^m)^n = a^{mn}$ $\quad$ $(ab)^m = a^mb^m$

$a^0 = 1$ (if $a \neq 0$) $\quad$ $\sqrt{a^m} = a^{\frac{m}{2}}$

$\frac{1}{a^m} = a^{-m}$ (if $a \neq 0$) $\quad$ $a^{\frac{m}{n}} = \sqrt[n]{a^m}$

$= \left(\sqrt[n]{a}\right)^m$

Coordinate geometry

Distance

$d = \sqrt{(x_1 - x_2)^2 + (y_1 - y_2)^2}$

Mid-point of a line

mid-point $\left(\frac{x_1 + x_2}{2}, \frac{y_1 + y_2}{2}\right)$

Gradient

$$m = \frac{y_2 - y_1}{x_2 - x_1}$$

Lines

$$y = mx + c$$

$$y - y_1 = m(x - x_1)$$

Perpendicular and parallel lines

$y = m_1x + c_1$ and $y = m_2x + c_2$ are parallel if $m_1 = m_2$ and perpendicular if $m_1 \times m_2 = -1$

Circles

$x^2 + y^2 = r^2$ circle with centre (0, 0) and radius r

$(x - a)^2 + (y - b)^2 = r^2$ circle with centre (a, b) and radius r

Perimeter

$C = \pi d$ or $2\pi r$ circumference of a circle

$\frac{\theta}{360} \times 2\pi r$ arc length of a sector (θ in degrees)

$r\theta$ arc length of a sector (θ in radians)

Trigonometry

$\sin \theta = \frac{o}{h}$ o = opposite

$\cos \theta = \frac{a}{h}$ h = hypotenuse

$\tan \theta = \frac{o}{a}$ a = adjacent side

$h^2 = a^2 + o^2$ Pythagoras' theorem

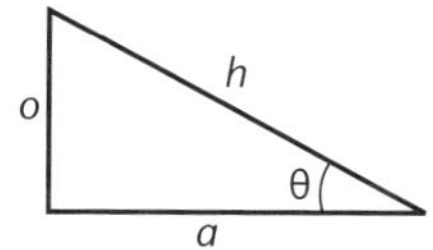

Triangles

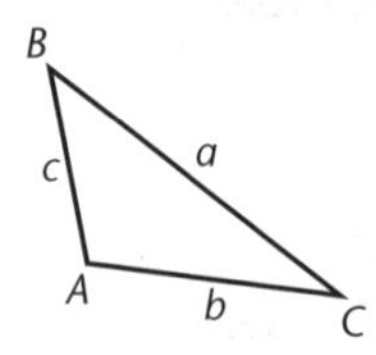

Sine rule: $\dfrac{a}{\sin A} = \dfrac{b}{\sin B} = \dfrac{c}{\sin C}$ or $\dfrac{\sin A}{a} = \dfrac{\sin B}{b} = \dfrac{\sin C}{c}$

Cosine rule: $a^2 = b^2 + c^2 - 2bc \cos A$ or $\cos A = \dfrac{b^2 + c^2 - a^2}{2bc}$

Area of triangle: $\text{Area} = \dfrac{1}{2}ab \sin C$

Sequences and series

Arithmetic series

$a + (a + d) + (a + 2d) + (a + 3d) + \ldots$

$t_n = a + (n - 1)d$

$S_n = \dfrac{n}{2}(2a + (n - 1)d)$

Geometric series

$a + ar + ar^2 + ar^3 + \ldots$

$t_n = ar^{n-1}$

$S_n = \dfrac{a(1 - r^n)}{(1 - r)},\ r \neq 1$

$S_\infty = \dfrac{a}{1 - r}$ for $|r| < 1$

Probability

$P(A) = 1 - P(A')$

$P(A \text{ or } B) = P(A \cup B) = P(A) + P(B) - P(A \cap B)$

A and B are independent events $\Leftrightarrow P(A \cap B) = P(A) \times P(B)$

Mean, variance and standard deviation

Statistic	Sample of ungrouped data	Samples of grouped data
Mean	$\bar{x} = \dfrac{\sum x_i}{n}$	$\bar{x} = \dfrac{\sum x_i f_i}{\sum f_i}$
Variance	$s^2 = \dfrac{\sum (x_i - \bar{x})^2}{n-1}$	$s^2 = \dfrac{\sum (x - \bar{x})^2 f_i}{\sum f_i - 1}$
Standard deviation	$s = \sqrt{s^2}$	$s = \sqrt{s^2}$

Random number tables

7898	8002	4418	2747	8079	7993	6863	9542	0849	4531	6955	5826	9971	6233	7887
8640	3204	6906	5719	1116	5982	9532	2422	8333	8828	9002	2680	1928	8532	3600
4431	3453	3070	5239	3168	6490	0275	8443	9984	7503	0263	8086	3372	5454	1599
5868	4764	0158	1225	5558	7840	9394	8126	6974	1561	4765	0758	8717	6979	6306
8214	6959	7775	5844	5149	9173	4558	9107	0453	6119	2915	6585	9670	6580	5202

3137	1170	0345	6099	6352	6074	6142	1898	3657	1924	5625	3556	8178	0103	6107
3490	3349	7010	2045	6123	6271	8981	5274	2183	9820	0957	3988	6747	3508	8914
0483	1041	3095	6028	7633	7984	8941	2598	4127	1487	3123	9720	8455	9499	5291
4991	7841	0264	9480	9545	3517	5701	6513	9644	5639	4982	0201	7013	3858	8432

0485	2823	0417	5787	7517	2305	1324	7674	2205	8107	1273	5851	0509	2403	2318
2185	1493	7751	6292	6392	9115	6708	0980	2518	7780	0180	0041	4175	1633	4543
6680	8525	7608	7492	1072	9013	2222	2655	1528	6854	2126	2954	9422	1264	8398
2711	7863	8449	4594	8509	4112	9334	8538	8800	5761	6052	7879	2268	0647	7176
9378	4934	0532	1703	1961	6017	8518	1789	3513	2545	0222	4235	1426	2712	8698

0720	6897	7989	3108	6782	2972	3982	1145	9561	6345	0042	5204	4273	2904	7626
2303	9453	7029	7403	7757	0934	4638	5117	2599	4583	9983	6193	4750	7787	6199
9100	3944	9685	9598	7455	0398	0154	7721	0889	7234	7540	1127	4652	6652	5149
8948	7973	9851	6475	8179	1170	0698	9187	6732	1511	3157	7865	0226	7819	2455
5479	2537	3835	0004	3631	0096	0328	2076	1531	9934	5949	2224	3741	1479	6848

7450	6649	2104	5027	8134	3711	6423	5454	1700	9432	9338	1118	0038	2102	2836
1641	0681	5279	6536	8638	1722	8024	1985	5826	0487	3885	0706	5881	5957	8632
5996	0850	4941	6741	8468	0509	1998	6628	5355	9472	3105	4636	3550	3055	3085
7764	6407	1877	6773	5155	4288	5227	3942	8333	8857	3170	7973	0182	1130	7404
8278	2681	3270	0876	6469	3394	4132	6135	6285	0697	8567	2445	7259	7768	0460

1643	2141	5319	9532	2747	8073	3078	1888	5321	1507	0029	4308	1576	6879	3392
2300	7236	9030	6359	1895	1145	9378	4578	5191	3625	5840	1386	9654	3590	6503
1798	1163	4399	7050	5236	0806	0470	1926	7014	5429	0638	9262	8578	1930	0264
9549	6484	5310	2096	8058	0133	0465	5875	1011	8260	7737	4699	5505	4164	2210
2126	3184	9694	6752	8043	7657	7984	8821	8263	5675	4934	1004	2897	2050	9451

5317	8453	8326	2426	1252	8703	5204	4472	7574	9286	8712	2409	7866	1073	0126
4499	9827	7273	9643	4425	8322	9292	3804	4031	2172	1347	6664	4806	4709	3817
4714	8115	8064	5362	5553	4187	1222	3776	0313	9845	1816	5596	9825	4995	1670
4468	4545	7860	6101	8248	8225	8958	6966	4867	5336	3847	9453	6938	2237	4599
2635	5117	2098	0040	3763	0209	3576	4736	6313	4032	3228	6721	1784	9024	1693

3987	5243	6602	3606	9716	3625	4271	1274	6265	4777	1486	2096	7900	9257	4275
9737	2280	0878	8515	9047	0694	6009	2213	4737	7041	7619	6192	3794	5568	6072
3789	1661	7351	4679	9194	2153	5535	8992	4561	1133	9312	9850	5867	3928	6658
0177	7522	6508	5820	5531	5730	0404	5546	8396	0592	1380	4072	6232	6974	1853
6855	3273	4109	7072	3059	6072	3962	5031	0787	2796	8335	0146	1069	6278	4729

0676	0592	0809	0952	9895	3187	2688	8481	1013	0065	4516	4566	5848	8797	3248
3096	7250	1051	1983	4204	5270	9077	5515	2626	7081	0940	9631	4813	0656	4504
4155	4791	2183	0168	9745	8766	7240	2350	8442	8486	5717	9720	7965	3354	0487
4185	9760	1278	9829	8625	0794	8291	3598	3376	9103	6903	3150	1463	3091	3214
5009	2586	4575	2772	8294	5857	5579	5182	5888	1277	8887	7466	9360	9978	2335

6173	7639	3204	6861	8238	9433	0046	8299	0590	9401	3923	2929	8513	8079	7824
6770	2756	5546	8285	8631	5390	8851	8126	6905	4418	3182	7937	0117	7137	6786
6310	1221	2243	9648	8824	5646	0694	5905	5781	3094	5319	0022	8688	3087	9175
6713	5493	4261	2659	4613	4568	7417	9365	3700	7445	2316	0695	6465	0299	8225
9219	2839	4183	8215	0782	8858	4013	7207	5366	9090	6301	3771	6698	2805	5640

5670	0607	3896	9938	9158	2140	4552	4405	2028	1414	2683	4282	0408	8250	3083
6332	9306	3431	2363	9067	6913	1010	7613	1740	2640	8973	8832	2073	8925	8944
5093	2746	6772	4528	4471	6628	5054	0419	7709	0764	3556	8257	5644	9107	9772
1242	0378	4004	9454	7491	0200	7430	0907	2145	8256	4973	3037	8232	4974	3567
7148	6277	4231	7976	2488	2657	3282	1032	5544	6714	6425	6614	3096	6613	2427

Answers

Answers for many questions include a 'Marking Guide':
A. ('Achievement', meaning 'satisfactory achievement').
M. ('Merit', meaning 'high achievement').
E. ('Excellence', meaning 'very high achievement').
The Marking Guide has been made by the authors and the publishers and is not an official guide but we hope it will be a help to students who are striving for the best possible results.

Unit 11.1 Number and Application

Unit 11.1 Activity 1A: The real number system (page 4)

1. a. $\frac{15}{1}$ **b.** $\frac{-4}{1}$ **c.** $\frac{0}{1}$ **d.** $\frac{125}{10}$ **e.** $\frac{45}{1\,000}$

f. $\frac{19}{8}$ **g.** $\frac{23224}{1000}$ **h.** $\frac{-62}{9}$ **i.** $\frac{-5}{1\,00000}$ **j.** $\frac{98}{25}$

2. a. $\frac{1}{3}$ **b.** $\frac{7}{9}$ **c.** $\frac{56}{99}$ **d.** $\frac{67}{99}$ **e.** $\frac{23}{9}$

f. $\frac{137}{11}$ **g.** $\frac{20920}{333}$

3. There exists a real number, 0, that for every real number, n, $n + 0 = n$

Example $5 + 0 = 5$

The additive identity is 0

4. There exists a real number, 1, that for every real number, n, $n \times 1 = n$

Example $7 \times 1 = 7$

The multiplicative identity is 1

5. The associative axiom of multiplication.

Unit 11.1 Activity 2A: Order of operations and integers (page 9)

Note: *These problems practise essential skills but only problems in context are given a Marking Guide.*

1. a. 10 **b.** 14 **c.** 32 **d.** 18 **e.** 10

f. 2 **g.** 28 **h.** 36 **i.** 9 **j.** 9

k. 4 **l.** 15 **m.** 6 **n.** 5 **o.** 2

p. 2 **q.** 21 **r.** 0 **s.** Undefined. **t.** Undefined.

2. K130 **(A)** **3.** 55 kg **(A)**

4. a. −4 **b.** 7 **c.** 12 **d.** −3 **e.** −20

f. 12 **g.** −24 **h.** 12 **i.** 0 **j.** 3

k. −7 **l.** −21 **m.** 5 **n.** −1 **o.** 12

p. 1 **q.** 25 **r.** −7 **s.** −3 **t.** 1

5. a. 2°C **(A)** **b.** −10°C **(A)** **c.** 12°C **(A)**

6. a. i. 7°C **(A)** **ii.** 7°C **(A)** **b.** 5°C **(A)** **c.** −2°C **(A)**

7. –K14.90, ie K14.90 overdrawn. **(A)**

8. a. 4 months **(E)** b. K418 **(E)**

Unit 11.1 Activity 2B: Powers and roots of numbers (page 13)

1. a. 25 b. 81 c. 625 d. 1 849 e. 6
   f. 13 g. 56 h. 98 i. 18.601 j. 35.128
   k. 29.155 l. 7.746 m. 3 n. 4 o. 5
   p. 9 q. –11 r. –20 s. –24.495 t. 5
2. a. –36 b. 1 c. 36 d. –75 e. 7 f. 0
3. a. 11–12 b. 9–10 c. 5–6 d. 20–21
4. a. 37 cm b. 15.326 cm longer
5. a. 81 m^2 b. 9 m
6. a. 512 b. 6 561 c. 243 d. 4 096 e. 3
   f. 11 g. 4 h. 2 i. 1.348 j. 1
   k. 1.479 l. 7.990 m. 2 n. 5 o. 7
   p. 4 q. –6 r. –6 s. Undefined. t. –1.369
7. a. 6 b. 81 c. 1.968 d. 1 e. 2 f. –1
8. a. 125 cm^3 b. 6.3 cm
9. a. 5.848 cm b. 7.368 cm c. $\frac{\text{long}}{\text{short}} = \frac{\sqrt[3]{400}}{\sqrt[3]{200}} = \sqrt[3]{2}$ so $k = 2$ **(E)**
10. 6 cm **(E)**

Unit 11.1 Activity 3A: Equivalent fractions and mixed numbers (page 17)

Note: *Only problems in context are given A, M, E ratings.*

1. a. $\frac{7}{15}$ b. $\frac{3}{8}$ c. $\frac{9}{16}$
2. a. $\frac{8}{17}$ b. $\frac{5}{17}$ c. $\frac{12}{17}$
3. a. $\frac{25}{50} = \frac{1}{2}$ b. $\frac{10}{50} = \frac{1}{5}$ c. $\frac{15}{50} = \frac{3}{10}$
4. a. $\frac{1}{2}$ b. $\frac{1}{4}$ c. $\frac{7}{15}$ d. $\frac{1}{8}$
   e. $\frac{6}{11}$ f. $\frac{1}{2}$ g. $\frac{3}{4}$ h. $\frac{3}{5}$
   i. $\frac{7}{10}$ j. $\frac{5}{8}$
5. a. $9\frac{1}{2}$ b. $9\frac{3}{5}$ c. $13\frac{1}{6}$ d. $8\frac{7}{10}$
   e. $8\frac{1}{3}$ f. $5\frac{1}{2}$ g. $5\frac{2}{3}$ h. $5\frac{3}{5}$
   i. $7\frac{5}{6}$ j. $9\frac{1}{2}$

6. a. $\frac{5}{2}$ **b.** $\frac{14}{3}$ **c.** $\frac{23}{4}$ **d.** $\frac{47}{8}$

e. $\frac{247}{20}$ **f.** $\frac{7}{3}$ **g.** $\frac{19}{5}$ **h.** $\frac{51}{20}$

i. $\frac{93}{8}$ **j.** $\frac{191}{15}$

7. a. $\frac{2}{7}, \frac{1}{3}, \frac{3}{8}$ **b.** $\frac{2}{15}, \frac{3}{20}, \frac{1}{6}$ **c.** $\frac{7}{8}, 1\frac{1}{5}, \frac{4}{3}$ **d.** $\frac{12}{5}, \frac{8}{3}, 2\frac{3}{4}$

8. a. Bob **(*A*)** **b.** Anne **(*A*)**

9. 60 **(*A*)**

10. a. $\frac{1}{3}$ **(*A*)** **b.** $\frac{4}{5}$ **(*A*)** **c.** $\frac{5}{8}$ **(*A*)**

Unit 11.1 Activity 3B: Operations with fractions (page 22)

Note: *Only problems in context are given A, M, E ratings.*

1. a. $\frac{4}{5}$ **b.** $\frac{7}{10}$ **c.** $\frac{6}{8} = \frac{3}{4}$ **d.** $\frac{5}{6}$

e. $\frac{13}{15}$ **f.** $\frac{41}{40} = 1\frac{1}{40}$ **g.** $\frac{9}{20}$ **h.** $\frac{17}{30}$

i. $\frac{31}{20} = 1\frac{11}{20}$ **j.** $1\frac{2}{5}$ **k.** $3\frac{3}{4}$ **l.** $4\frac{7}{12}$

2. a. $\frac{3}{8}$ **b.** $\frac{3}{10}$ **c.** $\frac{7}{12}$ **d.** $\frac{1}{6}$

e. $\frac{5}{12}$ **f.** $\frac{13}{72}$ **g.** $\frac{1}{20}$ **h.** $\frac{2}{48} = \frac{1}{24}$

i. $\frac{4}{96} = \frac{1}{24}$ **j.** $\frac{2}{3}$ **k.** $1\frac{2}{15}$ **l.** $1\frac{7}{8}$

3. a. $\frac{1}{6}$ **b.** $\frac{1}{6}$ **c.** $\frac{7}{30}$ **d.** $\frac{1}{20}$

e. $\frac{2}{3}$ **f.** $\frac{10}{27}$ **g.** $\frac{32}{3} = 10\frac{2}{3}$ **h.** $\frac{247}{10} = 24\frac{7}{10}$

i. $\frac{49}{4} = 12\frac{1}{4}$ **j.** 17 **k.** $\frac{20}{3} = 6\frac{2}{3}$ **l.** $\frac{172}{3} = 57\frac{1}{3}$

4. a. $\frac{3}{2} = 1\frac{1}{2}$ **b.** $\frac{3}{2} = 1\frac{1}{2}$ **c.** $\frac{15}{16}$ **d.** $\frac{5}{4} = 1\frac{1}{4}$

e. $\frac{11}{15}$ **f.** $2\frac{13}{18}$ **g.** 2 **h.** $\frac{1}{5}$

i. $\frac{5}{4} = 1\frac{1}{4}$ **j.** $\frac{2}{9}$ **k.** $\frac{26}{35}$ **l.** $\frac{98}{75} = 1\frac{23}{75}$

5. a. $\frac{4}{9}$ **b.** $\frac{125}{729}$ **c.** 16 **d.** $22\frac{1}{2}$

e. $2\frac{1}{4}$ **f.** $5\frac{4}{9}$ **g.** $54\frac{109}{125}$ **h.** $7\frac{58}{81}$

6. a. $14\frac{1}{16}$ m^2 b. $11\frac{25}{64}$ cm^2

7. 40 **(A)** 8. $\frac{1}{15}$ **(A)**

9. $\frac{17}{20}$ **(A)** 10. $\frac{3}{8}$ **(A)**

11. 16 **(A)** 12. $1\frac{1}{6}$ cups **(A)**

Unit 11.1 Activity 3C: Problem-solving involving fractions (page 25)

Note: *Only problems in context are given A, M, E ratings.*

1. a. $\frac{9}{14}$ b. $-16\frac{7}{10}$ c. $-12\frac{19}{40}$ d. $\frac{1}{7}$ e. $\frac{3}{25}$ f. $3\frac{3}{8}$ g. $5\frac{1}{16}$
   h. $\frac{9}{22}$ i. $4\frac{3}{4}$ j. $1\frac{329}{400}$ k. $1\frac{5}{8}$ l. $\frac{3}{4}$

2. 3 kg **(A)** 3. $\frac{1}{4}$ **(A)** 4. $\frac{9}{16}$ **(A)** 5. $\frac{11}{15}$ **(A)** 6. $\frac{1}{6}$ **(A)**

7. $\frac{21}{24} = \frac{7}{8}$ **(A)** 8. K800 **(M)** 9. 2 000 L **(M)** 10. 45 **(M)**

Unit 11.1 Activity 3D: Multiple choice—fractions (page 26)

1. D 2. B 3. B 4. A 5. D

6. C 7. A 8. C 9. B

Unit 11.1 Activity 4A: Calculations involving decimals (page 29)

Note: *Only problems in context are given A, M, E ratings.*

1. a. 0.8 b. 3.76 c. $0.8\dot{3}$ d. $18.58\dot{3}$ e. $0.\dot{1}\dot{3}$
   f. $\frac{273}{500}$ g. $3\frac{1}{50}$ h. $4\frac{1}{3}$ i. $\frac{4}{9}$ j. $-11\frac{1}{1\,000}$

2. a. 15.2 b. 1.82 c. 1.88 d. 0.02
   e. 4 f. 9.8 g. 30.29 h. 0.624

3. a. 54 b. 70 c. 420 d. 0.238
   e. 57 630 f. 0.002958 g. 13 581

4. 10.5 km **(A)** 5. K6.95 **(A)** 6. 3.52 **(A)** 7. Short answer. **(A)**

8. a. 3.2 m **(A)** b. Less. **(A)** 9. 77 **(A)** 10. K55.27 **(A)**

Unit 11.1 Activity 4B: Significant figures and rounding (page 33)

Note: *Only problems in context are given A, M, E ratings. Approximate answers will vary.*

1. **a.** 120 **b.** 5 910 **c.** 31 800 **d.** 14.7 **e.** 89.9 **f.** 0.049

2. **a.** 1.54 **b.** 11.7 **c.** 0.414 **d.** 89.86 **e.** 0.0041 **f.** 10.0

3. **a.** 13 500 **b.** 32 000 **c.** 280 **d.** 30 **e.** 20 **f.** 45 or 50

g. 200 **h.** 1 100 **i.** 4 000

4. **a.** 720 **b.** 665.25 **(A)**

5. **a.** The radius was measured to 2 significant figures and the area should then be given to 2 significant figures. 8 significant figures implies too much accuracy. **(A)**

b. 120.8 cm^2 **(A)**

6. 10 (9 packets would not be enough since she needs at least 184 tiles). **(A)**

7. **a.** 80 cm^3 **b.** 87.1 cm^3 **(A)**

8. **a.** 20 m^2 **b.** 18.5 m^2 (1 dp) **(A)**

Unit 11.1 Activity 4C: Standard form (page 37)

Note: *Only problems in context are given A, M, E ratings.*

1. **a.** 4.87×10^2 **b.** 9.12×10^3 **c.** 7.63×10^5 **d.** 7.43×10^{-2}

e. 5.18×10^{-3} **f.** 4.51×10^{-6} **g.** 6.95×10^7 **h.** 4.958×10^2

i. 1.6401×10^1 **j.** 1.265×10^4 **k.** 3.2×10 **l.** 4.859×10^{-2}

2. **a.** 251 **b.** 6 390 **c.** 84 120 **d.** 0.357

e. 0.0017 **f.** 0.000975 **g.** 34.8 **h.** 46 000

i. 0.0783 **j.** 18 450 **k.** 1.264 **l.** 0.00057

3. **a.** 3.65×10^2, 3.7×10^2, 2.4×10^3, 1.901×10^4

b. 0.0014, 0.002, 4.1×10^{-2}, 2.6×10^{-1}

c. 9.9×10^3, 12.6×10^3, 12 965, 1.3×10^4

4. **a.** 4.8×10^3 **b.** 5.665×10^{-3} **c.** 9.0×10^4

d. 7.8×10^{-2} **e.** 4×10^5 **f.** 1.0×10^{-2}

5. **a.** 2.7×10^6, 2.21×10^6 **b.** 6.5×10^5, 7.01×10^5

c. 4.9×10^9, 4.79×10^9 **d.** 700, 7.92×10^2

6. **a.** 4.07×10^6 **b.** 3.02×10^6 **(M)** **c.** 8.80×10^5 **(M)**

7. 1 837 **(M)**

8. 1.415×10^9 km **(M)**

9. a. 1.286×10^9. Answer to 3 sf would be 1.29×10^6, the original figure for the population of China. **(M)**

b. 327 **(M)**

10. a. Sweets by 1.40×10^7 **(M)** **b.** 3.52×10^5 **(M)** **c.** 1.61×10^5 **(M)**

Unit 11.1 Activity 4D: Multiple choice—decimals (page 38)

1. C **2.** B **3.** C **4.** D **5.** A

Unit 11.1 Activity 5A: The numbers π and e (page 40)

1. a. 3.140845 … correct to 3 d.p.

b. 3.142857 … correct to 2 d.p.

c. 3.14626 … correct to 2 d.p.

d. 3.141592920 … correct to 6 d.p.

e. 3.01707 … correct to 0 d.p. (including $-\frac{1}{31}$ correct to 1 d.p.)

f. 3.04936 … correct to 0 d.p.(including $\frac{1}{20^2}$; correct to 1 d.p.)

2. a. 2.593742 … correct to 0 d.p.

b. 2.704813 … correct to 1 d.p.

c. 2.716923 … correct to 2 d.p.

d. 2.716666 … 2 d.p.

e. Including $\frac{1}{7!}$: 2.7182539· ...correct to 4 d.p.

Unit 11.1 Activity 5B: Surds (page 41)

1. $\sqrt{101}, \sqrt{17}, \sqrt{8}, \sqrt{83.4}$

2. a. 8, 0.5, 4.33, 20, 63.25, –0.6, –0.6, 0.04, –0.2, 500

b. $\sqrt{64}, \sqrt{0.25}, \sqrt{400}, \sqrt{-0.36}, \sqrt{0.0016}$

3. a. Q **b.** Q' **c.** Q **d.** Q'

e. Q **f.** Q' **g.** Q **h.** Q

4. a. $5\sqrt{2}$ **b.** $4\sqrt{3}$ **c.** $7\sqrt{2}$ **d.** $8\sqrt{5}$

e. $15\sqrt{3}$ **f.** $6\sqrt{3}$ **g.** 18 **h.** $2\sqrt{2}$

5. a. $\sqrt{75}$ **b.** $-\sqrt{28}$ **c.** $\sqrt{1000}$ **d.** $\sqrt{150}$

e. $-\sqrt{45}$ **f.** $\sqrt{32}$ **g.** $\sqrt{44}$ **h.** $\sqrt{0.56}$

Unit 11.1 Activity 5C: Simplifying surds (page 42)

1. a. $10\sqrt{7}$ b. $2\sqrt{3}+3\sqrt{2}$ c. $\sqrt{6}+2\sqrt{3}+3$ d. Already simplified
2. a. $\sqrt{6}$ b. $13\sqrt{5}$ c. $-2\sqrt{3}$ d. $12\sqrt{3}$
   e. $\sqrt{3}$ f. $\sqrt{5}$ g. $11\sqrt{2}+3\sqrt{11}$ h. $6\sqrt{2}$
3. a. $2\sqrt{3}+2\sqrt{3}-1$ b. $3\sqrt{5}-5\sqrt{3}$ c. $16\sqrt{3}-6\sqrt{6}-8$
   d. $12\sqrt{2}-\sqrt{6}-8\sqrt{3}$ e. $14+14\sqrt{2}$ f. $13\sqrt{3}-39$

Unit 11.1 Activity 5D: Simplifying and expanding surds (page 44)

1. a. 5 b. $5\sqrt{3}$ c. $3\sqrt{6}$ d. $15\sqrt{2}$
   e. $6\sqrt{6}$ f. $30\sqrt{2}$
2. a. $9+4\sqrt{5}$ b. $30-12\sqrt{6}$ c. $49-6\sqrt{10}$
   d. $41+24\sqrt{2}$ e. $219-120\sqrt{3}$
3. a. $6-\sqrt{3}$ b. $10+2\sqrt{21}$ c. $\sqrt{6}$ d. $6+2\sqrt{3}-2\sqrt{6}-2\sqrt{2}$
   e. $8-15\sqrt{6}$ f. $15\sqrt{3}-9$
4. a. $2\sqrt{3}+4$ b. $2-2\sqrt{6}$ c. $5+3\sqrt{5}$ d. $5\sqrt{3}-8$
5. a. -1 b. 3 c. 4 d. 13
   e. 23 f. 13 g. 94

Unit 11.1 Activity 5E: Simplest form (page 46)

1. a. $\frac{1}{5}$ b. $\frac{4}{3}$ c. $\frac{6}{5}$ d. $\sqrt{7}$ e. $\frac{\sqrt{6}}{2}$
2. a. $\frac{\sqrt{5}}{5}$ b. $\frac{\sqrt{2}}{4}$ c. $\sqrt{3}$ d. $\frac{\sqrt{6}}{12}$ e. $\frac{5\sqrt{2}+4}{2}$
3. a. $\sqrt{3}+\sqrt{2}$ b. $5-2\sqrt{5}$ c. $3+\sqrt{3}$ d. $\frac{11-4\sqrt{7}}{3}$ e. $\frac{2\sqrt{5}+20}{19}$
   f. $\frac{29-6\sqrt{6}}{7}$

4. a. $17+12\sqrt{2}$ b. $\frac{12+6\sqrt{2}+2\sqrt{3}-\sqrt{6}}{11}$ c. $\frac{10-15\sqrt{2}-4\sqrt{3}+6\sqrt{6}}{-14}$

d. $\frac{61+7\sqrt{5}}{44}$ e. $\frac{2+\sqrt{2}}{2}$ f. $\frac{41+4\sqrt{10}}{39}$ g. $\frac{\sqrt{15}-3}{2}$

h. $\frac{8\sqrt{21}-28}{5}$ i. $\frac{2\sqrt{11}+2\sqrt{6}+\sqrt{66}+6}{5}$ j. $\frac{57+\sqrt{15}}{42}$

Unit 11.1 Activity 6A: Negative indices (page 49)

1. a. 0.0625 b. 0.4019 c. 7.3046 d. 5.0027 e. 4.1890 f. 9
g. 27.648 h. 0.0865 i. 1.06 j. −1.2022 k. 0.2712

2. a. $\frac{1}{9}$ b. $\frac{1}{64}$ c. $\frac{1}{16}$ d. $\frac{27}{64}$ e. $\frac{8}{25}$ f. 9
g. 16 h. $\frac{25}{9}$ i. $\frac{125}{64}$ j. $\frac{27}{4}$ k. $\frac{25}{27}$ l. $\frac{1}{24}$
m. $\frac{1}{10}$ n. $\frac{7}{5}$ o. $\frac{8}{3}$ p. $\frac{1}{6}$ q. $\frac{1}{15}$ r. $\frac{1}{12}$
s. 10 t. 12 u. $\frac{1}{36}$ v. $\frac{1}{175}$ w. 3 x. 5
y. $\frac{1}{14}$ z. $\frac{4}{3}$

3. a. $\frac{3}{2}$ b. $\frac{1}{36}$ c. 144 d. $\frac{1}{576}$ e. $\frac{5}{6}$ f. $\frac{5}{18}$
g. $\frac{33}{10}$ h. $\frac{1}{6}$ i. $\frac{1}{2}$ j. $\frac{17}{12}$ k. $-\frac{73}{18}$ l. $\frac{6}{5}$
m. $\frac{5}{36}$ n. $\frac{11}{16}$ o. $\frac{11}{4}$ p. $\frac{4}{5}$

4. a. $\frac{1}{2}$ b. $\frac{1}{6}$ c. 2 d. $\frac{5}{9}$ e. $\frac{3}{2}$ f. $\frac{2}{3}$
g. 9 h. $\frac{3}{2}$ i. $1\frac{3}{5}$

5. a. 3×2^2 b. $2^3 \times 5^2$ c. 5×3^2 d. $5^2 \times 2^{-4}$ e. 2×3^{-1} f. 22×3^{-3}
g. $2^4 \times 3^{-4}$ h. $2^6 \times 3^{-4}$ i. $3^2 \times 2^{-2}$

6. a. 2^4 b. 5^3 c. 3^4 d. 2^{-4} e. 2^6
f. 5^{-4} g. 3^{-2} h. 3^{-5}

7. a. 25 b. − 25 c. $\frac{1}{25}$ d. $\frac{5}{6}$ e. $\frac{1}{36}$
f. $-\frac{6}{7}$ g. − 1

8. **a.** 2^{C-2} **b.** 2^{2m} **c.** 2^{3+t} **d.** 2^{2x+2} **e.** 2^{n-1} **f.** 2^1

g. 2^{1+x} **h.** 3^{2+x} **i.** 3^{y-3} **j.** 3^{5a} **k.** 3^3 **l.** 3^{3a-1}

Unit 11.1 Activity 6B: Simplifying expressions with indices (page 51)

1. **a.** $\frac{1}{x^3}$ **b.** $\frac{1}{y^2}$ **c.** $\frac{1}{a^5}$ **d.** $\frac{1}{b^4}$ **e.** $\frac{1}{z^7}$ **f.** $\frac{1}{y^6}$

g. $\frac{1}{z^{11}}$ **h.** $\frac{1}{xy}$ **i.** $\frac{1}{abc}$ **j.** $\frac{1}{x^2y^2}$ **k.** $\frac{1}{2x}$ **l.** $\frac{1}{4x^2}$

m. $\frac{1}{9a^2}$ **n.** $\frac{1}{8x^3}$ **o.** $\frac{3}{x}$ **p.** $\frac{a}{b^3}$ **q.** $\frac{b^2}{a^3}$ **r.** $\frac{1}{a}$

s. $\frac{1}{2x}$ **t.** $\frac{1}{4x^3}$ **u.** $\frac{3}{5x^2}$ **v.** $\frac{2}{7a^3}$ **w.** $\frac{4}{9a^2}$ **x.** $\frac{3}{5x^4}$

y. $\frac{4}{5x^5}$ **(A)**

2. **a.** x^{-6} **b.** y^{-5} **c.** z^{-2} **d.** $4x^{-1}$ **e.** $5y^{-2}$ **f.** $7z^{-5}$

g. $\frac{x^{-2}}{2}$ **h.** $\frac{a^{-4}}{3}$ **i.** $\frac{y^{-4}}{5}$ **j.** $\frac{2x^{-2}}{5}$ **k.** $\frac{3y^{-3}}{4}$ **l.** $\frac{3x^{-2}}{4}$

m. $\frac{4y^{-2}}{25}$ **n.** $\frac{x^{-1}}{8}$ **o.** $\frac{x^{-3}}{8y^{-2}}$ **(A)**

3. **a.** $\frac{y^3}{2x^2}$ **b.** $\frac{2b^4}{3a^3}$ **c.** $\frac{x}{5}$ **d.** $\frac{2y^4}{x^2}$ **e.** $\frac{x^5}{y}$ **f.** $\frac{2y^3}{x}$

g. $\frac{8}{x^7y^2}$ **h.** $\frac{1}{25x^3}$ **i.** $\frac{1}{6xy}$ **j.** $\frac{1}{27x^4}$ **k.** $\frac{x}{2y^2}$ **l.** $\frac{x^2}{9y^3}$

m. $\frac{2}{3x^2}$ **n.** $\frac{4y^2}{5x^3}$ **o.** $\frac{x+y}{xy}$ **p.** $\frac{x+1}{x^2}$ **q.** $\frac{xy^2+2}{2y}$ **(A)**

Unit 11.1 Activity 6C: Fractional indices (page 53)

1. **a.** $a^{\frac{1}{2}}$ **b.** $c^{\frac{1}{2}}$ **c.** $x^{\frac{1}{3}}$ **d.** $a^{\frac{1}{5}}$ **e.** $u^{\frac{1}{6}}$ **f.** $x^{\frac{2}{5}}$

g. $x^{\frac{3}{7}}$ **h.** $a^{\frac{3}{8}}$ **i.** $x^{-\frac{1}{2}}$ **j.** $y^{-\frac{1}{2}}$ **k.** $a^{-\frac{1}{3}}$ **l.** $x^{-\frac{1}{5}}$

m. $x^{-\frac{2}{7}}$ **n.** $x^{-\frac{5}{9}}$ **o.** $a^{-\frac{3}{7}}$ **p.** $a^{-\frac{8}{9}}$ **q.** $4a^{\frac{1}{3}}$ **r.** $6y^{-\frac{1}{3}}$

s. $8x^{-\frac{7}{3}}$ **t.** $\frac{x^{-\frac{2}{3}}}{7}$ **u.** $\frac{2x^{-\frac{3}{4}}}{5}$ **v.** x^{-1} **w.** $x^{\frac{1}{6}}$ **x.** $a^{-\frac{5}{3}}$

y. $x^{-\frac{1}{12}}$ **(A)**

2. a. $\sqrt[7]{a^3}$ b. $\sqrt[4]{b^3}$ c. $\sqrt[5]{x^2}$ d. $\frac{1}{\sqrt[3]{x^2}}$ e. $\frac{1}{\sqrt[4]{y^3}}$ **(A)**

3. a. $(xy)^{\frac{1}{2}}$ or $(x)^{\frac{1}{2}}(y)^{\frac{1}{2}}$ b. $(x)^{\frac{1}{2}}y$ c. $x(y)^{-\frac{1}{3}}$ d. $x^2(y)^{-\frac{1}{3}}$ e. $2x(y)^{-\frac{3}{5}}$

f. $(x)^{-\frac{5}{4}}y$ g. $(x)^{\frac{1}{2}}(a)^{\frac{1}{3}}$ h. $(a)^{-\frac{1}{3}}(c)^{\frac{1}{2}}$ i. b **(A)**

4. a. $\frac{1}{8x^3}$ b. $\frac{1}{9x^6}$ c. $\frac{1}{4y^4}$ d. $27A^3$ e. $\frac{1}{32B^{10}}$

f. $\frac{1}{4x^4}$ g. $4A^2B^2$ h. $\frac{1}{25A^2B^4}$ i. $3x^{\frac{3}{2}}$ j. $256x^{10}$ **(A)**

Unit 11.1 Activity 6D: Calculating with indices (page 55)

1. 16 2. 0.11 3. 82.19 4. –32 5. 2.25 6. 1.32
7. –2.11 8. No sol. 9. 3.51 10. 1720.51 11. 2.73 12. 1.83
13. 74.30 14. –0.5109 15. 1.25 16. 2.1715 17. 14.64 18. 0.0833
19. No sol. 20. 0.0638

Unit 11.1 Activity 6E: Equations with indices (page 57)

1. a. 1.26 b. 8 c. 81 d. 25 e. $\frac{32}{243}$ f. 0.2 or $\frac{1}{5}$

g. $\frac{1}{4}$ or 0.25 h. 81 i. 32 j. 8 k. $\frac{1}{6}$ l. $\frac{1}{6}$

m. 4 n. 8 o. 5 p. 27 q. $\frac{1}{27}$ r. $\frac{1}{4}$

s. $\frac{1}{9}$ t. $\frac{1}{243}$ **(A)**

2. a. 64 b. 2 381 c. 0.20 d. 3.04 e. No Sol. f. –10.08

g. 6.55 h. 270.8 i. 1.67 **(A)**

3. a. $x = 3$ b. $x = -\frac{1}{2}$ c. $x = \frac{5}{2}$ d. $x = 0$ e. No solution

f. $x = 2$ g. $x = 5$ h. $x = -3$ i. $x = 3$ j. $x = 1\frac{1}{2}$

k. $x = -3$ l. $x = \frac{1}{2}$ m. $x = -\frac{1}{2}$ n. $x = -\frac{1}{4}$

4. a. $x = 3$ b. $x = 3$ c. $x = 2$ d. $x = -2$ e. $x = 2$ f. $x = -2$

5. a. K51 b. 81.51 m **(A)**

6. a. 14336 b. 4 c. 1.8 **(M)**

7. a. T gets smaller b. 0.01094 **(M)**

Unit 11.1 Activity 7A: Other logarithmic functions (page 59)

1. 2.10 **2.** 199.53 **3.** 2 508.52 **4.** 0.04 **5.** 3.68 **6.** 2.5

Unit 11.1 Activity 7B: Using logarithmic properties (page 63)

1. **a.** $\log_2 8 = 3$ **b.** $\log_3 81 = 4$ **c.** $\log_5 25 = 2$ **d.** $\log_6 216 = 3$ **e.** $\log_x b = a$

2. **a.** $3^2 = 9$ **b.** $(9)^{\frac{1}{2}} = 3$ **c.** $25^{1.5} = 125$ **d.** $(1000)^{\frac{1}{3}} = 10$ **e.** $3^{-1} = \frac{1}{3}$

3. **a.** 6 **b.** –1 **c.** 6 **d.** 7 **e.** –1 **f.** –3
g. $\frac{1}{2}$ **h.** 3 **i.** $\frac{1}{2}$ **j.** $\frac{1}{4}$ **k.** $\frac{1}{2}$ **l.** $\frac{3}{2}$
m. 2 **n.** $\frac{5}{6}$ **o.** $-\frac{1}{2}$ **p.** 1 **q.** –1 **r.** –3
s. $\frac{5}{3}$ **t.** $\frac{3}{4}$ **(A)**

4. **a.** 0.954 **b.** –2.117 **c.** 2.510 **d.** 2.885 **e.** 0.972 **f.** 1.262
g. 1.390 **h.** 1.710 **i.** 2.767 **j.** 2.484 **k.** 31 **(A)**

5. **a.** 1 **b.** 1 **c.** 3 **d.** $\frac{1}{2}$ **e.** $\frac{1}{4}$ **f.** $\frac{5}{4}$
g. $\frac{3}{4}$ **h.** $\frac{1}{2}$ **i.** $\frac{4}{5}$ **j.** –1

6. **a.** $\log_a \frac{1}{2}$ **b.** $\log_a \frac{1}{2}$ **c.** $\log_a \frac{x^2}{y}$ **d.** $\log_a 3(x+1)$ **e.** $\log_b\left(\frac{2x^2}{3}\right)$ **f.** $\log_c 2$
g. $\log_a(ac^2)$ **h.** $\log_a c^{\frac{3}{2}}$ **i.** $\log_a\left(\frac{pq^2}{r^2}\right)$ **j.** $\log_{10}\frac{1}{2}$

Unit 11.1 Activity 8A: Units of measurement (page 68)

1. **a.** x **b.** vii **c.** xix **d.** v **e.** xi **f.** i
g. xv **h.** xviii **i.** ix **j.** viii **k.** ii **l.** iii
m. xii **n.** iv **o.** xvi

2. **a.** 800 cm **b.** 40 mm **c.** 4 kg **d.** 3 500 mm **e.** 30 000 cm
f. 0.6 L **g.** 4 800 m **h.** 8.3 t **i.** 370 mg **j.** 40 cm
k. 750 mL **l.** 0.945 g **m.** 0.14 km **n.** 0.0137 t **o.** 0.045 km
p. 4 700 kg **q.** 0.12 kL **r.** 0.64 km **s.** 4 000 mm **t.** 7 300 000 g
u. 1.86 m **v.** 63 kg **w.** 3 800 g **x.** 9 370 mL

3. **a.** 18.50 m **b.** 240 cm or 2.4 m **c.** 38.5 cm

4. **a.** 1.34 kg **b.** 4 g **c.** 140 g

5. **a.** 1:22 pm **(A)** **b.** 2:12 pm **(A)** **c.** 0430 **(A)**
   d. April 4 **(A)** **e.** 1 hr 45 min **(A)**

6. **a.** 3:35 pm **(A)** **b.** 5 hr 32 min **(A)** **c.** 18 hr 10 min **(A)**
   d. 2 678 400 seconds **e.** 2325 **(A)**

Unit 11.1 Activity 8B: Reading measuring instruments (page 71)

This exercise gives practice in reading scales and other measuring equipment. While these skills are important, no A, M, E rating is given as these skills are usually part of a problem-solving exercise.

1. **a.** 0.8 kg **b.** 800 g **c.**

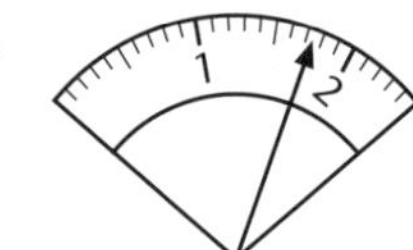

2. **a.** 116° **b.** 64°

3. **a.** **i.** 380 mL **ii.** 230 mL **iii.** 150 mL
   b. **i.** 75 mL **ii.** 225 mL **iii.** 450 mL
   c. **i.** 410 mm **i.** 425 mm **iii.** 437 mm
   d. **i.** 2 m 30 cm (2.3 m) **ii.** 2 m 32 cm (2.32 m) **iii.** 2 m 27.2 cm (2.272 m)
   e. **i.** 85 cm **ii.** 86.7 cm **iii.** 89.3 cm
   f. **i.** 500 g **ii.** 1 075 g **iii.** 1 250 g
   g. **i.** 41°C **ii.** 37.8°C **iii.** 36.6°C
   h. **i.** 0.3 Amp **ii.** 4 Amp **iii.** 6 Amp
   i. **i.** 4 **ii.** 30 **iii.** 400
   j. **i.** 540 kHz **ii.** 675 kHz **iii.** 850 kHz
   k. **i.** 52 **ii.** 54 **iii.** 58

Unit 11.1 Activity 8C: Limits of accuracy (page 73)

1. **a.** 5.5 m $\leq x <$ 6.5 m **(M)** **b.** 4.25 cm $\leq x <$ 4.35 cm **(M)**
   c. 5.235 m $\leq x <$ 5.245 m **(M)** **d.** 1.75 cm $\leq x <$ 1.85 cm **(M)**
   e. 23.5 mm $\leq x <$ 24.5 mm **(M)** **f.** 757.5 km $\leq x <$ 758.5 km **(M)**
   g. 1.035 L $\leq x <$ 1.045 L **(M)** **h.** 0.55 mL $\leq x <$ 0.65 mL **(M)**
   i. 6.95 mg $\leq x <$ 7.05 mg **(M)** **j.** 8.995 m $\leq x <$ 9.005 m **(M)**

2. 15.415 s ≤ time < 15.425 s **(M)**

3. 3.395 m ≤ length < 3.405 **(M)**

4. 10.4 m ≤ perimeter < 10.8 m **(M)**

5. 47 500 L ≤ capacity < 48 500 L **(M)**

6. The limits of accuracy for the old record are 12.35 seconds ≤ time taken < 12.45 seconds. The limits of accuracy for the recent runner's time are

   12.405 seconds ≤ time taken < 12.415 seconds.

   Within the limits of accuracy for the old record, there are times such as 12.42 seconds which are slower than the times within the recent runner's limit of accuracy. Therefore the recent runner could well have run faster that the record holder. However, 12.41 seconds as a single time would not break the record. **(E)**

Unit 11.1 Activity 8D: Conversion of units (page 76)

1. **a.** 2.7432 yd **b.** 321.860 km **c.** 1.0160 m
d. 5468.32 yds **e.** 1853 m **f.** 3.937 in.

2. **a.** 1196 yd^2 **b.** 7.7229 sq miles **c.** 247.104 acres
d. 0.1672 ha **e.** 1935.48 mm^2 **f.** 1416.415 m^2

3. **a.** 141.343 ft^3 **b.** 7.2083 ft^3 **c.** 1.7047 litres
d. 1136.6 cm^3 **e.** 0.5368 pints **f.** 26.3197 fl oz.

Unit 11.1 Activity 8E: Multiple choice—measurement (page 76)

1. A **2.** C **3.** A **4.** C
5. B **6.** D **7.** A

Unit 11.1 Activity 9A: Ratios (page 79)

Note: *Only problems in context are given A, M, E ratings.*

1. **a.** 1 : 2 **b.** 1 : 3 **c.** 5 : 3 **d.** 1 : 8 **e.** 3 : 2 **f.** 4 : 1
g. 10 : 1 **h.** 1 : 3 **i.** 20 : 1 **j.** 1 : 250 **k.** 1 : 500 **l.** 14 : 1

2. **a.** 15 : 25 **b.** 40 : 16 **c.** 9 : 36 : 45 **d.** K3.20 : K4.80
e. K13.50 : K18.90 **f.** 36 : 54 : 90

3. **a.** K3 000 **b.** K112.50 **(A)**

4. **a.** K20 000 **b.** 2 : 5 **c.** Mr Liklik, K50 000; Mr Bik, K125 000 **(A)**

5. **a.** 1 : 3 **b.** 6 L **c.** 4.5 L **(A)** **d.** 16 **(M)**

6. **a.** 900 g **(A)** **b.** 2.25 kg **(M)**

Unit 11.1 Activity 9B: Map scales (page 82)

1. **a.** 100

b.
0 100 200 300 400 km

c. **i.** 3.3 cm **ii.** 5.5 cm

d. **i.** 330 km **ii.** 550 km

2. **a.** 1 : 10 000

b. 1 : 100 000

c. 1 : 20 000 000

3. **a.**

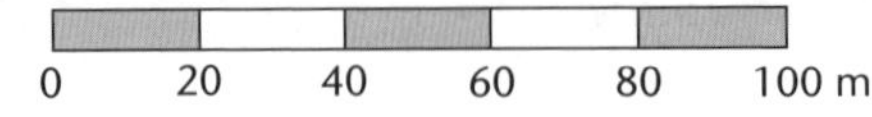

b. 0 3 6 km

c. 0 1 km

4. **a.** **i.** 507 m **ii.** 780 m

b. 6.7 cm

c. 0 130 260 390 520 m

5. 15.4 cm

6. 11.2 cm

7. 1 : 5 000 000

Unit 11.1 Activity 9C: Variation (page 85)

1. **a.** $y \propto x$; $y=kx$ **b.** 20 **c.** 2.8

2. **a.** 7.5 **b.** 252

3. **a.** 270 **b.** 10

4. **a.** and **b.**

5. **a.** $A = 0.421362P$ **b.** AUK33.71 **c.** PGK237.33

6. **a.** 13 cm **b.** 3.846 kg

7. K381.25

8. 8.92 cm

9. 10.102 sec.

10. 12.247 km

11. **a.** 0.57 sec **b.** 24.5 cm

Unit 11.1 Activity 9D: Inverse variation (page 87)

1. **a.** 11 **b.** 6.6

2. **a.**

x	2	3	8	12
y	60	40	15	10

b.

p	100	50	25	7.5
q	0.75	1.5	3	10

3. a. $V = \frac{k}{P}$ b. 1480 c. 2.96 L

4. 4 hours 10 minutes.

5. a. 53.0 cm b. 35.4 cm

6. a. i. 212.5 units ii. 34 units

   b. 3.54 m

Unit 11.1 Activity 10A: Algebraic expressions (page 91)

1. $9 + a$ 2. $b - 4$ 3. $5c$ 4. $x + 6$ 5. $y - 7$ 6. $4(x + y)$

7. $\frac{x}{6}$ 8. $\frac{x + y}{10}$ 9. $12 - y$ 10. $3(x + 4)$ 11. $\frac{y - 7}{2}$ 12. $\sqrt{x + 10}$

13. $n + n + 1$ or $2n + 1$ 14. $\frac{a + 10}{7b}$ 15. $2y - 3$ 16. $(p - 2)^3$ 17. $x - 5$

18. $t + 3l$ 19. $10x + 15y$ 20. $a + b + 4$

Unit 11.1 Activity 10B: Combining algebraic terms (page 93)

1. $8a$ 2. $6a$ 3. $3 + 2x$ 4. $x^2 + x + 1$

5. $4x^2 - 3x + 5$ 6. $4x^2y - 3xy^2$ 7. $10x$ 8. $3a$

9. $4 + 3x$ 10. $x^2 + 8x + 2$ 11. $3x^2 - x - 3$ 12. $x^2 - 9x + 9$

13. $24ab$ 14. $12a^2b$ 15. $24pqr$ 16. $10a^2b^2$

17. $3x^6 - x^4$ 18. $3x$ 19. $\frac{3bc}{2}$ 20. $\frac{2}{3y}$

Unit 11.1 Activity 10C: Algebraic fractions (page 96)

1. a. x^8 **(A)** b. $3y^8$ **(A)** c. $4a^3$ **(A)** d. $4b$ **(A)** e. $3x$ **(A)**

   f. $3a^4$ **(A)** g. $\frac{a^3b}{2}$ **(A)** h. $\frac{3x^2}{2y}$ **(A)** i. $\frac{2a^2b}{5}$ **(A)** j. $\frac{2x^5}{3z^2}$ **(A)**

2. a. $\frac{x + 1}{3}$ **(M)** b. $\frac{y - 2}{9}$ **(M)** c. $\frac{7x}{12}$ **(M)** d. $\frac{a}{10}$ **(M)** e. $\frac{7k}{15}$ **(M)**

   f. $\frac{y}{6}$ **(M)** g. $\frac{7}{x + 1}$ **(M)** h. $\frac{x^2 + 2}{2x}$ **(M)** i. $\frac{4b + 3a}{ab}$ **(M)**

   j. $\frac{4}{3x}$ **(M)** k. $\frac{1 + 2x}{x^2}$ l. $\frac{3 - 2y}{3y^2}$ **(M)**

3. a. $\frac{a^2}{12}$ b. 2 c. 1 d. $\frac{x^2}{10}$ e. $\frac{2}{5}$

   f. $\frac{4}{3}$ **(A)** g. $\frac{1}{9}$ **(A)** h. $\frac{8a^3}{45}$ **(A)** i. $\frac{9x}{2}$ **(A)**

   j. a^2b **(A)** k. $\frac{9xy^2}{10}$ **(A)** l. $\frac{2y^3}{3}$ **(A)**

4. **a.** $\frac{x+9}{20}$ **(M)** **b.** $\frac{9x+1}{20}$ **(M)** **c.** $\frac{3-x}{4}$ **(M)** **d.** $\frac{9x-5}{14}$ **(M)**

e. $\frac{8-5x}{12}$ **(M)** **f.** $\frac{4x+3}{12x}$ **(M)**

5. **a.** $\frac{7a}{12}$ **(M)** **b.** $\frac{-1}{x}$ **(M)** **c.** $\frac{a^2}{24}$ **(M)** **d.** $\frac{7y}{30}$ **(M)**

e. $\frac{w}{3}$ **(M)** **f.** $\frac{10a^2}{3}$ **(M)**

Unit 11.1 Activity 11A: Expanding brackets (page 98)

Many of these problems practise basic skills.

1. $3x + 3$ **2.** $4x + 8$ **3.** $ab - 2a$ **4.** $x^2 - 3x$

5. $-4x + 20$ **6.** $2x^3 + 6x^2 - 8x$ **7.** $2 - 6x$ **8.** $11 - x$

9. $2x + 10$ **10.** $11x + 6y$ **(A)** **11.** $10a + 21b$ **(A)** **12.** $2x - y$ **(A)**

13. $x^2 + 5x - 2$ **(A)** **14.** $3p^2 - 2pq + 2q^2$ **(A)** **15.** $-x^2 - x$ **(A)** **16.** $x^2 - 6x$ **(A)**

17. $3x^2 - 6x^3 - 3x^4$ **(A)** **18.** $-2x^2y + 2xy^2 + 6xyz$ **(A)** **19.** $x^2 + 3x$ **(A)**

20. a. $K(y - 10)$ **(A)** **b.** $K3(y - 10)$ or $K(3y - 30)$ **(A)**

Unit 11.1 Activity 11B: Further expansions (page 100)

1. $x^2 + 9x + 20$ **(A)** **2.** $x^2 - x - 6$ **(A)** **3.** $x^2 - x - 20$ **(A)**

4. $x^2 - 2x - 15$ **(A)** **5.** $x^2 - 11x + 28$ **(A)** **6.** $x^2 - 11x + 30$ **(A)**

7. $2x^2 + 11x + 12$ **(A)** **8.** $12x^2 - 7x - 12$ **(A)** **9.** $2x^2 - 15x + 18$ **(A)**

10. $6 + 7x - 3x^2$ **(A)** **11.** $35 + 3x - 2x^2$ **(A)** **12.** $6x^2 - 23x + 20$ **(A)**

13. $3x^2 + 10x + 3$ **(A)** **14.** $15x^2 + 14x - 8$ **(A)** **15.** $12 + x - 6x^2$ **(A)**

16. $x^2 + 12x + 36$ **(A)** **17.** $x^2 + 8x + 16$ **(A)** **18.** $x^2 - 14x + 49$ **(A)**

19. $x^2 - 100$ **(A)** **20.** $16x^2 - 1$ **(A)** **21.** $4x^2 - 25$ **(A)**

22. $4x^2 + 12x + 9$ **(A)** **23.** $9x^2 - 30x + 25$ **(A)** **24.** $x^3 + 3x^2 + 3x + 2$ **(M)**

25. $x^3 - x^2 - 10x + 12$ **(M)** **26.** $2x^3 - 9x^2 + 4x + 15$ **(M)** **27.** Length YZ = $\sqrt{2x^2 + 2}$ **(M)**

28. $A = 2x^2 + 3x - 18$ **(M)** **29.** $10x + 25$ **(M)** **30.** $x^3 + 6x^2 + 12x + 8$ **(M)**

Unit 11.1 Activity 11C: Basic factorising (page 102)

Many of these problems practise basic skills.

1. $4(a + 3)$ **2.** $7(x - 2)$ **3.** $5(y + 5)$ **4.** $3(3x + 5y)$

5. $x(y - 3)$ **6.** $y(y^2 + 5)$ **7.** $a(a + b)$ **8.** $x(x - 7)$

9. $2(3x + 4y)$ **10.** $a(b - a)$ **11.** $3p(p - 2)$ **12.** $2(2x + 7y)$

13. $3x(x^2 + 2x - 1)$ **14.** $4(2x + 3y - 5z)$ **15.** $5(3x + 2y - 5)$

16. $2xy^2(4xy + 6y^2 - 5)$ **(A)** **17.** $4(2x - 6y + 5)$ **(A)** **18.** $ab(3a + 5a^2b + 1)$ **(A)**

19. $2(2x^2 + 4x - 5)$ **(A)** **20.** $y(2 - 6y + y^2)$ **(A)** **21.** $3pq(p - 3q)$ **(A)**

22. $p^2(p^2 - 3)$ **(A)** **23.** $2a^2(2 - 9a)$ **(A)** **24.** $a^2b^2(b + a)$ **(A)**

25. $(x + 3)(2x + 5)$ **(A)** **26.** $(x - 2)(6x - 5)$ **(A)** **27.** $(a + x)(b + c)$ **(M)**

28. $(x + 2y)(3 + a)$ **(M)** **29.** $(x + 3y)(5 - p)$ **(M)** **30.** $(a - b)(x + y)$ **(M)**

Unit 11.1 Activity 11D: Factorising quadratic expressions (page 105)

1. $(x + 4)(x + 2)$ **(A)** **2.** $(x + 5)^2$ **(A)** **3.** $(x - 4)^2$ **(A)**

4. $(x + 5)(x - 3)$ **(A)** **5.** $(x + 8)(x - 3)$ **(A)** **6.** $(x - 7)(x + 2)$ **(A)**

7. $(x + 5)(x + 4)$ **(A)** **8.** $(x - 5)^2$ **(A)** **9.** $(x + 5)(x + 2)$ **(A)**

10. $(x - 3)(x - 1)$ **(A)** **11.** $(x - 7)(x - 4)$ **(A)** **12.** $(x + 5)(x - 4)$ **(A)**

13. $(x + 7)(x - 3)$ **(A)** **14.** $(x + 6)(x + 3)$ **(A)** **15.** $(x - 6)^2$ **(A)**

16. $(x - 3)(x + 3)$ **(A)** **17.** $(x - 7)(x + 7)$ **(A)** **18.** $(x + 8)(x - 8)$ **(A)**

19. $(5x - 4)(5x + 4)$ **(A)** **20.** $(2x + 9)(2x - 9)$ **(A)** **21.** $5(x - 2)(x + 2)$ **(A)**

22. $3(x - 3)(x + 7)$ **(A)** **23.** $2(x + 1)(x - 2)$ **(A)** **24.** $4(x - 3)(x - 1)$ **(A)**

25. $(2x - 1)(x + 3)$ **(M)** **26.** $(2x + 5)(x + 2)$ **(M)** **27.** $(3x - 1)(x - 2)$ **(M)**

28. $(5x + 2)(x - 3)$ **(M)** **29.** $(3x + 7)(x - 1)$ **(M)** **30.** $(3x - 4)(2x + 3)$ **(M)**

Unit 11.1 Activity 11E: Simplifying rational expressions by factorising (page 106)

1. 9 **(M)** **2.** 7 **(M)** **3.** $x + 3$ **(M)** **4.** 3 **(M)** **5.** x **(M)**

6. $x + 6$ **(M)** **7.** $x + 4$ **(M)** **8.** $\frac{x+4}{x-2}$ **(M)** **9.** $\frac{x-8}{x+2}$ **(M)** **10.** $\frac{x+4}{x-2}$ **(M)**

11. $\frac{x+2}{x+3}$ **(M)** **12.** $\frac{x+4}{2(x+5)}$ **(M)**

Unit 11.1 Activity 11F: Multiple choice—factorising quadratic expressions (page 106)

1. A **2.** B **3.** D **4.** C **5.** D **6.** C **7.** A

8. B **9.** A **10.** B **11.** D **12.** C **13.** A **14.** B

Unit 11.1 Activity 12A: Solving quadratic equations (page 111)

1. **a.** $-2, -3$ **(A)** **b.** $2, 4$ **(A)** **c.** $-1, 3$ **(A)** **d.** $5, -2$ **(A)**

e. $0, 4$ **(A)** **f.** $0, -5$ **(A)** **g.** $\frac{3}{2}, -6$ **(A)** **h.** $-\frac{1}{2}, 1$ **(A)**

i. $-\frac{3}{5}, -\frac{5}{2}$ **(A)** **j.** $\frac{3}{4}, \frac{9}{2}$ **(A)** **k.** $\frac{4}{3}, 5$ **l.** $\frac{2}{3}, -\frac{2}{3}$ **(A)**

2. **a.** 0, –3 **(M)** **b.** 0, 5 **(M)** **c.** –3, 2 **(M)** **d.** –1, –4 **(M)**
e. 8, –5 **(M)** **f.** 3 **(M)** **g.** 2, 6 **(M)** **h.** –5, 2 **(M)**
i. ±2 **(M)**

3. **a.** 5, –3 **(M)** **b.** –6, 3 **(M)** **c.** 1, 3 **(M)** **d.** 0, 6 **(M)** **e.** 0, $\frac{5}{2}$ **(M)**
f. ±7 **(M)** **g.** –2, 4 **(M)** **h.** ±2 **(M)** **i.** $\frac{3}{2}$, –2 **(M)**

4. **a.** ±8 **(M)** **b.** –3, 9 **(M)** **c.** ±12 **(M)** **d.** ±5 **(M)**
e. $\frac{-9}{2}$, 4 **(M)** **f.** $\frac{-19}{6}$, 2 **(M)**

Unit 11.1 Activity 12B: Solving word problems with quadratic equations (page 113)

1. **a.** Sam's age. **b.** Sam = 14, Sam's sister = 11, Sam's brother = 16. **(M)**

2. Day 3 and Day 9. **(M)** **3.** 12 m by 32 m **(M)** **4.** 2 m square **(M)**

5. 5 **(M)** **6.** 8 cm **(M)** **7.** 4, 14 or –4, –14 **(M)**

8. 16 cm square **(E)** **9.** (18 and 20) or (–10 and –36) **(E)** **10.** 8, 12, 18 points **(E)**

Unit 11.1 Activity 12C: Multiple choice—solving quadratic equations (page 114)

1. C **2.** D **3.** B **4.** A **5.** D **6.** B **7.** C **8.** C

Unit 11.1 Activity 13A: Quadratic graphs (page 119)

1. **a.**

x	–2	–1	0	1	2
x^2	4	1	0	1	4
$y = x^2 - 2$	2	–1	–2	–1	2

b.

x	0	1	2	3	4
$x - 2$	–2	–1	0	1	2
$y = (x - 2)^2$	4	1	0	1	4

c.

x	–5	–4	–3	–2	–1
$x + 3$	–2	–1	0	1	2
$y = (x + 3)^2$	4	1	0	1	4

d.

x	–2	–1	0	1	2
x^2	4	1	0	1	4
$y = 2 - x^2$	–2	1	2	1	–2

e.

x	–2	–1	0	1	2
x^2	4	1	0	1	4
$y = x^2 + 3$	7	4	3	4	7

f.

x	–2	–1	0	1	2
x^2	4	1	0	1	4
$y = 2x^2$	8	2	0	2	8

2. **a.** $y = x^2 + 3$ **b.** $y = (x + 3)^2$ **c.** $y = 2x^2$ **d.** $y = 2 - x^2$
e. $y = x^2 - 2$ **f.** $y = (x - 2)^2$

3. a.

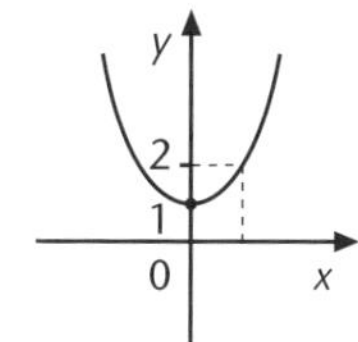

(A)

b.

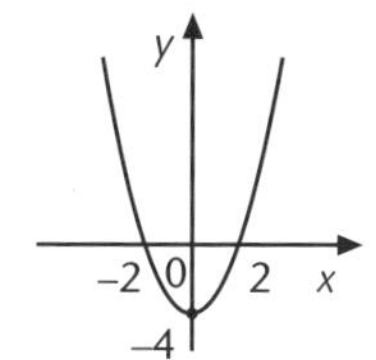

(A)

c.

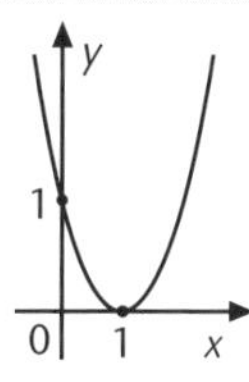

(A)

d.

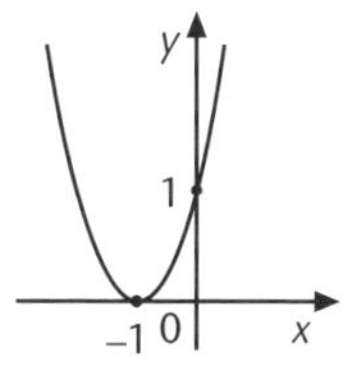

(A)

e.

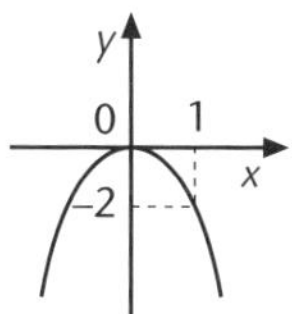

(M)

f.

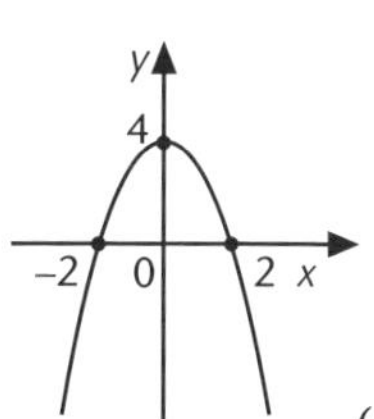

(A)

4. a.

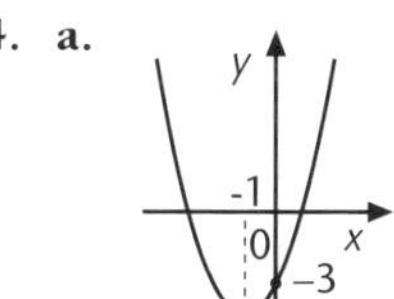

(M)

b.

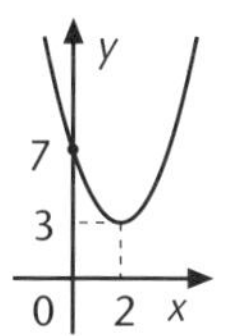

(M)

c.

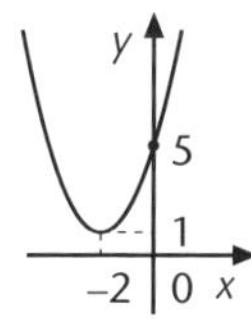

(M)

d.

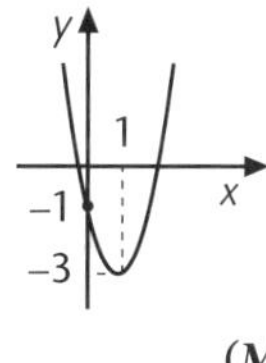

(M)

e.

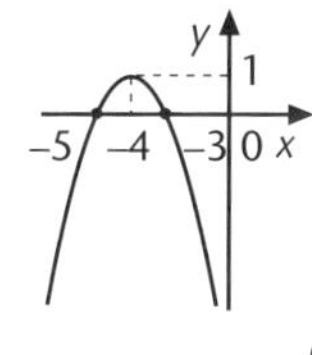

(M)

f.

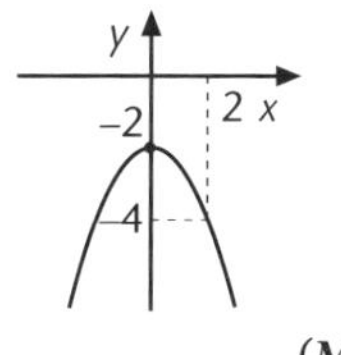

(M)

Unit 11.1 Activity 13B: Quadratic graphs (page 123)

1. a.

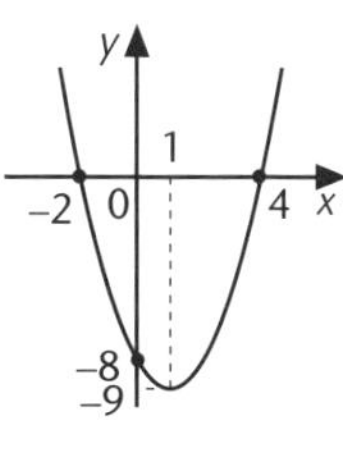

(A)

b.

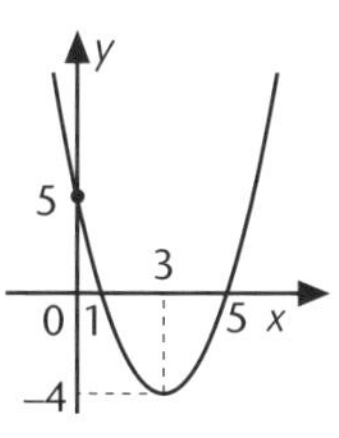

(A)

c.

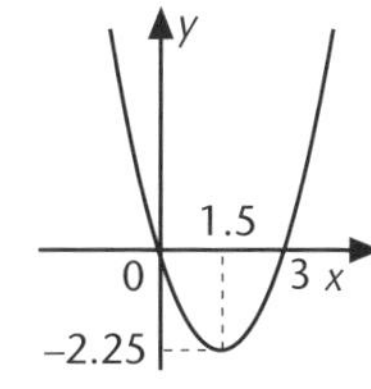

(A)

d.
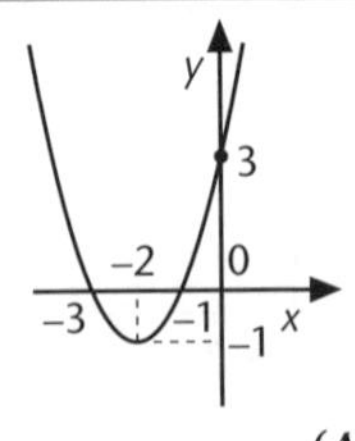

(A)

e.
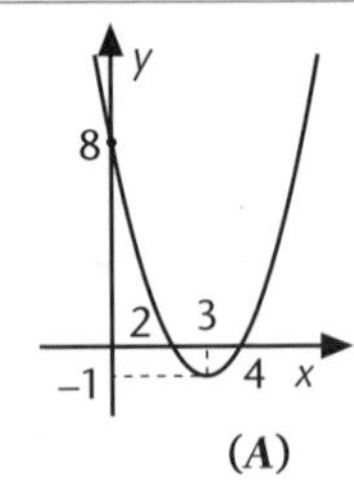

(A)

f.
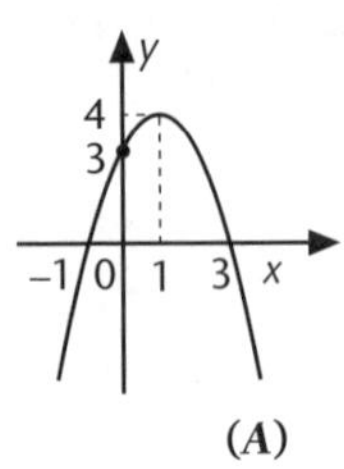

(A)

g.
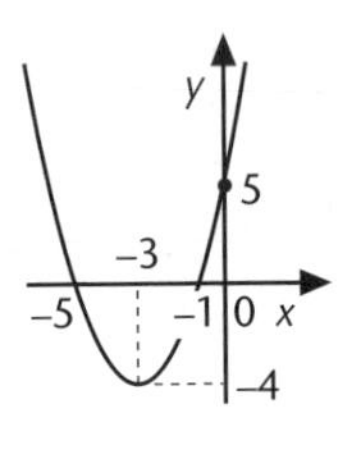

(A)

h.
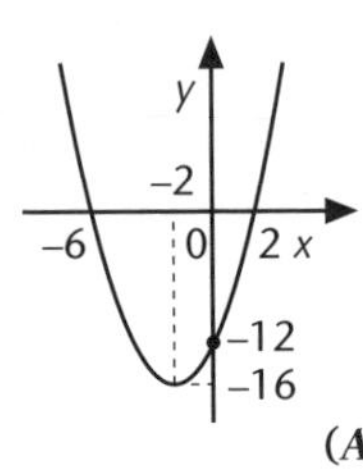

(A)

i.
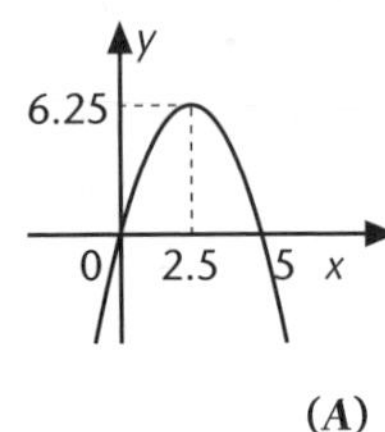

(A)

2. a. i. $(-4, 0)$ **(A)** ii. $(2, 0)$ **(A)** iii. $(-1, 0)$ **(A)** iv. $(-1, -9)$ **(A)**

b. -9 c. $y = (x + 1)^2 - 9$ **(M)**

3. a. $y = (x - 15)(x - 55)$ *or* $y = (x - 35)^2 - 400$ **(E)**

b. $y = 3(x - 5)(x - 9)$ *or* $y = 3(x - 7)^2 - 12$ **(E)**

c. $y = 2(3 - x)(x - 15)$ *or* $y = -2(x - 3)(x - 15)$ *or* $y = -2(x - 9)^2 + 72$ **(E)**

d. $y = \frac{-1}{2}x^2 + 800$ *or* $y = \frac{-1}{2}(x + 40)(x - 40)$ **(E)**

e. $y = -(x - 2)^2 + 9$ *or* $y = -(x + 1)(x - 5)$ **(E)**

f. $y = \frac{1}{3}(x - 9)^2 + 3$ **(E)**

Unit 11.1 Activity 13C: Using quadratic graphs to solve problems (page 126)

1. a.

x	0	1	2	3	4	5	6	7	8	9	10	11	12
A	0	22	40	54	64	70	72	70	64	54	40	22	0

b.
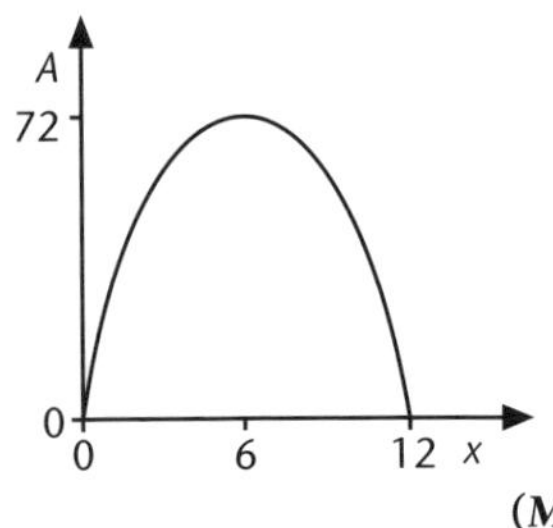

(M)

c. 72 m^2 **(M)**

d. Cannot have a negative length. **(A)**

e. 3.6 m or 8.4 m **(M)**

2. a.

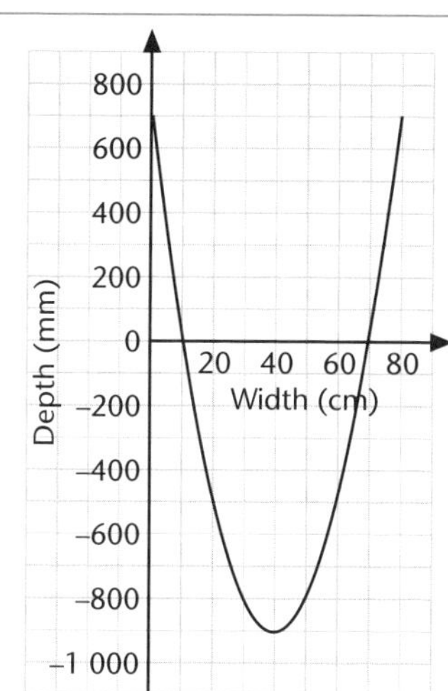

(A)

b. i. 40 cm **(*A*)**

ii. 900 mm **(*A*)**

c. i. 60 cm **(*A*)**

ii. Distance between intercepts on x-axis. **(*A*)**

d. 20 cm and 60 cm **(*A*)**

e. For $10 \leq W < 70$ (a depth of +700 cm at the pole is inapplicable). **(*E*)**

f. Cross-section of the drain may not exactly 'fit' the parabola and the sides and bottom of the drain may be uneven. **(*E*)**

3. a.

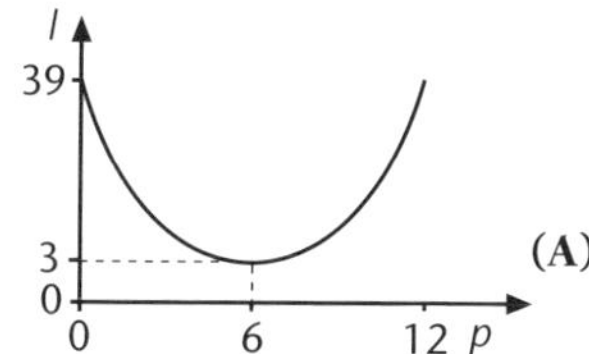

(A)

b. 3 m **(*A*)**

c. i. 12 m **(*A*)**

ii. Symmetry and distance from first pole to the lowest point is 6 m. **(*A*)**

4. a. $y = \frac{1}{8}(x-2)^2 + 1.5$ *or* $y = \frac{1}{8}x^2 + 1.5$

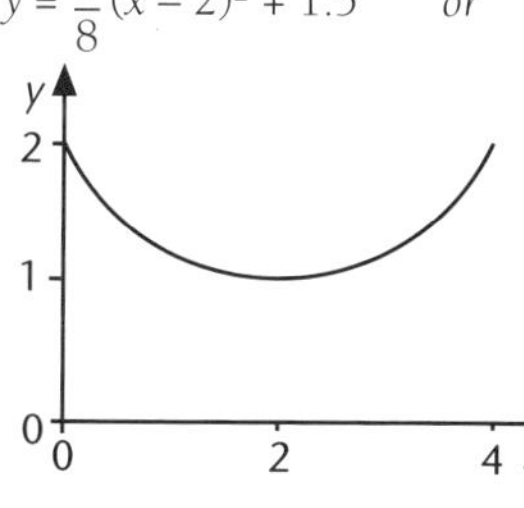

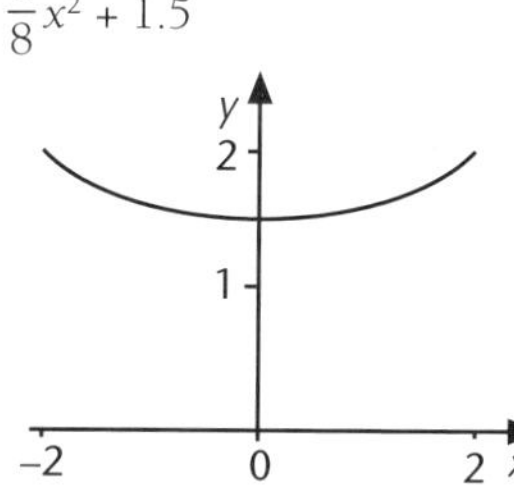

(E)

b. No, height of string is 1.68 m. **(*E*)**

5. a. $h = 50d + 10$ **(*M*)**

b. $h = -40(d-4)(d-10)$ or $h = -40(d-7)^2 + 360$ or $h = -40d^2 + 560d - 1\,600$ **(*E*)**

6. a.

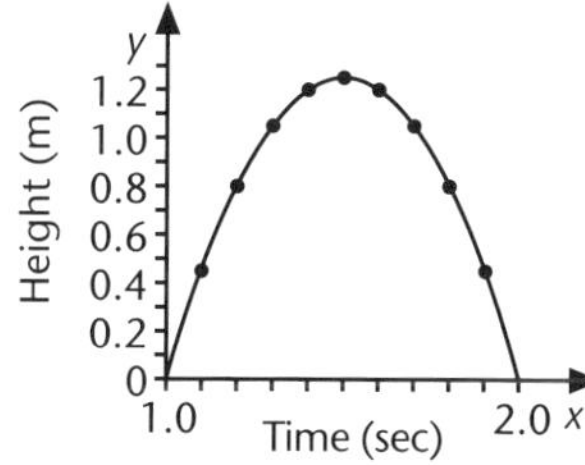

b. 1.25 m **(*A*)**

c. $y = 5(x-1)(2-x)$ or $y = -5x^2 + 15x - 10$ **(*E*)**

d. 0.9375 m **(*A*)**

7. 7.5 cm **(*E*)**

Unit 11.1 Activity 14A: Simultaneous equations (page 133)

1. a. $x = 14, y = 19$ **(M)** b. $x = 11, y = 15$ **(M)** c. $x = 7, y = 17$ **(M)**
   d. $x = 12, y = 27$ **(M)**
2. a. $x = 4, y = 3$ **(M)** b. $x = 5, y = 4$ **(M)** c. $x = 3, y = 5$ **(M)**
   d. $x = 7, y = -3$ **(M)** e. $x = 2\frac{1}{2}, y = 1$ **(M)** f. $x = 1, y = -5$ **(M)**
   g. $x = 4, y = 5$ **(M)** h. $x = 5, y = 0$ **(M)**
3. a. $x = 3, y = 4$ **(M)** b. $x = 2, y = 7$ **(M)** c. $x = 5, y = 5$ **(M)**
   d. $x = -3, y = 2$ **(M)** e. $x = 8, y = 3$ **(M)** f. $x = -\frac{1}{2}, y = 2$ **(M)**
   g. $x = 2, y = -1$ **(M)** h. $x = -1, y = 10$ **(M)**
4. a. $x = 2, y = 0$ **(M)** b. $x = 0, y = -1$ **(M)** c. $x = 3, y = -1$ **(M)**
   d. $x = 4, y = 5.5$ **(M)**
5.

a.

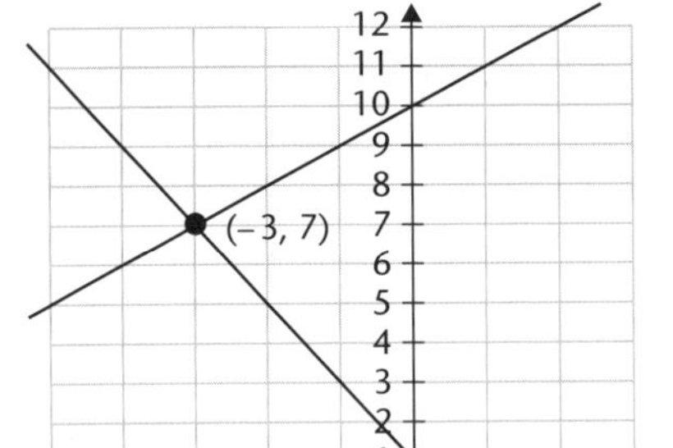

b.

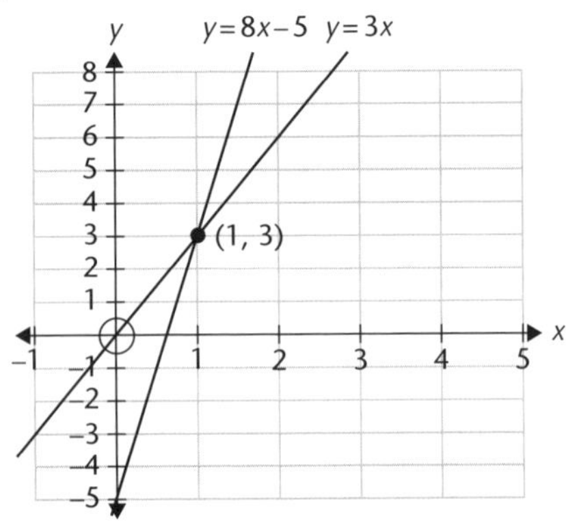

c.

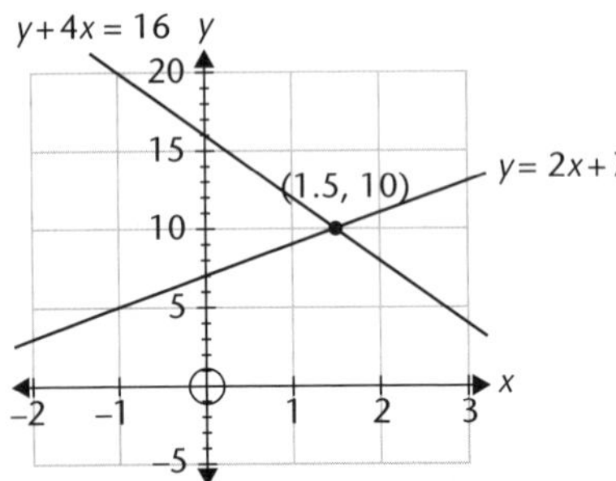

d.

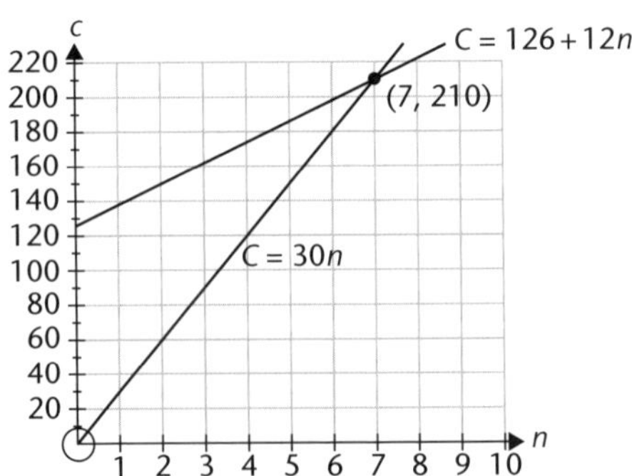

Unit 11.1 Activity 14B: Word problems with simultaneous equations (page 134)

1. K4.40 **(M)** **2.** 3.5 hours **(M)** **3.** Artemius 32, Brown 18. **(M)**

4. 6 beans, 9 aibika. **(M)**

5. **a.** $7s + 3h = 50$ and $3s + 2h = 24$ where s = Kcost socks, h = Kcost handkerchiefs **(*E*)**

b. Socks cost K5.60 a pair. **(M)**

6. Flower garden is 10 m by 13 m; vegetable garden is 12 m by 28 m. **(*E*)**

Unit 11.1 Activity 15A: Inequations (page 136)

1. **a.** **i.** $x \leq 2$ **ii.** x: –5 –4 –3 –2 –1 0 1 2 3 4 5 **iii.** $(-\infty, 2)$

b. **i.** $x \leq -2$ **ii.** x: –5 –4 –3 –2 –1 0 1 2 3 4 5 **iii.** $(-\infty, -2)$

c. **i.** $x \geq -4$ **ii.** x: –5 –4 –3 –2 –1 0 1 2 3 4 5 **iii.** $(-4, \infty)$

d. **i.** $x \geq \frac{3}{2}$ **ii.** x: –5 –4 –3 –2 –1 0 1 2 3 4 5 **iii.** $\left(\frac{3}{2}, \infty\right)$

e. **i.** $x > -7.5$ **ii.** x: –10 –9 –8 –7 –6 –5 –4 –3 –2 –1 0 1 2 3 **iii.** $(-7.5, \infty)$

f. **i.** $x > \frac{7}{2}$ **ii.** x: –2 –1 0 1 2 3 4 5 6 7 8 **iii.** $\left(\frac{7}{2}, \infty\right)$

2. **a.** $x > \frac{-5}{3}$ **b.** $x \leq 3$ **c.** $x \leq 21$ **d.** $x \geq 2$

e. $x \geq \frac{9}{2}$ **f.** $x < -11$ **g.** $x \leq 14$ **h.** $x \geq \frac{5}{2}$

3. **a.** $a > \frac{-3}{2}$ **b.** $a > \frac{4}{3}$ **c.** $a \leq \frac{1}{3}$ **d.** $a \geq 1$

e. $a > \frac{5}{2}$ **f.** $a \leq \frac{-1}{2}$ **g.** $a < 2$ **h.** $a \leq \frac{9}{5}$

4. **a.** $b > -11$ **b.** $b \leq -13$ **c.** $b < 12$

d. $b \leq \frac{13}{7}$ **e.** $b \geq 1$ **f.** $b \leq \frac{11}{5}$

Unit 11.1 Activity 15B: Regions (page 139)

1. a.

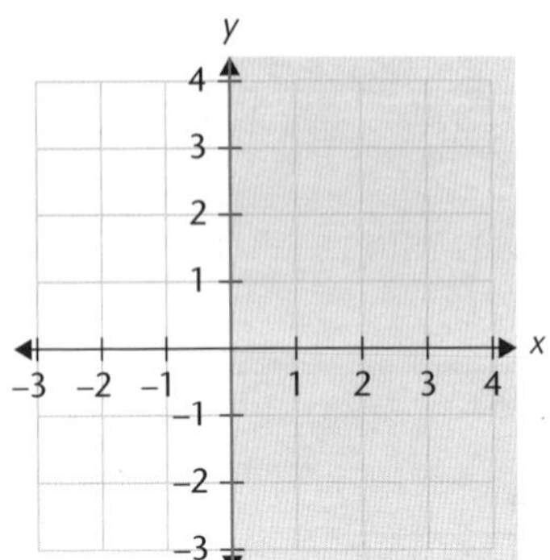

b.

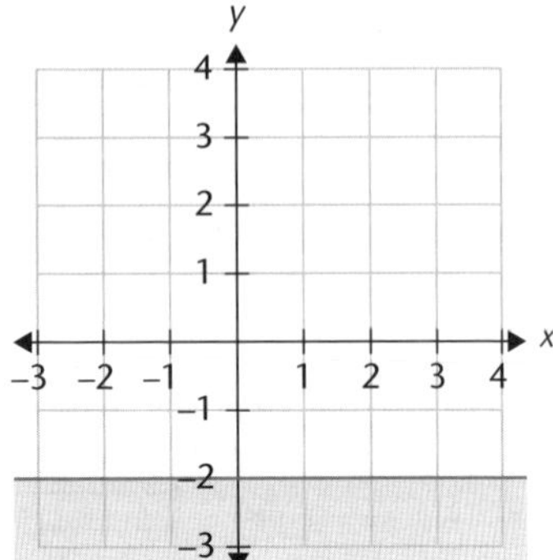

c.

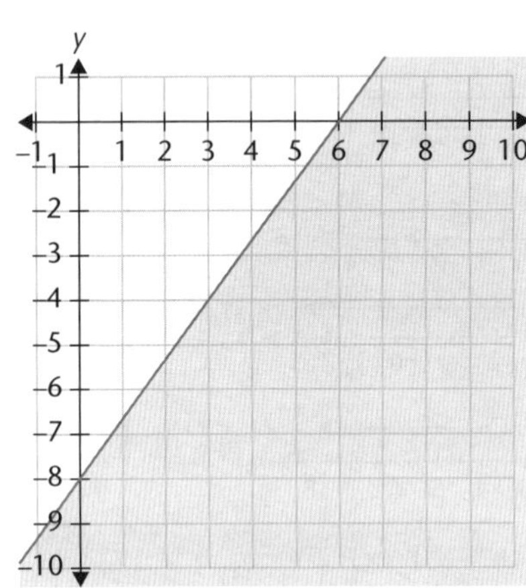

d.

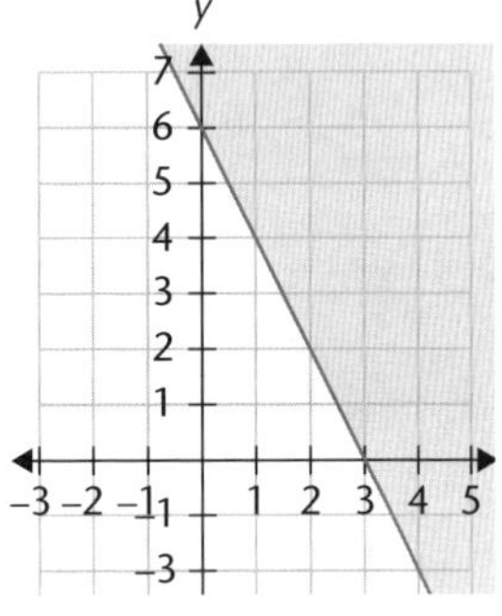

e.

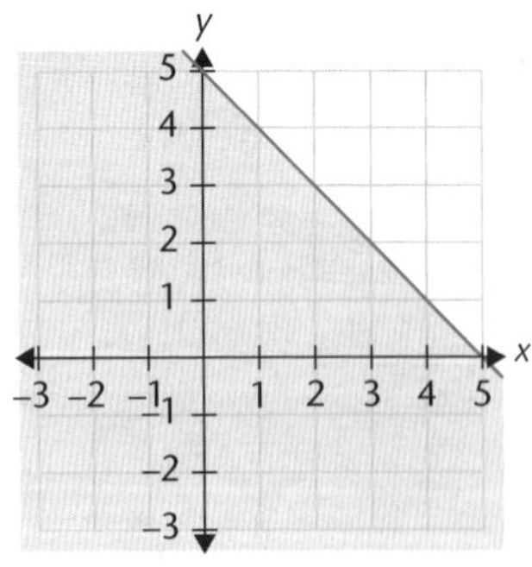

f.

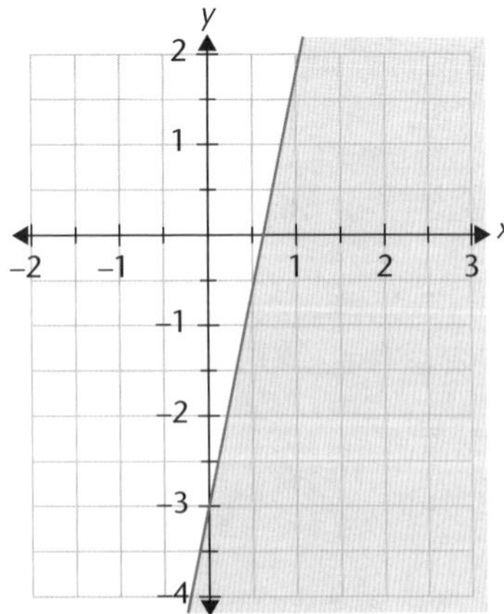

g.

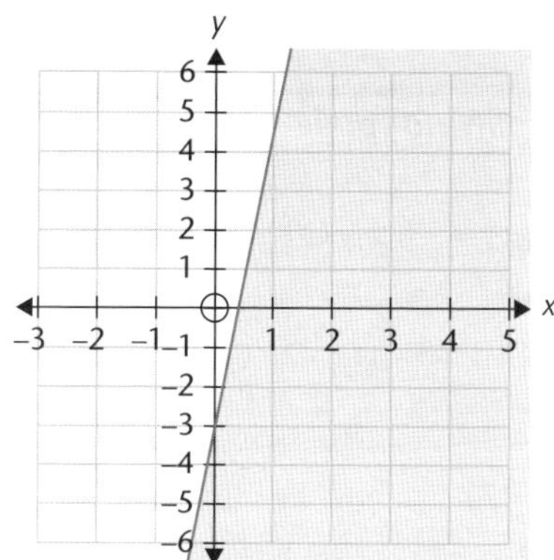

2. a. $x > -1$ b. $y \leq 0$ c. $x + y < 4$

d. $y \leq -2x + 4$ e. $y \leq \frac{2}{3}x + 2$ f. $y \leq 3x$

Unit 11.2 Graphs and Functions

Unit 11.2 Activity 1A: Solving word problems (page 141)

1. $5x + 6(x + 1) = 32.4$, K2.40 **(M)**
2. $x + 2(x + 1) + 6x = 15.5$, T K1.50; B, D K2.50, rest K3 each **(M)**
3. $2x + x - 20 = 86.5$, K35.50, K15.50 **(M)**
4. $2x + 2(x - 3.3) = 50$, 14.15 m **(M)**
5. $4x + 6 + 110$, 26 **(M)**
6. $17.35x = 1\,000 - 687.70$, K18 **(M)**
7. $\frac{x+2}{2} = \frac{117}{2}$, $\frac{x+2}{2} = \frac{117}{26}$, K7 **(M)**
8. $x - 20 = \frac{7x}{11}$, 55 **(M)**
9. $12(x - 5) = 5x + 45$, 15 **(M)**
10. $1.15(x + 5) = 13.8$, K7 **(M)**
11. $36(1 + \frac{x}{100}) = 40.86$, 13.5% **(M)**
12. $\frac{x}{x+44} = \frac{11}{17}$, 143 **(M)**
13. $x - 30 = 0.25x$, 40 **(M)**
14. $4x - 12 = x + 12$, K8, K32 **(M)**
15. $\frac{x + 3\,000}{x + 5\,000} = \frac{13}{17}$, K3500 **(M)**

Unit 11.2 Activity 1B: Expanding (page 144)

1. $15x + 5$ **(A)**
2. $10x^2 + 15x$ **(A)**
3. $2x^2 + 3xy + y^2$ **(A)**
4. $6x^2 + 11x + 4$ **(A)**
5. $12A^2 + 7A - 10$ **(A)**
6. $2x^2 + x - 21$ **(A)**
7. $12PQ + 18P - 16Q - 24$ **(A)**
8. $6A^2 + AB - B^2$ **(A)**
9. $3x^2 + y^2 - 4xy + 2x - 2y$ **(A)**
10. $3x^2 + 4y^2 + 8xy - 3x - 6y$ **(A)**
11. $2x^2 + y^2 - 3xy + 8x - 6y + 8$ **(A)**
12. $x^2 + 2x + 1$ **(A)**
13. $9x^2 - 24xy + 16y^2$ **(A)**
14. $P^2 + R^2 - 4P - 4R + 4 + 2PR$ **(A)**
15. $A^3 + 3A^2B + 3AB^2 + B^3$ **(A)**
16. $x^4 - 8x^2 + 16$ **(A)**
17. 8 **(A)**
18. $3x^3 - 26x^2 + 35x + 100$ **(A)**
19. $12A^4 - A^3B + 13A^2B^2 + 7AB^3 + 5B^4$ **(A)**
20. $x^4 - 4x^3 + 10x^2 - 12x + 9$ **(A)**
21. $8x^3 + 36x^2y + 54xy^2 + 27y^3$ **(A)**

22. $-x^3 + 6x^2 - 11x + 6$ **(A)23.**

23. $6x^3 - 43x^2 + 22x + 120$ **(A)**

24. $-18x^3 + 114x^2 + 240x$ **(A)**

25. $-30x^3 + 44x^2 - 16x$ **(A)**

Unit 11.2 Activity 1C: Expanding and simplifying (page 144)

1. $7x + 11$ **(A)**
2. $7A + B$ **(A)**
3. $6x + 3y$ **(A)**
4. $8x + 8y + 14$ **(A)**
5. $12x + 19y$ **(A)**
6. $3A - 17B$ **(A)**
7. $x^2 + 5x - 2$ **(A)**
8. $2x^2 - 12$ **(A)**
9. $2x^2 + 6x + 5$ **(A)**
10. $2x+ 7$ **(A)**
11. $-8x^4 - 12x^3 + 17x^2 + 17x - 11$ **(A)**
12. $6A^3 + 16A^2 - 1$ **(A)**
13. $48A^2 + 48A + 28$ **(A)**
14. $4x^2 - 3x + 1$ **(A)**

Unit 11.2 Activity 1D: Factorising (page 147)

1. $a(x - b)$ **(A)**
2. $a(b - a)$ **(A)**
3. $x(11x - 1)$ **(A)**
4. $4a(x + 2y)$ **(A)**
5. $(x + 4)(x + 3)$ **(A)**
6. $(x + 9)(x + 4)$ **(A)**
7. $(P + 3)(P + 9)$ **(A)**
8. $(p - 2q)(p - q)$ **(A)**
9. $(R - 2P)(r - 4P)$ **(A)**
10. $(x - 7y)(x + 6y)$ **(A)**
11. $(p + 5q)(p - q)$ **(A)**
12. $(p - 5q)(p + 2q)$ **(A)**
13. $(R + 6S)(R + 4S)$ **(A)**
14. $(x - 7)(x + 7)$ **(A)**
15. $P(P - 36)$ **(A)**
16. $(L - 3)(L - 2)$ **(A)**
17. $(M + 10)(M + 6)$ **(A)**
18. $(M - 8)(M + 1)$ **(A)**
19. $25(M - 2)(M + 2)$ **(A)**
20. $(2x + 1)(x + 1)$ **(A)**
21. $(x - 5b)(x - b)$ **(A)**
22. $(2x + 1)(x + 3)$ **(A)**
23. $(4P + 1)(P + 5)$ **(A)**
24. $(3z + 7)(z - 1)$ **(A)**
25. $(4P - 7)(P + 3)$ **(A)**
26. $(4x - 1)(3x - 2)$ **(A)**
27. $(LP - 6)(LP + 4)$ **(A)**
28. $(7 - 5x)(2 + 3x)$ **(A)**
29. $(7p - q)(p + q)$ **(A)**
30. $8(L - 2x)(L + 2x)$ **(A)**
31. $3P(3x - y)(3x + y)$ **(A)**
32. $2(5 - x)(5 + x)$ **(A)**
33. $A(3P - 1)(P - 1)$ **(A)**
34. $x^2(9 + x^5)$ **(A)**
35. $x(18y - 21x - 14)$ **(A)**
36. $8p^3q^3(1 + 2p + 3p^2 q)$ **(A)**
37. $(p^3q^2r^2 - 3x^5y^6)(p^3q^2r^2 + 3x^5y^6)$ **(A)**
38. $(x - b)(x + b)(x^3 + b^3)$ **(A)**
39. $(A - B)(C - D)$ **(A)**
40. $(x + 1)(x + 1 + a)$ **(A)**

Unit 11.2 Activity 1E: Simplifying quotients (page 148)

1. $\dfrac{x-5}{x+1}$ **(A)**
2. $\dfrac{x+2}{x+3}$ **(A)**
3. $\dfrac{x}{x+1}$ **(A)**
4. $\dfrac{x+7}{x+2}$ **(A)**
5. $\dfrac{1}{x+1}$ **(A)**
6. $\dfrac{x+5}{x-4}$ **(A)**
7. $\dfrac{p+7}{p+3}$ **(A)**
8. $\dfrac{5}{p+1}$ **(A)**
9. $\dfrac{a-b}{b}$ **(A)**
10. $\dfrac{y-3}{4y}$ **(A)**
11. $\dfrac{2y+1}{y+1}$ **(A)**
12. $\dfrac{a-b+c}{b-a+c}$ **(A)**
13. $\dfrac{a^2}{a+1}$ **(A)**
14. $\dfrac{x-y}{z}$ **(A)**
15. $\dfrac{x+b}{x+a}$ **(A)**
16. Likely to be correct (both substitutions give 18). **(A)**
17. Incorrect ($27 \neq 17$). **(A)**

Unit 11.2 Activity 2A: Multiplication of rational expressions (page 150)

1. a. $\dfrac{3a}{4b}$
b. $\dfrac{5ac}{bd}$
c. $\dfrac{3ac}{4db}$
d. $\dfrac{a^2}{bc}$
e. $\dfrac{b}{2}$
f. $3p$
g. $15a$
h. $\dfrac{a}{6}$
i. $\dfrac{2}{3}k^2$
j. $\dfrac{13b}{8}$
k. $\dfrac{y^3}{x^3}$
l. $\dfrac{y}{3x}$

2. **a.** $\frac{x}{x+1}$ **b.** $\frac{1}{x-1}$ **c.** $\frac{x-2}{x+2}$ **d.** $\frac{4}{x+4}$ **(A)**

Unit 11.2 Activity 2B: Division of rational expressions (page 151)

1. **a.** $\frac{2c}{3a}$ **b.** $\frac{abe}{cd}$ **c.** $\frac{3x^2}{y^2}$ **d.** $\frac{a}{5}$ **e.** $\frac{b}{c}$ **f.** $\frac{3b}{a}$

g. $2a$

2. **a.** $\frac{2a+2}{15}$ **b.** $\frac{5(x^2-1)}{7x}$ **c.** $\frac{4b-4}{9}$ **d.** $\frac{2x+2y}{3y}$ **e.** $\frac{9x-3}{4}$ **f.** $\frac{6x}{5x+5}$ **(A)**

3. **a.** 5 **b.** $\frac{x+2}{x+1}$ **c.** $(x+1)(x-2)$ **d.** $\frac{x+3}{x+1}$ **e.** $\frac{x-2}{x+5}$ **(A)**

Unit 11.2 Activity 2C: Addition and subtraction of rational expressions (page 152)

1. **a.** $\frac{3a+b^2}{3b}$ **b.** $\frac{3a}{8}$ **c.** $\frac{a+1}{b}$ **d.** $\frac{4a+b}{4b}$ **e.** $\frac{4a-3b}{12}$ **f.** $\frac{a^2-bc}{ac}$

g. $\frac{a+2}{a^2}$ **h.** $\frac{2a+3}{3b}$ **i.** $\frac{10+b}{4}$ **j.** $\frac{21p+2q}{7}$

2. **a.** $\frac{2a+3}{2}$ **b.** $\frac{3b+4a}{3}$ **c.** $\frac{7a+17b}{6}$ **d.** $\frac{4a+4}{3}$ **e.** $\frac{15b+31}{6}$ **f.** $\frac{9a+2}{2a}$

3. **a.** $\frac{a+7}{12}$ **b.** $\frac{13b+6}{15}$ **c.** $\frac{-p-7q}{12}$ **d.** $\frac{11p+8q}{20}$ **e.** $\frac{14a-12}{15}$ **f.** $\frac{21-5p}{6}$

g. $\frac{26+5p}{10}$ **h.** $\frac{1+7r}{12}$ **i.** $\frac{6p-4q}{3}$

4. **a.** $\frac{2x+3}{(x+1)(x+2)}$ **b.** $\frac{3p+5}{(p+3)(p+1)}$ **c.** $\frac{3p-5}{(p+1)(p-3)}$ **d.** $\frac{5R+2}{(R+4)(R-2)}$

e. $\frac{2R+9}{R(R+3)}$ **f.** $\frac{R-A-1}{(A+3)(R+2)}$ **g.** $\frac{3A-1}{(A-3)(A-1)}$ **h.** $\frac{2A^2+2A-2}{(A+1)(A-1)}$

i. $\frac{3x-2}{6(x+1)}$ **j.** $\frac{xy-3x^2-4y^2}{12(x+y)(x-y)}$ **k.** $\frac{5x+9}{(x+1)(x+2)(x+3)}$ **l.** $\frac{x+8}{x(x+1)(x+2)}$

m. $\frac{3x-y+1}{(x-y+2)(x+y-3)}$ **n.** $\frac{x-5y+10}{x(x+y-2)(x-y+2)}$ **o.** $\frac{4y-3x}{xy(x-y)}$ **(A)**

5. **a.** $\frac{a}{3b^3}$ **b.** $\frac{4y^2z^4}{9x^2}$ **c.** $\frac{3a(a+c)}{c^2}$ **d.** $\frac{x+2}{x(x-3)}$ **e.** $\frac{(u-6)(x-1)}{(x+4)(u-1)}$

f. $\frac{4(a+2)}{3(a-2)}$ **g.** $\frac{5}{2x-2y}$ **h.** $\frac{5x+1}{x^2-1}$ **i.** $\frac{x+6y}{x^2-4y^2}$ **j.** $\frac{x}{x^2-y^2}$

k. $\frac{2x}{(x+1)(x+2)}$ **l.** $\frac{2}{(a-4)(a-5)(a-6)}$ **m.** $\frac{4x+2}{(x-1)(x+1)(x+3)}$

n. $\frac{(a-1)(a-2)(7a-12)}{12}$ **(A)**

Unit 11.2 Activity 2D: Simplifying algebraic expressions—multiple choice (page 154)

1. C **2.** D **3.** D **4.** A **5.** B

6. C **7.** D **8.** A **9.** B

Unit 11.2 Activity 3A: Writing formulae (page 156)

1. $100D$ cents **2. a.** $\frac{x}{20}$ hours **b.** $180x$ seconds

3. a. $(1000x + y)$ grams **b.** $(x + \frac{y}{1\,000})$ kg

4. a. $100LW$ cm^2 **b.** $\frac{LW}{100}$ m^2 **c.** $(2000L + 20W)$ mm

5. a. $800L$ cm **b.** $8000L$ mm **6. a.** $\frac{1}{2}ab$ **b.** $\sqrt{a^2 + b^2}$

7. a. $\frac{\pi L^3}{4}$ **b.** $\frac{3\pi L^2}{2}$ **8. a.** $2y + 2z$ **b.** $yz - xy + x^2$

9. a. $2(L + M) + 8x$ **b.** LMx **10. a.** $16L^3$ **b.** $40L^2$

Unit 11.2 Activity 3B: Changing the subject of formulae (page 158)

1. $\frac{T}{3}$ **(A)** **2.** $\frac{T}{A}$ **(A)** **3.** PT **(A)** **4.** P^2 **(A)**

5. $AP + BP$ **(A)** **6.** $\frac{5P}{2}$ **(A)** **7.** $\frac{T^2}{5}$ **(A)** **8.** $A - B$ **(A)**

9. $A + B$ **(A)** **10.** $2B$ **(A)** **11.** $2B^2 + 3B$ **(A)** **12.** B^2 **(A)**

13. AB^2C **(A)** **14.** $B(C + 3)$ **(A)** **15.** $\frac{AC}{B}$ **(A)** **16.** $\pm\sqrt{A}$ **(A)**

17. $\pm\sqrt{\frac{A + B}{C}}$ **(A)** **18.** $\frac{B^2 + D}{A}$ **(A)** **19.** $\frac{B^2}{A - D}$ **(A)** **20.** $\frac{ABC}{A + B}$ **(A)**

21. $\frac{ED + A}{1 - E}$ **(A)** **22.** C^2 **(A)** **23.** C^2D^2 **(A)** **24.** $\frac{9L^2G}{4A^2}$ **(A)**

25. $\frac{A^2G}{C^2X^2}$ **(A)** **26.** $\sqrt{\frac{D}{3 - A}}$ **(A)** **27.** $\frac{D^2}{(A - B)^2}$ **(A)** **28.** $\frac{(A + C)^2}{E^2}$ **(A)**

29. $\frac{4y - 5}{By - A}$ **(A)** **30.** $\sqrt[3]{\frac{Fy - E}{4y - 3}}$ **(A)**

Unit 11.2 Activity 3C: Using formulae (page 159)

1. a. 45.08 m^2 **b.** 71.13 m^2 **c.** 3.44 m **d.** 1.94 m **e.** 10.82 m

2. a. 817.24 cm^3 **b.** 106.55 cm^3 **c.** 3.44 cm **d.** 5.70 cm **e.** 5.33 cm

3. a. K2 750 **b.** 8.5% **c.** 4 years

Unit 11.2 Activity 4A: Polynomials (page 163)

1. $\frac{2}{x+2}$ **i.** $3\sqrt{x^2-1}$; $x^2 + x + x^{-1}$; not positive powers of x

2. **a.** **i.** 4, $2x^2$ **ii.** 6

b. **i.** 3, $-14x^3$ **ii.** -14

c. **i.** 3, $15x^3$ **ii.** 8

d. **i.** 4, $7x^4$ **ii.** -2

e. **i.** 5, $-7x^5$ **ii.** 1

3. **a.** -22

b. -90

c. 0

d. -32

e. $P(m) = m^3 - 4m^2 - 7m + 10$

4. **a.** $2x^2 + 12x - 12$

b. $x^3 + x^2 - 8x + 9$

c. $2x^3 + 5x^2 - 18x + 9$

d. $4x^2 - 12x + 9$

e. 25

Unit 11.2 Activity 4B: Polynomials and the remainder theorem (page 165)

1. **a.** i. -12 **ii.** -30 **iii.** 0

b. $x - 3$

2. **a.** **i.** 6 **ii.** 30 **iii.** 0

b. $x + 4$

3. **a.** $x - 2$

b. $x + 4$

c. $x + 8$

d. $x + 3$ or $x - 4$

Unit 11.2 Activity 4C: Factorising polynomials (page 167)

1. **a.** $x^2 - 2x^2 - 5x + 6 = x^2(x - 1) - x(x - 1) - 6(x - 1)$

$= (x - 1)(x^2 - x - 6)$

b. $x^2 - 3x - 2 = x^2(x+1) - x(x + 1) - 2(x + 1)$

$= (x+1)(x^2 - x - 2)$

c. $2x^3 + x^2 - 2x - 1 = 2x^2(x + 1) - x(x + 1) - (x + 1)$

$= (x + 1)(2x^2 - x - 1)$

2. **a.** $x - 1$

b. $(x - 1)(x^2 + 2x - 15)$

c. $(x - 1)(x + 5)(x - 3)$

3. **a.** $x - 4$

b. $(x - 4)(x^2 + 3x + 2)$

c. $(x - 4)(x + 2)(x + 1)$

4. **a.** $x - 2$

b. $(x - 2)(x^2 + 4x - 12)$

c. $(x - 2)(x - 2)(x + 6) = (x - 2)^2(x + 6)$

5. **a.** $x - 3$

b. $(x - 3)(x^2 - 9)$

c. $(x - 3)(x - 3)(x + 3) = (x - 3)^2(x + 3)$

6. **a.** $(x - 1)(x - 4)(x + 3)$

b. $(x - 2)(x - 2)(x + 1) = (x - 1)^2(x + 1)$

c. $(x - 1)(x + 4)(x + 7)$

d. $(x - 4)(x + 2)(x + 2) = (x + 2)^2(x - 4)$

e. $(x + 1)(x + 2)(x + 3)$

f. $(x - 2)(x^2 + x + 2)$

g. $(x - 5)(3x - 1)(x + 11)$

h. $(x + 1)(2x + 1)(x - 5)$

i. $(x - 4)(3x + 1)(2x - 1)$

j. $(x - 1)(x + 1)(x^2 + 1)$

k. $(x - 9)(x + 1)(x^2 - x + 1)$

l. $(x - 1)(x - 5)(x + 4)^2$

Unit 11.2 Activity 5A: Working with indices (page 171)

1. **a.** 0.0625 **b.** 0.4019 **c.** 7.3046 **d.** 5.0027 **e.** 4.1890 **f.** 9

g. 27.648 **h.** 0.0865 **i.** 1.06 **j.** –1.2022 **k.** 0.2712

2. **a.** $\frac{1}{9}$ **b.** $\frac{1}{64}$ **c.** $\frac{1}{16}$ **d.** $\frac{27}{64}$ **e.** $\frac{8}{25}$ **f.** 9

g. 16 **h.** $\frac{25}{9}$ **i.** $\frac{125}{64}$ **j.** $\frac{27}{4}$ **k.** $\frac{25}{27}$ **l.** $\frac{1}{24}$

m. $\frac{1}{10}$ **n.** $\frac{7}{5}$ **o.** $\frac{8}{3}$ **p.** $\frac{1}{6}$ **q.** $\frac{1}{15}$ **r.** $\frac{1}{12}$

s. 10 **t.** 12 **u.** $\frac{1}{36}$ **v.** $\frac{1}{175}$ **w.** 2 **x.** 4

y. $\frac{1}{14}$ **z.** $\frac{4}{3}$

3. a. $\frac{3}{2}$ **b.** $\frac{1}{36}$ **c.** 144 **d.** $\frac{1}{576}$ **e.** $\frac{5}{6}$ **f.** $\frac{5}{18}$

g. $\frac{33}{10}$ **h.** $\frac{1}{8}$ **i.** $\frac{1}{2}$ **j.** $\frac{17}{12}$ **k.** $-\frac{73}{18}$ **l.** $\frac{6}{5}$

m. $\frac{5}{36}$ **n.** $\frac{11}{16}$ **o.** $\frac{11}{4}$

4. a. $\frac{1}{2}$ **b.** $\frac{1}{6}$ **c.** 2 **d.** $\frac{5}{9}$ **e.** $\frac{3}{2}$ **f.** $\frac{2}{3}$

g. 9 **h.** $\frac{3}{2}$ **i.** $1\frac{3}{5}$

5. a. 3×2^2 **b.** $2^3 \times 5^2$ **c.** 5×3^2 **d.** $5^2 \times 2^{-4}$ **e.** 2×3^{-1} **f.** 22×3^{-3}

g. $2^4 \times 3^{-4}$ **h.** $2^6 \times 3^{-4}$ **i.** $3^2 \times 2^{-2}$

Unit 11.2 Activity 5B: Simplifying expressions with indices (page 173)

1. a. $\frac{1}{x^3}$ **b.** $\frac{1}{y^2}$ **c.** $\frac{1}{a^5}$ **d.** $\frac{1}{b^4}$ **e.** $\frac{1}{z^7}$ **f.** $\frac{1}{y^6}$

g. $\frac{1}{z^{11}}$ **h.** $\frac{1}{xy}$ **i.** $\frac{1}{abc}$ **j.** $\frac{1}{x^2y^2}$ **k.** $\frac{1}{2x}$ **l.** $\frac{1}{4x^2}$

m. $\frac{1}{9a^2}$ **n.** $\frac{1}{8x^3}$ **o.** $\frac{3}{x}$ **p.** $\frac{a}{b^3}$ **q.** $\frac{b^2}{a^3}$ **r.** $\frac{1}{a}$

s. $\frac{1}{2x}$ **t.** $\frac{1}{4x^3}$ **u.** $\frac{3}{5x^2}$ **v.** $\frac{2}{7a^3}$ **w.** $\frac{4}{9a^2}$ **x.** $\frac{3}{5x^4}$

y. $\frac{4}{5x^5}$ **(A)**

2. a. x^{-6} **b.** y^{-5} **c.** z^{-2} **d.** $4x^{-1}$ **e.** $5y^{-2}$ **f.** $7z^{-5}$

g. $\frac{x^{-2}}{2}$ **h.** $\frac{a^{-4}}{3}$ **i.** $\frac{y^{-4}}{5}$ **j.** $\frac{2x^{-2}}{5}$ **k.** $\frac{3y^{-3}}{4}$ **l.** $\frac{3x^{-2}}{4}$

m. $\frac{4y^{-2}}{25}$ **n.** $\frac{x^{-1}}{8}$ **o.** $\frac{x^{-3}}{8y^{-2}}$ **(A)**

3. a. $\frac{y^3}{2x^2}$ **b.** $\frac{2b^4}{3a^3}$ **c.** $\frac{x}{5}$ **d.** $\frac{2y^4}{x^2}$ **e.** $\frac{x^5}{y}$ **f.** $\frac{2y^3}{x}$

g. $\frac{8}{x^7y^2}$ **h.** $\frac{1}{25x^3}$ **i.** $\frac{1}{6xy}$ **j.** $\frac{1}{27x^4}$ **k.** $\frac{x}{2y^2}$ **l.** $\frac{x^2}{9y^3}$

m. $\frac{2}{3x^2}$ **n.** $\frac{4y^2}{5x^3}$ **o.** $\frac{x+y}{xy}$ **p.** $\frac{x+1}{x^2}$ **q.** $\frac{xy^2+2}{2y}$ **(A)**

Unit 11.2 Activity 5C: Fractional indices (page 174)

1. a. $a^{\frac{1}{2}}$ **b.** $c^{\frac{1}{2}}$ **c.** $x^{\frac{1}{3}}$ **d.** $a^{\frac{1}{5}}$ **e.** $u^{\frac{1}{6}}$ **f.** $x^{\frac{2}{5}}$

g. $x^{\frac{3}{7}}$ **h.** $a^{\frac{3}{8}}$ **i.** $x^{-\frac{1}{2}}$ **j.** $y^{-\frac{1}{2}}$ **k.** $a^{-\frac{1}{3}}$ **l.** $x^{-\frac{1}{5}}$

m. $x^{-\frac{2}{7}}$ **n.** $x^{-\frac{5}{9}}$ **o.** $a^{-\frac{3}{7}}$ **p.** $a^{-\frac{8}{9}}$ **q.** $4a^{\frac{1}{3}}$ **r.** $6y^{-\frac{1}{3}}$

s. $8x^{-\frac{7}{3}}$ **t.** $\frac{x^{-\frac{2}{3}}}{7}$ **u.** $\frac{2x^{-\frac{3}{4}}}{5}$ **v.** x^{-1} **w.** $x^{\frac{1}{6}}$ **x.** $a^{-\frac{5}{3}}$

y. $x^{-\frac{1}{12}}$ **(A)**

2. a. $\sqrt[7]{a^3}$ **b.** $\sqrt[4]{b^3}$ **c.** $\sqrt[5]{x^2}$ **d.** $\frac{1}{\sqrt[3]{x^2}}$ **e.** $\frac{1}{\sqrt[4]{y^3}}$ **(A)**

3. a. $(xy)^{\frac{1}{2}}$ or $(x)^{\frac{1}{2}}(y)^{\frac{1}{2}}$ **b.** $(x)^{\frac{1}{2}}y$ **c.** $x(y)^{-\frac{1}{3}}$ **d.** $x^2(y)^{-\frac{1}{3}}$ **e.** $2x(y)^{-\frac{3}{5}}$

f. $(x)^{-\frac{5}{4}}y$ **g.** $(x)^{\frac{1}{2}}(a)^{\frac{1}{3}}$ **h.** $(a)^{-\frac{1}{3}}(c)^{\frac{1}{2}}$ **i.** b **(A)**

4. a. $\frac{1}{8x^3}$ **b.** $\frac{1}{9x^6}$ **c.** $\frac{1}{4y^4}$ **d.** $27A^3$ **e.** $\frac{1}{32B^{10}}$

f. $\frac{1}{4x^4}$ **g.** $4A^2B^2$ **h.** $\frac{1}{25A^2B^4}$ **i.** $3x^{\frac{3}{2}}$ **j.** $256x^{10}$ **(A)**

Unit 11.2 Activity 5D: Calculating with indices (page 176)

1. 16 **2.** 0.11 **3.** 82.19 **4.** –32 **5.** 2.25 **6.** 1.32

7. –2.11 **8.** No sol. **9.** 3.51 **10.** 1720.51 **11.** 2.73 **12.** 1.83

13. 74.30 **14.** –0.5109 **15.** 1.25 **16.** 2.1715 **17.** 14.64 **18.** 0.0833

19. No sol. **20.** 0.0638

Unit 11.2 Activity 5E: Equations with indices (page 177)

1. a. 1.26 **b.** 8 **c.** 81 **d.** 25 **e.** $\frac{32}{243}$ **f.** 0.2 or $\frac{1}{5}$

g. $\frac{1}{4}$ or 0.25 **h.** 81 **i.** 32 **j.** 8 **k.** $\frac{1}{8}$ **l.** $\frac{1}{8}$

m. 4 **n.** 8 **o.** 5 **p.** 27 **q.** $\frac{1}{27}$ **r.** $\frac{1}{4}$

s. $\frac{1}{9}$ **t.** $\frac{1}{243}$ **(A)**

2. a. 64 **b.** 2 381 **c.** 0.20 **d.** 3.04 **e.** No Sol. **f.** –10.08

g. 6.55 **h.** 270.8 **i.** 1.67 **(A)**

3. a. K51 **b.** 81.51 m **(*A*)** **4. a.** 14 336 **b.** 4 **c.** 1.8 **(*M*)**

5. a. T gets smaller **b.** 0.01094 **(*M*)**

Unit 11.2 Activity 6A: Further logarithmic problems (page 180)

1. a. 571 429 **(*M*)** **b.** 2 711 864 **(*M*)** **c.** 1.63 years **(*M*)** **d.** 3 million (***E***)

2. a. 1 000 **b.** 1 911 **c.** 2000 **d.** 1 year **(*M*)**

3. a. K4 500 **b.** K1 824 **c.** 3.65 years **d.** 2.3 years **(*M*)**

4. a. $A = 100\,000$, $k = 0.0395$ **b.** K148 043 **c.** 58.3 years later **(*M*)**

5. 10 m (***E***)

6. a. 64 **b.** 81 **c.** 2 **d.** 216 **e.** 1.73 **f.** 64
g. 81 **h.** 3 **i.** 2 **j.** 36 **k.** 5 **l.** 1
m. –3 **n.** $\frac{1}{3}$ **o.** $\frac{2}{3}$ **(*A*)**

7. a. $\log xyz$ **b.** $\log x^2y$ **c.** $\log xy^2$ **d.** $\log \frac{xy}{z}$ **e.** $\log \frac{ab^3}{c^2}$ **f.** $\log \frac{(p)^{\frac{1}{2}}}{(q)^{\frac{3}{2}}}$
g. $\log xyz$ **h.** $\log y$ **i.** $\log xy^4$ **j.** $\log y$ **k.** $\log \frac{1}{y^2}$ **l.** $\log y^2$
m. $\log xy$ **n.** $\log \frac{1}{z^2}$ **o.** $\log xy^3$ **p.** $\log_a x^2a^5$ **(*A*)**

8. a. log 60 **b.** log 12 **c.** log 75 **d.** log 12 **e.** log 24 **f.** $\log \frac{6}{7}$
g. log 1 **h.** $\log \frac{25}{3}$ **i.** log 96 **j.** log 12 **(*A*)**

9. a. 3 **b.** $\frac{2}{3}$ **c.** $\frac{2}{3}$ **d.** $\frac{4}{3}$ **e.** $\frac{6}{5}$ **f.** 2
g. $\frac{11}{5}$ **h.** 1 **i.** 1 **j.** 2 **(*A*)**

10. a. 1.285 **b.** 0.715 **c.** 0.717 **d.** 1.781 **(*M*)**

Unit 11.2 Activity 7A: Relations and functions (page 185)

1. a. c. d. e. h. i.

2. c.

3. a. i. {2, 3, 4, 5 } **ii.** {3, 4, 5, 6 }
b. i. $x \in R$ **ii.** $y \le 4$
c. i. $x \in R$ **ii.** $y \in R$
d. i. $-4 \le x \le 4$ **ii.** $-4 \le y \le 0$
e. i. $x \in R$ **ii.** $y \le 3$
f. i. $x \in R$ **ii.** $y \in R$
g. i. $x \in R$ **ii.** $y > 0$

Unit 11.2 Activity 7B: Functions and graphs (page 189)

1. a. (–3, 0)

   b. (3, 0)

   c. $x \in R$

2. a.

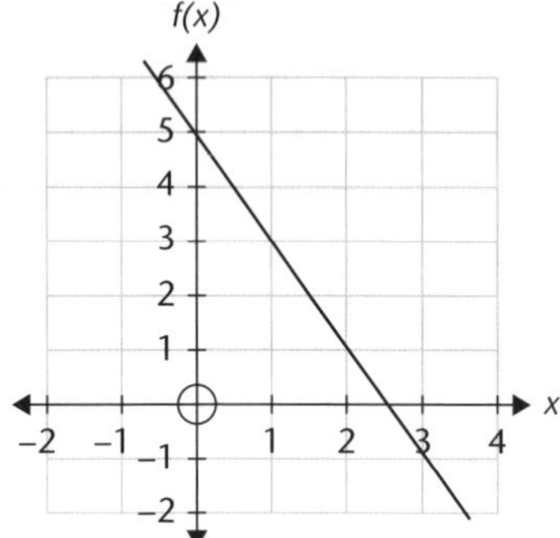

   b. i. (2.5, 0)

      ii. (0, 5)

      iii. $x \in R$

      iv. 3

3. a. (–3, 0), (1, 0)

   b. (0, –3)

   c. (–1, –4)

   d. minimum point

   e. $x > -1$

4. a. –2

   b. 2

   c. 0

   d. (–1, 2)

   e. (1, –2)

   f. i. $x < -1, x > 1$

      ii. $-1 < x < 1$

5. a. i. minimum

      ii. maximum

   b. $-2 < x < 1$

   c. $x < -2, x > 1$

d. $x \in \mathrm{R}$

e. $x \in \mathrm{R}$

6. a. –2, 2

b. 2

c. (0, 2), maximum

d. [–2, 2]

e. [0, 2]

Unit 11.2 Activity 8A: Distance between points (page 192)

1. a. $\sqrt{32}$ **b.** 10 **c.** 13 **d.** $\sqrt{40}$ **e.** $\sqrt{41}$

2. a. 2.2 km **b.** 6.1 km **c.** 3.6 km **d.** 8.6 km **e.** 5.8 km **(A)**

3. 24.166 km **(A)** **4.** 72.691 km **(A)**

5. b. $AB = \sqrt{26}$, $AC = \sqrt{26}$, $BC = \sqrt{32}$ **(A)** **6.** Side lengths are $\sqrt{40}$, $\sqrt{40}$, $\sqrt{32}$ **(A)**

7. (–1, 2) **(A)** **8. a.** 5 cm **b.** (1, 6) **(A)**

9. $x = -2$ or -6 (**E**) **10.** $a = 1$ or -5 (**E**)

Unit 11.2 Activity 8B: Midpoints of line segments (page 193)

1. a. (3.5, 6) **b.** (–2, 2) **c.** (4, –3) **d.** (1.5, 0)

e. (0.5, 4) **f.** $(2a + 1, a - 1)$

2. (4, 1) **(A)** **3.** (7, –7) **(A)** **4.** 1 km east and 2 km north of camp **(A)**

5. a. 40 km west and 20 km north of island. **b.** 60 km west and 40 km north of port **(A)**

6. 13 **(M)** **7.** $a = -4$, $b = -9$ **(M)**

Unit 11.2 Activity 9A: The gradient of a straight line (page 197)

1. a. $m = \frac{1}{3}$ **b.** $m = \frac{3}{4}$ **c.** $m = 1$ **d.** $m = -2$ **e.** m is undefined **f.** $m = 0$

2. Both lines have gradient 2 **3.** Both lines have gradient 3

4. Gradients are 1 and 3 **5.** Not parallel

6. Parallel **7.** $k = 3$ **8.** $k = 4$ **9.** $k = -3$

10. $k = 11$ **11.** $x = 1$ **12.** $x = 13$ **13.** $x = -13$

14. 4 **15.** –4 **16.** Proof (grad AB = grad BC = 2) **(M)**

17. Proof (grad AB = grad BC = 3) **(M)**

Unit 11.2 Activity 9B: Drawing straight lines (page 198)

Solutions must be straight lines which go through the following pairs of points:

1. (0, 1), (5, 3) **2.** (–2, 4), (2, 7) **3.** (3, 2), (7, 3) **4.** (4, 3), (7, 1)

5. (2, 1), (4, 0) **6.** (–2, –2), (1, –2) **7.** (–3, 0), (–3, 2)

Unit 11.2 Activity 10A: Straight line equations (page 200)

1. The straight lines should go through the indicated points:

a. (0, 1), (1, 3) **b.** (0, –4), (1, –1) **c.** (0, 1), (4, 2) **d.** (0, 2), (3, 4)

e. (0, 5), (1, 4) **f.** (0, 3), (4, 0) **g.** (0, 3), (5, 1) **h.** (0, 2), (1, 2)

i. (–3, 0), (–3, 1) **j.** (0, –5), (1, –5) **k.** (–1, 1), (–1, 2) **l.** (0, –1), (–1, 0)

2. **a.** 2 **b.** 3 **c.** $\frac{1}{4}$ **d.** $\frac{2}{3}$ **e.** –1 **f.** $-\frac{3}{4}$

g. $-\frac{2}{5}$ **h.** 0 **i.** Undefined **j.** 0 **k.** Undefined **l.** –1

3. **a.** 5 **b.** –9 **c.** 3 **d.** 1 **e.** –4 **f.** –3 **g.** $3\frac{1}{6}$ **h.** $\frac{2}{3}$ **i.** –1.5 **j.** $-7\frac{1}{2}$

4. **a.** $x = 2, y = -4$ **b.** $x = -2, y = 6$ **c.** $x = 6, y = -4$ **d.** $x = 2, y = 2$

e. $x = 7\frac{1}{2}, y = 5$ **f.** $x = 0, y = 0$

Unit 11.2 Activity 10B: Line equations using gradient and point (page 202)

1. $y = x + 1$ **(A)** **2.** $y = 2x + 3$ **(A)** **3.** $y = 3x - 2$ **(A)** **4.** $y = 4x + 3$ **(A)**

5. $y = \frac{1}{2}x - 4$ **(A)** **6.** $y = \frac{2}{3}x + 2$ **(A)** **7.** $y = -2x + 3$ **(A)** **8.** $y = -\frac{3}{4}x + 2$ **(A)**

9. $y = -4$ **(A)** **10.** Let $(x, y) \pm (x_1, y_1)$ be any point on line $m = (y - y_1)/(x - x_1)$. Rearrange to get $y - y_1 = m(x - x_1)$ **(E)**

Unit 11.2 Activity 10C: Line equations using two points (page 204)

1. $y = x - 1$ **(A)** **2.** $y = 2x + 3$ **(A)** **3.** $y = \frac{3}{4}x + 2\frac{1}{2}$ **(A)** **4.** $y = -x + 4$ **(A)**

5. $y = \frac{2}{3}x - 4$ **(A)** **6.** $y = 4$ **(A)** **7.** $y = -\frac{1}{2}x$ **(A)** **8.** $y = -\frac{3}{4}x + 2$ **(A)**

9. $y = 3x - 1$ **(A)** **10.** $y = 0.5x + 2$ **(M)** **11.** $y = 2x - 6$ **(M)**

12. Median is $y = -x$ and (0, 0) lies on the line. **(M)**

Unit 11.2 Activity 10D: Other forms of the straight line equation (page 205)

1. **c**, **d**, **h**

2. **a.** $y = \frac{2}{3}x + \frac{5}{3}$ **b.** $y = \frac{3}{2}x - \frac{7}{2}$ **c.** $y = \frac{3}{4}x - \frac{5}{4}$

d. $y = \frac{3}{2}x - 2$ **e.** $y = \frac{4}{3}x + \frac{19}{18}$ **f.** $y = -\frac{3}{2}x + 3$

3. **a.** $(A - D)x + (B - E)y + (C - F) = 0$ is of the form $ax + by + c = 0$ and so is a straight line.

b. **ii** and **vi** are straight line equations.

Unit 11.2 Activity 10E: Linear modelling (page 206)

1. a.

Force	0	1	2	3	4
Length	20	20.125	20.25	20.375	20.5

b.

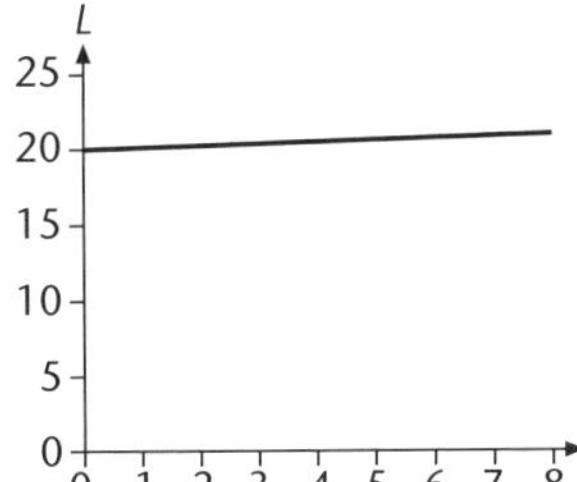

$L = 20 + 0.125F$ **(A)**

c. 60.8 N

2. a. **b.** $V = 10 - 0.12t$ **c.** After 40 minutes

3. a.

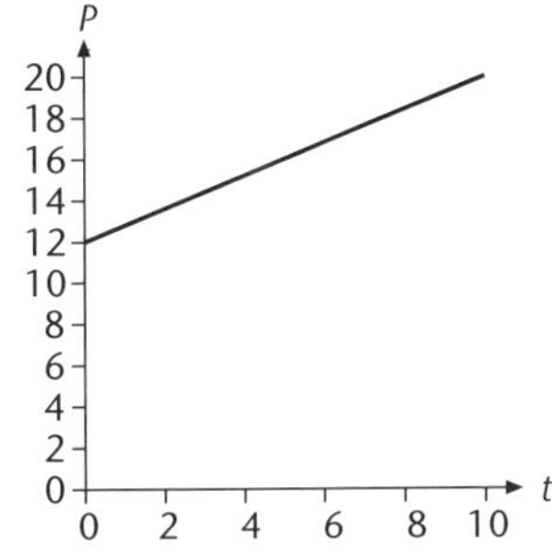

b. $P = 12 + 0.8t$ **c.** 60 pages

4. a. $V = 20\,000 - 0.25d$ **b.** 20 000 km **(A)**

5. a. 0.1 L per minute **b.** 3.8 litres **c.** 562 minutes or 9 hours 22 minutes

6. $y = 4x + 3$ **(A)**

7. a. $D = 85t + 15.83$ **b.** 15.83 km **(M)**

8. a. 2nd, 4th days **b.** Dec 30, Dec 26, Dec 23, Dec 21, Dec 17 **(E)**

Unit 11.2 Activity 10F: Equation of a straight line—multiple choice (page 207)

1. C **2.** C **3.** A **4.** A **5.** B

6. D **7.** D **8.** B **9.** A **10.** B

Unit 11.2 Activity 11A: Parallel and perpendicular lines (page 210)

1. **a.** $\frac{-1}{3}$ **b.** $\frac{-1}{10}$ **c.** $\frac{1}{2}$ **d.** $\frac{1}{5}$ **e.** -2 **f.** $\frac{-4}{3}$

g. $\frac{3}{2}$ **h.** 6 **i.** $\frac{-2}{3}$ **j.** $\frac{5}{16}$ **k.** Undefined **l.** -1000

2. **a.** $y = 2x - 2$ **(A)** **b.** $y = \frac{1}{2}x + 4\frac{1}{2}$ **(A)** **c.** $y = -3x + 7$ **(A)**

d. $y = -\frac{2}{3}x + 6$ **(A)** **e.** $2x - 3y - 7 = 0$ **(A)** **f.** $y = 7x + 7$ **(A)**

3. **a.** $y = -\frac{1}{3}x + 6$ **(A)** **b.** $y = -\frac{1}{2}x + 7$ **(A)** **c.** $y = 2x - 5$ **(A)**

d. $y = -\frac{3}{2}x + 3$ **(A)** **e.** $2y + 3x - 2 = 0$ **(A)** **f.** $3x - 4y - 15 = 0$ **(A)**

g. $5x - 2y + 10 = 0$ **(A)** **h.** $7x + 4y + 7 = 0$ **(A)**

4. **a.** $k = \frac{8}{3}$ **b.** $k = -\frac{3}{2}$ **(A)**

Unit 11.2 Activity 11B: Coordinate geometry and straight lines (page 212)

1. $y = 2x - 2$ **(M)** **2.** $y = x + 2$ **(M)** **3.** $y = \frac{x}{2} + 0.5$ **(M)** **4.** $y = \frac{x}{2} + 4.5$ **(M)**

5. $y = x + 4$ **(M)** **6.** $y = 5$ **(M)** **7.** $x = 2$ **(M)** **8.** $y = x$ **(M)**

9. $y = 4x - 1$ **(M)** **10.** $a^2 = b^2$ **(E)**

Unit 11.2 Activity 11C: Altitudes and distance of a point from a line (page 215)

1. **a.** Proof (AB = AC = $\sqrt{50}$) **(M)** **b.** $y = -x$ **(M)**

2. $y = \frac{3}{4}x - 1\frac{3}{4}$ **(M)**

3. **a.** Proof (sides are 5, 12, 13, so not isosceles although right-angled) **(M)**

b. $y = -\frac{12}{5}x + 36$ **(M)**

4. $y = -2x + 6$ **(M)** **5.** 7 **(E)** **6.** 27 **(E)**

7. Substitute into formula for distance of a point from a line to get result. **(E)**

8. $3\sqrt{2}$ (By use of the distance formula in finding the perpendicular distance of (–3, 2) from $2y + 3x = 0$ the result is easily established.) **(E)**

Unit 11.2 Activity 11D: Parallel and perpendicular lines—multiple choice (page 215)

1. B **2.** C **3.** D **4.** A **5.** C

6. B **7.** B **8.** B **9.** A **10.** B

Unit 11.2 Activity 12A: Intersection of straight lines (page 218)

1. a. (1, 2) **(A)** **b.** (0, 3) **(A)** **c.** (2, 3) **(A)** **d.** (2, 1) **(A)** **e.** (–2, 0) **(A)**

f. (1, 1) **(A)** **g.** (2, 1) **(A)** **h.** (3, 1) **(A)** **i.** (3, 1) **(A)** **j.** $(\frac{33}{14}, -\frac{1}{28})$ **(A)**

2. a. (1, 4) **(A)** **b.** (1, 5) **(A)** **c.** (2, 2) **(A)** **3.** $\frac{5}{2}$ **(M)**

4. Solving any two different pairs from each set will give different solution **(M)**

5. Parallel lines $(m = \frac{2}{3})$ **(M)** **6.** (4, 8) **(M)** **7.** 22.62 **(M)**

8. a. $y = x + 3$ and $y = 3 - x$ are perpendicular as the product of their gradients is –1. **(E)**

b. 3 **(E)**

9. The claim is true. **(E)**

Unit 11.2 Activity 13A: Inequations (page 220)

1. a. i. $x \le 2$ **ii.** **iii.** $(-\infty, 2)$

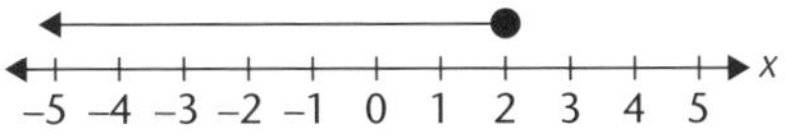

b. i. $x \le -2$ **ii.** **iii.** $(-\infty, -2)$

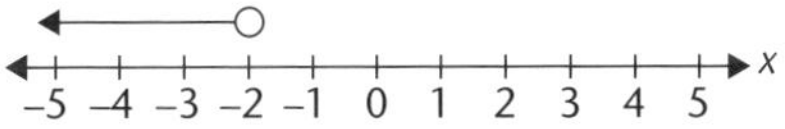

c. i. $x \ge -4$ **ii.** **iii.** $(-4, \infty)$

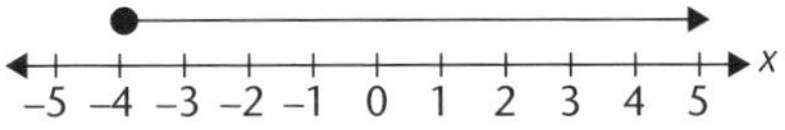

d. i. $x \ge \frac{3}{2}$ **ii.** **iii.** $\left(\frac{3}{2}, \infty\right)$

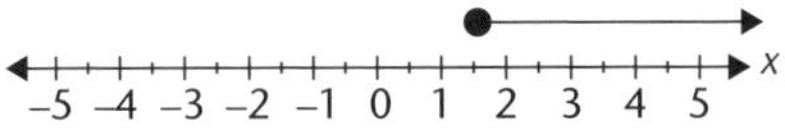

e. i. $x > -7.5$ **ii.** **iii.** $(-7.5, \infty)$

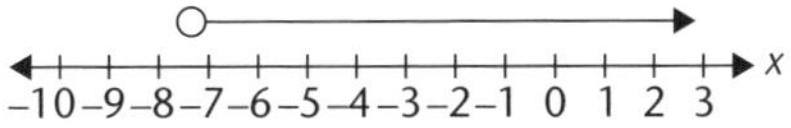

f. i. $x > \frac{7}{2}$ **ii.** **iii.** $\left(\frac{7}{2}, \infty\right)$

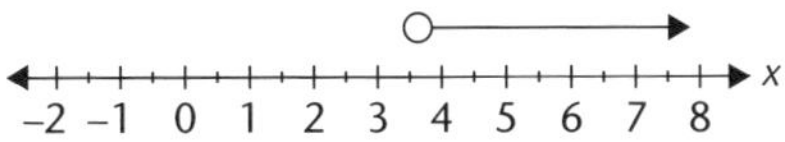

2. a. $x > \frac{-5}{3}$ **b.** $x \le 3$ **c.** $x \le 21$ **d.** $x \ge 2$

e. $x \ge \frac{9}{2}$ **f.** $x < -11$ **g.** $x \le 14$ **h.** $x \ge \frac{5}{2}$

3. a. $a > \frac{-3}{2}$ **b.** $a > \frac{4}{3}$ **c.** $a \le \frac{1}{3}$ **d.** $a \ge 1$

e. $a > \frac{5}{2}$ **f.** $a \le \frac{-1}{2}$ **g.** $a < 2$ **h.** $a \le \frac{9}{5}$

4. a. $b > -11$ **b.** $b \le -13$ **c.** $b < 12$

d. $b \le \frac{13}{7}$ **e.** $b \ge 1$ **f.** $b \le \frac{11}{5}$

Unit 11.2 Activity 13B: Regions (page 223)

1. a.

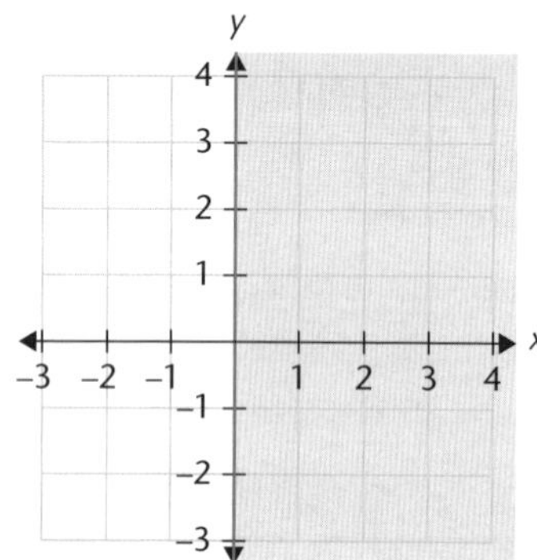

b.

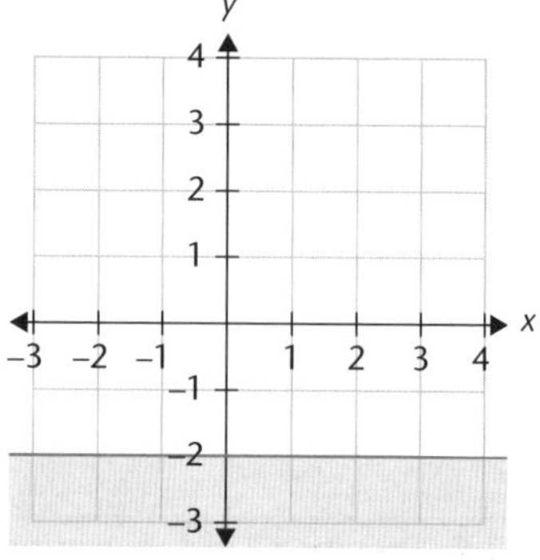

c.

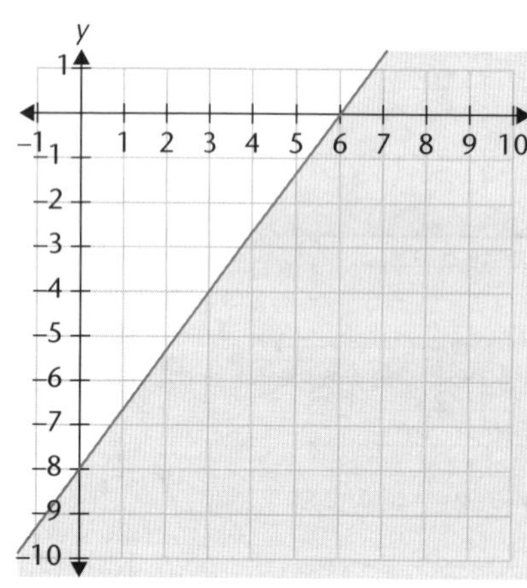

d.

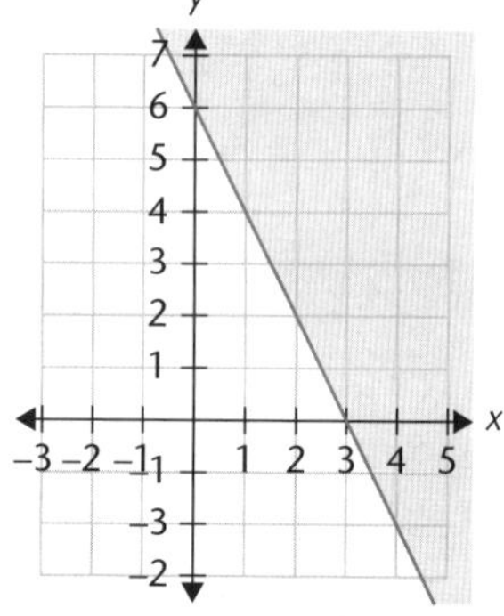

e.

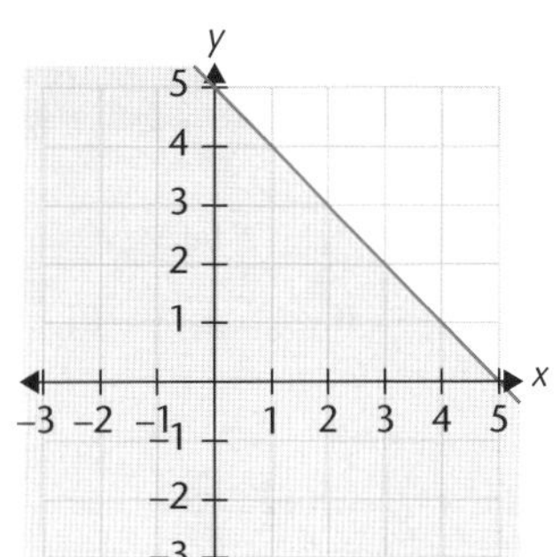

f.

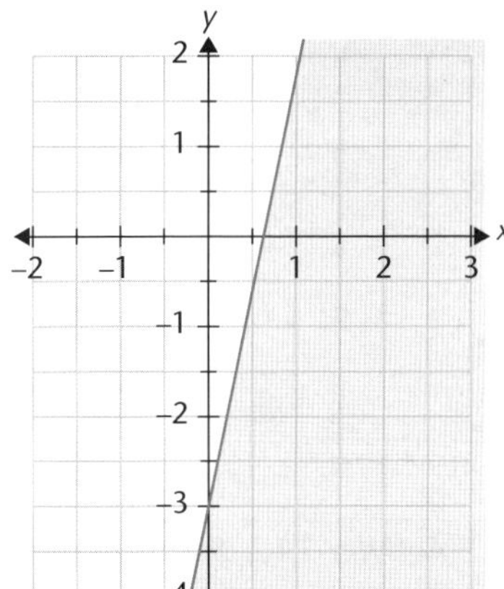

g.

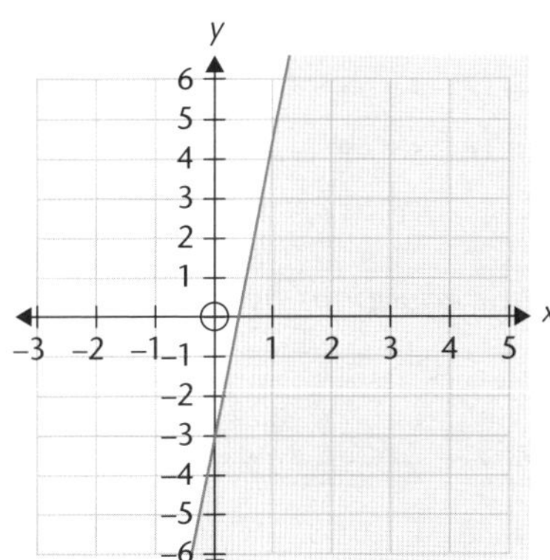

2. **a.** $x > -1$ **b.** $y \le 0$ **c.** $x + y < 4$

d. $y \le -2x + 4$ **e.** $y \le \frac{2}{3}x + 2$ **f.** $y \le 3x$

Unit 11.2 Activity 14A: Polynomials (page 225)

1. *a*, *b*, *c*, *f*, *h*, *i*, *j*

2. **a.** 3 **b.** 5 **c.** 6 **d.** 4 **e.** 1 **f.** 2

g. 3 **h.** 4 **i.** 2 **j.** 0

3. *a*, *c*, *e*, *f*, are quadratics; *b*, *g*, are cubics

Unit 11.2 Activity 14B: Substitution and solving equations (page 226)

1. **a.** **i.** 4 **ii.** 18 **iii.** –2 **iv.** 0 **v.** $1\frac{3}{4}$

b. $x = -2, 1$ **c.** $x^2 + x - 2$ **(A)**

2. **a.** –6 **b.** 30 **c.** 0 **d.** $-10\frac{10}{27}$ **e.** –40.176 **f.** $x = \frac{1}{2}, 4, -1$ **(A)**

3. **a.** –1, 3 **b.** 3, –4, 5 **c.** 0, 1 **d.** –2, 1 **e.** –4 **f.** 3, –4

g. 1, –2 **h.** –1, 1, 4 **i.** 0, –1 **j.** 0, 1 **k.** 0, 7, 3, $-\frac{4}{5}$ **(A)**

4. **a.** **i.** 6 **ii.** –4 **iii.** $-2\frac{1}{4}$ **iv.** $8\frac{4}{9}$ **v.** $-6\frac{8}{81}$

b. **i.** 12 **ii.** $8\frac{3}{4}$ **iii.** 56 **c.** **i.** 22 **ii.** 12 **iii.** $-1\frac{1}{3}$

d. **i.** –4, 1 **ii.** –4, –3 **iii.** –, 1 **(A)**

5. **a.** **i.** 0 **ii.** $-\frac{9}{8}$ **iii.** 8 **iv.** 5.247 **v.** –2.288

b. **i.** 1, 2, 4 **ii.** 1, 2, –1 **iii.** –1, 2, –3

Unit 11.2 Activity 14C: Solution of quadratic equations by factorisation (page 228)

1. –1, 4 **(A)** **2.** –5, 2 **(A)** **3.** –2, 4 **(A)** **4.** –4, –9 **(A)** **5.** –3, 6 **(A)**

6. ±4 **(A)** **7.** –3, 4 **(A)** **8.** –8, 1 **(A)** **9.** –2, 0 **(A)** **10.** –2, 6 **(A)**

11. –5, 1 **(A)** **12.** –1, $\frac{1}{3}$ **(A)** **13.** $-1\frac{1}{2}, -\frac{1}{2}$ **(A)** **14.** ±2 **(A)** **15.** ±2 **(A)**

Unit 11.2 Activity 14D: Solution of quadratic equations by formula (page 229)

1. **a.** –3.73, –0.27 **b.** –7.61, –0.39 **c.** 3.35, 0.15 **d.** –2.83, 0.83 **e.** –2.27, 1.77

f. 0.23, –1.33 **g.** 1.81, –0.29 **h.** –8.24, 0.24 **i.** 1.55, –0.22 **j.** –4.44, –0.06 **(M)**

2. **a.** –5.83, –0.17 **b.** 0.18, 2.82 **c.** 2.29, –0.29 **d.** –2.30, 1.30 **e.** –2.26, 0.59

f. 0.70, 4.30 **g.** –1.87, 5.87 **h.** –0.58, 0.58 **i.** –0.83, 4.83 **j.** 1 **(M)**

Unit 11.2 Activity 14E: Word problems (page 230)

1. width is 9 m 5 cm, length is 11 cm 5 cm **(M)** **2.** width is 10 m, length is 20 m **(M)**

3. 3.58 m **(M)** **4.** 7.83 m **(M)** **5.** 24, 53 **(M)** **6.** 61, 62 **(M)** **7.** 2.25 m **(M)**

8. 9.44 **(M)** **9.** 0.73 m, 1.73 m, 2.73 m **(M)** **10.** 4.79 or 0.21 **(M)**

11. **a.** 31 **b.** 45 **(M)** **12.** height = 3.28 m, base = 2.28 m **(M)** **13.** K8.50 **(M)**

14. 4.3 m **(M)**

Unit 11.2 Activity 14F: Simultaneous linear and non-linear equations (page 232)

1. (0, –1), (1, 0) **(M)** **2.** (0, 1), $(-\frac{4}{5}, -\frac{3}{5})$ **(M)** **3.** (0, 2), $(-\frac{4}{3}, \frac{2}{3})$ **(M)**

4. (0, –3), $(\frac{12}{5}, \frac{9}{5})$ **(M)** **5.** (0, 3), $(-\frac{36}{13}, -\frac{15}{13})$ **(M)** **6.** (–2, –1), (1, 2) **(M)**

7. (5, 2), (–2, –5) **(M)** **8.** (2, –2), (72, 12) **(M)** **9.** (4, 3), $(\frac{9}{4}, \frac{16}{3})$ **(M)**

10. $(\frac{3}{2}, 4), (2, 3)$ **(M)**
11. $(-3, -2), (2, 3)$ **(M)**
12. $(2, 1), (1, 2)$ **(M)**
13. $(4, 2), (-2, -4)$ **(M)**
14. $(\frac{16}{3}, 4), (\frac{4}{3}, -2)$ **(M)**
15. $(1, \frac{1}{2}), (-\frac{2}{3}, 3)$ **(M)**
16. $(4, 3), (-4, 3)$
17. $(4, 3), (1.5, -0.75)$
18. $(-3, 1), (4.4, -0.48)$

Unit 11.2 Activity 14G: The nature of the roots of a quadratic equation (page 234)

1. **a.** **i.** 1 **ii.** 2
b. **i.** –7 **ii.** 0
c. **i.** –15 **ii.** 0
d. **i.** 0 **ii.** 1
e. **i.** –20 **ii.** 0
f. **i.** –23 **ii.** 0
g. **i.** 85 **ii.** 2
h. **i.** –60 **ii.** 0
i. **i.** 0 **ii.** 1
j. **i.** 40 **ii.** 2

2. **a.** $\frac{1}{4}$ **b.** $-5\frac{1}{3}$ **c.** $\pm\frac{4\sqrt{2}}{3}$ **d.** $k < \frac{25}{24}$ **e.** $p > \frac{9}{4}$ **(E)**

3. answers include $k = 0, 2, 4, 10$ **(E)**
4. 9.61 **(E)**

Unit 11.2 Activity 15A: Graphs of $y = \pm(x - a)^2 + b$ (page 236)

1. **a.** $y = x^2 + 4x + 8$ $a = 1, b = 4, c = 8$
b. $y = x^2 - 6x + 8$ $a = 1, b = -6, c = 8$
c. $y = -x^2 + 0x + 8$ $a = -1, b = 0, c = 8$
d. $y = -x^2 + 4x + 1$ $a = -1, b = 4, c = 1$
e. $y = x^2 - 8x + 5$ $a = 1, b = -8, c = 5$

2. **a.**

x	–3	–2	–1	0	1	2	3
y	11	6	3	2	3	6	11

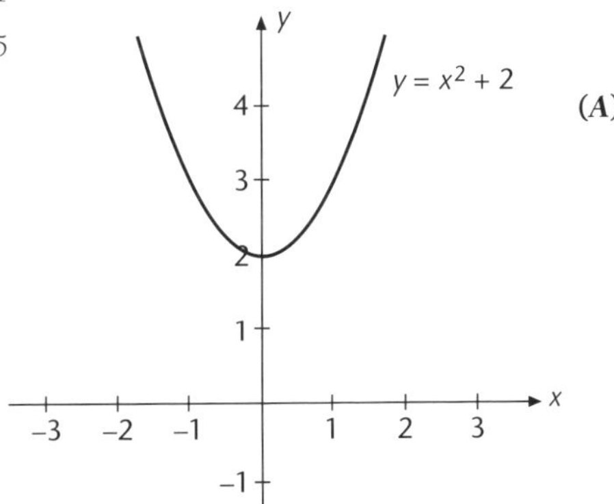

(A)

b.

x	–3	–2	–1	0	1	2	3
y	–10	–5	–2	–1	–2	–5	–10

(A)

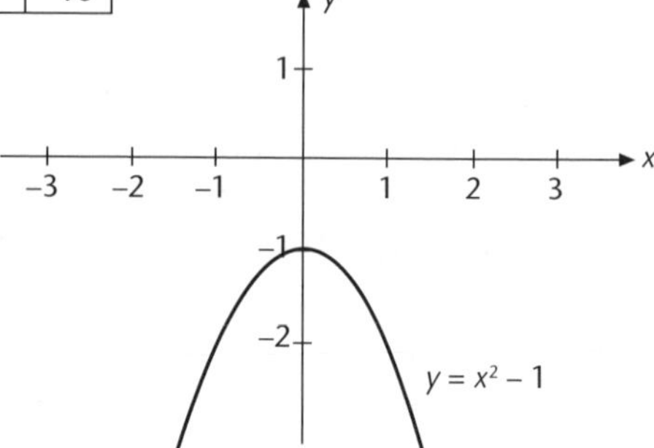

c.

x	–3	–2	–1	0	1	2	3
y	16	9	4	1	0	1	4

(A)

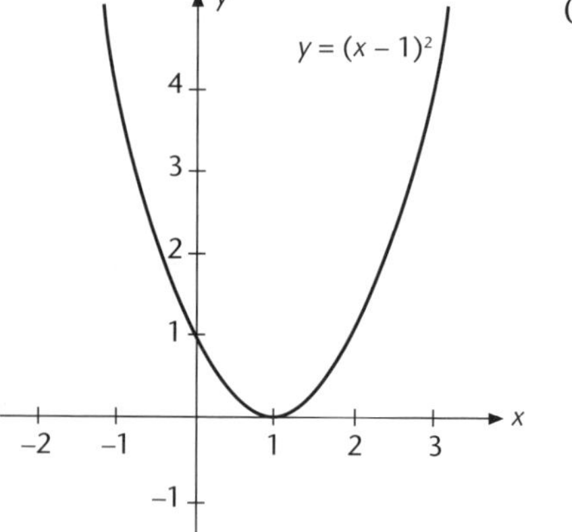

d.

x	–5	–4	–3	–2	–1	0	1
y	–9	–4	–1	0	–1	–4	–9

(A)

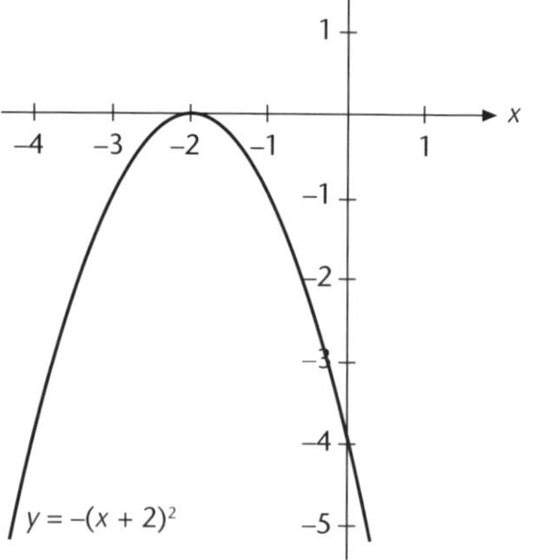

3. a.

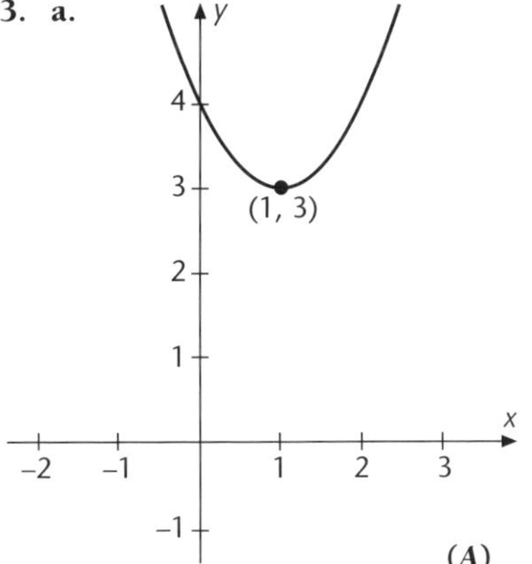

b.

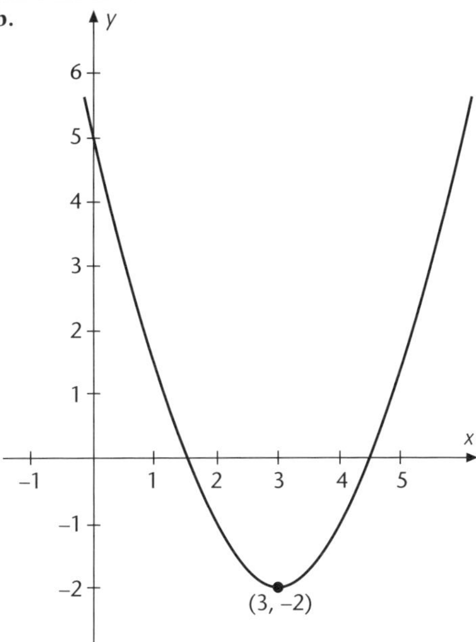

(A)

c.

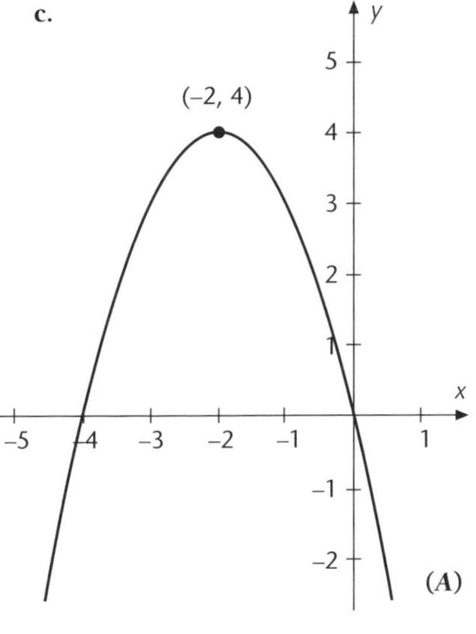

d.

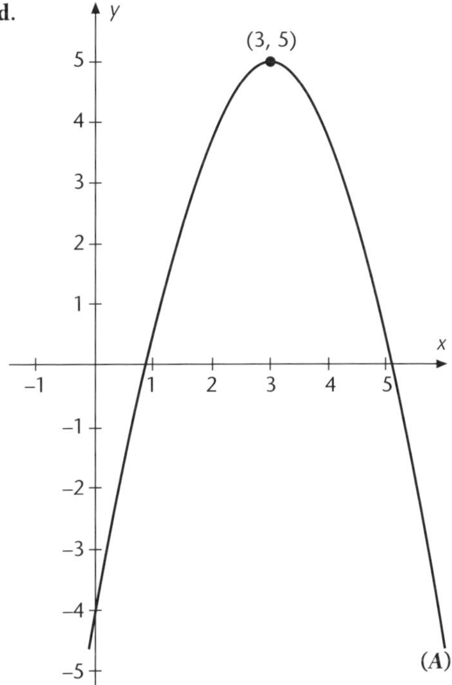

(A)

4. a. (1, 3) minimum

b. (3, –2) minimum

c. (–2, 4) maximum

d. (3, 5) maximum **(A)**

5. Axis of symmetry

a. $x = 1$ b. $x = 3$ c. $x = -2$ d. $x = 3$ **(A)**

6. I. $y = (x + 3)^2 + 2$ II. $y = (x + 3)^2$ III. $y = -(x + 3)^2 - 2$

IV. $y = -(x - 3)^2 - 2$ V. $y = (x - 2)^2$ **(M)**

7. a. $y = (x - 2)^2$ b. $y = -x^2 + 5$ or $y = 5 - x^2$

c. $y = (x - 3)^2 + 2$ d. $y = (x + 4)^2 + 1$

e. $y = -(x + 3)^2 + 7$ or $y = 7 - (x + 3)^2$ f. $y = -(x - 2)^2 + 5$ or $y = 5 - (x - 2)^2$ **(M)**

Unit 11.2 Activity 15B: Graphs of quadratics in factorised form (page 238)

1. a.

x	–4	–3	–2	–1	0	1	2
y	5	0	–3	–4	–3	0	5

b.

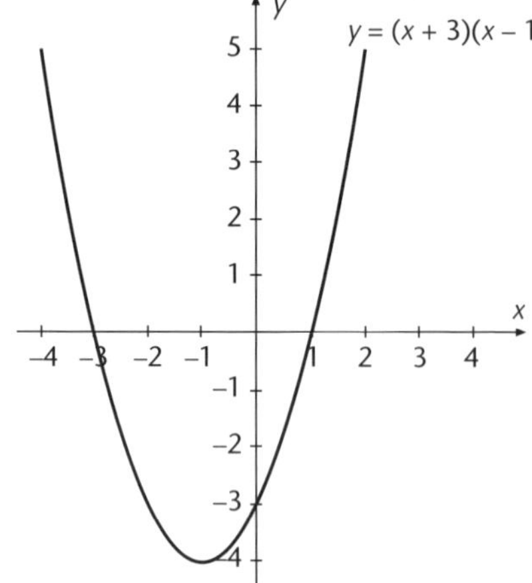

c. (–3, 0), (1, 0)

d. (0, –3)

e. $x = -1$

f. (–1, –4)

g. $y = x^2 + 2x - 3$ **(A)**

2. a.

x	–3	–2	–1	0	1	2	3
y	–5	0	3	4	3	0	–5

b. $x = 0$

c. (0, 4)

d. (–2, 0), (2, 0)

e. $y = -x^2 + 4$ **(A)**

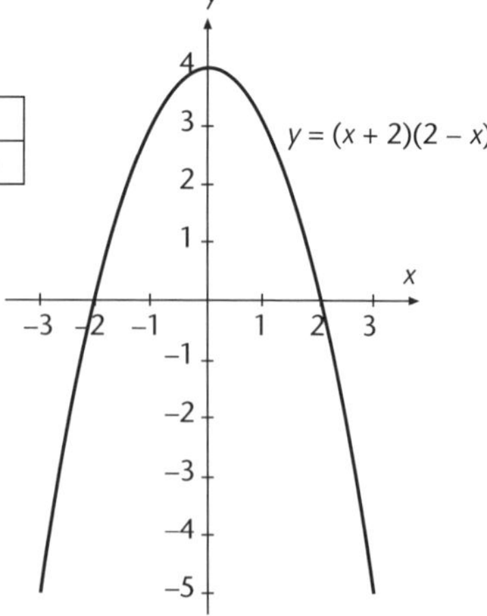

3. **I.** $y = (x + 2)(x + 4)$

II. $y = (x + 1)(x - 3)$

III. $y = -(x - 1)(x - 5)$ or $y = (1 - x)(x - 5)$ or $y = (x - 1)(5 - x)$ **(M)**

Unit 11.2 Activity 15C: Graphs of quadratics which can be factorised (page 240)

1.

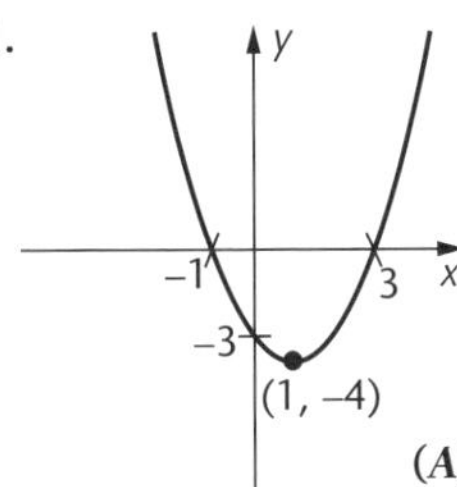

2.

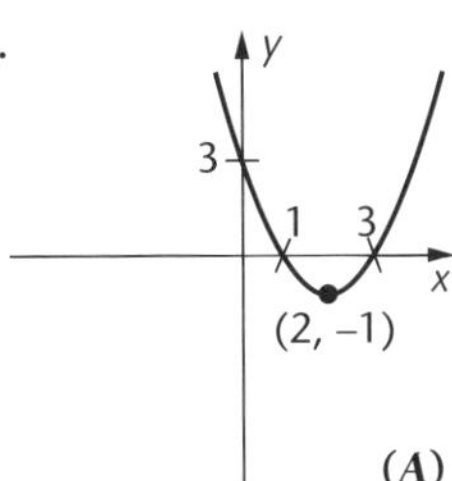

3.

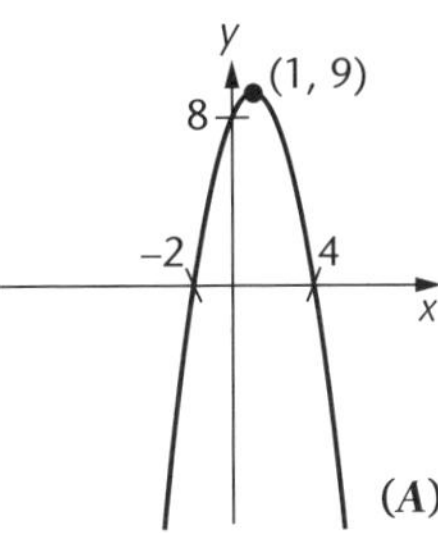

4.

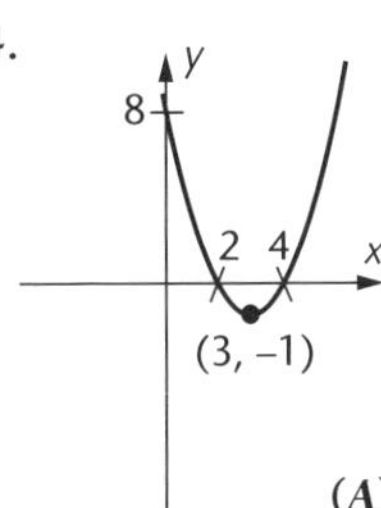

5.

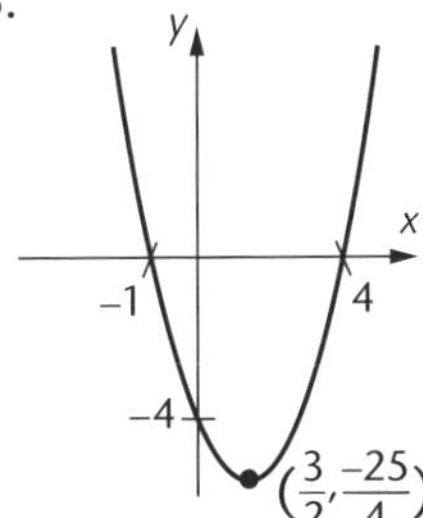

6.

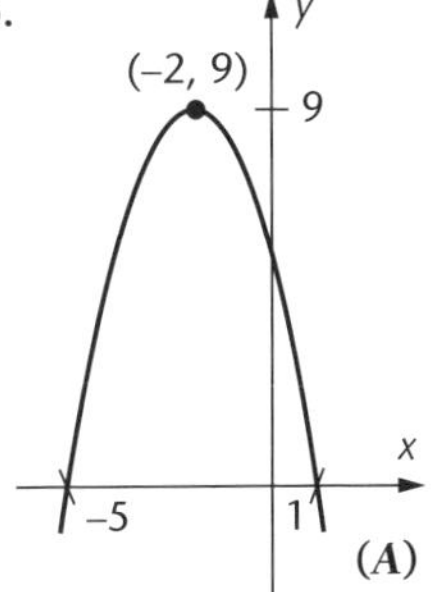

Unit 11.2 Activity 15D: Graphs of $y = ax^2$ (page 240)

1.

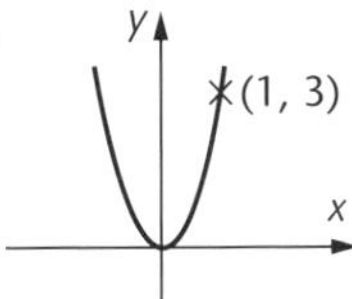

2.

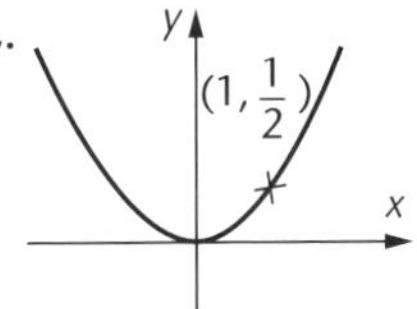

3.

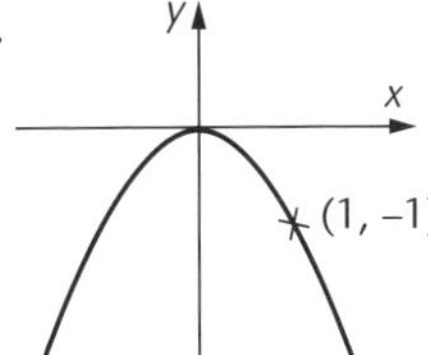

4.

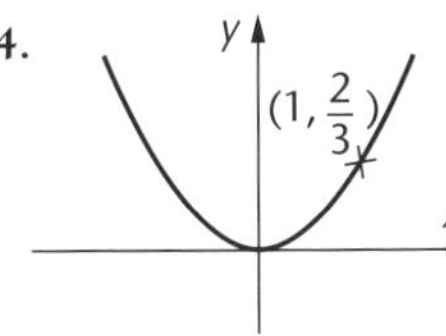

5.

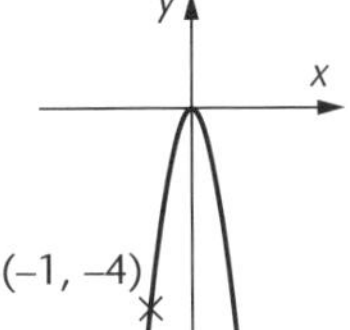

6. **a.** $y = 4x^2$ **b.** $y = \frac{1}{4}x^2$ **c.** $y = \frac{3}{4}x^2$ **d.** $y = \frac{-1}{3}x^2$ **(E)**

Unit 11.2 Activity 15E: Graphs of $y = ax^2 + bx + c$ (page 243)

1. a.

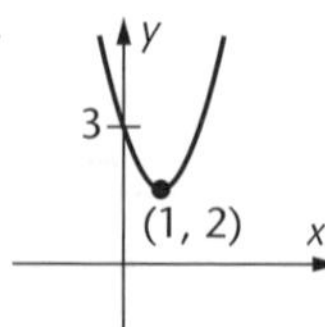

b.

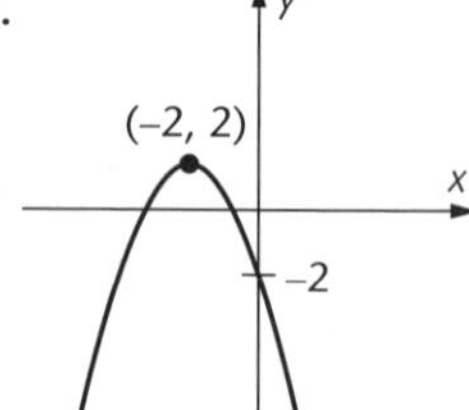

c.

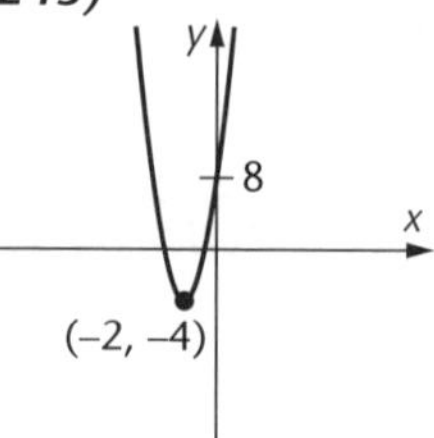

d.

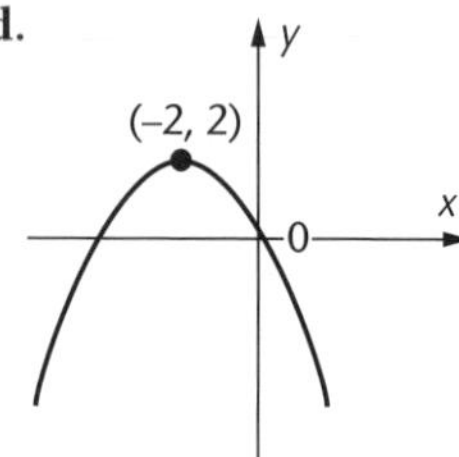

e.

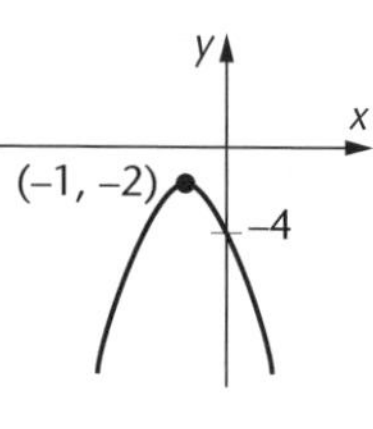

2. a. $y = (x - 3)^2$ **b.** $y = (x - 2)^2 + 3$ **c.** $y = 2(x - 1)^2 + 2$

d. $y = 3(x + 3)^2 - 2$ **e.** $y = \frac{-1}{2}(x - 2)^2 + 5$ or $y = 5 - \frac{1}{2}(x - 2)^2$ **(E)**

3. a. $y = 2x^2 - 12x + 10$ **b.** $y = 0.5x^2 - x - 4$ **c.** $y = 3x^2 - 6x - 9$ **(E)**

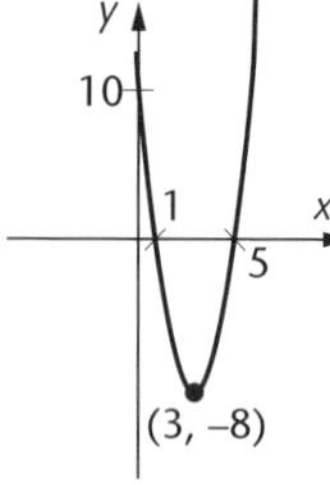

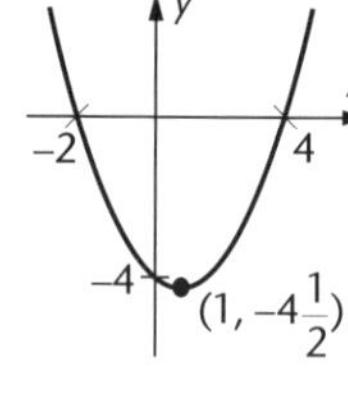

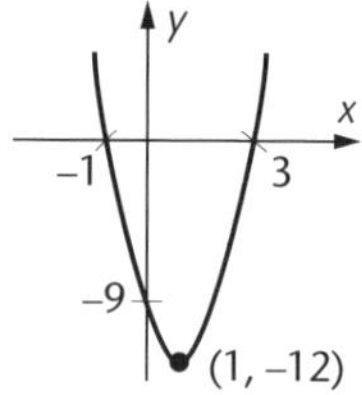

4. a. i. $k = 2$ **ii.** $(3, -2)$ **b.** $\frac{1}{4}$ **c.** $L = 4$ **(M)**

5. a. $y = (x + 3)(x - 1)$ **b.** $y = \frac{2}{3}(x - 1)(x - 6)$ **c.** $y = -2(x - 1)^2 + 5$

d. $y = 3(x + 1)^2 + 2$ **(M)**

Unit 11.2 Activity 15F: Quadratic modelling (page 245)

1.

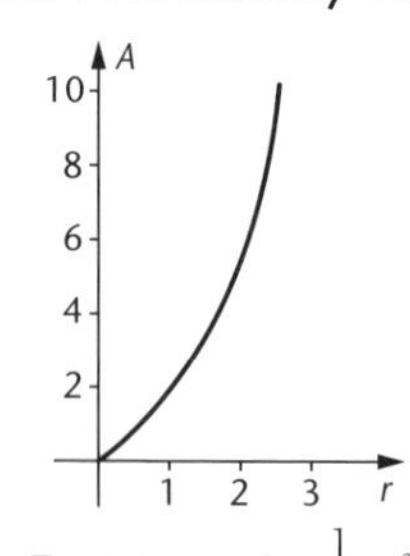

Equation is $A = \frac{1}{2}\pi r^2$

2.

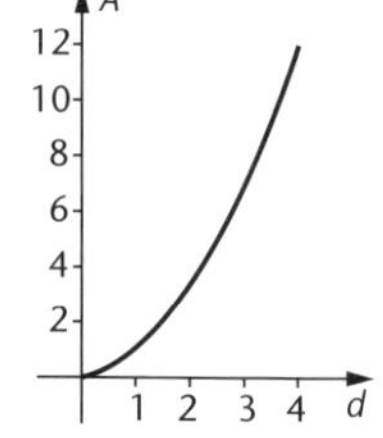

Equation is $A = \frac{1}{4}\pi d^2$

3.

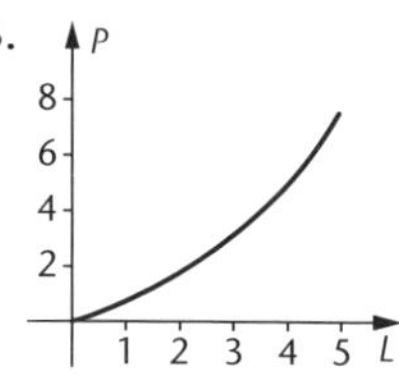

Equation is $P = 0.3L^2$

4.

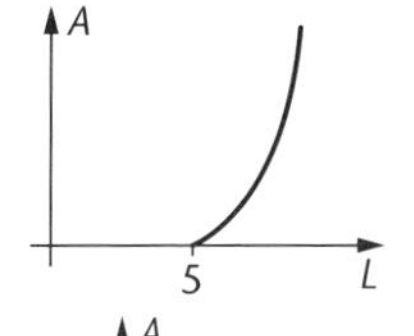

Equation is $A = L^2 - 25$ **(E)**

5.

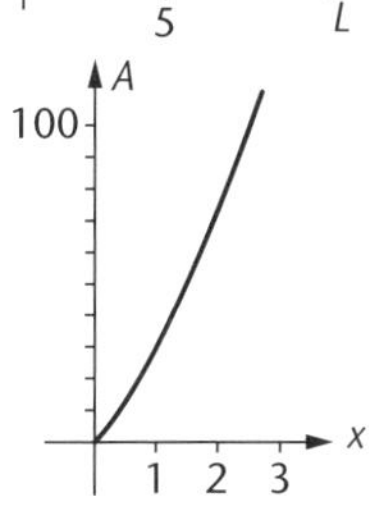

Equation is $A = (2x + 8)^2 - 64$, where x is distance from fence **(E)**

6. $D = 5 + \frac{1}{2}(t - 2)^2$ **(E)**

7. a.

BD(x)	AB	BC	Area of rectangle
3	5	0	0
$3\frac{1}{2}$	$4\frac{1}{2}$	$\frac{1}{2}$	2.25
4	4	1	4
$4\frac{1}{2}$	$3\frac{1}{2}$	$1\frac{1}{2}$	5.25
5	3	2	6
6	2	3	6
7	1	4	4
8	0	5	0

b.

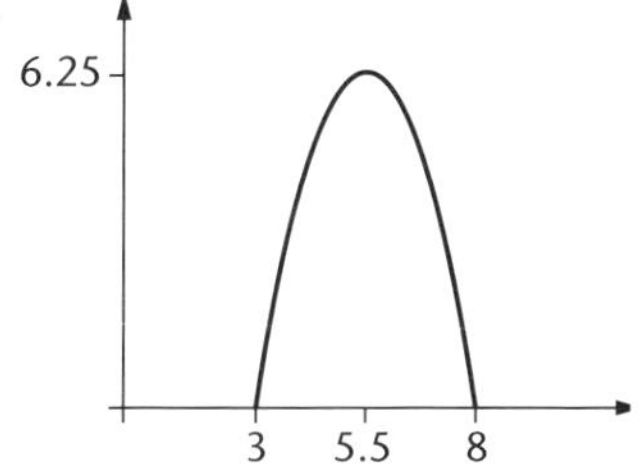

c. $A = (8 - x)(x - 3)$

d. 6.25 **(E)**

8. a.

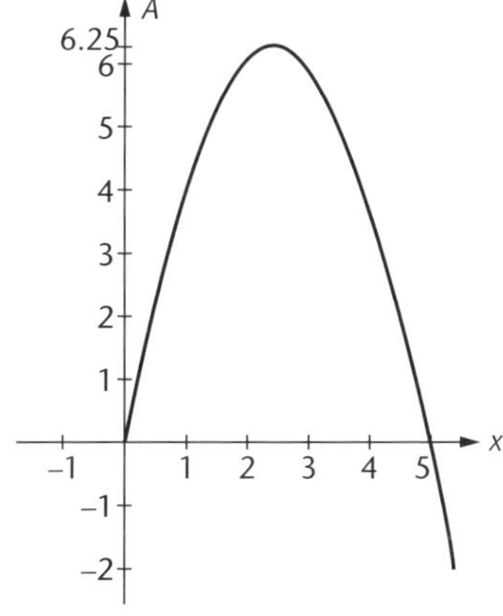

b. 6.25

c. Equation would be $A = x(4 - x)$ which has a maximum of $A = 4$ at $x = 2$. **(E)**

9. a. $C = 50\pi r^2$

b.

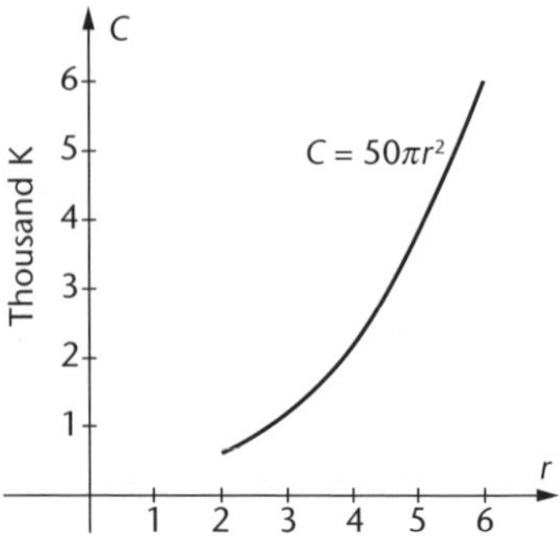

c. a little more than 6 m

d. about K600 (from graph) **(E)**

Unit 11.2 Activity 15G: Intersection of parabola and line (page 247)

1. a. (2, 4), (–2, 4) b. (2, 4) c. No intersection d. (0, 0), (1, 1)
   e. (0, 0), (2, 4) f. (0, 0), (3, 9) g. (3, 9), (–1, 1) h. (1, 1)
   i. (–4, 16), (1, 1) j. (1.62, 2.62), (–0.62, 0.38) **(E)**
2. (1, 0), (0, –1) **(M)** 3. (3, 0), (11, 34) **(M)** 4. (2, 7) **(M)** 5. (1, 4), (–3, 8) **(M)**
6. Proof (point of intersection is (1, 6)) **(M)**

Unit 11.2 Activity 15H: Graphing quadratic functions—multiple choice (page 248)

1. B 2. A 3. C 4. C
5. D 6. D 7. B 8. A

Unit 11.2 Activity 16A: Cubic graphs (page 251)

1. a. **(A)**

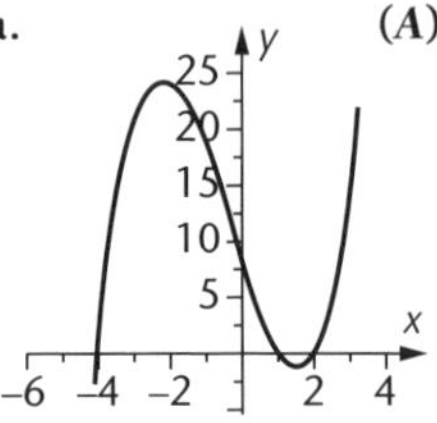

b. **(A)**

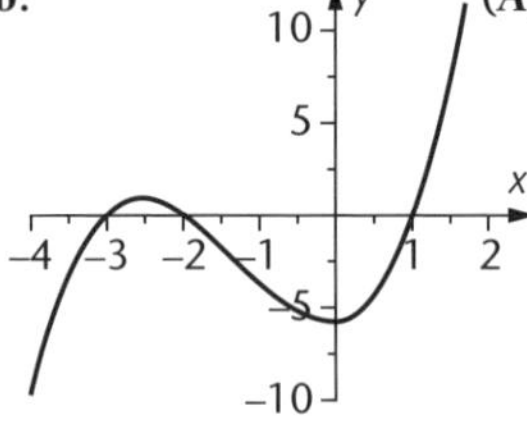

c. **(A)**

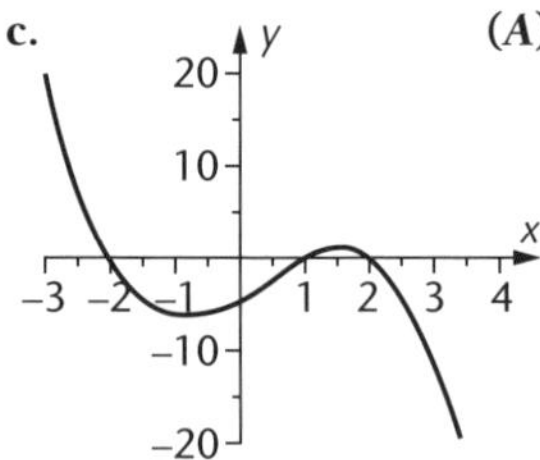

d. **(A)**

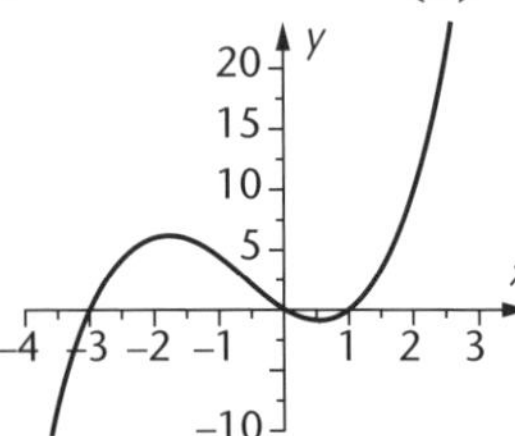

e. **(A)**

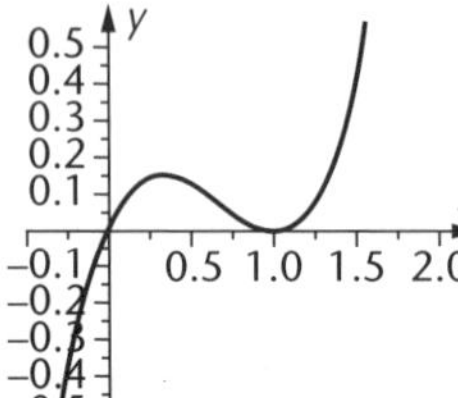

f. **(A)**

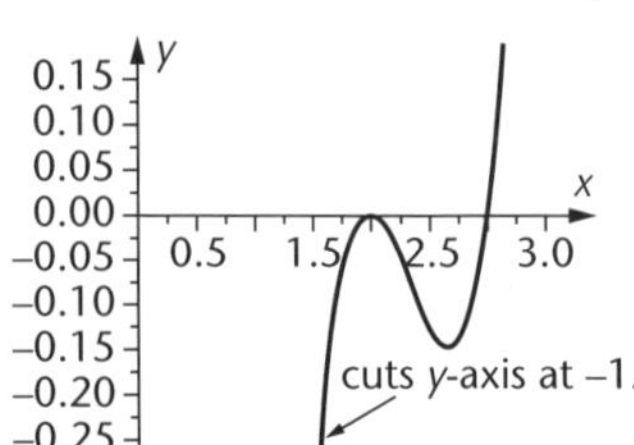

g. **(A)**

cuts y-axis at 2

h. **(A)**

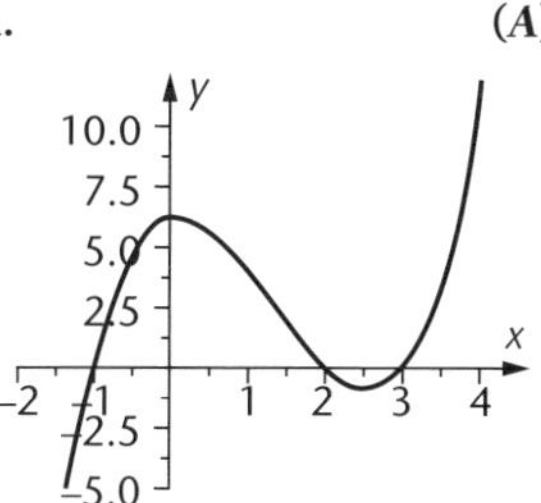

2. **a.** $y = (x + 2)(x - 1)(x - 3)$ **(M)**

b. $y = (x + 2)(x + 1)(x - 2)$ **(M)**

c. $y = -(x + 1)(x - 1)(x - 3)$ **(M)**

d. $y = (x + 3)(x - 2)^2$ **(M)**

3. **(E)**

$V = (L - 2)(L - 1)L$

4. **(E)**

$V = L^2(L + 1)$

5. **a.** April, November 1 **(M)**

b. K10.50 **(M)**

c. Before February, then early July until nearly the end of September. **(M)**

d. $y = -0.1(x - 3)^2(x - 10)$

Where x is the number of months since the beginning of the year, y is the Kvalue above K1.50 of 1 kg of the commodity. **(E)**

Unit 11.2 Activity 16B: Cubic graphs—multiple choice (page 252)

1. B **2.** D **3.** B **4.** B

5. A **6.** D **7.** A **8.** C

Unit 11.2 Activity 17A: Absolute value of a number (page 253)

1. **a.** 5 **b.** 3.4 **c.** $1\frac{5}{6}$ **d.** 0.0056

e. 2.5 **f.** $\sqrt{2}$ **g.** 0 **h.** 1024

2. **a.**

$\lvert a\rvert + \lvert b\rvert$	$\lvert a + b\rvert$
9	9
9	5
9	5
9	9

b. No

3. a.

$\lvert a\rvert - \lvert b\rvert$	$\lvert a - b\rvert$
5	5
5	5
5	11
5	5

b. No

4. a.

$\lvert a\rvert \times \lvert b\rvert$	$\lvert a \times b\rvert$
21	21
21	21
21	21
21	21

b. Yes

5. a.

$\lvert a\rvert \div \lvert b\rvert$	$\lvert a \div b\rvert$
3	3
3	3
3	3
3	3

b. Yes

6. a. 35 b. 10 c. 36 d. 36
e. 11 f. −11 g. 0 h. 0

Unit 11.2 Activity 17B: Graphs of absolute value functions (page 258)

1.

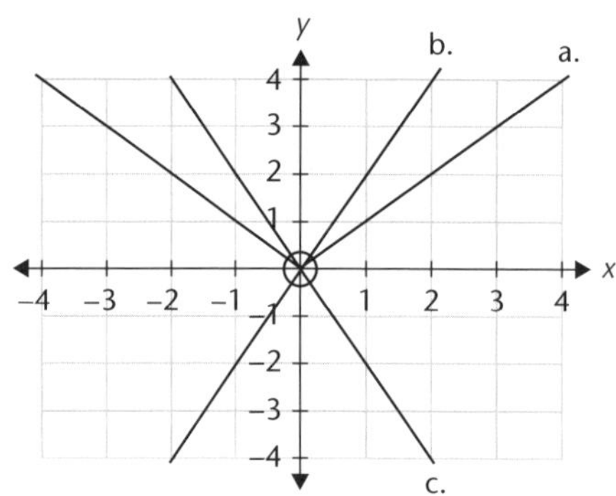

2.

y
b.
a.
c.
x
4
3
2
1
−1
−2
−3
−4
−4 −3 −2 −1 1 2 3 4

3.

a.
b.
y
c.
x
4
3
2
1
−1
−2
−3
−4
−4 −3 −2 −1 1 2 3 4

4.

y
c.
a.
b.
x
8
6
4
2
−2
−4
−4 −3 −2 −1 1 2 3 4

5. a.

y
x
6
5
4
3
2
1
−1
−2 −1 1 2 3 4

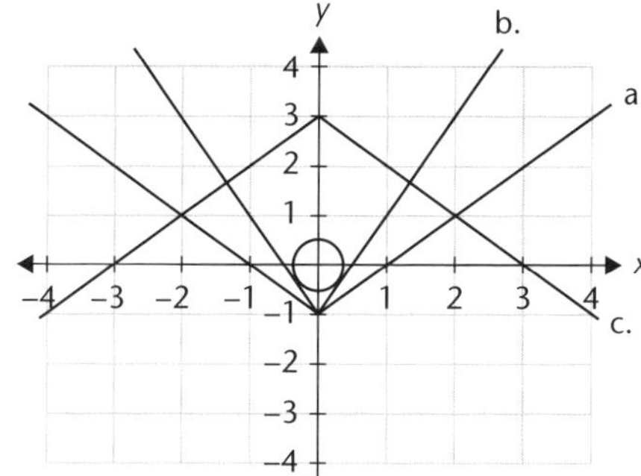

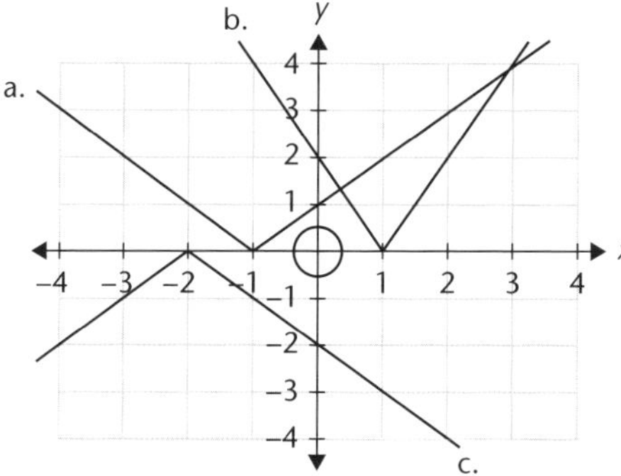

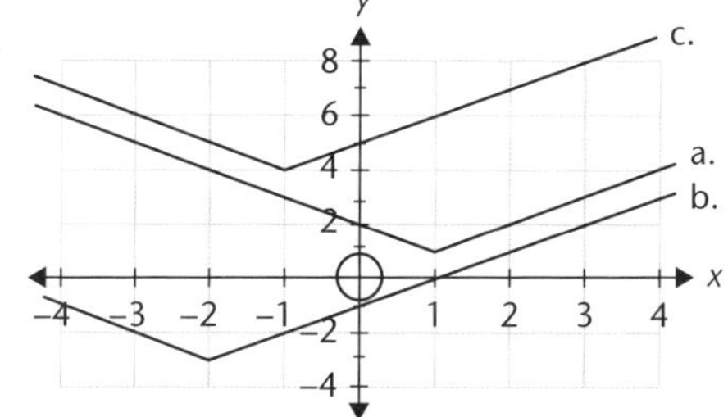

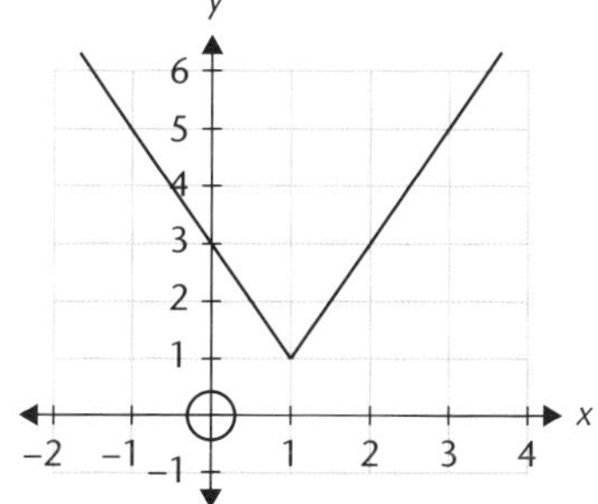

b. (0, 3)

6. a.

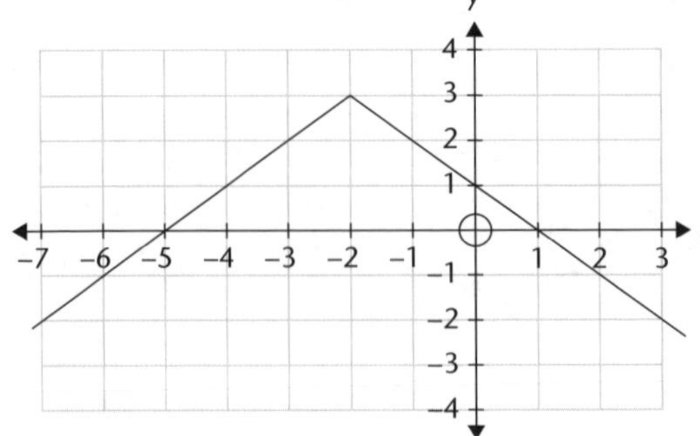

b. (0, 1)

c. (–5, 0), (1, 0)

d. $y = 1 - x$

e. $y = x + 5$

7. a.

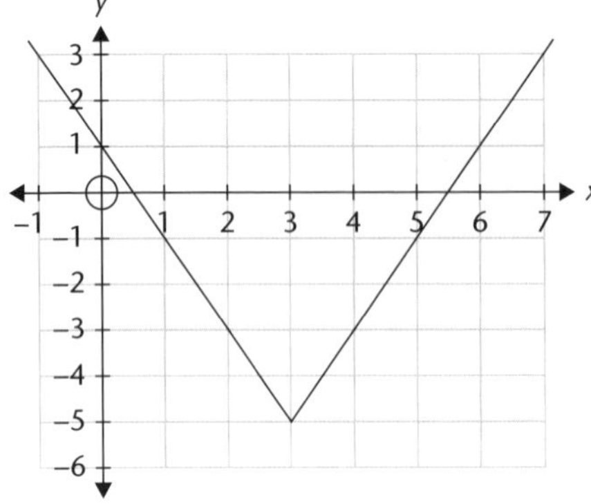

b. (0, 1)

c. $\left(\frac{1}{2}, 0\right), \left(\frac{11}{2}, 0\right)$

d. $y = 2x - 11$

e. $y = 1 - 2x$

8.

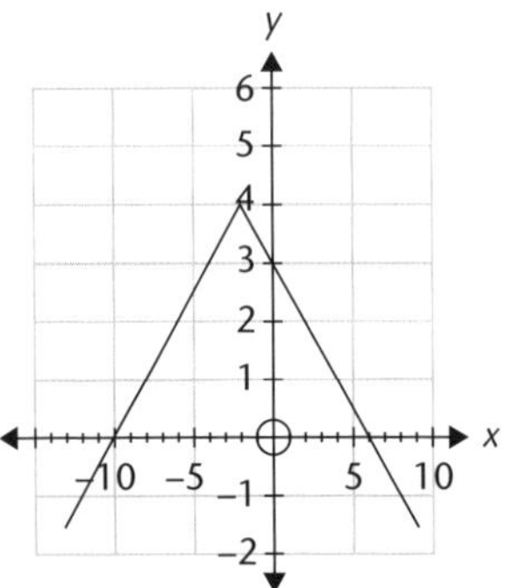

b. (0, 3)

c. (–10, 0), (6, 0)

d. $y = 3 - \frac{1}{2}x$

e. $y = 5 + \frac{1}{2}x$

Unit 11.3 Managing Data

Unit 11.3 Activity 1A: Basic data display (page 267)

1.

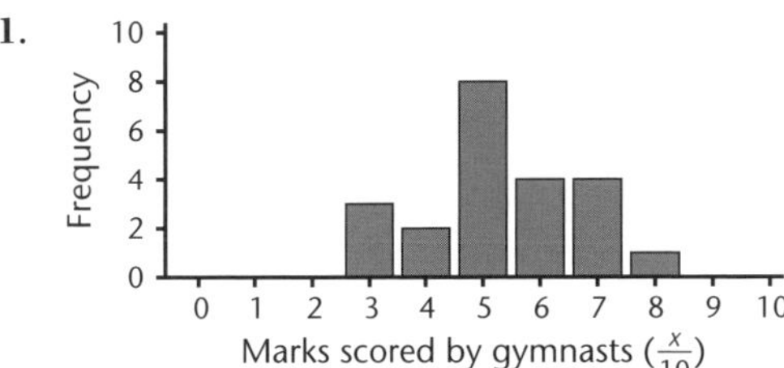

2. a.

Number	Tallies	Frequency
20	|	1
21	|||	3
22		0
23	|	1
24	||	2
25	|||	3
26	||	2
27	||	2
28	||	2
29	||	2
30	||	2
31	|||	3
32	|	1
33	||	2

Number	Tallies	Frequency
34	||||\	5
35	||	2
36	||||	4
37	||	2
38	|	1
39	|	1
40	||||\	5
41	|	1
42	|	1
43	||||	4
44		0
45	|	1
46	|	1

b.

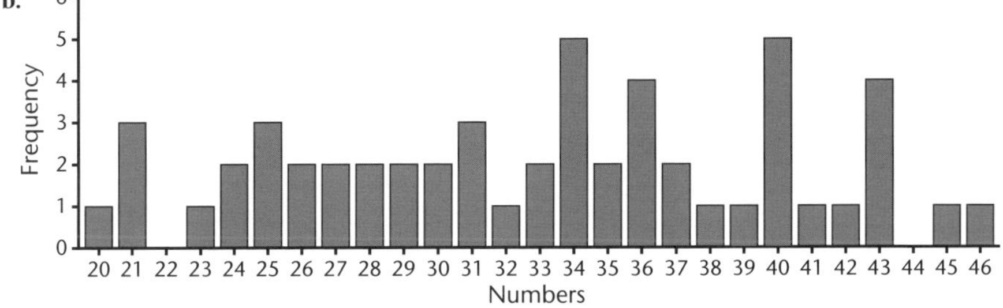

c.

Number	Tallies	Frequency
20–24	~~IIII~~ II	7
25–29	~~IIII~~ ~~IIII~~ I	11
30–34	~~IIII~~ ~~IIII~~ III	13
35–39	~~IIII~~ ~~IIII~~	10
40–44	~~IIII~~ ~~IIII~~ I	11
45–49	II	2
		54

i.

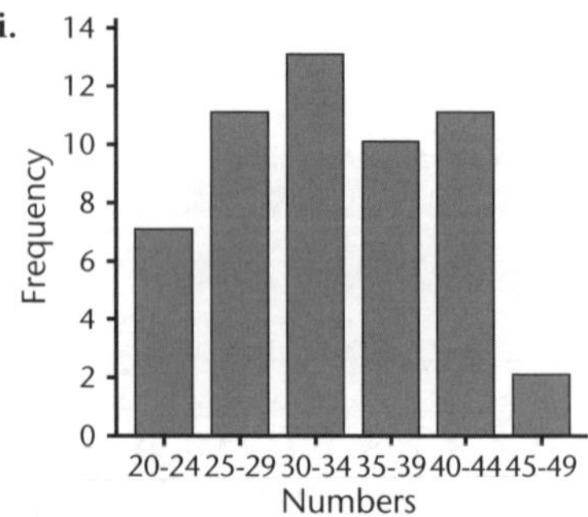

ii. The pattern of the above graph demonstrates that the frequency is smaller at the two ends, while the majority of the numbers are situated in the middle region where the frequency peaks. ***(A)***

iii. The more suitable graph is the second graph, with the data grouped in fives. This is because we get a general idea of the spread of the numbers by using groups of five. Using the frequency of each **individual** number does not reveal any **major** trends. ***(M)***

3.

Vowel	Tallies	Frequency	Relative frequency
a	~~IIII~~ ~~IIII~~ I	11	11 ÷ 66 = 0.167
e	~~IIII~~ ~~IIII~~ ~~IIII~~	15	15 ÷ 66 = 0.227
i	~~IIII~~ ~~IIII~~ ~~IIII~~ ~~IIII~~ III	23	23 ÷ 66 = 0.348
o	~~IIII~~ ~~IIII~~ III	13	13 ÷ 66 = 0.197
u	IIII	4	4 ÷ 66 = 0.061
		66	1.0

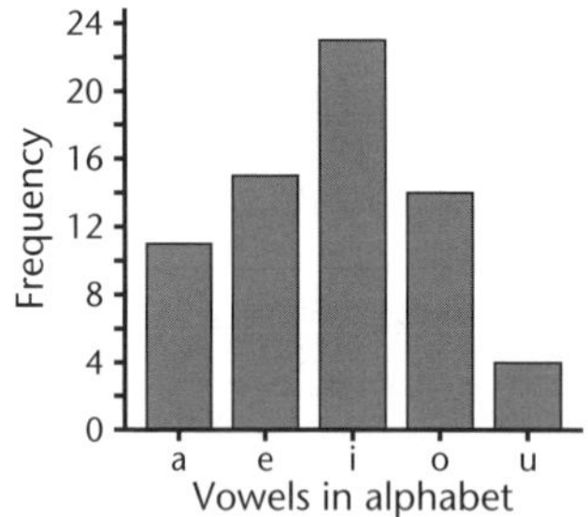

4. **a.** Discrete **b.** Continuous

c. Discrete **d.** Continuous

e. Discrete **f.** Discrete

g. Discrete **h.** Continuous

i. Discrete **j.** Discrete

5.

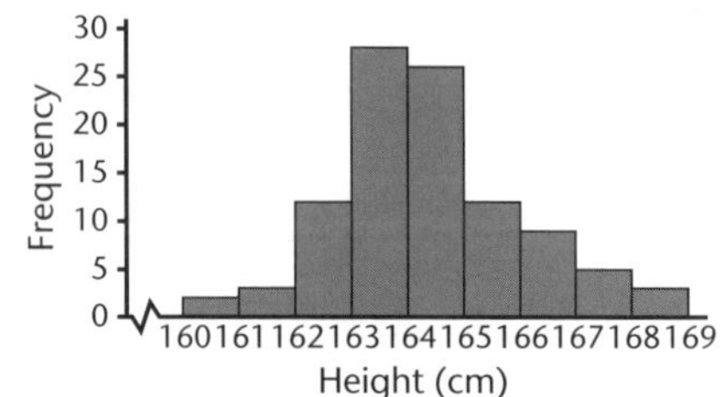

6. Histogram showing distribution of catch

Number of fish

Weight (kg)

7. a. i. No labels on axes

ii. No title

iii. Incorrect scale on horizontal axis

iv. Incorrect values drawn for histogram

b.

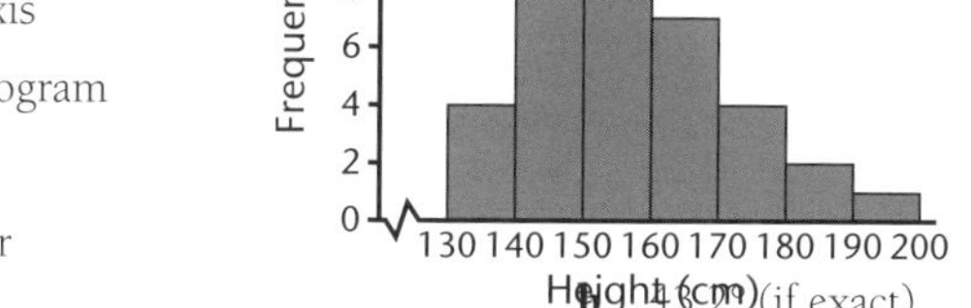

8. a. Interest cost = K27 295 970, labour cost = K15 869 750

b. 43.2° (if exact)

c.

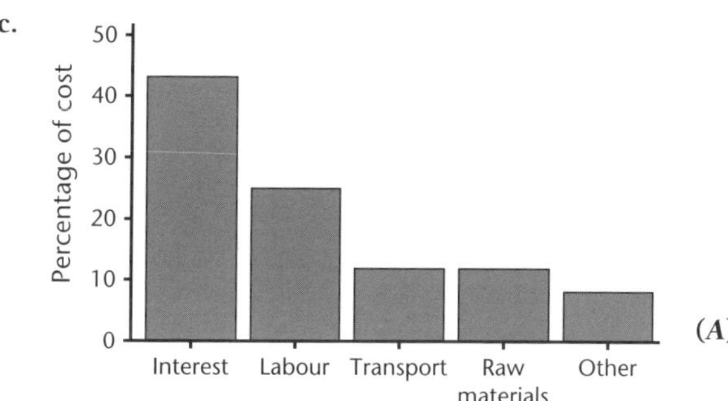

(A)

9.

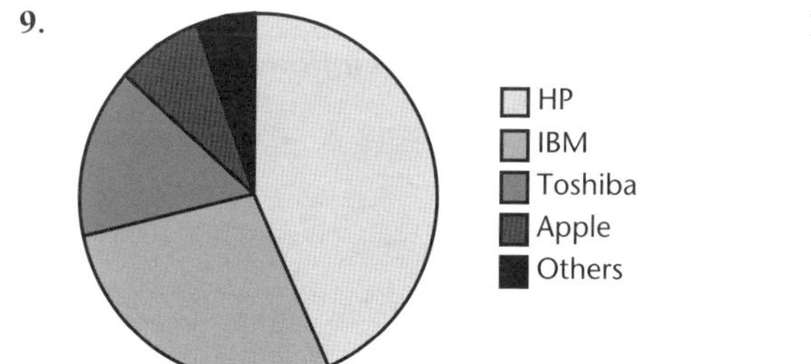

10.

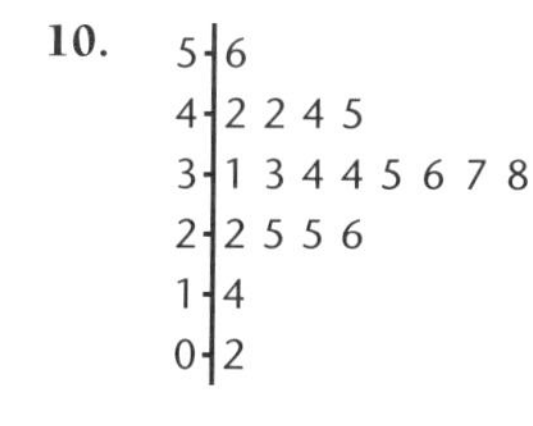

11. (*A*)

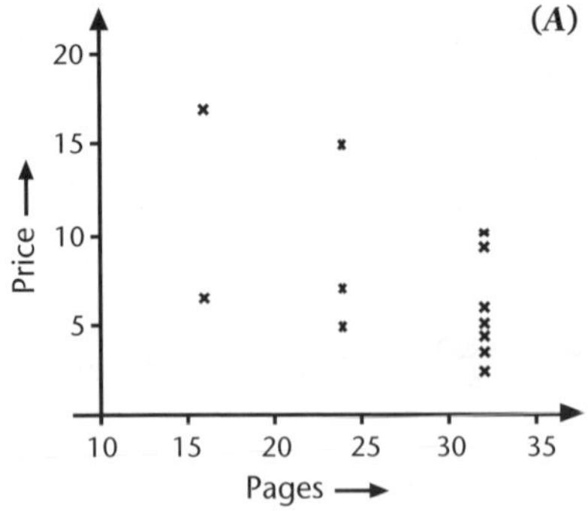

Some points have been omitted.

There appears to be no correlation.

12. (*A*)

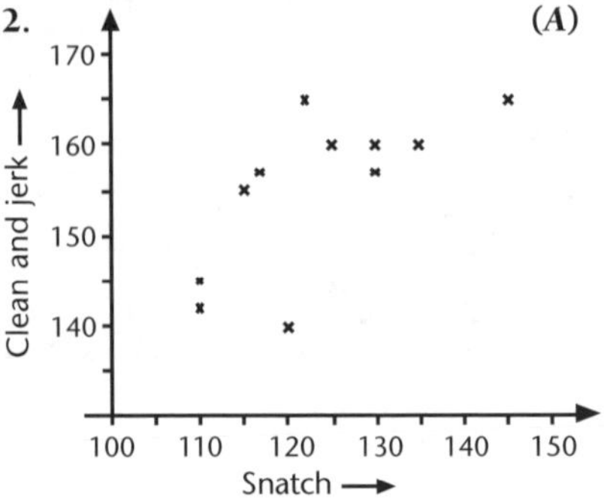

There is a tendency for performance in the 'clean and jerk' to rise as performance in the 'snatch' rises.

Some correlation.

Unit 11.3 Activity 2A: The mean (page 273)

1. a. 10.125 **b.** $77\frac{1}{7}$ or 77.14 (2 dp) **c.** 4.79 (2 dp)

2. 80

3. There are many such sets, eg 21, 22, 23, 24, 25.

Unit 11.3 Activity 2B: Medians, modes and quartiles (page 275)

1. a.	Median = 27.5	LQ = 16	UQ = 42	There is no mode.
b.	Median = 4	LQ = 2.91	UQ = 4.3	Mode = 4.3
c.	Median = 3	LQ = 2	UQ = 3	Mode = 3
d.	Median = 48.5	LQ = 32	UQ = 70	There is no mode.

2. There are many such sets.

Unit 11.3 Activity 2C: Estimating means and standard deviations (page 278)

1. a. $\bar{x} = 11.5$, $s = 7.9$ **b.** $\bar{x} = 4.9$, $s = 5.2$

2. a. $\bar{x} = 65.1$, $s = 30.2$ **b.** $\bar{x} = 57.4$, $s = 41.2$

3. a. First has mean of about 2.3, the second has mean of about 4.7, standard deviation is about 1.5 for both.

b. i. For first, $\bar{x} = 3.3$, $s = 1.5$ For second, $\bar{x} = 5.7$, $s = 1.5$

ii. For first, $\bar{x} = 4.6$, $s = 3.0$ For second, $\bar{x} = 9.4$, $s = 3.0$

iii. For first, $\bar{x} = 2.3$, $s = 1.5$ For second, $\bar{x} = 4.7$, $s = 1.5$

Unit 11.3 Activity 2D: Box plots (page 280)

1. **2. a.**

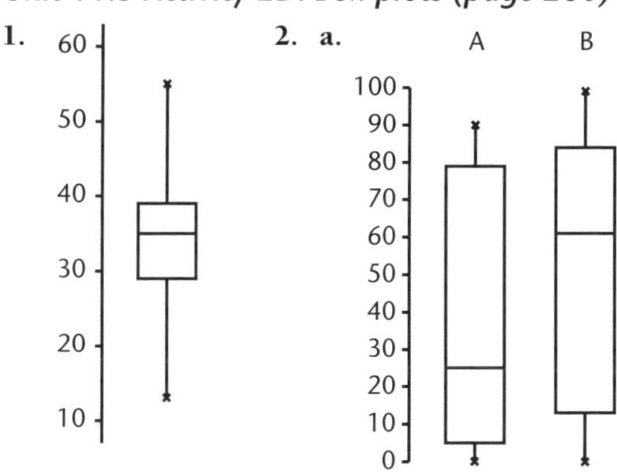

b. Group B is higher in quartiles, median and maximum than Group A. Both have same minimum. Group B has slightly greater spread.

Group A is skewed towards values above its median; Group B is skewed towards values below its median. **(A)**

Unit 11.3 Activity 3A: Cumulative frequency graphs (page 284)

1. **a.** **i.** 183 cm

ii. 189 cm

iii. 180 cm

iv. 191 cm

v. 174 cm

b. **i.** 183 cm

ii. 180 cm

iii. 80%

iv. 62%

2. **a.**

Maximum daily temperature (°C)	Cumulative frequency
< 15	0
< 20	1
< 25	6
< 30	20
< 35	25
< 40	28
< 45	30

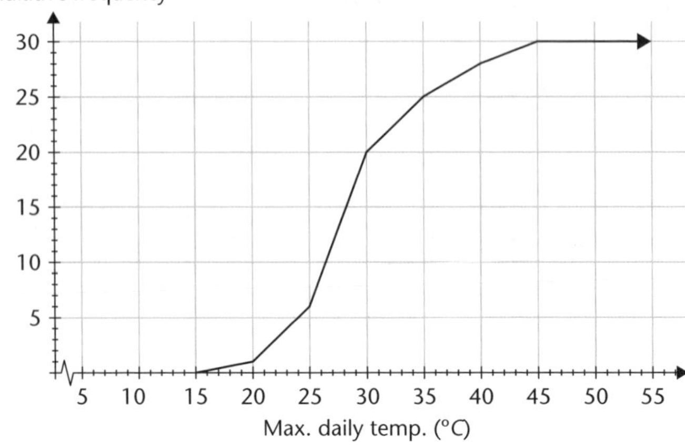

b. **i.** 28°C **ii.** 33°C

iii. 22°C **iv.** 6.5°C

c. **i.** On half of the days the maximum daily temperature (MDT) was 28°C or more.

ii. 80% of the days had a MDT of 33°C or less.

iii. 10% of the days had a MDT of 22°C or less.

iv. the middle 50% of the MDTs were spread over 6.5°C

3. **a.** 600 **b.** 162

c.

Speed (km/hr)	% cumulative frequency
< 50	0
< 60	1.3
< 70	4.3
< 80	18.3
< 90	37.7
< 100	73
< 110	98
< 120	99.7
< 130	100

d.

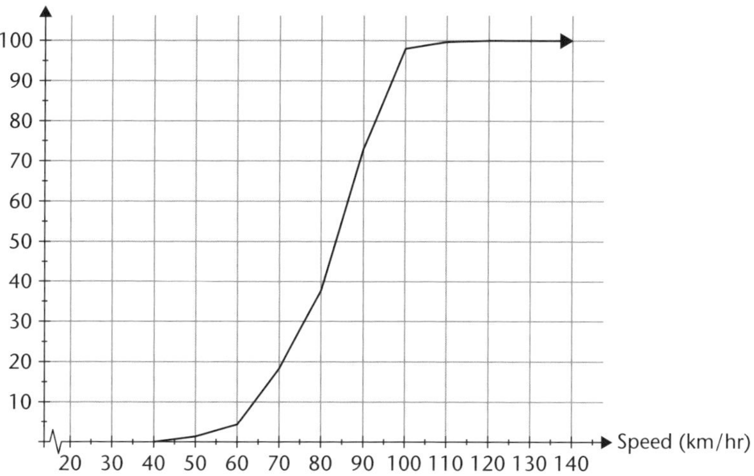

e. **i.** 83 km/hr; half of the cars are travelling at 83km/hr or more.

ii. 74 km/hr; 25% of the cars are travelling at 74 km/hr or less.

iii. 91 km/hr; 75% of the cars are travelling at 91 km/hr or more.

iv. 97 km/hr; 90% of the cars are travelling at 97km/hr or less.

v. 64 km/hr; 10% of the cars are travelling at 64 km/hr or less.

vi. 17 km/hr; the speed of the middle 50% of the cars is spread over 17 km/hr.

4. **a.**

Swim time (secs)	% cumulative frequency
< 50	0
< 55	2.5
< 60	20
< 65	62.5
< 70	90
< 75	95
< 80	100

b.

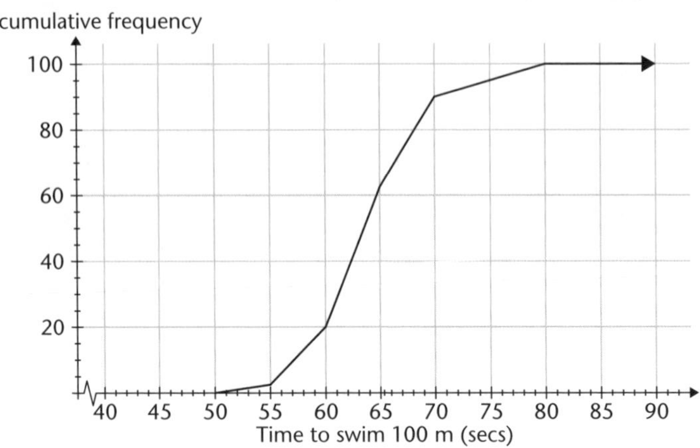

c. **i.** 63 secs; half of the swimmers had a time of 63 seconds or less

ii. 60.5 secs; 25% of the swimmers had a time of 62.5 secs or more

iii. 67.5 secs; 75% of the swimmers swam a time of 67.5 secs or less.

iv. 70 secs; 90% of the swimmers swam a time of 70 seconds or less.

v. 57.5%; the fastest 10% of the swimmers swam a time of 57.5 secs. or less.

vi. 7 secs; the middle 50% of the swimmers had swim times spread over 7 secs.

Unit 11.3 Activity 4A: Time series graphs (page 288)

1. a. Estimated volume of cured vanilla bean exports, 1997–2006

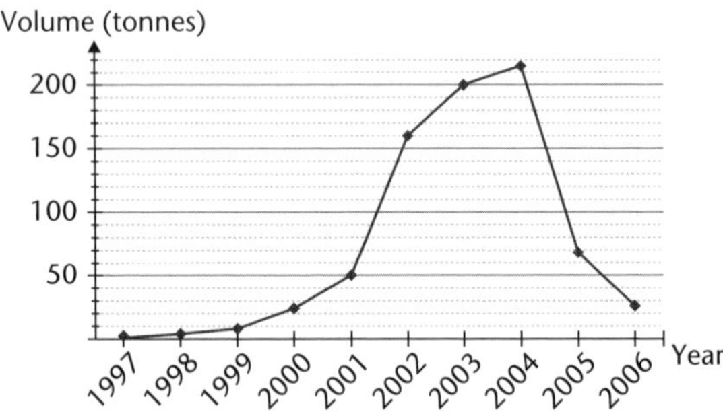

b. The vanilla bean exports showed an increase from 1997 to 2001, then there was a large increase in the year 2002, from 50 to 160 tonnes. The exports continued to increase in 2003 and 2004 then there was a large decrease in 2005, from 215 down to 68 tonnes. The decrease continued into 2006 with only 26 tonnes exported.

2. a. **i.** Yes **ii.** No **iii.** Yes, a downward trend

b. **i.** 1985 **ii.** 1997

c. There is a general downward trend over the years 1980 to 1997 although large fluctuations in production occur in this time; the highest production (950 tonnes) is in 1985 and the lowest in 1987(50 tonnes). After 1997 there appears to be a steady upward trend.

d. Approx. 450 tonnes.

3. a. i. Fruit and vegetable imports from Australia showed an increasing trend for the years 1983 to 1997. For the years 1997 to 2000 there was a dramatic decline in imports(750 tonnes to 350 tonnes) and then imports were steady for two years before declining slightly in 2003.

ii. Fruit and vegetable imports from New Zealand showed random fluctuations over the years 1983 to 2003 and a slight downward trend.

b. The imports from Australia and New Zealand show the same pattern over the years 2000 to 2003; steady for the first three years and then a decline in 2003. Australia imported approximately 500 tonnes more of fruit and vegetables.

c. Approximately 2500 tonnes and 2000 tonnes.

4. a. The coffee production shows a yearly seasonal trend with high production in June to August and the lowest production in February. Production was higher in 1983 than in 1982.

b. The food sales show a yearly seasonal trend with high sales in June to August and the lowest sales in February. Food sales were similar in both 1982 and 1983.

c. The patterns follow the same pattern over each year.

5. a. There appears to be an upward trend in the price of sweet potato in Port Moresby for 1998 to 2001, no increase in 2003 then a decline in the price(from 130t to 105 t) in 2003. The price increased again in 2004 and 2005 to 130t.

b. The price of sweet potato was steady for 1998 to 2000 and then there was a large increase in the price (from 35t to 110t) over the next two years. In 2003 the price decreased again to 55t and remained about that price until 2005.

c. The price of sweet potato appears to follow the same pattern over the years 1998 to 2005. However, the price was always less in Madang than in Port Moresby and the fluctuations were more pronounced.

6. a. The data shows seasonal fluctuations with a seven-day repeat of the pattern. The highs generally occur on Wednesdays and the lows on Sundays. There also appears to be an upward trend in the data.

b.

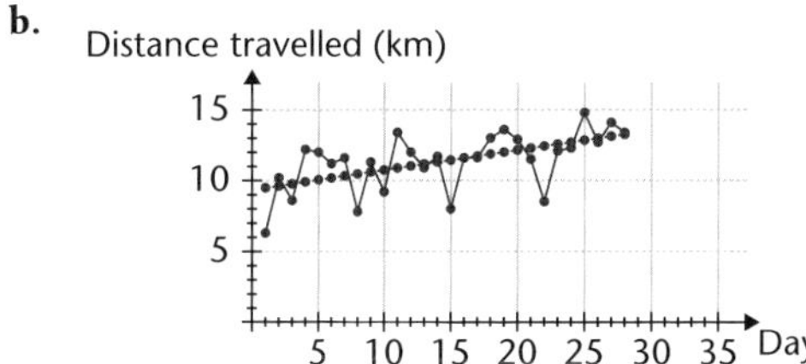

c. i. 17 km ii. 18 km iii. 19 km

d. Saturday's prediction. Saturday's values are always closer to the prediction line than Sundays and Wednesdays.

7. a. Yes, there is a pattern repeating at regular intervals.

b. A seasonal pattern

c. Every four quarters which is every year.

d. In the first quarter of each year....January to March.

e. In the 2nd and 3rd quarters; April to September.

f. A slight upward trend is apparent; the highest rates each year are increasing.

g. About 6.9%

Unit 11.3 Activity 5A: Interpreting graphs and statistics (page 296)

1. a. That for every 1000 children under 5, in 2009 in PNG, 68 will die.

b. India

c. South Africa

d. i. PNG, Indonesia, India, Turkey and South Africa.

ii. PNG, India, South Africa

e. 25.3% decrease

2. a.

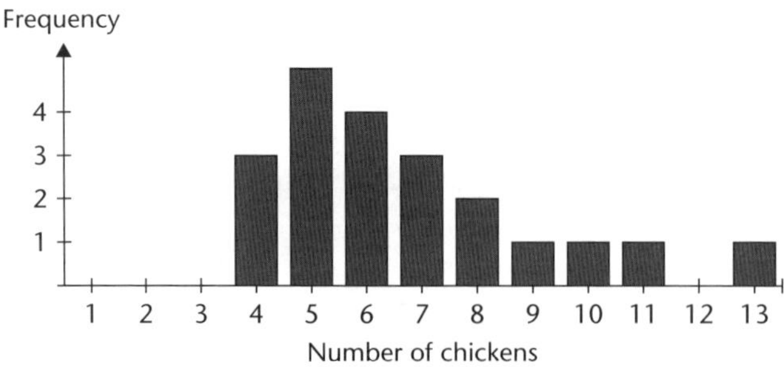

b. Median = 6, upper quartile = 8, lower quartile = 5

c.

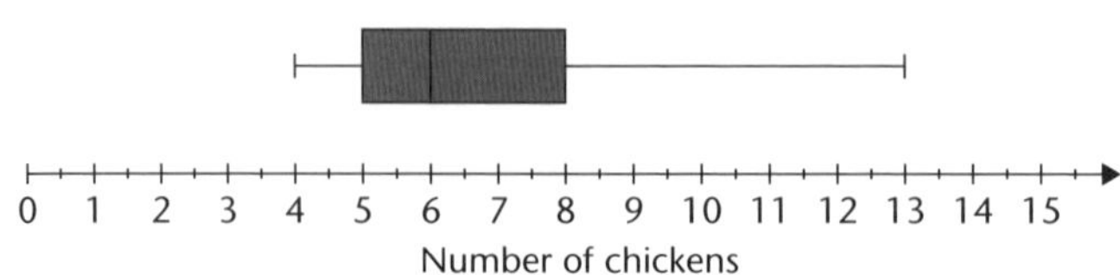

d. The data is positively skewed with a median of 6 chickens per household. (The mean is 6.7 chickens, suggesting that the data is positively skewed, and the mode is 5.) The number of chickens per household above the median is spread from 6 to 13 whereas the number of chickens per household below the median is spread from 4 to 6.

The middle 50% of the households have between 5 and 8 chickens. the minimum number of chickens in a household is 4 and the maximum number is 13.

3. a. In 1975 the production of Palm oil was about 17 000 tonnes and in 2005 it was about 330 000 tonnes; the production in 2005 was about 20 times the production in 1975.

b. The production of coffee has about doubled over the thirty years from 1975 to 2005.

c. The production of Copra has declined from about 90 000 tonnes in 1975 to 20 000 tonnes in 2005.

4. Both the graph and the boxplot show that the data is negatively skewed; the spread of the number of avocados per tree below the median is from 192 to 221 whereas the spread of the number of avocados per tree above the median is from 221 to 235. The median number of avocados per tree was 221 with 25% of the trees producing 211 or fewer avocados and 25% producing 226 or more avocados. The minimum number of avocados produced on a tree was 192 and the maximum number was 235.

5. The distributions of scores in 2009 and 2010 were symmetric about a median score of 18, however the scores in 2009 were more variable. The range for 2009 was 28 compared with a range of 19 for 2010.

The scores for 2011 were positively skewed with a median of 19. The range of scores in 2011 was 21 so the scores were not as variable as 2009 but slightly more variable than 2010.

Overall the scores in 2011 were higher; the median was higher than the other two years and 2011 had the highest minimum score(11) and the highest maximum score (32).

6. a. It means that 65% of the males in PNG can read and write.

b. PNG, Malaysia, South Africa and the Philippines.

c. India, Afghanistan and Egypt.

d. Afghanistan

e.

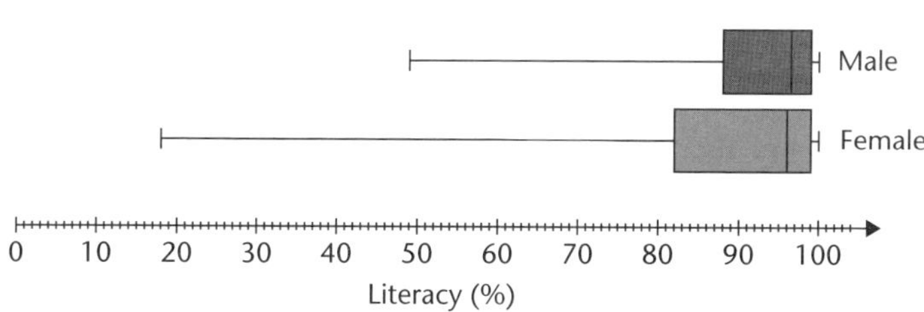

f. The distribution of the data for both males and females is negatively skewed indicating a much larger spread of literacy rates below the median literacy rate. The spread of the data for females is larger with a range of 82 compared to a range of 51 for males. The median literacy rate for males was slightly higher than that for the females; 97% compared to 96%. For females the middle 50% of the countries had a literacy rate that was more variable; an IQR of 17 compared to an IQR of 11 for males.

Overall it can be said (based on this sample of 14 countries) that female literacy is slightly lower than male literacy and that the female literacy is more variable.

Unit 11.3 Activity 6A: Statistical investigation (page 305)

1. **a.** 280

   b. Consult teacher.

   c.

Output (avocadoes)	Frequency
0–40	0
41–80	4
81–120	11
121–160	9
161–200	5
201–240	4
241–280	2

   d.

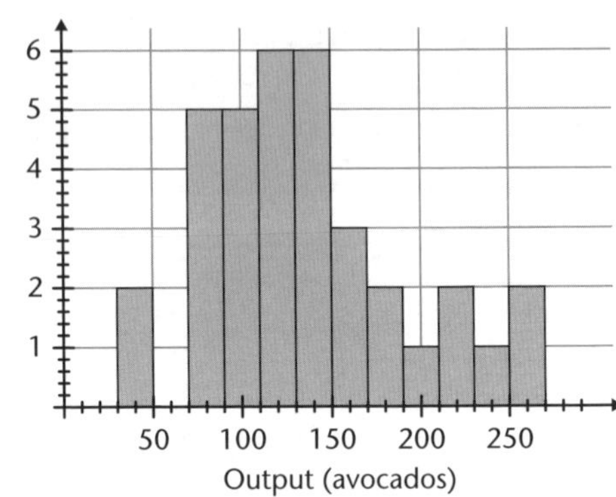

   e. Mean = 140; median = 128

   The average number of avocadoes per tree was 140. The median number of avocadoes per tree was 128 indicating that half of the trees produced 128 avocadoes or more. The difference in value of these two statistics indicates that the distribution of avocadoes per tree was positively skewed with some large values distorting the mean.

   f. The range of the data = 260 – 44 = 216

   The inter-quartile range is 169 – 97 = 72

   There is a difference of 216 avocadoes between the tree that produced the most and the tree that produced the least number of avocadoes.

   The middle 50% of trees produced between 97 and 167 avocadoes; a spread of 72.

   g.

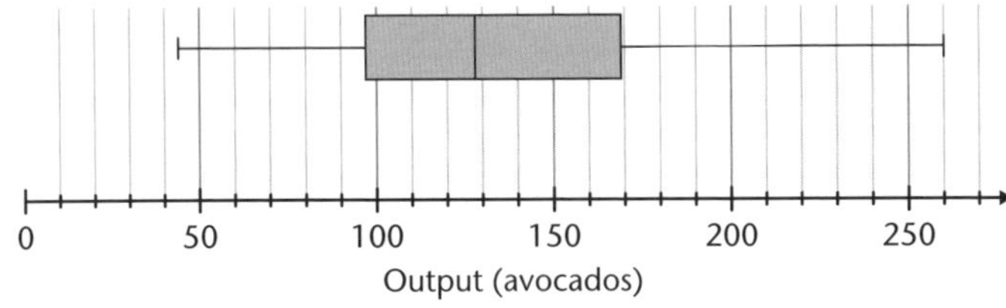

   The boxplot shows that the data is positively skewed; the higher half of the data having a greater spread than the lower half of the data. This means that there was greater variety in the number of avocadoes produced in the half of the trees that produced more avocadoes.

   h.

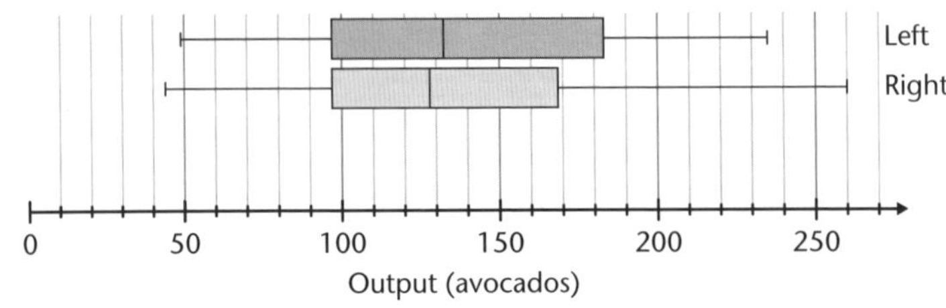

The median for the right side of the road is slightly less than the median for the trees on the left side. This does not support the agricultural adviser's claim.

The spread of the output from the trees on the right of the road is larger; the highest yield from a tree in the sample was from a tree on the right side of the road as was the tree with the lowest yield.

There is no substantial evidence to support the advisor's claim.

i.

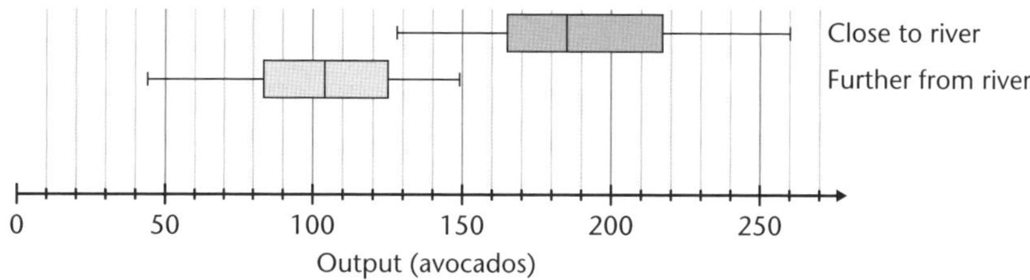

The trees that were planted closer to the river (numbered at least 190) have a median output of 185 compared to the median output of the other trees (numbers less than 190) of 104.

More than 75% of the trees planted close to the river had an output that was greater than the maximum output (149) of the trees planted further from the river.

These statistics substantially support the villager's statement that the closer to the river a tree is planted the greater its output.

Unit 11.3 Activity 7A: Factorial notation and permutations (page 310)

1. **a.** 40320 **b.** 3603600 **c.** 4838400

2. **a.** 42 **b.** $\frac{4}{3}$ **c.** $\frac{1}{6}$ **d.** 9900

e. 56 **f.** $\frac{1}{24}$ **g.** 120 **h.** 55

3. **a.** 210 **b.** 24 **c.** 4 **d.** 7

4. **a.** **i.** W, O, R, D

ii. WO, WR, WD, OW, RW, DW, OR, RO, OD, DO, RD, DR

b. **i.** ${}^4P_1 = 4$ **ii.** ${}^4P_2 = 12$

c. 24

5. 120 **6.** 720 **7.** 120

8. **a.** 210 **b.** No

Unit 11.3 Activity 7B: Further arrangements (page 313)

1. MOON, MONO, MNOO, NOOM, NOMO, NMOO, OMON, ONOM, OOMN, OONM, OMNO, ONMO

2. 180 **3.** 35

4.

A	A	A	A	A	A
D ◯ B	D ◯ C	B ◯ D	C ◯ D	C ◯ B	B ◯ C
C	B	C	B	D	D

5. 120

6. a. i. 720 **ii.** 48 **iii.** 240 **b.** 480

7. a. 360 **b.** 1296 **c.** 60 **d.** 180 **e.** 120

8. a. 5040 **b.** 576 **c.** 288 **d.** 144

9. a. 67600 **b.** 17576000 **c.** 17558424

10. a. 1-2-3-4, 3-3-3-6 for examples **b.** 1296 **c.** 6

Unit 11.3 Activity 7C: Combinations (page 317)

1. a. 45 **b.** 126 **c.** 20 **d.** 165

e. 364 **f.** 35 **g.** 495 **h.** 70

2. a. 1 **b.** 9 **c.** 6 **d.** 55

e. 78 **f.** 21 **g.** 12 **h.** 28

3. Missing subscript is 5 **4.** 1365 **5.** 10

6. 8145060 **7.** 2598960

8. a. 5 **b.** 4 **c.** 6 **d.** 8 **e.** 6

Unit 11.3 Activity 7D: Combinations with restrictions (page 319)

1. 211680 **2.** 72 **3.** Yes

4. a. 220 **b.** 60 **c.** 10 **d.** 4

e. 30 **f.** 80

5. a. 5400 **b.** 4320 **c.** 5076

6. a. 2598960 **b.** 65780 **c.** 1287 **d.** 192

e. 24 **f.** 193 **g.** 2860

7. a. 1287 **b.** 495 **c.** 525

d. 1056 **e.** 825

8. 26

Unit 11.3 Activity 8A: Probability and sample spaces (page 323)

1. a. $\frac{9}{16}$ **b.** $\frac{7}{16}$ **c.** 1 **d.** 0

2. a. $\frac{9}{16}$ **b.** $\frac{9}{15}$ **c.** $\frac{7}{16}$ **d.** $\frac{6}{15}$

3. a. $\frac{8}{20}$ **b.** $\frac{5}{20}$ **c.** $\frac{7}{20}$ **d.** $\frac{12}{20}$

4. a. $\frac{5}{20}$ **b.** $\frac{5}{19}$ **c.** $\frac{7}{20}$ **d.** $\frac{6}{19}$

5. a. $\frac{2}{7}$ **b.** $\frac{1}{7}$ **c.** $\frac{3}{7}$ **d.** $\frac{4}{7}$ **e.** 1 **f.** 0

6. a. $\frac{7}{12}$ **b.** $\frac{1}{12}$ **c.** $\frac{2}{12}$ **d.** $\frac{4}{12}$ **e.** $\frac{2}{12}$ **f.** $\frac{5}{12}$

7. a. $\frac{4}{9}$ **b.** $\frac{5}{9}$ **c.** 1 **d.** 0 **e.** $\frac{3}{9}$ **f.** $\frac{3}{9}$ **g.** $\frac{1}{9}$ **h.** $\frac{5}{9}$

8. a. $\frac{4}{52}$ **b.** $\frac{13}{52}$ **c.** $\frac{1}{52}$ **d.** $\frac{16}{52}$ **e.** $\frac{36}{52}$ **f.** $\frac{3}{52}$ **g.** $\frac{12}{52}$

9. a. i. $\{M_1, M_2, M_3, M_4, W_1, W_2\}$ **ii.** $\frac{1}{3}$ **iii.** $\frac{1}{3}$ **b. i.** $\frac{4}{9}$ **ii.** $\frac{8}{9}$ **iii.** $\frac{2}{3}$

c. i. $\frac{2}{5}$ **ii.** $\frac{2}{5}$ **iii.** $\frac{1}{3}$ **iv.** $\frac{1}{15}$

10. a. 53 **b.** 13 **c.** 160 **d.** 121 **e.** $\frac{121}{160}$ **f.** $\frac{60}{160}$

g. $\frac{40}{160}$ **h.** $\frac{39}{160}$ **i.** $\frac{21}{160}$ **j.** $\frac{128}{160}$

Unit 11.3 Activity 8B: Probability trees (page 327)

1. a. i. $\frac{4}{9}$ ***(M)*** **ii.** $\frac{2}{9}$ ***(M)*** **iii.** $\frac{5}{9}$ ***(M)***

b. i. $\frac{1}{2}$ ***(M)*** **ii.** $\frac{1}{12}$ ***(M)*** **iii.** $\frac{1}{2}$ ***(M)***

2. a.

start
$\frac{1}{2}$ 10
$\frac{1}{2}$ 10 — $\frac{1}{2}$ 10, $\frac{1}{2}$ 5
$\frac{1}{2}$ 5 — $\frac{1}{2}$ 10, $\frac{1}{2}$ 5
$\frac{1}{2}$ 5
$\frac{1}{2}$ 10 — $\frac{1}{2}$ 10, $\frac{1}{2}$ 5
$\frac{1}{2}$ 5 — $\frac{1}{2}$ 10, $\frac{1}{2}$ 5 ***(M)***

b. i. $\frac{1}{8}$ ***(M)*** **ii.** $\frac{3}{8}$ ***(M)*** **iii.** $\frac{7}{8}$ ***(M)***

iv. $\frac{2}{8}$ ***(M)*** **v.** $\frac{7}{8}$ ***(M)***

3. a.

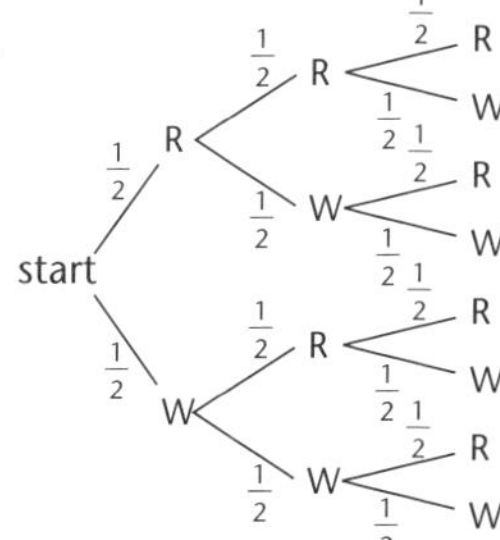

b. i. $\frac{1}{8}$ ***(M)*** **ii.** $\frac{1}{8}$ ***(M)***

iii. $\frac{2}{8}$ ***(M)*** **iv.** $\frac{4}{8}$ ***(M)***

4. a. $\frac{4}{9}$ **(M)** **b.** $\frac{1}{3}$ **(M)** **c.** $\frac{5}{9}$ **(M)** **d.** $\frac{2}{3}$ **(M)**

5. a. $\frac{240}{702}$ **(M)** **b.** $\frac{132}{702}$ **(M)** **c.** $\frac{272}{702}$ **(M)** **d.** $\frac{492}{702}$ **(M)** **e.** $\frac{312}{702}$ **(M)**

Unit 11.3 Activity 9A: Probabilities of one event and/or another event (page 331)

1. 0.92 **2.** 0.77 **3.** 0.5 **(M)**

4. a. 0.55 **(M)** **b.** 0.15 **(M)**

5. 0.37 **6.** 0.55 **7.** 0.6

8. a. 0.7 **b.** 0.3 **c.** 0.21 **d.** 0.06

e. 0.27 **f.** 0.76 **g.** 0.79

9. a. A = {4, 5, 6}; B = {1, 2, 3}; C = {5}; D = {2, 4, 6}; E = {1, 3, 5}; F = {3}

b. i. Y **ii.** N **iii.** N **iv.** N

v. Y **vi.** N

c. i. $\frac{1}{2}$ **ii.** $\frac{1}{2}$ **iii.** $\frac{1}{6}$ **iv.** $\frac{1}{6}$

v. 1 **vi.** $\frac{1}{2}$ **vii.** $\frac{2}{3}$ **viii.** $\frac{1}{2}$

10. a. i. Y **ii.** N **iii.** N **iv.** Y

b. i. $\frac{1}{4}$ **ii.** $\frac{1}{4}$ **iii.** $\frac{1}{13}$ **iv.** $\frac{1}{2}$

c. i. $\frac{1}{13}$ **ii.** $\frac{1}{2}$ **iii.** $\frac{1}{13}$ **iv.** 0

v. $\frac{2}{13}$ **vii.** $\frac{1}{13}$ **viii.** $\frac{1}{2}$

Unit 11.3 Activity 10A: Independent events (page 333)

1. **a, b** **2.** none are independent **3.** There are many possible answers for each.

4. a. **i, ii** are independent **b. i.** $\frac{4}{100}$ **ii.** $\frac{16}{100}$

5. The probability of A and B occurring is not equal to the product of the probability of A and the probability of B.

Unit 11.3 Activity 10B: Conditional probability (page 335)

1. a. 0.3 **(M)** **b.** 0.95 **(M)** **c.** 0.6 **(M)** **d.** 0.035 **(M)**

2.

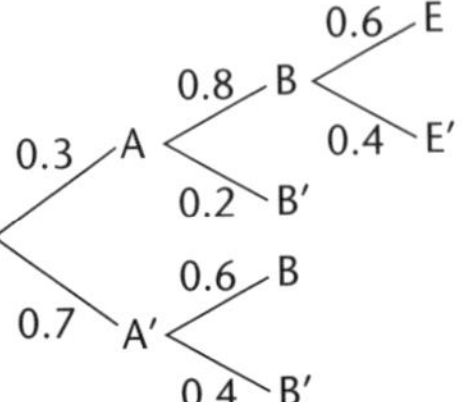

(M)

a. 0.4 **(M)**

b. 0.4 **(M)**

c. 0.42 **(M)**

d. 0.144 **(M)**

3. a.

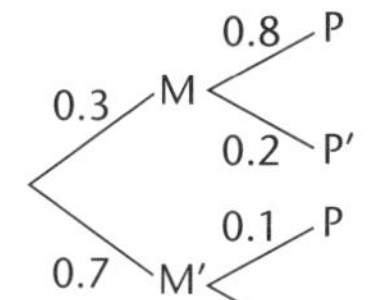

(*M*)

b. 0.9 **(*M*)**

c. 0.59 **(*M*)**

4. a. 0.1 **(*M*)** **b.** 0.4 **(*M*)** **c.** 0.9 **(*M*)** **d.** 0.81 **(*M*)** **e.** 0.06 **(*M*)**

5. a. 0.51 **(*M*)** **b.** 0.4 **(*M*)**

6. a. 0.45 **(*M*)** **b.** $\frac{32}{45}$ **(*M*)** **c.** 0.18 **(*M*)** **d.** $\frac{18}{55}$ **(*M*)**

Unit 11.3 Activity 10C: Probability and expected values (page 338)

1. a. 0.02 **(*M*)** **b.** 0.28 **(*M*)** **c.** 14 **(*M*)**

2. a. $\frac{1}{4}$ **(*M*)** **b.** $\frac{5}{18}$ **(*M*)**

3. a.

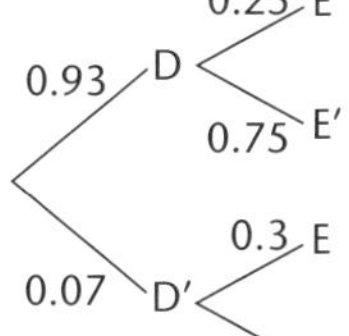

(*M*)

b. 0.2325 **(*M*)** **c.** 0.2535 **(*M*)**

4. a.

0.75 H
0.5 G
0.25 H′
0.4 H
0.5 G′
0.6 H′ **(*M*)**

b. 0.375 **(*M*)** **c.** 0.3 **(*M*)**

Unit 11.4 Geometry

Unit 11.4 Activity 1A: Angles and triangles (page 341)

1. *In question 1, the finding of the second lettered angle is Achieved without reasons and Merit with reasons.*

a. a = 92° (∠ sum Δ) **b.** b = 46° (∠ sum isos Δ)

c. c = 74° (ext ∠ Δ) **d.** d = 70° (∠ sum isos Δ), e = 125° (ext ∠ Δ)

e. f = 81° (∠ sum Δ), g = 45° (ext ∠ Δ) **(*A*)**

f. h = 84° (corr ∠s, // lines), i = 22° (ext ∠ Δ)

g. d = 70° (∠ sum isos Δ), e = 110° (ext ∠ Δ)

h. f = 71° (∠ sum Δ), g = 51° (ext ∠ Δ)

i. h = 40° (alt ∠s, // lines), i = 148° (ext ∠ Δ)

2. ∠ sum Δ	**3.** x	**4.** Let ∠BFC = c
QT	x	∴ ∠DFE = c (vert opp ∠s)
40°	base ∠s isos Δ **(M)**	$x = a + c$ (ext ∠ Δ)
alt ∠s = **(M)**		$y = b + c$ (ext ∠ Δ)
		∴ $x = y$ **(E)**
		(Other possible solutions exist.)

Unit 11.4 Activity 2A: Similar triangles (page 345)

1. **a.** B & C **b.** B & C **c.** A & B **d.** A & B
2. **a.** AAA **b.** AAA **c.** SSS **d.** SAS
3. Discuss with your teacher.
4. $x = \frac{800}{21} \approx 28.1\text{cm}$; $y = 45$ cm
5. **a.** $x = 3\frac{8}{9}$ m; $y = 7.2$ m
   b. $a = 12.5$ cm; b = 9.6 cm
   c. $p = 4$ m; $q = 7$ m
   d. $x = 6$ cm; $y = 6$ cm
   e. $x = 8$ m
   f. $x = 16$ cm; $y = 7.2$ cm
   g. x = 3.6 m
   h. $x = 9\frac{1}{3}$ cm; $y = 4.5$ cm

Unit 11.4 Activity 3A: Congruent triangles (page 351)

1. b.
2. a. and c.; b. and d.; e. and f.
3. Discuss with your teacher.

Unit 11.4 Activity 3B: Congruent triangles and other figures (page 353)

No answers are given for this Activity—discuss your proofs with your teacher.

Unit 11.4 Activity 4A: Pythagoras' Theorem (page 357)

1. **a.** 4.47 **(A)** **b.** 8.94 m **(A)** **c.** 7.62 cm **(A)**
   d. 9.90 **(A)** **e.** 6.93 m **(A)** **f.** 30.74 **(A)**
   g. 8.31 cm **(A)** **h.** 17 mm **(A)** **i.** 7.55 cm **(A)**
2. 7.86 m **(A)**
3. 139.6 cm **(A)**
4. 13.27 mm **(A)**
5. **a.** Right-angled. **b.** Not right-angled. **c.** Right-angled.

Unit 11.4 Activity 4B: Solving problems using Pythagoras' Theorem (page 359)

1. 240 m **(M)**
2. 28.84 m **(A)**
3. **a.** When the walls are at right angles, the lengths of the walls and the diagonal will obey Pythagoras' Theorem. **(M)**

   b. 9.44 m **(A)**
4. 7.16 m **(M)**
5. 82 cm **(M)**
6. Yes—maximum possible length is 1.08 m. **(M)**
7. 38.47 m **(A)**
8. 9.17 m **(M)**
9. 66.4 m **(A)**
10. **a.** 5.3 cm **(M)** **b.** 3.6 cm **(M)**

Unit 11.4 Activity 5A: Trigonometric ratios (page 362)

1. a. 13 **b.** 12 **c.** 12 **d.** $\frac{12}{13}$ **e.** $\frac{5}{12}$ **f.** $\frac{12}{13}$ **g.** $\frac{5}{13}$

2. a. 10 **b.** 8 **c.** 6 **d.** $\frac{8}{6}$ **e.** $\frac{6}{10}$ **f.** $\frac{8}{10}$ **g.** $\frac{8}{10}$

Unit 11.4 Activity 5B: Using your calculator (page 364)

1. a. 0.3420 **b.** 0.9848 **c.** 2.7475 **d.** 0.7880 **e.** 1.1106 **f.** 0.4384
g. 0.8320 **h.** 4.8716 **i.** 0.2147

2. a. 40.0° **b.** 40.0° **c.** 60.0° **d.** 50.0° **e.** 38.4° **f.** 26.9°
g. 28.9° **h.** 51.0° **i.** 6.7°

Unit 11.4 Activity 5C: Finding unknown sides (page 365)

1. a. 5 **(A)** **b.** 7.2 **(A)** **c.** 6.7 **(A)** **d.** 14 **(A)** **e.** 14.3 **(A)**
f. 13.5 **(A)** **g.** 14.2 **(A)** **h.** 7.4 **(A)** **i.** 16.2 **(A)**

2. a. 13.7 **(M)** **b.** 4.4 **(M)** **c.** 14.8 **(M)** **d.** 41.5 **(M)** **e.** 6.57 **(M)**
f. 12.7 **(M)** **g.** 19.2 **(M)** **h.** 13.8 **(M)** **i.** 52.9 **(M)**

Unit 11.4 Activity 5D: Finding unknown angles (page 367)

1. a. 34.8° **(A)** **b.** 53.1° **(A)** **c.** 29.9° **(A)** **d.** 56.3° **(A)** **e.** 36.9° **(A)**
f. 53.1° **(A)** **g.** 38.9° **(A)** **h.** 51.3° **(A)** **i.** 60.3° **(A)**

2. a. $s = 58.0°$, $t = 32.0°$ **(A)** **b.** $u = 74.5°$, $v = 15.5°$ **(A)**
c. $w = 52.3°$, $x = 37.7°$ **(A)** **d.** $y = 19.0°$, $z = 71.0°$ **(A)**
e. $a = 35.7°$, $b = 54.3°$ **(A)** **f.** $c = 55.4°$, $d = 34.6°$ **(A)**

Unit 11.4 Activity 5E: Problem solving using trigonometry (page 369)

1. 56.4° **(A)**
2. 1.33 m **(A)**
3. 508 m **(M)**
4. **a.** 176 m **(M)** **b.** 19.4° **(M)**
5. 2.16 m **(A)**
6. 119 m **(M)**
7. **a.** 10.4 m **(M)**

   b.
   - Measurement error.
   - Mistake in calculation when working out height.
   - Equipment used not suitable for accurate measurement. **(M)**
8. **a.** 5.0 m **(M)** **b.** 9.05 m **(M)**
9. 49.3° **(E)** **10.** 139.5° **(E)**

Unit 11.4 Activity 6A: 3-D trigonometry (page 374)

1. 1.27 m **(M)**
2. **a.** ∠AHE **b.** ∠CEG **c.** ∠HBD **d.** ∠DFC
3. **a.** 6.7 **(M)** **b.** ∠CDF = 26.6° **(M)**
4. **a.** 20 **(M)** **b.** 21.54 **(M)** **c.** ∠CEG = 21.8° **(M)**
5. **a.** 56.3° **(M)** **b.** 16.97 m or 1 697 cm **(M)**
6. 5.7 cm **(E)**
7. 6.9 cm **(M)**
8. 29.8° **(E)**

Unit 11.4 Activity 6B: Further 3-D trigonometry (page 377)

1. **a.** ∠AHE or ∠BGF **b.** ∠ADE or ∠BCF **c.** ∠DEH or ∠CFG **d.** ∠ACB or ∠EGF
2. 56.3° **(M)**
3. 90° **(M)**
4. ∠BCA = 32.0° **(M)**
5. ∠EXH = 71.6° **(M)**
6. 52.7° **(E)**
7. 15 m (other alternative was 17.46 m) **(E)**
8. **a.** 20.83 m (to nearest cm) **(E)** **b.** 15.89 m along roof/wall line from end at T. **(E)**

Unit 11.4 Activity 7A: The cosine rule and applications (page 380)

1. **a.** 5.0 **(A)** **b.** 60° **(A)**
2. **a.** 12.4 **(A)** **b.** 51.0° **(A)**

3. **a.** 6.0 **(A)** **b.** 60.5° **(A)**
4. **a.** 6.5 **(A)** **b.** 132.5° **(A)**
5. **a.** 78.54° **(A)** **b.** 1.9 m **(A)**
6. **a.** 328 m **(A)** **b.** 79° **(A)**

Unit 11.4 Activity 7B: The sine and area rules and applications (page 382)

1. **a.** 7.3 **b.** 9.7 **c.** 25 **(A)**
2. **a.** 6.0 **b.** 8.7 **(A)**
3. **a.** 22.4 **b.** 14.4 **c.** 233.2 **(A)**
4. **a.** 187.86 km **b.** E48.8°N **(A)**
5. 162 cm **(A)**
6. **a.** 3.5 cm **b.** 7.3 cm^2 **c.** 374 cm^3 **(A)**
7. **a.** K1 114 **b.** 1 282.6 m^2 **(M)**
8. **a.** 5.27 km **b.** 5.44 km **(M)**
9. **a.** W46.2°N **b.** 15.07 km **(M)**
10. $\frac{2}{\sqrt{3}}$ **(M)**
11. **a.** 53.1° **b.** 199.2 m **c.** 2 220 m^2 **(M)**
12. 5.2 km **(M)**

Unit 11.4 Activity 8A: Circle geometry (page 388)

Exercises with one step of logic are included for practice and are ungraded (A, M, E).

1. **a.** a = 35° (∠s same arc) **b.** b = 92° (∠ at centre) **c.** c = 21° (∠ at centre)
   d. d = 90° (∠ in semi-circle), e = 70° (base ∠ isos Δ), f = 140° (ext ∠ Δ)
   e. g = 90° (∠ in semi), h = 25° (∠ in semi), i = 65° (base ∠ isos Δ), j = 130° (∠ sum Δ)
   f. k = 32° (base ∠ isos Δ), l = 64° (∠ on line), m = 90° (∠ in semi), n = 50° (∠ sum Δ)
2. **a.** a = 90° (∠ in semi-circle), b = 28° (∠s same arc), c = 62° (∠ sum Δ) **(M)**
   b. d = 26° (base ∠ isos Δ), e = 52° (∠ at centre) **(M)**
   c. f = 56° (∠s same arc), g = 34° (∠ in semi-circle, sum ∠s Δ) **(M)**
   d. r = 80° (∠ at centre), s = 50° (base ∠ isos Δ) **(M)**
   e. t = 42° (alt ∠s // lines), u = 42° (∠s same arc), v = 96° (∠ sum Δ, vert opp ∠s) **(M)**
   f. w = 40° (∠s same arc), x = 90° (∠ in semi), y = 130° (ext ∠ Δ),
   z = 25° (base ∠ isos Δ) **(M)**
3. AO = OD (radii)
   ∠AOD = 52° (∠s in isos Δ)
   ∠BOC = 52° (symmetry)
   ∠DOC = 76° (∠s on line) **(M)**
4. ∠OQS = 35° (base ∠ isos Δ OQS)
   ∠QOS = 110° (∠ sum Δ)
   ∠QPS = 55° (∠ at centre) **(M)**

5. **a.** $\angle OBA = 63°$ (base $\angle$ isos Δ), $\angle OBC = 27°$ ($\angle$ in semi) **(M)**

b. $\angle AOD = 126°$ ($\angle$s on line), $\angle ODA = 27°$ (base $\angle$ isos Δ) **(M)**

c. Alternate angles $\angle OBC$ and $\angle ODA$ are equal so AD is parallel to BC. **(M)**

6.

$$x = 2a \text{ (}\angle\text{ at centre)}$$
$$x = 2b \text{ (}\angle\text{ at centre)}$$
$$\therefore 2a = 2b$$
$$\therefore a = b$$ **(E)**

7.

$$\angle BAO = a \text{ (base }\angle\text{ isos }\Delta)$$
$$\angle AOD = 2a \text{ (ext }\angle\ \Delta)$$
$$\angle BCO = b \text{ (base }\angle\text{ isos }\Delta)$$
$$\angle DOC = 2b \text{ (ext }\angle\ \Delta)$$
$$\angle AOC = \angle AOD + \angle DOC$$
$$= 2a + 2b$$
$$= 2(a + b)$$
$$= 2\angle ABC$$ **(E)**

8.

$$x + y = 180° \quad (\angle\text{s on on line})$$
$$\angle ACO = \frac{1}{2}x \quad (\angle\text{ at centre})$$
$$\angle BCO = \frac{1}{2}y \quad (\angle\text{ at centre})$$
$$\angle ACB = \angle ACO + \angle OCB$$
$$= \frac{1}{2}x + \frac{1}{2}y$$
$$= \frac{1}{2}(x + y)$$
$$= \frac{1}{2} \times 180°$$
$$= 90°$$ **(E)**

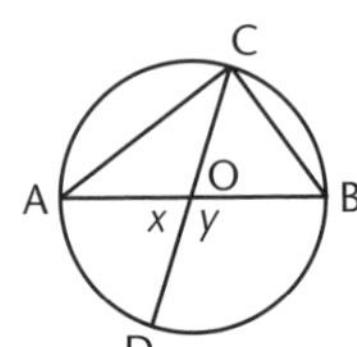

Unit 11.4 Activity 8B: Cyclic quadrilaterals (page 392)

No grading given for one-step problems.

1. **a.** $i = 80°$ (opp $\angle$s cyc quad), $e = 84°$ (ext $\angle$ cyc quad)

b. $j = 58°$ (rad $\perp$ tan), $k = 122°$ (opp $\angle$s cyc quad)

c. $m = 65°$ (rad $\perp$ tan), $n = 130°$ ($\angle$ sum isos Δ)

d. $t = 75°$ ($\angle$s same arc), $u = 61°$ (ext $\angle$ cyc quad)

e. $v = 48°$ (rad $\perp$ tan), $w = 144°$ ($\angle$ sum isos Δ), $x = 72°$ ($\angle$ at centre),
$y = 108°$ (opp $\angle$s cyc quad)

f. $a = 90°$ (rad $\perp$ tan), $b = 48°$ ($\angle$s Δ), $c = 42°$ (symmetry), $d = 48°$ (symmetry)

2. **a.** $\angle ADC = 40°$ (alt $\angle$s // lines), $x = 50°$ (rad $\perp$ tan) **(M)**
$w = 80°$ ($\angle$ at centre), $\angle ODB = 50°$ (base $\angle$ isos Δ), $z = 40°$ (rad $\perp$ tan) **(M)**

b. $\angle OAC = 40°$ (symmetry), $\angle OCA = 90°$ (rad $\perp$ tan), $a = 50°$ ($\angle$ sum Δ) **(M)**
$\angle COB = 100°$ (symmetry), $b = 50°$ ($\angle$ at centre) **(M)**

c. $\angle OBT = \angle OAT = 90°$ (rad ⊥ tan), $c = 130°$ (∠ sum quad) **(M)**

$\angle BCA = 65°$ (∠ at centre), $d = 30°$ ($\angle BCO + d = 65°$) **(M)**

d. $\angle ADC = 80°$ (∠ at centre), $e = 80°$ (ext ∠ cyclic quad) **(M)**

e. $\angle SQR = 55°$ (∠s on same arc), $f = 35°$ (∠ in semi) **(M)**

$\angle PSQ = 55°$ (alt ∠s // lines), $g = 35°$ (angles on line) **(M)**

f. $\angle TWU = 48°$ (alt ∠s // lines), $h = 42°$ (∠ in semi (TWV)) **(M)**

$\angle WUV = 90°$ (corr ∠s and rad ⊥ tan), $i = 48°$ (∠ sum Δ) **(M)**

3. Reflex angle $DOB = 216°$ (∠ at centre)

$a = 144$ (∠s at point) **(M)**

$\angle DAB = 72°$ (opp ∠s cyclic quad)

$\angle OAB = 60°$ (Δ OAB is equilateral)

$b = 12°$ **(M)**

4. a. A, B, D, E **(M)**

b. A, B, C, D and A, B, D, E **(M)**

c. B, C, E, F and A, B, D, F **(M)**

d. A, B, C, D **(M)**

e. A, B, C, D **(M)**

f. A, B, F, E or B, C, D, E **(M)**

5. a. and **b.**

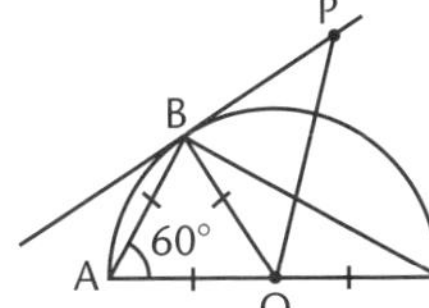

c. i. Right-angled. **ii.** Isosceles.

iii. Right-angled. **iv.** Equilateral.

d. No, because $\angle PBO = 90°$ no matter what position P is. Hence the other two angles of ΔOBP are acute. **(M)**

6. $\angle RCB = \angle BQR$ (∠s same arc)

$\angle BQR = \angle PAB$ (ext ∠ cyc quad)

$\angle RCB = \angle PAB$

PA is parallel to CR (alt ∠s =) **(E)**

7. $\angle OAC = 90° - x$ (rad ⊥ tan)

$\angle OCA = 90° - x$ (OAC isos Δ)

$\angle AOC = 180° - 2(90° - x)$ (∠s Δ)

$= 180° - 180° + 2x$

$= 2x$

$\angle AOC = 2y$ (∠ at centre)

$\therefore\ 2x = 2y$

$x = y$ **(E)**

8. Obtuse $\angle BOD = 2C$ (∠ at centre)

Reflex $\angle BOD = 2A$ (∠ at centre)

$2A + 2C = 360°$ (∠s at point)

$A + C = 180°$ **(E)** [dividing by 2]

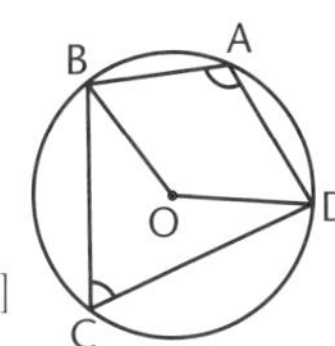

Unit 11.4 Activity 9A: Circles centred at (a, b) (page 395)

1. 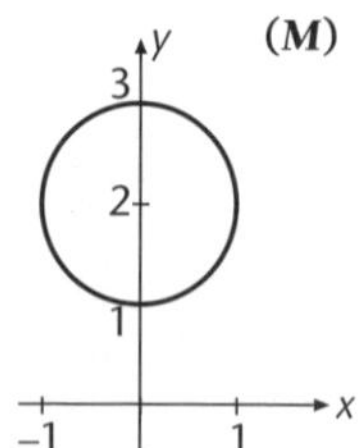

(M)

2. 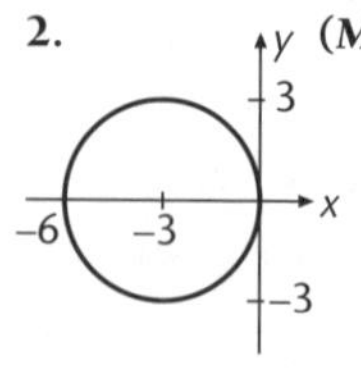

(M)

3. 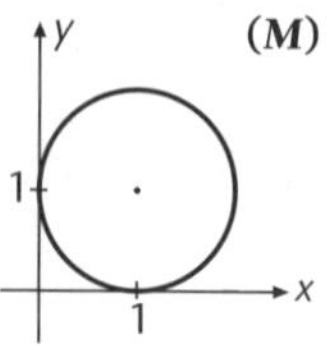

(M)

4. 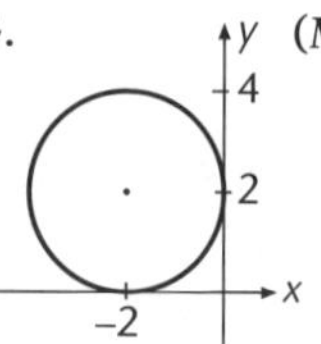

(M)

5. 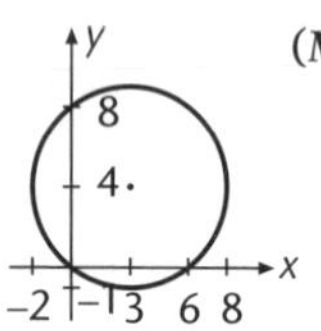

(M)

6. a. $(x-2)^2 + (y-2)^2 = 4$ b. $(x+1)^2 + (y-3)^2 = 16$
   c. $(x-4)^2 + y^2 = 25$

7. a. $(x-2)^2 + (y-3)^2 = 25$ **(M)** b. $(x+3)^2 + (y-1)^2 = 16$ **(M)**
   c. $(x+2)^2 + (y+3)^2 = 4$ **(M)** d. $(x-1)^2 + (y-4)^2 = 9$ **(M)**
   e. $(x-1)^2 + (y-2)^2 = 17$ **(M)** f. $(x-4)^2 + (y-2)^2 = 13$ **(M)**
   g. $(x-\frac{1}{2})^2 + (y-1)^2 = \frac{50}{4}$ **(M)** h. $(x-1)^2 + y^2 = 8$ **(M)**

8. a. (–1, 3), 4 b. (2, 4), 5
   c. (4, –1), 6 d. (3, 4), 7

9. a. $(x-1)^2 + (y-2)^2 = 25$ **(M)** b. $(x-3)^2 + (y-7)^2 = 100$ **(M)**
   c. $(x+5)^2 + (y-2)^2 = 169$ **(M)** d. $(x-2)^2 + (y-3)^2 = 2$ **(M)**

10. Shane **(M)**

Unit 11.4 Activity 9B: Intersection of a circle and a line (page 398)

1. a. (0, 2) b. (1.73, 1), (–1.73, 1) c. (2, 0)
   d. (–1, 1.73), (–1, –1.73) e. (1.32, 1.5), (–1.32, 1.5) f. No intersection
   g. (1.41, 1.41), (–1.41, –1.41) h. (0.89, 1.79), (–0.89, –1.79)
   i. (0.63, 1.90), (–0.63, –1.90) j. (1.79, 0.89), (–1.79, –0.89) **(E)**

2. a. (0.82, 1.82), (–1.82, –0.82) b. (1.82, 0.82), (–0.82, –1.82)
   c. (–1.27, –1.54), (0.47, 1.94) d. (0, 2), (–2, 0)
   e. (0, –2), (1.6, 1.2) **(E)**

3. Each is a tangent. **(E)**

4. $k = \pm\sqrt{18}$ [a very accurate graph will give $k = \pm 4.2$] **(E)**

5. There is no intersection. **(M)**

6. (–3, –3), (3, 5) **(M)**

Unit 11.4 Activity 10A: Arc length and area of circles (page 400)

1. **a.** 4.19 cm **b.** 41.89 cm **c.** 90.06 cm

2. **a.** 122.17 m^2 **b.** 698.13 m^2

3. **a.** 576.12 cm^2 **b.** 4450.42 cm^2

4. **a.** 98 cm, 599.96 cm^2 **b.** 93.44 m, 491.48 m^2

5. 9.72 cm

6. **a.** 6.66 m^2 **b.** 13.00 m

7. 1658.30 cm^2 **8.** 168.02 m^2

Unit 11.4 Activity 11A: Spheres, latitude and longitude (page 408)

1. 13.23 cm **2.** 11.62 cm **3.** 5.28 m

4. **a.** 3863 km **b.** 9124 km **c.** 735 km **d.** 3202 km

5. **a.** **i.** 1390 km **ii.** 1390 km

b. **i.** 8193 km **ii.** 7047 km

c. **i.** 17323 km **ii.** 13698 km

d. **i.** 9026 km **ii.** 8645 km

Glossary/Index

12-hour clock (68): Instrument for recording the passing of time in hours, minutes and seconds from midnight or noon. Morning times use am and afternoon times use pm.

24-hour clock (68): Instrument for recording the passing of time in hours, minutes and seconds from midnight, eg 1430 is 30 minutes after 1400 hours, ie half-past two in the afternoon.

absolute value ($|x|$) 214, 253–8): The size of a number (positive value). Also called the modulus.

acute angle (341): An angle between 0° and 90°.

acute-angled triangle (341): A *triangle* in which all angles are *acute*.

addition (+): An arithmetic operation. Adding rational expressions **(151–2)**.

adjacent side (339, 343, 361–2): The side of a triangle, not the hypotenuse, which is next to a particular angle.

algebra (89–140): The use of numbers and *variables* to solve problems.

algebraic equation: An *equation* with at least one *variable*.

algebraic expression (50–1, 89–92, 97, 141–8, 172–4): A collection of *algebraic* (and maybe *numerical*) *terms*, linked by + and –.

algebraic fraction (93–6, 105, 149): A rational expression with variables.

algebraic method (89, 97, 109, 131, 202–3, 396): Solving a problem using algebraic modelling and techniques.

algebraic term (89–93): An algebraic term consists of a combination of a *coefficient*, *variable*, and *exponent*. Addition and subtraction **(91–2)**; multiplication and division **(92)**.

alternate angles: The angles between parallel lines and on different sides of the *transversal*.

altitude (213–14): A line from the *vertex* of a triangle which meets the opposite side at *right angles*.

analysis (293–8): The process of deriving information from *data* in a *statistical investigation*.

and (329): The event *A and B* means the outcomes which are in *both* event *A* and event *B*.

angle (∠ or ^): A turning, usually measured in degrees.

angle between a line and plane (372–3): The *angle* between a line and its projection on the plane.

angle between two planes (375–6): The largest *angle* between two *planes*.

angle of depression (369): Angle from the horizontal *down* to an object.

angle of elevation (369): Angle from the horizontal *up* to an object.

apex (373): *Vertex* of pyramid, or other solid.

approximate answer (32): A less accurate answer which is close in value to the actual answer.

arc (349–50, 385–7, 399): Part of the circumference of a circle; arc length **(399)**.

area (74–5, 125, 155–6, 244, 410): The amount of 2-dimensional space an object covers. Units are m^2, cm^2 or *hectare*; Area of a *sector* **(399)**. Of a triangle **(381)**. Of a circle **(244, 399)**. Of a segment **(400)**.

area rule (381): A relationship used to find the area of any triangle given the length of two sides and the angle between them.

average (272, 305): A measure of the central or middle value of a *data* set, eg the *median*, *mean* and *mode*.

axis (plural **axes**) **(126, 137):** One of a pair of *number lines* which intersect at right angles on a 2-dimensional graph.

axis of symmetry (115, 238–9, 340, 358, 390): A line about which an object may be folded so that one half sits exactly on the other half. Also called a mirror line.

bar graph (259–61, 266, 271): A data display using bars to represent the frequency of the data. Also called a *column graph*.

base (47, 54, 59): The number being raised to a power in an index expression, eg in 5^2, the number 5 is the base. The unequal side of an *isosceles triangle* **(192, 340–1)**.

base angles (340–1): The equal angles on the *base* of an *isosceles triangle*.

base changing rule (62): Used to change *logarithms* from one base to another.

base unit (65): The fundamental units (eg metre) from which other units (eg centimetre and millimetre) are derived.

bearing: An *angle* measured clockwise from North, expressed using three digits, used for describing direction.

BEMA (or **BEDMAS) (5, 12, 24, 90, 95, 98):** A mnemonic used to indicate the order in which the operations are done.

bias (299–300): When the conclusions from a survey do not describe the characteristics of the whole population.

box plot (279–80): A form of data display. Also called *box-and-whisker plot*.

box-and-whisker plot (279): Data display showing five key features of the data distribution.

brackets (( ) or [ ]) **(5–7, 11–12, 42, 90, 94, 97–9, 135, 143):** Punctuation used to establish the order of basic operations.

calculators (6), use of with: integers **(9)**; squares and square roots **(11)**; higher powers and roots **(13)**; fractions **(20, 27–8)**; standard form **(35–6)**; graphs **(243)**; logarithms **(59)**; simultaneous equations **(129, 217)**; indices **(54, 169–70, 175)**; standard deviation **(277)**; random numbers **(302)**; factorials **(308–9, 315)**; trigonometric functions **(363–8)**.

cancelling (16, 21, 47, 94): Simplification of a fraction by dividing the numerator and denominator by the same factor.

capacity (67, 70): Amount of liquid a container holds, measured in *litres*, *millilitres*, etc.

Cartesian plane (137): The (x, y) coordinate number plane on which graphs are drawn.

Celsius (67): See *degrees Celsius*.

census (299): Investigation of an entire *population*.

cent (c) **(85):** A unit of *money*. 100 c = $1.

centigrade (67): See *degrees Celsius*.

centimetre (**cm**) **(66–7, 74, 80–1):** A smaller unit of length than the *metre*. 100 cm = 1 m.

central tendency (271–80): The central value of a sample or population.

centre (385, 399): The middle of a *circle*. See also *central tendency*.

changing the subject (157–8): Rearranging a *formula* to make a different *variable* the *subject*.

chord (386–7, 390, 400): A straight line joining any two points on the circumference of a circle.

circle (39, 184, 385–401): A set of points equidistant from a fixed point (its *centre*). Equation of **(395)**; intersecting with a line **(232, 396–7)**.

circumference (385, 399, 403, 411): The *perimeter* of a *circle*.

class interval (294): Range of values for a group in a table of *grouped data*.

coefficient (89, 91–2, 102): A number which multiplies the *variable*.

collinear (197): Three or more points which lie on the same straight line.

column graph (304): See *bar graph*.

common denominator (19, 94, 151–2): A denominator common to two or more fractions.

common factors (16, 77, 92, 101–5, 145, 149, 165): Factors which divide into each term of an expression without remainder.

comparing fractions (16): Ordering fractions by size.

compass bearing: Method of describing direction based on the directions North, South, East and West.

complement (A') (409): A' is the set of elements *not* in set A.

compress (242): Reduce, by a scale factor.

compression (242, 261): Shortening of the horizontal axis.

concertina (261): The broken line on the horizontal axis of a statistical graph to indicate that there is nothing plotted between zero and the first value.

conclusions (299): Opinions arrived at and justified by statistical analysis.

concurrent (217): Concurrent lines pass through the same point.

concyclic points (391): Points which all lie on the circumference of the same circle.

conditional probability (334–5): The probability of an event assuming another event has taken place.

cone (155, 159, 410): A circle-based *pyramid*.

congruent (235–9, 349–54): Identical in shape and size.

constant (39, 84–7, 89): A fixed amount which does not vary.

continuous data (261, 303): Data obtained by measuring (no gaps between possible values).

continuous population: A population of values which are continuous. See *continuous data*.

conversion (66, 74–5): Changing from one unit to another in the metric system.

coordinates (191): *Ordered pair* of numbers (x, y) describing the position of a point on a *Cartesian plane*.

correlation (265–6): Two variables are correlated if there is a pattern in a scatter diagram of ordered pairs. See *positively correlated*, and *negatively correlated*.

corresponding angles (343): The angles on the same side of the *transversal* and on the corresponding side of each parallel line.

cosine (cos) (361–3): A *trigonometric ratio*. Relative to a particular angle it is the ratio $\frac{\text{adjacent side}}{\text{hypotenuse}}$.

cosine rule (379–80): A relationship between three sides and one angle of a triangle.

counting numbers (1, 2, 5): The numbers 1, 2, 3, 4, . . . Also called the *natural numbers* (N).

cross-section (126, 377, 403): The view when a plane 'cuts' an object at right angles to the base.

cube (13, 410): Dice-shaped solid with 6 congruent (equal) square faces. Also called a regular *hexahedron*.

cube numbers (12, 13, 40): The cube of a number such as 5 is $5^3 = 5 \times 5 \times 5 = 125$.

cube root (13, 40, 52, 173): A number which when cubed gives the number under the $\sqrt[3]{\ }$ sign.

cubic centimetre (cm^3): Unit of measurement for *volume*. 1 cm^3 = 1 000 mm^3.

cubic functions (161, 188, 249): An equation of the form y = cubic expression.

cubic metre (m^3) (75): Base unit of measurement for *volume*. 1 m^3 = 1 000 000 cm^3.

cubic millimetre (mm^3): Unit of measurement for *volume*. 1 000 mm^3 = 1 cm^3.

cubics: Drawing graphs of **(249–50)**.

cuboid (371–2, 375–7, 410): A solid with 6 rectangular faces meeting at right angles (opposite faces congruent). Also known as a rectangular prism.

cyclic quadrilateral (390–1): A quadrilateral with all four corners on the *circumference* of a circle.

cylinder: Circle-based solid with uniform cross-section. Surface area **(410)**. Volume **(33, 84–5, 410)**.

data (259–338): Pieces of information; collected by observation, interviews or questionnaires.

data display (259–70, 279): Discrete, ungrouped data **(259–60)**; discrete, grouped data **(260)**; continuous data **(261)**; data with non-equal intervals **(262)**.

decagon (339): A 10-sided *polygon*.

decimal (1–3, 27–9): A number which has a *decimal point*.

decimal places (31–2, 39): The number of *digits* after the decimal point.

decimal point (3, 27): A full stop used to separate the whole part from the fractional part in a *decimal*.

decreasing function (187): A function whose 'y' values decrease as its 'x' values increase.

deduction (259): A conclusion drawn from a calculation or survey.

degree (161–3, 225): The highest power of the variable in any of the terms of a polynomial.

degrees (405): Unit used to measure angles. 1 degree (1°) is $\frac{1}{360}$ revolution.

degrees Celsius (°C) **(67):** A unit for *temperature*. Water freezes at 0°C and boils at 100°C. Also called *degrees centigrade*.

denominator (15): The integer or variable below the horizontal line in a *fraction*.

diagonal (339): A line segment joining two vertices of a polygon and which is not a side.

dial (70): A circular *scale* used in measurement.

diameter (39, 385): A straight line passing through the *centre* of a circle and joining two points on the *circumference*.

die (plural **dice) (301, 323, 330, 333):** A 6-sided cube usually labelled 1, 2, 3, 4, 5, 6.

difference: The result of subtracting two terms or numbers.

difference of two squares (104, 146): *Quadratic expression* of the form $a^2 - b^2$. Factorising **(104)**.

digits: The ten symbols, 0, 1, 2, 3, 4, 5, 6, 7, 8, 9, used in numbers.

direction: Line or course of a *vector's* movement.

discrete data (259–60): Data where there are no other values between possible values obtained in the data (discrete data is usually obtained by counting).

discriminant (233–4): The value $(b^2 - 4ac)$ calculated from the coefficients of a quadratic (part of the quadratic formula).

dispersion (271): See *spread*.

distance between two points (191, 241, 376): A formula in coordinate geometry.

distance of a point from a line (214): Formula for finding the shortest distance of a point from a line.

distribution (272, 294–6): The spread of data values between their lowest and highest values.

distributive law (42, 97, 143): A law describing how *brackets* are expanded, $a(b + c) = ab + ac$.

division (÷): An arithmetic operation. Dividing rational expressions **(150)**.

divisor (17): The term 'doing' the division.

dodecagon (339): A 12-sided *polygon*.

dollar ($) **(287):** A unit for money. $1 = 100 cents.

domain (184–5): The set of first elements in the ordered pairs of a relation. Of a trigonometric function **(187)**.

elimination method (130–2): A method for solving *simultaneous equations* where the two equations are added to or subtracted from each other, in order to create an equation with only one variable.

enlarge: Make larger, by a *scale factor*.

equally likely outcomes (321–2): *Outcomes* which all have an equal chance of occurring.

equation: A mathematical sentence involving an equals sign and at least one *variable*. Trigonometric **(361–70)**; linear **(135–6)**; simultaneous **(133–4, 231–2)**; quadratic **(109–28)**; circle **(395–401)**; involving indices **(55–7)**.

equation of a straight line (199–208): Gradient intercept form is $y = mx + c$. General form $ax + by + c = 0$, 178.

equidistant: The same distance away from another point or line.

equilateral triangle (340): A *triangle* in which all three sides are of equal length.

equiprobable (321–2): Events with equally likely outcomes.

equivalent fraction (15): A different way of representing the same *fraction*.

estimate (32–3): An approximate value. A number *rounded* to fewer *significant figures* than it originally had.

even numbers: *Multiples* of 2.

event (321–2, 329–33): A possible outcome or set of outcomes resulting from a statistical trial.

expanding (13, 19, 24): Removing *brackets* from an expression using the rule $a(b + c) = ab + ac$. When antidifferentiating **(42–4, 94, 97–100)**.

expected number or value (337): The average number of occurrences of an event after a set number of trials.

exponent (5, 35, 47, 54): A superscript (raised) number or variable used to indicate repeated multiplication of a *factor*. Also called a *power* or *index*

exponential function (59): A function of the form $y = a^x$, whose inverse is the logarithmic function $y = \log_a x$.

exterior angles (339–40, 391): The outside angles of a polygon formed by extending the sides of a *polygon*.

factor (16, 41, 77): Any *natural number* (or *algebraic expression*) which divides into another number or expression without a remainder.

factorising (96, 101, 145, 163, 166): Writing an *expression* as a product of its *factors*. Factorising quadratics **(102–38, 145, 228)**.

FOIL (99): A mnemonic to indicate the order in which to expand a pair of brackets.

formula (plural **formulae**) **(155–60):** An *equation* which shows the relationship between different quantities. Area **(74–5, 125, 410)**. Volume **(13, 75, 410)**. Changing the subject of **(157–8)**; using formulae **(158–9)**; and sequences **(412)**.

fraction (1–4, 15–26): A number expressed as a quotient. The top number is called the numerator, the bottom the denominator. An algebraic fraction or rational expression is a fraction with variables **(93–6)**. Addition and subtraction **(18–20)**. Multiplication and division **(20–2)**.

free of bias (299–300): When each member of the *population* is equally likely to be included in the sample.

frequency (259–61): The number of times a value occurs in a survey.

frequency table (259, 282, 304): A table which lists data values and their frequencies. (Also called a *frequency distribution table* or a *frequency distribution*.)

function (59, 115, 124, 161, 183–5): A relation in which each element of the *domain* appears in only one of the ordered pairs of the relation.

general form of equation of line (204–5): $ax + by + c = 0$ is the general form of the equation of a straight line.

gradient (m) (195–8, 201): A measure of the steepness of a line or curve. Also called *slope*.

gram (g) (65–7): The base unit of *mass* in the *metric system*. 1 000 g = 1 kg.

graph (155): Visual display of information, often involving axes and a grid. Of parabolas **(115–18)**; of cubic **(184–5, 249–50)**; of circles **(395–8)**.

graphical method (396–7): Solving a problem using a graph. For straight lines **(132, 196, 201, 209, 217)**; for straight lines and circles **(396–8)**; for straight lines and parabolas **(246–7)**.

grouped data (261): *Data* values that are put into groups or classes.

hectare (ha) (75): A unit for *area* in the *metric system*. 1 ha = 10 000 m^2.

hexagon (339): A 6-sided *polygon*.

highest common factor (HCF) (16, 101): The largest *factor* shared by two or more numbers or expressions.

histogram (261, 304): Bar graph on which continuous data are displayed (bars touch).

horizontal line (200, 210, 369): Line parallel to the horizon.

hypotenuse (191, 349, 355, 361): The side of a right-angled triangle opposite the right angle.

improper fraction (2–3, 16–17, 20–2): A fraction where the *size* of the *numerator* is larger than the *size* of the *denominator*.

increasing function (187): A function whose 'y' values increase as its 'x' values increase.

independent (333–7, 412): Uninfluenced by preceding outcomes.

index, indices (47): An alternative expression for *exponent*. Working with indices **(47–57)**.

inequality (73, 135–9, 219–22): One of the four *relations*, $<$, $>$, $\leq$ or $\geq$.

inequality symbols ($<$, $\leq$, $>$, $\geq$) (135): Symbols used in inequations.

inequation (135–9, 219–22): A mathematical expression which has an *inequality* instead of an equals sign.

inference (271): Process of using sample statistics to estimate population parameters.

integers (I) (1–2, 5, 7–11): The set of *whole numbers* and their opposites: …, −3, −2, −1, 0, 1, 2, 3, …

intercepts (117–25): Points where a graph cuts the x-axis or y-axis.

interior angles (399–40): The angles inside a *polygon* and between adjacent sides.

interpret: Make sense of. Understand the implications of.

interquartile range (275): The difference between the *upper* and *lower quartiles* of a data set.

intersection (329): The set containing the elements common to two or more other sets; a point where the graphs of two (or more) functions meet **(396–7)**.

interval notation (135, 185, 219): a notation for subsets of the real numbers, using brackets.

inverse function (59–60): The function which reverses the effect of another function.

inverse operation (12, 157): An operation which reverses the effect of another operation.

inverted (116): Turned upside down.

irrational numbers (1–2, 5, 39–46): Numbers which are not *rational*, ie which cannot be written as a *fraction*.

isosceles trapezium (388): A *trapezium* in which the non-parallel sides are equal in length.

isosceles triangle (192, 340–1, 352, 358, 387): A triangle with two sides of equal length.

kilogram (**kg**) **(65, 67):** A larger unit for *mass* than the *gram*. 1 kg = 1000 g.

kilometre (**km**) **(66–7, 74):** A longer unit of length than the metre. 1 km = 1000 m.

kite (354, 410): A *quadrilateral* with two pairs of equal adjacent sides.

latitude (403–7): Position of a point on earth with respect to the equator.

laws of indices (93, 95, 179): Rules for operations on algebraic terms with *exponents*.

leaf (263): The single-digit 'right part' of data values in a *stem and leaf* diagram.

length (65, 67): The distance from one end of an object to the other. The base unit is the *metre*.

like terms (91): Terms that have the same *variable*(s) and *exponents*.

limit: The value that a function approaches.

limits of accuracy (72–3): Range of values within which the true value of a measurement falls.

line: Applications of straight-line equations **(199–208)**; intersection with another line **(217–18)**; intersection with a circle **(232)**.

line segment (193, 211–12, 339, 385–6): Part of a line (with two end points). Also called an interval.

line symmetry: A figure has line symmetry if it has an *axis of symmetry*.

linear: An equation or polynomial in which the highest power of any variable is 1.

linear equation (132, 221, 231): An algebraic equation in which the power of the variable is 1.

linear graphs: Straight-line graphs.

linear inequation (135–8, 219–22): An algebraic inequation in which the power of the variable is 1.

line of best fit (264): Fairest line for a *scatter diagram* showing a narrow linear band pattern of points.

litre (**L**) **(65, 67, 75):** A unit for the *capacity* (or *volume*) of an object.

local maximum (235–6): A turning point where a function changes from increasing to decreasing.

local minimum (235–6): A turning point where a function changes from decreasing to increasing.

logarithm (log) (59–64, 179–82, 410): The inverse of $y = a^x$ is $y = \log_a x$, a function with domain $(x > 0)$.

longitude (404–7): Position of a point on earth with respect to the Greenwich meridian.

lower limit (73): The smallest value possible for a particular measurement.

lower quartile (274–5): The value for which 25% of the sample or population is less than or equal to. Also called the 25th percentile. On a box plot **(279)**.

lowest common multiple (LCM) (16, 94, 151): The smallest possible *multiple* shared by two or more numbers.

m (195): Commonly used symbol for the gradient of a straight line.

major arc (385): Arc of length more than a semi-circle.

map (79–84): A representation of a geographical area, drawn to scale often with grid lines for reference.

mass (65, 67): The amount of material in an object. Often referred to as weight.

mathematical modelling (141, 155–6): The process where problems described in words are changed into equations; linear modelling **(205–6)**; quadratic modelling **(244–5)**.

mathematical sentence (90): A description of a problem using numbers, *variables* and *operations*.

maximum (116): The y value for a function that has a peak at its turning point.

maximum turning point (187): A turning point where a function changes from increasing to decreasing.

mean (272–8): Central value in data, often called the *average*. The mean is found by adding all the scores and then dividing by the number of scores.

measurement (65–76): The process of assigning a numerical value to properties of an object, such as *mass*.

measures of central tendency (271–2): See *average*.

measures of spread (275): A measure of spread describes how data are distributed about a central value, eg the *range* and the *interquartile range*.

median (203): A line joining a *vertex* of a triangle to the midpoint of the opposite side. The central (or middle) value found by *ranking* data and selecting the middle number **(273–4, 282–3, 294–6)**.

mediator (390): See *perpendicular bisector*.

memory (6, 277): Function of calculator used for storing the answer to a calculation.

metre (m) (65–7): The base unit for *length* in the metric system.

metric system (65–8): A base 10 system of measurement.

midpoint (193, 203, 211): The point halfway between the end-points of a line segment.

milligram (mg) (67): A unit of *mass* smaller than the *gram*. 1000 mg = 1 g.

millilitre (mL) (67): A unit of *capacity* (or *volume*) smaller than the *litre*. 1000 mL = 1 L.

millimetre (mm) (66–7, 74): A unit of *length* smaller than the centimetre and metre.

minimum (115): The *y* value for a function that has a trough at its turning point.

minimum turning point (118, 187): A turning point where a function changes from decreasing to increasing.

minor arc (385): *Arc* of length less than a semi-circle.

mixed number (3, 16–28): An *integer* and *fraction* written together.

mode (273): The most frequently occurring value in a data set.

modulus (214): See *absolute value*.

money: Currency used to exchange goods and services. The basic unit is the *kina*.

multiplication principle (307): A multiplying technique to work out the number of ways sequence of events can occur.

multipliers (66): Numbers used to multiply by.

multiplying (×)**:** An arithmetic operation. Multiplying rational expressions **(149)**.

natural numbers (**N**) **(1, 2, 5):** The set of counting numbers: 1, 2, 3, . . .

nature (233): The number and type of roots of a quadratic equation.

nature of a turning point (187): Whether a turning point is a maximum or a minimum.

negatively correlated (265): Plotted ordered pairs of values showing a linear trend with negative gradient.

net (376): Flat shape that folds up to make a three-dimensional figure.

***n*th root (13):** The inverse of raising a number to the *n*th power. Raising the *n*th root (of a number) to the *n*th power results in the original number.

numerator (15): The integer or variable above the horizontal line in a *fraction*.

numerical expression (89): Numbers combined using operations such as +, –, × etc.

obtuse angle (341): An angle between 90° and 180°.

obtuse-angled triangle (341): A triangle which has one obtuse angle.

octagon (339): An 8-sided *polygon*.

odd numbers: The numbers 1, 3, 5, 7...

operation (5): Process applied to numbers, such as squaring, or adding.

opposite side (361): The side of the triangle which is directly opposite a particular angle.

or (329): The event *A or B* means the outcomes in *either* event *A or* event *B or* in both events.

ordered pairs (183): Two numbers (*a*, *b*) which give the position of a point relative to axes in the *cartesian plane* (the order of the numbers is important).

ordered stem and leaf plot (263–4): *Stem and leaf plot* in which leaves are arranged in order of size.

origin (138, 222, 253): The point (0, 0), where the *x*- and *y*-axes intersect on the *Cartesian plane*.

outcome (299, 321–3): The result of a trial in a statistical experiment.

outlier (274–5): A value in a data set which is unusually high or low.

parabola (115): The curve which is the graph of a *quadratic equation*.

parallel lines (209): Two lines are parallel if they are always the same distance apart. Construction of a line parallel to another line **(211)**.

parallelogram (353): A *quadrilateral* in which the opposite sides are parallel to each other.

pattern (287–8): A repeating feature, often expressed as a *formula*, found by looking at simple examples of a problem.

pentagon (339): A polygon with 5 sides.

percentage (%) **(263, 282, 304):** A percentage means 'out of every 100'.

perfect square (11, 40–1, 100): A *square number* such as 25 or an algebraic expression such as $(a + b)^2$, formed by multiplying a factor by itself.

perimeter (73, 411): The distance around the outside of an object.

perpendicular bisector (212, 352, 390): A line cutting a line segment in half and at right angles to it. Also called *mediator*.

perpendicular line (209–10): A line at *right angles* to a given line. Construction of the perpendicular at a point on a line **(214)**. Construction of a perpendicular to a line from a point **(213)**.

pie graph (262–3): Circular statistical graph in which sectors show data distribution. Also called pie chart.

place holder (29, 31, 35): A zero used to fill a position in a number.

plane (372–5): A flat (2-D) surface.

plotting (115–18, 238, 240, 249): Transferring the ordered pairs of a relation accurately onto a graph.

point symmetry (249): A graph which can be rotated through 180° about a point onto itself has point symmetry.

polygon (339): A closed figure with many straight sides.

polynomial (161–7): A function which is the sum of several terms. Each term is of the form ax^n, where n is a whole number.

population (259, 271–4): The entire group of interest in a *statistical investigation*.

population mean (μ) (272): A measure of central tendency in a population.

population parameters (271): Measures (such as the mean) of a population.

population standard deviation (σ) (277): A measure of spread of a population (the square root of the *variance*).

positively correlated (265): Plotted ordered pairs of values showing a linear trend with positive gradient.

power (5, 12, 21, 29): See *exponent*. Power key on a calculator **(13)**.

predict (287–8): Use patterns to work out future values.

prefix (65): Part words, such as kilo, which precede other words, such as metre, to change their meaning, eg a kilometre is 1000 metres. The prefixes *milli* and *centi* are also commonly used in the metric system.

prism (410): A solid object with the same *cross-section* throughout.

probability (321–38): The likelihood or chance of an event occurring.

probability tree (325–7): A branching diagram showing combinations of events and probabilities.

product (4): The result of multiplying two or more terms together.

pronumeral (89): A letter or symbol representing an unknown quantity or number.

proof (52, 60, 173, 321, 352–3): Statements showing that a relationship is correct.

proper fraction (2, 16): A fraction where the *numerator* is smaller than the denominator.

proportion (77–88, 261): A comparison of one quantity with another (often a part with a whole).

proportional (84): Of equal proportions; in the same ratio.

pyramids (373–6): Solid objects with all but the base face meeting at the same point (the *apex*). Volume **(410)**.

Pythagoras' theorem (2, 191, 355–60): A relationship ($a^2 + b^2 = c^2$) involving the three sides of a *right-angled triangle*. 3D trigonometry and Pythagoras **(371–3)**.

quadratic (102): A polynomial of degree 2. Drawing graphs of **(115–26)**.

quadratic equation (109): An equation of the form $ax^2 + bx + c = 0$, where a, b, c are constants. Solution of **(109–12)**; roots of **(229, 233–4)**.

quadratic expressions (102): *Expressions* of the form $ax^2 + bx^2 + c$. Factorising **(102–3)**.

quadratic formula (229): A formula for solving a quadratic equation.

quadratic function (115): A function with equation of the form $y = ax^2 + bx + c$.

quadratic graph (115): *Parabola*.

quadrilateral (155, 339, 390–1): A 4-sided *polygon*.

quartiles (274): Two numbers associated with the central tendency of samples and populations, eg 25% of the data values lie above the *upper quartile*.

quotient (17, 89, 148): An expression with a *denominator* and *numerator*; the result of dividing two numbers or variables.

radius (plural **radii) (155, 159, 244, 385, 389, 395, 403–7):** A straight line from the *centre* to a point on the *circumference* of a *circle* or *sphere*.

random numbers (300–1): A list of digits in which there are no predictable patterns. Random numbers on a calculator **(302)**.

random sample (272, 300): A sample in which every member of the *population* had an equal chance of being selected.

range (275): The difference between the highest and lowest scores in a set of data.

range (184–7): The set of second elements from a relation.

ranked (259): Ranked data are listed in order of size, eg from lowest value to highest value.

rate (205–6): A comparison of two quantities which have different units, one often being time.

rate of change (206): The gradient of a line or curve.

ratio (1–3, 39, 77–88): A comparison of two or more like quantities (measured using the same unit). Sharing quantities in a given ratio **(78)**.

rational expression: Any quotient consisting of *algebraic expressions*.

rational numbers (40–5): Numbers which can be written as a *fraction* (an integer over an integer).

raw data (259): Data at an early stage of analysis (unsorted).

real numbers (R) (1–2, 5–14): The set of all possible numbers on the number line.

rearrange (110): Change the order of terms in an expression, often so that another *variable* becomes the subject.

reciprocal (55–6, 95, 150, 176–7): The multiplicative inverse. The reciprocal of x is $\frac{1}{x}$. The reciprocal of $\frac{a}{b}$ is $\frac{b}{a}$.

reciprocal powers (55–6, 176–7): Used for solving equations with indices.

rectangle: A *quadrilateral* in which all the *interior angles* are right angles.

recurring decimal (2, 3, 28): A decimal with the same number (or sequence of numbers) repeated infinitely. Also called a repeating decimal.

reflex angle: An angle between 180° and 360°.

regular (339): A regular *polygon* has the lengths of all sides equal and the sizes of all *interior angles* equal.

relative frequency (261): The relative frequency of an event is the *proportion* of times it occurs.

repeating decimal (28): See *recurring decimal*.

representative (299): A representative (*unbiased*) sample has features similar to the population from which it was selected.

rhombus (354, 410): A *parallelogram* in which all the sides are equal.

right-angled triangle (191, 341, 355): A *triangle* which has a *right angle*.

right angle (∟) (209): An angle of 90°.

root (229, 233–4): A *solution* to an equation.

rounding (31–2): The writing of a number with fewer significant figures.

rule: *Formula* which expresses a relationship.

sample (271): Part of a *population*.

sample mean ($\bar{x}$) (272): The mean value of a sample.

sample space (321): A set whose elements describe every possible outcome of a trial.

sample standard deviation (s) (277): The standard deviation of a sample.

sample statistics (271): Measures calculated from data in a sample, such as the sample mean, $\overline{x}$, or the sample standard deviation, *s*.

sampling (300): Taking a subset of a population.

satisfy: Values which, when substituted, make an equation true, are said to satisfy the equation.

scale (67, 70): A line of values at set intervals marked on a measuring device.

scale factor (**SF**) **(343):** The *ratio* of the image size to the object size.

scalene triangle (340): A *triangle* which has no equal sides and no equal angles.

scatter diagram (264–6): Graph used with paired data, to investigate relationships between the variables.

sector (263, 386, 399): A fraction of a whole *circle* formed by two radii and the arc in between. Area **(399–400)**.

segment (386, 400): The region of a *circle* enclosed by a chord and an *arc*.

semi-circle (385, 387): A shape enclosed by a diameter of a circle and half the *circumference* of the circle.

sequence (412): A function which associates a real number with each natural number.

sign (+ or –) **(7–8):** The positive or negative nature of a number.

significant figure (**sig fig** or **sf**) **(30–2):** Any digit in a number which is not a zero at the beginning of a decimal or a zero at the end of a whole number. The number of significant figures shows how accurate a number is.

simplest form (2, 16, 28, 77, 93): *Simplified* as far as possible.

simplify: To write an expression in a briefer form (by combining like terms, etc). Simplifying fractions **(16)**. Simplifying algebraic fractions **(50)**.

simultaneous equations (129–34): Equations, each of which contains the same variables, and which usually have a common solution when solved together. Linear and non-linear **(231–2)**.

sine (**sin**) **(361–2):** A *trigonometric ratio*. Relative to a particular angle it is the ratio $\frac{\text{opposite side}}{\text{hypotenuse}}$.

sine rule (381): The relationship between one side of a triangle and its opposite angle, and a second side and its opposite angle.

size (7): The distance (positive) of a number from 0 on the number line.

sketch: A freehand drawing showing the main features of an object.

slope (195): Another expression for *gradient*.

SOH CAH TOA (362): Mnemonic relating the *trigonometric ratios* to the sides of a *right-angled triangle*.

solution: A value which substitutes into an *equation* to make it true; an answer.

solving: Finding the values of the *variables* which make an *equation* or *inequation* true.

speed (v): The positive rate of change of distance with respect to time. (Also called velocity.)

sphere (403–7): Completely rounded three-dimensional figure; ball-shaped.

spread (271): A description of the *distribution* of data values between their highest and lowest values.

spreadsheet (266, 274, 302): Numerical solution method for solving equations etc by computer.

square: A *quadrilateral* in which all four sides are equal and meet at *right angles*.

square (11): The square of a number is the number multiplied by itself, eg $4^2 = 4 \times 4 = 16$.

square centimetre (cm^2): Unit of measurement of *area* in the *metric system*. 10 000 cm^2 = 1 m^2.

square kilometre (km^2): Unit of measurement of *area* in the *metric system*. 1 km^2 = 1 000 000 m^2.

square metre (m^2): Base unit of measurement of *area* in the *metric system*.

square millimetre (mm^2): Unit of measurement of *area* in the *metric system*. 100 mm^2 = 1 cm^2.

square root (11): A number which when squared gives the number under the $\sqrt{}$ sign.

standard deviation (s or σ) (276–8): A measure of *spread*. The *square root* of the *variance*.

standard form (34): A number between 1 and 10, multiplied by a power of 10. Used to write and use large or small numbers. Standard form and calculators **(35–6)**.

statistical graphs (260–7): Data displays. See *graph*.

statistical investigation (299–305): A sequence of activities involving data collection and analysis in order to reach conclusions about a question or hypothesis.

statistics: The collection, display and analysis of data.

stem (263): See *stem and leaf plots*.

stem and leaf diagrams (263): Diagrams used in data display, similar to tally charts. The stem is the 'left part' of the data values, and the leaf is the final digit.

stem and leaf plots (295): Data display which splits data into a *stem* (the first significant figure(s) of a data value) and a *leaf* (last significant figure of a data value).

subject (130, 155, 157–8, 231, 379): The *variable* appearing on its own on one side of the equals sign in an equation or formula.

substitution (129–31): Replacing a *variable* with a number or with another equivalent expression.

substitution method (130): A method for solving simultaneous equations. One unknown is the subject of one equation and is then substituted in the other equation.

subtend (386–7, 399–400, 407): Lie under.

subtracting (–): An arithmetic operation.

sum: The result of adding two or more numbers or like terms.

summary statistics (293): Values calculated from a data set which show key features or characteristics of the data set, eg the *mean* or the *range*.

supplementary (391): Adding to 180°.

surd (40–6): The root of a rational number which is not rational itself, such as $\sqrt{2}$.

surface area (403, 410): The sum of the *areas* of the faces of a solid.

table: Information arranged in columns and rows.

tallies (259): Vertical and diagonal lines used to record frequencies.

tally chart (259, 263): A simple recording system for the number of times an outcome occurs.

tangent (tan) (361): A *trigonometric ratio*. Relative to a particular angle it is the ratio $\frac{\text{opposite side}}{\text{adjacent side}}$.

tangent to a circle (385, 389–90, 397): A straight line that touches a circle at one point on the circumference.

temperature (65, 67): How hot or how cold an object is. The unit used is *degrees Celsius* or *centigrade*.

term (60, 77, 89, 91, 97–101, 287): One member of a sequence or expression.

time (67–8): How long an event takes to occur. The basic unit is the second (s).

time series (287): Data recorded as time passes.

time-series graph (287–92): Line graph associated with a *time series*, used to show trends.

tonne (67): A unit of *mass*. 1000 kg = 1 tonne.

transformation (116–18, 122, 249): A change made to an object. *Translation* **(117–18, 122, 236)**.

transformed (116–19, 249): Changed in position or size.

transforming: See *rearranging*.

translation (117–18, 122, 236): A *transformation* which changes the position of an object by sliding.

transversal: A line that crosses a pair of lines.

trapezium (155–6): A *quadrilateral* that has one pair of parallel sides.

trend (260, 265–6, 287–8): A pattern of changes over time.

trend line (288): See also *line of best fit*.

trial (321, 327): One performance of an *experiment*.

triangle (339–54): A 3-sided *polygon*.

triangles: Convention for labelling of **(361)**.

trigonometric functions (361, 379–81): The functions *sine*, *cosine* and *tangent* etc.

trigonometric ratios (361–9, 379–81): Relationships between side lengths and angle sizes in a *right-angled triangle*. The three basic trigonometric ratios are *sine*, *cosine* and *tangent*.

turning point (TP) (115): A point where a function changes from increasing to decreasing (or vice versa) and the gradient of a tangent at the point is zero. Nature of **(187)**; finding turning points **(188, 239)**. See also *vertex*.

unbiased data (300): Data which are *representative* of the *population* from which they are sampled.

uncorrelated (265): Without a *pattern* in a *scatter diagram* of *ordered pairs*.

ungrouped data (259–60, 272, 275, 413): Data for which each possible value is listed individually.

union (∪) (329, 409): $A \cup B$ is the set of elements in A or B or both.

univariate data: Data with only one variable.

unlike terms (91–2): Terms which do not have the same *variable*(*s*) and *exponents*.

upper limit (73): The greatest possible value for a particular measurement.

upper quartile (274, 282–3): The value for which 75% of the sample or population is less than or equal to. On a box plot **(279)**.

variable (89, 91): A letter or symbol representing an unknown number.

variance (σ^2 or s^2) (275–7, 413): A measure of spread.

vector (122, 236, 241): A quantity with both *magnitude* and *direction* which is represented by an arrowed line.

Venn diagrams (322, 329–30): A diagram used to illustrate the relationships between two or more sets.

vertex (plural **vertices**) **(115–22, 235–8, 255–7):** The *turning point* of a parabola. The junction where two sides of a polygon *intersect* **(350)**.

vertical line (184, 200, 210): Line at *right angles* to a *horizontal line*.

vertical line test (184): The test applied to a graph to find if a relation is a function.

volume (13, 67, 75, 155): The amount of 3-dimensional space an object occupies. Units are m^3, cm^3 etc. Volume and *capacity* **(67, 70)**.

whole numbers (**W**) **(1–4, 5, 7):** The set of numbers 0, 1, 2, 3, 4, 5, 6, …

word problem (24, 112, 141, 230, 356): A description of a mathematical problem in words.

x-axis (186, 199): Horizontal axis in *Cartesian plane*.

x-coordinate (137, 193, 221): First number of a point's *coordinates* on a graph.

***x*-intercept (117, 120–3, 186, 199–200, 233, 238):** The point where a graph cuts the *x-axis*.

y-axis (186, 193, 221): Vertical axis in *cartesian plane*.

y-coordinate (121, 193): Second number of a point's *coordinates* on a graph.

***y*-intercept (117–20, 186, 199–202, 238–9):** The points where a graph cuts the *y-axis*.

ZFS (111): A mnemonic for the strategy required to solve *quadratic equations* (zero, factorise, solve).